PRACTICAL SHOP MATHEMATICS

Volume I—Elementary

PRACTICAL SHOP MATHEMATICS

Volume I—Elementary

John H. Wolfe and Everett R. Phelps

4th edition

McGRAW-HILL BOOK COMPANY

New York St. Louis San Francisco Dallas
London Toronto Sydney

ABOUT THE AUTHORS

JOHN H. WOLFE Apprentice and journeyman toolmaker, inspector, draftsman, shop mathematics teacher, and Director of Apprentice Training and Engineering Schools for the Ford Motor Company have been important milestones in John Wolfe's industrial experience. He has always put theoretical knowledge to the practical test. During World War II, he supervised the preliminary training for thousands of the Navy and aircraft recruits in the Ford schools. Later, he was granted an honorary degree of Doctor of Science in recognition of this work. In this new edition of *Practical Shop Mathematics*, John Wolfe continues his teaching. Several hundred thousand practical men have already learned their shop mathematics from earlier editions of this book.

EVERETT R. PHELPS Behind every successful textbook there is usually a master teacher. In this case the teacher who tutored the practical man in mathematics collaborated with his student to write this book. For many years Dr. Everett Phelps has been Professor of Physics and Astronomy at Wayne State University in Detroit. Unwilling to limit his interests to the theoretical part of a subject, he is always working with men who put theory to work. *Practical Shop Mathematics* is just one example of the numerous books and articles he has published.

PRACTICAL SHOP MATHEMATICS
VOLUME I—ELEMENTARY

Library of Congress Catalog Card Number: 58-10016

161718 VBVB 654

07-071358-8

PREFACE

This book is a direct outcome of classroom and shop experience. For several years the senior author taught a course in shop mathematics at the Ford Apprentice School of the Ford Motor Company. To provide content for this course, he developed a set of loose-leaf printed lesson sheets. To gain the benefits of his classroom experience, he revised the sheets periodically and then tested the revised lessons with his classes. The material in this book and its companion, Volume II, is the outgrowth of these classroom tested lessons. This edition provides new and important material that has never before been presented.

In writing this text, the authors have kept in mind its use not only in factory schools, trade schools, vocational high schools, etc., but also in all high schools to replace the usual geometry course for those students not intending to go to college. By using fifty propositions, they present the geometry necessary for the solution of practical shop problems, together with the necessary work to give continuity. These fifty propositions are proved in a formal manner in order that the training value of rigorous proofs may not be lost. The authors feel that the geometry as presented and the numerous practical problems that require a combined application of geometry and plane trigonometry are of much greater value to the high school student who is not going to college than is the usual geometry course consisting of about one hundred and fifty theorems and the usual more or less artificial and stereotyped exercises.

The value of this text in teaching students the shop mathematics they need to solve actual shop problems would be difficult to overestimate. The exposition of the principles involved, the solution of many practical problems, and the presentation of hundreds of problems for solution (many of which are accompanied by hints for the solution) teach the student the general methods he needs to apply in solving problems that arise in a shop. John Wolfe's fourteen years of machine-shop experience preceding his seventeen years of teaching shop mathematics has made it possible for the authors to select from the customary and regular run of actual shop problems those problems that are most helpful to students or workers who are pursuing a course in tool- and die-making, machine construction, machine repair, or draftsmanship.

After the student or mechanic has thoroughly mastered the general type of industrial shop problems by studying the theory and the worked-out typical problems and by solving the similar problems given in this text, he should be capable of handling the many shop problems involving plane geometry and trigonometry regardless of their difficulty. This is true because the problems of this text have been carefully chosen to give experience over the wide range of geometric propositions and trigonometric relations that are involved in actual shop problems.

The chapter on the use of the slide rule gives a brief discussion of the theory on which the slide rule is based and is followed by a detailed explanation of how to use a slide rule for the simple mathematical processes. This material is finally summarized in several formulas that will enable the student quickly to carry out multiplication, division, squares, square roots, proportion, and various combinations of these including those that involve the use of trigonometric functions. The authors have given this information on the slide rule with the object of saving the student hours of laborious computations. Actual shop problems usually require an accuracy to five significant figures, whereas the slide rule can be depended on for only three significant figures. However, in order to acquire more experience with the geometrical and trigonometrical phases of the practice problems, instead of concentrating on the numerical results, the use of the slide rule is very valuable.

and a great time saver. Furthermore a slide-rule solution may be used as a quick check on any problem.

One of the features of this book is the use of what the authors call the "variable system." Instead of all dimensions of a problem being given, one has been omitted and its value represented by a letter called the variable. Adjacent to the problem, or immediately after each exercise, six or seven values for the variable are given, any one of which may be used for the omitted dimension. Thus the instructor can, by using the six given values for the variable, assign separate problems of the same type to six students, each of whom will obtain a different answer. This helps greatly in preventing students from comparing solutions and answers. The students' own abilities are consequently developed to a much fuller extent. Of course, if the instructor does not care to make use of this method, he may assign the same value of the variable to all members of the class. Whenever seven values for the variable are given, the answer for the seventh one accompanies the figure. (See *Use of the Variable System*, page xiii.)

To this present edition the authors have added several new topics and have expanded several others included in previous editions. First, they have included additional material on the theory involved in checking of multiplication and division by the method of "excess of nines" and have given greater emphasis to the accuracy and advantages of this very useful checking device.

Second, the authors have inserted in this edition of Volume I the simple theory of gear ratios and lead screws since it is of an elementary nature and very useful in the lower brackets of machine-shop training. It is also helpful to understand this theory when operating production machines and other devices that require a definite movement with the use of a combination of gears and lead screws. The knowledge of gear and lead screw theory is also beneficial for the time study man who must estimate the time it should take to finish a certain type of machine operation before the job enters the machine shop. The authors have included sufficient problems to lead to an understanding of gear ratios and lead screws and to the ability to solve problems in this field.

Third, to this edition of Volume I, the authors have added the study of plain and differential indexing for the benefit of mechanics working in tool and die rooms.

Fourth, for the benefit of those who would rather use the law of sines and the law of cosines than the method of subdividing the oblique triangles into right triangles for the solution to a problem, they have added this theory.

The explanations in this text are presented very completely so that the student or mechanic can profitably use the book for home study or as a reference source. The solutions that accompany many of the problems are also of great help to such students.

In addition to being an excellent textbook for classroom use and for individual home study, this volume can be used very effectively as a reference book. All toolmakers, diemakers, machinists, draftsmen, apprentices, engineers, and vocational students will benefit greatly by owning this text for quick and easy reference.

The second volume of this text continues with the application of trigonometry and geometry to shop mathematics. It contains such subjects as solid trigonometry (commonly known as "compound angles"), the common types of gearing, compound gear ratios, continued fractions as applied to the cutting of leads and cams, and many special types of problems that occur in machine practice. There is a special course, never before given in any text, on the theory of duplicating helical gears. This course is very beneficial to those working in the field of gear cutting. The knowledge of the material contained in the second volume of *Practical Shop Mathematics* is very important and necessary in order to become a fully rounded-out toolmaker, diemaker, machine construction worker, machine repair worker, gear worker, draftsman, or engineer.

The authors wish to thank John W. Busman and William F. Mueller of the Ford Apprentice School faculty for their assistance in proofreading.

JOHN H. WOLFE
EVERETT R. PHELPS

CONTENTS

USE OF THE VARIABLE SYSTEM

In all problems, with a very few exceptions, one number or dimension is represented by a letter that is called the variable. This letter or variable has six or seven different numerical values which may be substituted for it to complete the problem as stated. This makes six or seven similar problems, each of which has a different answer. These six or seven values of the variable are given in tabular form at the end of a group of problems. For problems that are stated diagrammatically, the six values of the variable are usually placed to the right of the diagram. A seventh value of the variable and the corresponding answer are usually placed directly under the diagram.

To illustrate the use of the variable system when the variables are given in tabular form, consider problem 1 of page 4 which reads: "Reduce to a mixed number $\frac{191}{A}$." On page 5 immediately following this group of problems is a table of variables which gives for A the values: 16, 18, 20, 22, 24, 26. Thus the $\frac{191}{A}$ of problem 1 becomes $\frac{191}{16}, \frac{191}{18}, \frac{191}{20}, \frac{191}{22}, \frac{191}{24}, \frac{191}{26}$. Each student is to work with only one of these values, that is, one student may work with $\frac{191}{16}$, another with $\frac{191}{18}$, etc.

To illustrate the use of the variable system when variables are given to the right of the diagram, consider problem 1 at the bottom of page 19. x is the dimension to be computed and A represents another dimension, six values of which are given at the right of the figure. Any one of these values of A may be used to complete the statement of the problem. Thus for one student the dimension A is 2.1, for another 2.5, etc.

Throughout the text, x, y, and z are used to represent unknown distances, and any other letter of the English alphabet appearing in the problem is the variable. If the variable dimension is an angle, the Greek letter θ is generally used as the variable. Other Greek letters are used to represent the angular quantities to be computed.

A suggested plan for the use of the variable system in classrooms is presented in the chart below:

Name	No. of variable for 1st set of problems	No. of variable for 2d set of problems	No. of variable for 3d set of problems
Brown, John........	5	3	1
Collins, Ray.........	3	1	5
Grant, Peter.........	1	4	2
Hale, George........	2	6	4
Miller, Henry........	6	5	3
Smith, Williams......	4	2	6

The foregoing plan may be repeated for each group of six students.

The authors upon request will give further information regarding the use of the variable system.

PRACTICAL SHOP MATHEMATICS

CHAPTER I

COMMON FRACTIONS

DEFINITIONS

A **fraction** is a number expressing one or more of the equal parts of any whole quantity, as: $\frac{2}{7}$ bu., $\frac{5}{8}$ ft., $\frac{1}{2}$ mile.

The **terms** of a fraction are the denominator and numerator, which constitute a common fraction.

The **denominator** is the number below the line and shows the number of parts into which the whole is divided.

The **numerator** is the number above the line and shows how many parts are taken.

Example: $\frac{3}{4}$ of a foot shows that a foot has been divided into four equal parts, and three of the parts have been taken.

Common fractions are divided into the following classes: proper and improper fractions; mixed numbers; compound and complex fractions.

A **proper fraction** is one whose numerator is less than its denominator or whose value is less than unity, as $\frac{2}{5}$, $\frac{4}{7}$, $\frac{5}{8}$.

An **improper fraction** is one whose numerator equals or exceeds its denominator and whose value is equal to or greater than unity, as $\frac{5}{5}$, $\frac{7}{3}$, $\frac{9}{4}$.

A **mixed number** is a number expressed by an integer and a fraction, as $2\frac{2}{3}$, $4\frac{3}{4}$, $5\frac{5}{8}$.

A **compound fraction** consists of the indicated products of two or more proper or improper fractions, as $\frac{5}{8} \times \frac{3}{7}$, $\frac{1}{4} \times \frac{2}{3} \times \frac{7}{5}$.

A **complex fraction** is one in which one or both of its terms is a fraction or mixed number, as:

$$\frac{\frac{3}{5}}{8}, \quad \frac{\frac{9}{2}}{7}, \quad \frac{5\frac{1}{2}}{2\frac{7}{11}}.$$

REDUCTION OF FRACTIONS

To reduce an improper fraction to a whole or mixed number, divide the numerator by the denominator. The quotient will be the whole number. If there is a remainder, it will be the numerator of the fractional part, while the denominator will be the same as the denominator of the improper fraction.

Example: Reduce $\frac{21}{4}$ to a mixed number.

Solution: 21 contains 4 five times with one remaining. Thus: $\frac{21}{4} = 5\frac{1}{4}$.

To reduce a mixed number to an improper fraction, multiply the whole number by the denominator of the fraction, add the numerator to this product, and place the denominator under the result.

Example: Reduce $8\frac{3}{4}$ to an improper fraction.

Solution: $8 \times 4 = 32$, $32 + 3 = 35$; this result written over the denominator is $\frac{35}{4}$.

To reduce a fraction to higher or lower terms, multiply or divide the numerator and denominator by the same number. This does not change the value of the fraction.

LEAST COMMON DENOMINATOR

A **common denominator** of a group of fractions is a number which contains each of the denominators a whole number of times.

The **least common denominator** (L.C.D.) of a group of fractions is the *least* number which contains each of the denominators a whole number of times.

To Find the L.C.D. of a Group of Fractions.—Rewrite the denominators in a column, neglecting those denominators which are contained by others a whole number of times. Separate the remaining denominators into their prime factors.[1]

[1] A prime number is a number which is divisible only by itself and one, as 2, 3, 5, 7, 11, 13, 17, etc. A prime factor is one of two or more prime numbers which when multiplied together produce a given product.

The L.C.D. is the product of the different prime factors each taken the greatest number of times that it occurs in any one of the expressions.

Example: Find the L.C.D. of $\frac{1}{9}$, $\frac{1}{8}$, $\frac{1}{24}$, $\frac{1}{18}$, $\frac{1}{10}$. Since 24 and 18 contain 8 and 9, respectively, 8 and 9 are neglected. The remaining denominators separated into prime factors:

$$\begin{cases} 24 = 2 \times 2 \times 2 \times 3. \\ 18 = 2 \times 3 \times 3. \\ 10 = 2 \times 5. \end{cases}$$

The greatest number of times that 2 occurs is three; the greatest number of times that 3 occurs is two; the greatest number of times that 5 occurs is one. Therefore, the L.C.D. is the product of 2 used as a factor three times, 3 twice, and 5 once, or $2 \times 2 \times 2 \times 3 \times 3 \times 5 = 360$.

To reduce fractions to equivalent fractions having a L.C.D., divide the L.C.D. by the denominator of the fraction and multiply this quotient by the numerator of the fraction, then write this product as the numerator of the reduced fraction.

Example: Find the L.C.D. of $\frac{2}{3}$, $\frac{7}{9}$, $\frac{1}{2}$, $\frac{3}{4}$. By the foregoing method, the L.C.D. is equal to 36. $36 \div 3 = 12$, $12 \times 2 = 24$; therefore, $\frac{2}{3} = \frac{24}{36}$. Similarly $\frac{7}{9} = \frac{28}{36}$; $\frac{1}{2} = \frac{18}{36}$; $\frac{3}{4} = \frac{27}{36}$.

ADDITION OF FRACTIONS

Rule for Addition of Fractions.—*Reduce the fractions to equivalent fractions having a least common denominator, add their numerators, and write their sum over the common denominator.*

When fractions, mixed numbers, and whole numbers occur in addition of fractions, add the whole numbers and fractional parts separately and unite their sums. If the fractional part of the result is an improper fraction, it should be changed to a mixed number, the whole number part of which should be added to the rest of the whole numbers.

Example a: Add $\frac{2}{5}$, $\frac{4}{15}$, $\frac{4}{9}$, $\frac{3}{4}$. Reducing these to a L.C.D., $\frac{72}{180} + \frac{48}{180} + \frac{80}{180} + \frac{135}{180}$. Adding the numerators: $72 + 48 + 80 + 135 = 335$. Therefore the sum is $\frac{335}{180}$. Reducing to a mixed number in lowest terms, the sum is $1\frac{31}{36}$.

The **sum** is the result of addition of two or more quantities.

The **difference** is the result of subtraction of two quantities.

Example b: Add $4\frac{3}{8}$, $2\frac{5}{6}$, $5\frac{1}{2}$, $1\frac{1}{4}$.

Reducing the fractional parts to a common denominator, which is 24:

$$\frac{3}{8} = \frac{9}{24}, \qquad \frac{5}{6} = \frac{20}{24}, \qquad \frac{1}{2} = \frac{12}{24}, \qquad \frac{1}{4} = \frac{6}{24}.$$

Adding the numerators: $9 + 20 + 12 + 6 = 47$.

Then the sum of the fractional parts is $\frac{47}{24}$ or $1\frac{23}{24}$.

Adding all of the whole numbers: $1 + 4 + 2 + 5 + 1 = 13$.

Uniting the whole number and fractional part results in a total of $13\frac{23}{24}$.

PROBLEMS

Reduce to a mixed number, expressing the fractional part in its lowest terms:

1. $\frac{191}{A}$. **2.** $\frac{B}{42}$. **3.** $\frac{835}{C}$. **4.** $\frac{D}{29}$.

Reduce to an improper fraction:

5. $5\frac{7}{E}$. **6.** $F\frac{6}{7}$. **7.** $13\frac{13}{G}$. **8.** $15\frac{H}{35}$.

9. Determine the least common denominator of: $\frac{2}{J}$, $\frac{5}{34}$, $\frac{7}{8}$, $\frac{3}{24}$, and $\frac{8}{12}$.

10. Determine the least common denominator of: $\frac{1}{6}$, $\frac{1}{8}$, $\frac{1}{9}$, $\frac{1}{K}$, and $\frac{1}{3}$.

11. Reduce the following fractions to 72nds: $\frac{L}{36}$, $\frac{L}{12}$, $\frac{L}{8}$, $\frac{L}{9}$, and $\frac{L}{4}$.

12. Determine the sum of the following fractions: $\frac{5}{7}$, $\frac{M}{21}$, $\frac{20}{63}$, and $\frac{10}{21}$.

13. Reduce to the lowest terms: $\frac{36}{N}$.

14. How many thirds in P?

15. How many sixths are in the sum of: $5\frac{5}{6} + 6\frac{1}{6} + R\frac{2}{6} + 8\frac{7}{6}$?

16. Determine the sum of: $S\frac{4}{5} + 3\frac{9}{10} + 1\frac{2}{3} + \frac{19}{20}$.

17. Determine the sum of: $3\frac{9}{70} + 2\frac{12}{35} + T\frac{1}{7} + 10\frac{1}{10}$.

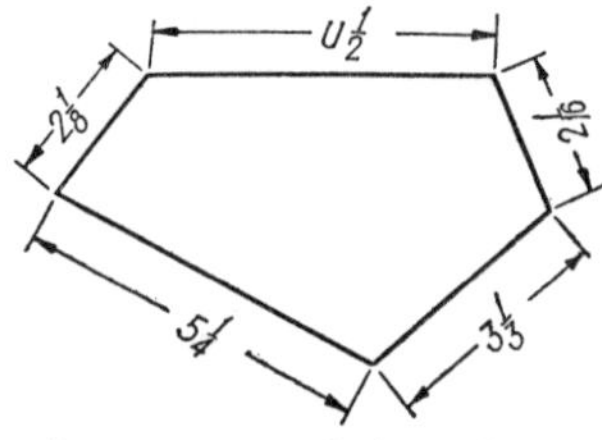

18. Determine the distance around the polygon.

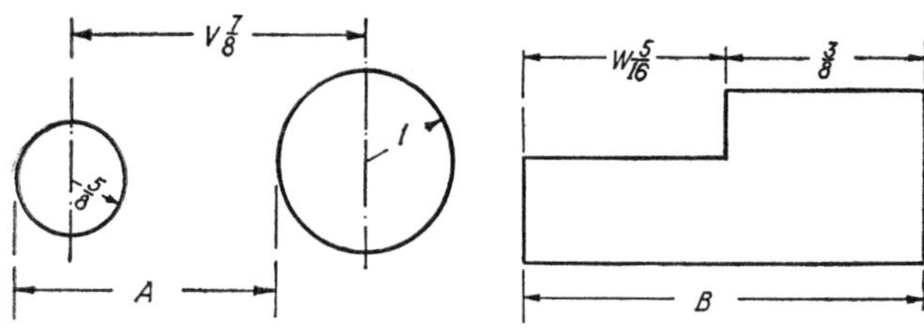

19. Determine the distance A. **20.** Determine the distance B.

VARIABLES

Prob.	Sym.	No. 1	No. 2	No. 3	No. 4	No. 5	No. 6
1	A	16	18	20	22	24	26
2	B	1511	1531	1551	1571	1591	1611
3	C	10	12	14	16	18	20
4	D	6608	6638	6668	6698	6728	6758
5	E	8	10	12	14	16	18
6	F	49	52	55	58	61	64
7	G	91	88	85	82	79	76
8	H	8	9	10	11	12	13
9	J	16	15	14	13	12	11
10	K	10	11	12	13	14	15
11	L	2	3	5	6	7	10
12	M	14	13	12	11	10	9
13	N	42	44	46	48	50	52
14	P	16	15	14	13	12	11
15	R	9	11	13	15	17	19
16	S	44	42	40	38	36	34
17	T	12	14	16	18	20	22
18	U	1	2	3	4	5	6
19	V	14	13	12	11	10	9
20	W	1	2	3	4	5	6

SUBTRACTION OF FRACTIONS

Rule for Subtraction of Fractions.—*Reduce the fractions to equivalent fractions having a least common denominator. Subtract the numerator preceded by the negative sign from the numerator preceded by the positive sign and place this difference over the common denominator.*

The first term in any expression is understood to be a plus quantity, unless otherwise specified.

Example: $\frac{7}{8} - \frac{3}{4} = \frac{7}{8} - \frac{6}{8} = \frac{1}{8}$.

To subtract a mixed number from a mixed number, subtract the whole numbers and fractional parts separately and unite these results.

Example: $3\frac{1}{2} - 1\frac{1}{4}$.

Solution: $\frac{1}{2} - \frac{1}{4} = \frac{2}{4} - \frac{1}{4} = \frac{1}{4}$; $3 - 1 = 2$.

Answer $= 2 + \frac{1}{4} = 2\frac{1}{4}$.

Sometimes in subtracting a mixed number from a mixed number, the fractional part in the subtrahend (the part to be subtracted) is greater than the fractional part in the minuend (the part that is subtracted from), and in this case it becomes evident that one unit must be borrowed from the whole number in the minuend and added to its fractional part.

Example: $4\frac{2}{7} - 2\frac{5}{8}$.

Solution: $\frac{2}{7} - \frac{5}{8} = \frac{16}{56} - \frac{35}{56}$; 35 cannot be subtracted from 16. Therefore one unit ($\frac{56}{56}$) is borrowed from the whole number of the minuend. This added to $\frac{16}{56}$ will be $\frac{72}{56}$. Then $\frac{72}{56} - \frac{35}{56} = \frac{37}{56}$. Next subtract the whole numbers: $3 - 2 = 1$. This united with the fractional part gives the final result, or $1\frac{37}{56}$.

To subtract a fraction from a whole number, borrow one unit from the whole number and express it as a fraction having the same denominator as the fraction to be subtracted. Subtract the fractional parts and annex this remainder to the remaining whole number.

Example: $7 - \frac{2}{5} = 6\frac{5}{5} - \frac{2}{5} = 6\frac{3}{5}$.

After addition or subtraction has been performed, it is customary to reduce the final fraction to its lowest terms.

PROBLEMS

1. What is the value of $\frac{21}{31}$ less $\frac{2}{A}$?

2. What is the value of $\frac{4}{C}$ less $\frac{2}{13}$?

3. What is the difference between $5\frac{1}{3}$ and $3\frac{23}{E}$?

4. What is the difference between $7\frac{2}{F}$ and $5\frac{1}{2}$?

5. Subtract $\frac{2}{3}$ from $3\frac{3}{G}$.

6. $2\frac{5}{8} - \frac{1}{K} - \frac{2}{3} - \frac{1}{8}$ is equal to what fraction in its reduced form?

7. $5\frac{1}{2} - 1\frac{2}{L} - 2\frac{1}{8} - \frac{2}{7}$ is equal to what fraction in its reduced form?

8. $7 - \frac{7}{8} - \frac{1}{10} - \frac{2}{M} - \frac{3}{11}$ is equal to what fraction in its reduced form?

9. $3\frac{1}{8} - \frac{6}{7} - \frac{5}{8} - \frac{3}{N} - \frac{1}{4}$ is equal to what fraction in its reduced form?

10. If $\frac{3}{R}$ is subtracted from a whole quantity, what part of the quantity remains?

11. If $\frac{3}{8}$ is subtracted from a whole quantity, and later $\frac{2}{S}$ is subtracted, what part of the quantity remains?

12. If $\frac{2}{T}$ is subtracted from a quantity, and later $\frac{3}{8}$ is subtracted, what part of the whole quantity remains?

13. Which fraction is the greater in value: $\frac{21}{U}$ or $\frac{15}{67}$?

14. Subtract $\frac{5}{V}$ from $2\frac{2}{3}$ and from the difference take away $\frac{2}{7}$.

Variables

Prob.	Sym.	No. 1	No. 2	No. 3	No. 4	No. 5	No. 6
1	*A*	13	14	15	16	17	18
2	*C*	10	9	8	7	6	5
3	*E*	25	26	27	28	29	30
4	*F*	8	7	6	5	4	3
5	*G*	10	11	12	13	14	15
6	*K*	3	4	5	6	7	8
7	*L*	15	14	13	12	11	10
8	*M*	8	9	10	11	12	13
9	*N*	21	18	15	12	9	6
10	*R*	14	15	16	17	18	19
11	*S*	13	12	11	10	9	8
12	*T*	15	16	17	18	19	20
13	*U*	36	35	34	33	32	31
14	*V*	12	13	14	15	16	17

ADDITION AND SUBTRACTION OF FRACTIONS

In any expression, where the plus and minus signs both occur, it is customary to add all of the plus quantities first, next to add all of the minus quantities, and then to subtract the sum of the minus quantities from the sum of the plus quantities.

Example: $\frac{2}{5} - \frac{7}{8} + \frac{3}{4} - \frac{2}{3} + \frac{1}{2} = ?$

Solution: Reducing the fraction to a common denominator:

$$\frac{48}{120} - \frac{105}{120} + \frac{90}{120} - \frac{80}{120} + \frac{60}{120}.$$

Adding all the plus quantities: $\frac{48}{120} + \frac{90}{120} + \frac{60}{120} = \frac{198}{120}$.

Adding all the minus quantities: $\frac{105}{120} + \frac{80}{120} = \frac{185}{120}$.

Subtracting the sum of the minus quantities from the sum of the plus quantities: $\frac{198}{120} - \frac{185}{120} = \frac{13}{120}$.

PROBLEMS

1. From the sum of $\frac{7}{9}$ and $\frac{5}{6}$ subtract the sum of $\frac{1}{3}$ and $\frac{2}{D}$.

2. From the sum of $4\frac{1}{2}$ and $6\frac{3}{5}$ subtract the sum of $2\frac{1}{E}$ and $3\frac{3}{4}$.

3. Add $5\frac{2}{9}$ and $6\frac{7}{F}$ and from the sum take away $4\frac{8}{9}$.

4. A man did $\frac{5}{G}$ of his work one day and $\frac{1}{3}$ of it the next. (*a*) What part of his work did he finish? (*b*) What part of his work was unfinished?

5. A truck drew $5\frac{3}{7}$ and $3\frac{5}{8}$ tons of pig iron on two successive days; another truck drew $6\frac{1}{2}$ and $7\frac{3}{H}$ tons on the same days. How many more tons did the latter draw than the former?

Simplify the following expressions by performing the operations indicated:

6. $5\frac{2}{5} - 7\frac{1}{8} + 4\frac{4}{7} + 7\frac{3}{4} - 4\frac{5}{J} = ?$

7. $\frac{7}{8} + \frac{6}{7} - \frac{7}{9} + \frac{8}{K} + \frac{3}{4} = ?$

8. $\frac{3}{4} - \frac{7}{8} - \frac{8}{L} + \frac{9}{10} + \frac{1}{3} = ?$

9. $6\frac{1}{2} - 7\frac{1}{3} + 8\frac{1}{3} + 5\frac{1}{2} - 4\frac{1}{M} - 3\frac{1}{2} = ?$

10. $\frac{6}{7} - \frac{4}{5} + \frac{3}{8} - \frac{8}{56} + \frac{5}{7} - \frac{7}{N} = ?$

11. If a man works $15\frac{1}{2}$ hr. on Monday, $10\frac{1}{3}$ hr. on Tuesday, $9\frac{3}{4}$ hr. on Wednesday, $7\frac{5}{6}$ hr. on Thursday, $11\frac{5}{P}$ hr. on Friday, and $8\frac{2}{3}$ hr. on Saturday, how many hours does he work during the week?

12. What is the perimeter (distance around) of a triangular piece, the sides of which measure $28\frac{2}{3}$, $45\frac{2}{R}$, and $67\frac{20}{21}$ in., respectively?

13. What is the perimeter of an irregular polygon, the sides of which measure $7\frac{1}{2}$, $5\frac{2}{3}$, $8\frac{3}{S}$, $2\frac{6}{7}$, and $2\frac{1}{3}$ in., respectively?

14. The perimeter of a four-sided irregular polygon is $24\frac{2}{3}$, three sides of which are $3\frac{2}{T}$, $5\frac{5}{7}$, and $4\frac{7}{8}$ in., respectively. Determine the length of the fourth side.

VARIABLES

Prob.	Sym.	No. 1	No. 2	No. 3	No. 4	No. 5	No. 6
1	*D*	4	5	6	7	8	9
2	*E*	13	12	11	10	9	8
3	*F*	14	15	16	17	18	19
4	*G*	13	12	11	10	9	8
5	*H*	8	9	10	11	12	13
6	*J*	20	19	18	17	16	15
7	*K*	10	12	14	16	18	20
8	*L*	34	32	30	28	26	24
9	*M*	3	4	5	6	7	8
10	*N*	19	18	17	16	15	14
11	*P*	10	11	12	13	14	15
12	*R*	5	6	7	8	9	10
13	*S*	15	14	13	12	11	10
14	*T*	21	22	23	24	25	26

MULTIPLICATION OF FRACTIONS

When multiplying fractions, do *not* reduce the fractions to a common denominator. Fractions should be reduced to a common denominator *only* in addition and subtraction.

In the multiplication of fractions, multiply together the numerators for the numerator of the product, and the denominators for the denominator of the product.

Examples: $\frac{3}{5} \times \frac{4}{7} = \frac{12}{35}$; $\frac{3}{4} \times \frac{5}{8} = \frac{15}{32}$.

CANCELLATION

Cancellation, which is used only in the multiplication of fractions, is the process of dividing a numerator and a denominator of an expression by a common factor. Should a plus or minus symbol occur in the numerator or in the denominator (or both) with other multiplication or division symbols, cancellation cannot be performed. Cancel whenever it is convenient, as it eliminates unnecessary work and also reduces the chances of errors.

Example: $\frac{6}{25} \times \frac{8}{3} \times \frac{15}{16}$.

Solution: 6 and 3 have 3 as a common factor; 15 and 25 have 5 as a common factor; 16 and 8 have 8 as a common factor. Reducing the foregoing by these common factors produces:

$$\frac{\overset{2}{\cancel{6}}}{\underset{5}{\cancel{25}}} \times \frac{\cancel{8}}{\cancel{3}} \times \frac{\overset{3}{\cancel{15}}}{\underset{2}{\cancel{16}}}.$$

This can be reduced further by dividing both numerator and denominator by a common factor 2 and the product becomes: $\frac{1 \times 1 \times 3}{5 \times 1 \times 1} = \frac{3}{5}$.

To multiply a fraction by an integer or an integer by a fraction, multiply the numerator of the fraction by the integer.

Examples: $\frac{3}{4} \times \frac{7}{1} = \frac{21}{4}$ or $5\frac{1}{4}$; $\frac{5}{6} \times \frac{11}{1} = \frac{55}{6}$ or $9\frac{1}{6}$; $6 \times \frac{3}{5} = \frac{18}{5}$ or $3\frac{3}{5}$.

To multiply a whole number by a mixed number, the usual form is to reduce the mixed number to an improper fraction and then proceed as in multiplying a whole number by a fraction.

Example: $42 \times 3\frac{5}{6} = \frac{\overset{7}{\cancel{42}}}{1} \times \frac{23}{\cancel{6}} = 161.$

If several fractions and mixed numbers are to be multiplied together, it is usually best to reduce all to improper fractions, for then the work may be shortened by cancellation.

Example: $\frac{5}{8} \times 4\frac{2}{3} \times 6 \times 5\frac{1}{2} \times \frac{1}{3} = \frac{5}{\underset{4}{\cancel{8}}} \times \frac{\overset{7}{\cancel{14}}}{3} \times \frac{\cancel{6}}{1} \times \frac{11}{\cancel{2}} \times \frac{1}{\cancel{3}} =$

$\frac{385}{12}$ or $32\frac{1}{12}$.

PROBLEMS

Solve:

1. $\frac{8}{D} \times \frac{16}{17} = ?$ **2.** $\frac{21}{22} \times \frac{E}{33} = ?$ **3.** $\frac{18}{F} \times \frac{30}{37} = ?$ **4.** $\frac{G}{28} \times \frac{51}{60} = ?$

5. $\frac{21}{22} \times \frac{55}{7} \times \frac{63}{11} \times \frac{72}{81} \times \frac{2}{H} = ?$ **6.** $\frac{5}{8} \times \frac{3}{J} \times \frac{32}{35} \times \frac{7}{9} \times \frac{36}{48} = ?$

7. $3\frac{2}{3} \times 1\frac{1}{2} \times 5\frac{3}{8} \times 4\frac{1}{6} \times 3\frac{1}{K} = ?$ **8.** $\frac{5}{8} \times 4\frac{2}{7} \times \frac{8}{L} \times 3\frac{1}{2} \times \frac{2}{3} = ?$

9. $\frac{9 \times 6}{4} \times \frac{M}{9} \times \frac{6 \times 3 \times 9}{27} = ?$

10. $\frac{N \times 18 \times 14}{6 \times 3 \times 2} \times \frac{42 \times 6 \times 1}{186 \times 9 \times 2} = ?$

11. $\frac{5}{P} \times \frac{4}{1} = ?$ **12.** $3\frac{1}{2} \times \frac{R}{8} = ?$

13. $4\frac{1}{S} \times 2\frac{2}{5} = ?$ **14.** $\frac{3}{7} \times \frac{T}{1} = ?$

15. $\frac{2}{3} \times \frac{5}{8} \times \frac{7}{9} \times \frac{3}{D} \times \frac{5}{12} \times \frac{6}{25} = ?$

16. $2\frac{3}{8} \times 3\frac{5}{6} \times 4\frac{1}{9} \times 1\frac{2}{5} \times 3\frac{1}{E} = ?$

17. $\frac{11}{13} \times \frac{26}{17} \times \frac{F}{55} \times \frac{25}{39} \times \frac{26}{27} \times \frac{9}{16} = ?$

18. $\frac{5}{8} \times \frac{2}{7} \times \frac{3}{25} \times \frac{21}{G} = ?$

VARIABLES

Prob.	Sym.	No. 1	No. 2	No. 3	No. 4	No. 5	No. 6
1	D	11	12	13	14	15	16
2	E	18	17	16	15	14	13
3	F	36	38	40	42	44	46
4	G	13	11	9	7	5	3
5	H	19	21	23	25	27	29
6	J	24	20	16	12	8	4
7	K	27	30	33	36	39	42
8	L	14	16	18	20	22	24
9	M	7	6	5	4	3	2
10	N	54	56	58	60	62	64
11	P	18	16	14	12	10	8
12	R	9	10	11	12	13	14
13	S	12	11	10	9	8	7
14	T	44	46	48	50	52	54
15	D	13	14	15	16	17	18
16	E	27	24	21	18	15	12
17	F	43	45	47	49	51	53
18	G	21	22	23	24	25	26

DIVISION OF FRACTIONS

In division of fractions the **dividend** is that number or fraction which is divided by some other number or fraction.

The **divisor** is that number or fraction which the dividend is divided by, *i.e.*, the **divisor** is the number or fraction which follows the division symbol (÷).

The **quotient** is the result obtained by dividing the dividend by the divisor.

RECIPROCALS

The **reciprocal** of a number is 1 divided by that number. Thus the reciprocal of 8 is $\frac{1}{8}$.

The reciprocal of a fraction is 1 divided by the fraction.

Example: The reciprocal of $\frac{5}{8}$ is $\frac{1}{\frac{5}{8}}$ or $\frac{1 \times 8}{\frac{5}{8} \times 8}$ or $\frac{8}{5}$.

From the foregoing it can be seen that the reciprocal of a fraction is the fraction inverted.

The reciprocal of a mixed number is the mixed number reduced to an improper fraction and then inverted.

Example: The reciprocal of $3\frac{2}{5}$ is $\frac{1}{\frac{17}{5}}$ or $\frac{5}{17}$.

Instead of dividing by a number or a fraction, one can multiply by the reciprocal of the number or fraction and get the same result. Therefore, in division of fractions, invert the divisor and multiply.

$$\begin{array}{lccccc} & \text{Dividend} & & \text{Divisor} & & \text{Quotient} \\ \textit{Example a:} & \frac{5}{8} & \div & \frac{3}{4} & = & \frac{5}{\cancel{8}_2} \times \frac{\cancel{4}}{3} = \frac{5}{6}. \end{array}$$

Example b: $$\frac{5}{7} \div \frac{3}{5} \div \frac{4}{9} \div \frac{5}{14} = \frac{\cancel{5}}{\cancel{7}} \times \frac{5}{\cancel{3}} \times \frac{\overset{3}{\cancel{9}}}{\underset{2}{\cancel{4}}} \times \frac{\overset{2}{\cancel{14}}}{\cancel{5}} = 7\frac{1}{2}.$$

PROBLEMS

1. Divide the product of 6, 9, 10 by the product of D, 3, 5.
2. Divide the product of E, 9, 12 by the product of 6, 8, 21.

3. $\frac{F}{8} \div \frac{3}{5} = ?$ **4.** $\frac{6}{7} \div \frac{J}{5} = ?$ **5.** $\frac{5}{8} \div \frac{2}{15} \div \frac{5}{24} \div \frac{45}{M} \div \frac{94}{100} = ?$

6. $\frac{3}{7} \div \frac{2}{3} \div \frac{5}{N} \div \frac{45}{21} \div \frac{5}{6} = ?$ **7.** $2\frac{1}{2} \div 3\frac{2}{3} \div 5\frac{6}{P} \div 4\frac{2}{5} \div 2\frac{6}{7} = ?$

8. $\frac{7}{12}$ of the distance from A to B is $R\frac{2}{3}$ in. What is the distance from A to B?

9. If a man chops $1\frac{1}{6}$ cords of wood a day, in what time can he chop $S\frac{4}{5}$ cords?

10. If $\frac{3}{4}$ of a ton of coal costs $5, how much will $T\frac{2}{3}$ tons cost?

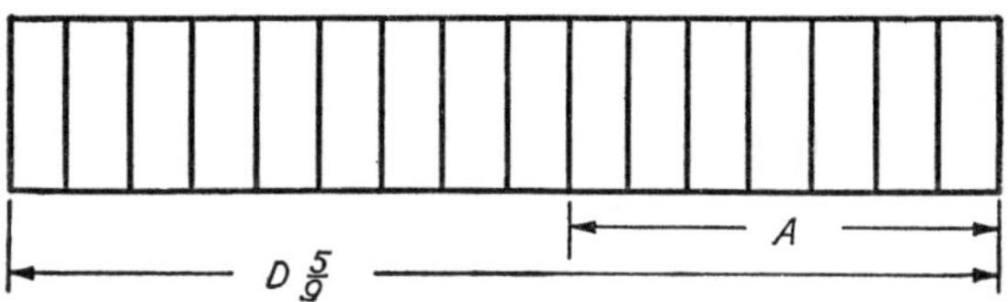

11. Determine the distance A.

12. $\frac{1}{7}$ is $\frac{5}{E}$ of what number? **13.** $\frac{3}{7}$ is $\frac{1}{G}$ of what number?

14. $\frac{5}{8}$ is how many times greater than $\frac{1}{J}$?

15. $\frac{6}{7}$ is how many times greater than $\frac{1}{K}$?

16. $\frac{2}{L}$ is what part of $\frac{5}{9}$? **17.** $\frac{5}{8} \div \frac{4}{7} \div \frac{2}{5} \div N \div 2 \div \frac{3}{4} \div \frac{1}{8} = ?$

18. $\frac{4}{7} \div 2\frac{3}{5} \div \frac{7}{9} \div 3\frac{2}{3} \div \frac{3}{P} \div \frac{9}{14} = ?$

VARIABLES

Prob.	Sym.	No. 1	No. 2	No. 3	No. 4	No. 5	No. 6
1	*D*	11	12	13	14	15	16
2	*E*	8	7	6	5	4	3
3	*F*	3	4	5	6	7	8
4	*J*	7	8	9	10	11	12
5	*M*	46	48	50	52	54	56
6	*N*	8	12	16	20	24	28
7	*P*	18	17	16	15	14	13
8	*R*	62	60	58	56	54	52
9	*S*	5	6	7	8	9	10
10	*T*	19	18	17	16	15	14
11	*D*	20	22	24	26	28	30
12	*E*	6	7	8	9	10	11
13	*G*	2	3	4	5	6	7
14	*J*	3	4	5	6	7	8
15	*K*	14	13	12	11	10	9
16	*L*	3	4	5	6	7	8
17	*N*	34	32	30	28	26	24
18	*P*	21	22	23	24	25	26

CHAPTER II

CHECKING MULTIPLICATION AND DIVISION BY THE EXCESS OF NINES

In all branches of mathematics, accuracy is of great importance and therefore one must have some convenient method of checking multiplication and division. A simple method of checking multiplication and division is by a process involving the excess of nines. The excess of nines is the remainder of a number which has been divided by nine.

Example: 38 divided by 9 has 2 as a remainder which is the excess of nines in 38. The excess of nines can also be found by dividing the *sum* of the digits by nine; then the remainder becomes the excess of nines. The excess of nines can be found with greater ease by eliminating the nines as soon as the sum of the digits is equal to, or immediately after it exceeds, nine. The excess of nines of any two digits whose sum is greater than nine is always one more than the last digit of the sum.

Examples: The excess of nines of 10 is $0 + 1 = 1$; the excess of nines of 14 is $4 + 1 = 5$, etc.

Example: Find the excess of nines of 74,685.

Solution: This carried out in detail form is as follows: Begin eliminating nines from left to right: $7 + 4 = 11$ where 2 is the excess of nines. Add this excess of nines to the next digit on the right, $2 + 6 = 8$. In cases like this, keep on adding the successive digits to the right until the sum equals or exceeds nine. Then $8 + 8 = 16$ where 7 is the excess of nines. Add this excess of nines to the next digit on the right and continue this process until all of the digits have been considered. Thus, $7 + 5 = 12$ where the excess of nines is 3. This final excess of nines is called the excess of nines of 74,685.

In order to determine the excess of nines with the greatest ease, the grouping method is recommended. The eye should be trained to recognize groups of two or three figures whose sums are nine, such as (1 and 8), (2 and 7), (3 and 6), (4 and 5), (2, 3, and 4), (5, 6, and 7). In order that the eye may be able to recognize the foregoing groups quickly, they should be memorized. These groups of digits need not be considered when finding the excess of nines by the foregoing process.

Example: Find the excess of nines of 5,762,382 by the grouping method.

Solution: At a glance it can be seen that 7 and 2, and 6 and 3, are each equal to nine. These groups should be neglected immediately. The remaining figures to be considered are 5, 8, and 2. Since the value of 8 is so close to that of nine it is best to take one from the 5 which completes another group of nine, leaving only 4 and 2 to be considered. 4 and 2 are 6, which is the excess of nines of the number 5,762,382. The following numbers are a few examples of grouping. The different groups will be indicated by the connecting lines: (6,768,654), (75,634), (68,423), (467,523).

To check multiplication by the excess of nines, multiply the excess of nines in the multiplicand by the excess of nines in the multiplier. The excess of nines in the product must equal the product of the excess of nines of the two original numbers.

Illustrative Problem a:

Multiply 54,876 by 87,542 and check by excess of nines.

```
               54876.................  3
               87542.................  8
                                      ---
6.........109752..2 × 3 = 6           24 or ⑥
3........219504  ..4 × 3 = 12 or 3      |
6.......274380   ..5 × 3 = 15 or 6      |
3......384132    ..7 × 3 = 21 or 3      |
6.....439008     ..8 × 3 = 24 or 6      |
      4803954792.................       ⑥
```

If an error has been made in the process of multiplication, in order to check the location of the error, the check by nines may be applied separately to each step. Check each step separately until the error is found. If the error is not in the multiplication phase of the problem, the error then without doubt is in the addition or subtraction. To check each step separately, multiply the excess of nines in the multiplicand by the excess of nines in the partial multiplier; the product must equal the excess of nines in the product represented by these two quantities.

For example, in multiplying 54,876 by 2 the excess of nines are 3 and 2, respectively and the product 109,752 gives an excess of 6 (which equals 3 × 2).

To check division by the excess of nines, proceed as follows: Multiply the excess of nines in the quotient by the excess of nines in the divisor. Add to this product the excess of nines in the remainder. The result must be equal to the excess of nines in the dividend.

Illustrative Problem b:

Divide 594,870 by 78,646 and check by excess of nines.

```
                    7.5638.........2
4......78646/594870..................... 6
1............550522......7 × 4 = 28 or 1
             443480
2............393230.....5 × 4 = 20 or 2
              502500
6.............471876....6 × 4 = 24 or 6
              306240
3..............235938...3 × 4 = 12 or 3
               703020
5...............629168..8 × 4 = 32 or 5
7................73852..remainder
```

The final check is 2 × 4 = 8; 8 + 7 = 15. The excess of nines of 15 is 6. This 6 is equal to the 6 that represents the excess of nines in the dividend, and since these two quantities are equal, the check is complete and indicates that the quotient is correct. The multiplication phase of division may also be checked step by step as indicated above in order to locate a possible error.

This check by the excess of nines has been proven by frequent class room experimentations to be extremely close to 100% accurate. It must be understood that this system of checking by the excess of nines will only check errors that are strictly accidental. The check loses its degree of accuracy if the figures are willfully rearranged. This check applied to arithmetical figures is extremely useful in the search for accidental errors made in the process of multiplication, division, addition, and subtraction. This check offers the only simple assurance that the multiplication or division has been correctly carried out. The check does not apply if the arrangement of the figures are not according to arithmetical rules.

PROBLEMS

What is the excess of nines of:

1. D. **2.** E. **3.** F. **4.** G. **5.** H.

6. Multiply 16,425 by J and check each step by the excess of nines. What is the excess of nines of the sum of the excess of nines of each of the results obtained by multiplying the multiplicand by the first four digits (starting from left to right) of the multiplier?

7. Multiply K by 4295 and check each step by the excess of nines. What is the excess of nines of the sum of the excess of nines of each of the results obtained by multiplying the multiplicand by the four digits of the multiplier?

8. Divide 438,569 by L and check the final result and each step of multiplication by the excess of nines. What is the excess of nines in the remainder? Quotient to consist of five significant figures.

9. Divide M by 5783 and check the final result and each step of multiplication by the excess of nines. What is the excess of nines in the remainder? Quotient to consist of five significant figures.

VARIABLES

Prob.	Sym.	No. 1	No. 2	No. 3	No. 4	No. 5	No. 6
1	D	587634	238674	396872	457635	843926	537462
2	E	237674	568943	487632	896543	487542	865437
3	F	395827	456843	785326	956472	234589	456787
4	G	475652	589542	678762	324578	567823	235678
5	H	894263	754326	821254	756234	724235	678756
6	J	24345	45678	56278	45673	54678	45638
7	K	5892	6785	7368	9375	8537	6278
8	L	34567	75623	65892	54237	68752	87542
9	M	6782	7368	9375	8537	7564	6785

CHAPTER III

DECIMALS

A **decimal fraction** is a fraction whose denominator is 10 or some multiple of 10. The denominator of a simple decimal fraction is always omitted but is expressed by a dot called the decimal point, placed in different positions of a number corresponding to the magnitude of the denominator. One figure to the right of the decimal point indicates that the denominator is 10; two figures to the right of the decimal point indicates that the denominator is 100; three figures to the right of the decimal point indicates that the denominator is 1000; etc.

A decimal number, or **decimal,** is a number involving a decimal fraction. Thus: .237 or 6.346.

The nomenclature of the decimal system is as follows:

3	5	7	9	.	5	8	3	4	7	2
thousands	hundreds	tens	units		tenths	hundredths	thousandths	ten-thousandths	hundred-thousandths	millionths

The decimal quantity is always read from left to right, annexing the name corresponding to the last decimal figure.

Example: The figure 35.6437 is read thirty-five and six thousand, four hundred thirty-seven ten-thousandths.

To change a decimal to a common fraction: The numerator will be the same as the original figure omitting the decimal point; the denominator will always be one followed by as many ciphers as there are figures to the right of the decimal point.

Examples: $.73 = \frac{73}{100}$; $5.496 = \frac{5496}{1000}$.

PROBLEMS

Change the following decimal numbers to common fractions:

1. *A*. **2.** *B*. **3.** *C*.

Change the following common fractions to decimal numbers:

4. *D*. **5.** *E*. **6.** *F*.

VARIABLES

Prob.	Sym.	No. 1	No. 2	No. 3	No. 4	No. 5	No. 6
1	*A*	.0327	.0433	.0569	.0671	.0723	.0837
2	*B*	2.427	3.567	4.287	5.367	6.487	8.367
3	*C*	.0037	.0033	.0039	.0041	.0043	.0047
4	*D*	$\frac{235}{1000}$	$\frac{356}{1000}$	$\frac{467}{1000}$	$\frac{897}{1000}$	$\frac{752}{1000}$	$\frac{689}{1000}$
5	*E*	$\frac{41}{10000}$	$\frac{51}{10000}$	$\frac{61}{10000}$	$\frac{71}{10000}$	$\frac{81}{10000}$	$\frac{91}{10000}$
6	*F*	$\frac{23}{100}$	$\frac{48}{100}$	$\frac{56}{100}$	$\frac{67}{100}$	$\frac{73}{100}$	$\frac{84}{100}$

ADDITION AND SUBTRACTION OF DECIMALS

Since it is necessary to have a common denominator when adding or subtracting common fractions, and since decimal fractions are only a modified form of common fractions, it becomes evident that to add or subtract decimals the decimal points must be placed in a column directly under each other.

Examples: Add

```
      2.6875
       .0789
     35.3000
      6.4789
Sum = 44.5453
```

Subtract

```
             7.6300
             2.1682
Remainder =  5.4618
```

PROBLEMS

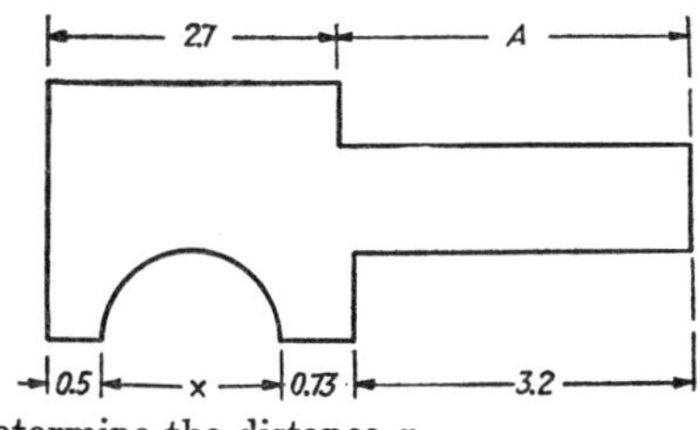

VARIABLE

No.	Sym.	Value
1	*A*	2.1
2	*A*	2.5
3	*A*	2.8
4	*A*	2.9
5	*A*	2.3
6	*A*	2.7

1. Determine the distance x.

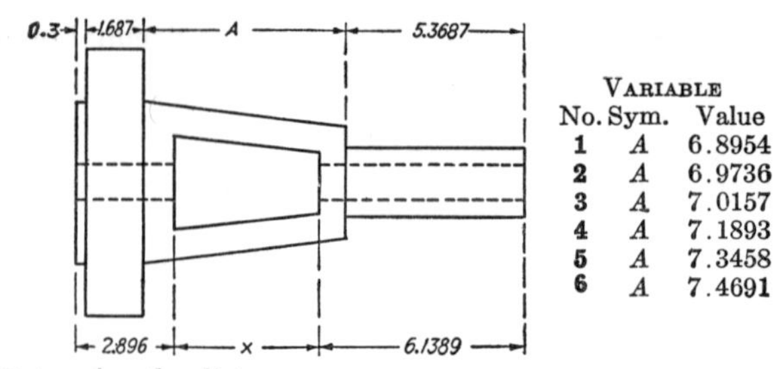

VARIABLE

No.	Sym.	Value
1	A	6.8954
2	A	6.9736
3	A	7.0157
4	A	7.1893
5	A	7.3458
6	A	7.4691

2. Determine the distance x.

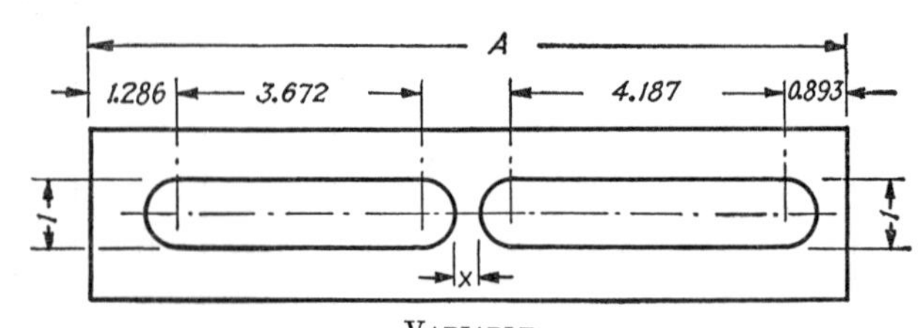

VARIABLE

1. $A = 11.503$ **2.** $A = 11.548$ **3.** $A = 11.586$
4. $A = 11.609$ **5.** $A = 11.637$ **6.** $A = 11.652$

3. Determine the distance x.

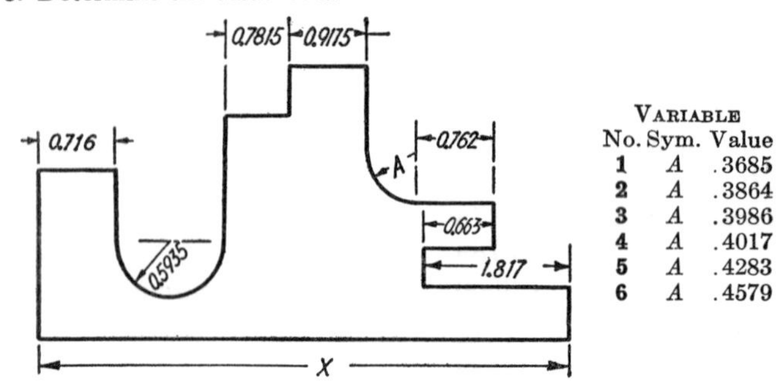

VARIABLE

No.	Sym.	Value
1	A	.3685
2	A	.3864
3	A	.3986
4	A	.4017
5	A	.4283
6	A	.4579

4. Determine the distance x.

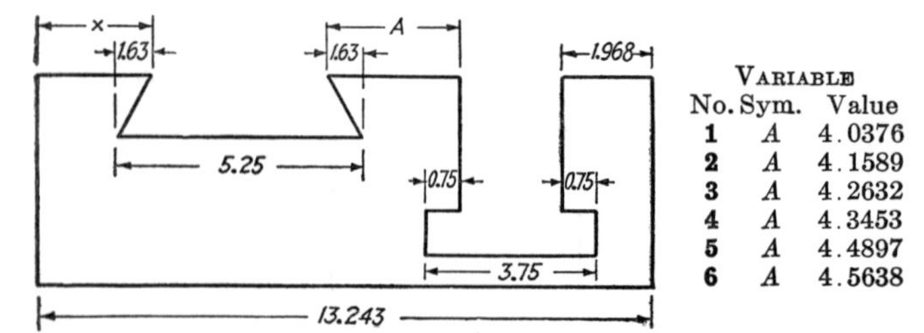

VARIABLE

No.	Sym.	Value
1	A	4.0376
2	A	4.1589
3	A	4.2632
4	A	4.3453
5	A	4.4897
6	A	4.5638

5. Determine the distance x.

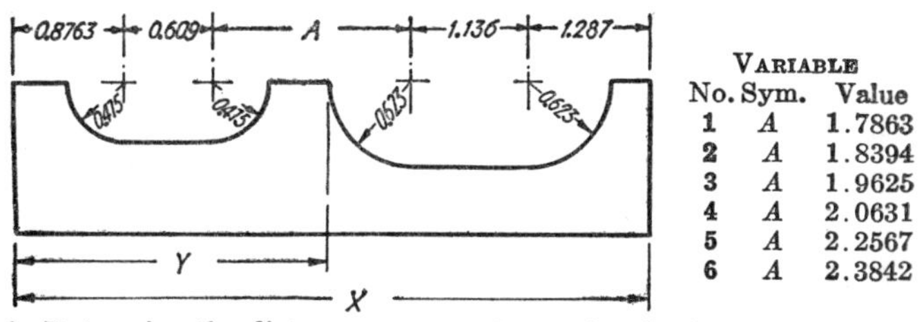

VARIABLE		
No.	Sym.	Value
1	*A*	1.7863
2	*A*	1.8394
3	*A*	1.9625
4	*A*	2.0631
5	*A*	2.2567
6	*A*	2.3842

6. Determine the distance x. **7.** Determine the distance y.

MULTIPLICATION OF DECIMALS

Multiplication is the process of adding a number as many times as there are units in the quantity by which it is multiplied.

The **multiplicand** is the number which is to be multiplied.

The **multiplier** is the number by which the multiplicand is to be multiplied.

The **product** is the result of the multiplication.

The multiplicand and multiplier are both **factors** of the product.

Example: $13 \times 11 = 143$.

In this example 13 is the multiplicand, 11 is the multiplier, and 143 is the product. 13 and 11 are both factors of 143.

In the multiplication of decimal quantities, proceed as in the multiplication of whole numbers. Since the product of two common fractions whose denominators are 10 will produce a fraction whose denominator is 100, the product of two decimal fractions stated in tenths will produce a decimal fraction stated in hundredths. From this it is evident that the number of decimal places in the product is equal to the sum of the decimal places in the multiplicand and multiplier.

Example: Multiply 7.8546 by 487.69.

```
          7.8546...  4 decimal places
          487.69...  2 decimal places
         --------    -
         706914      6 decimal places
        471276       Therefore place the decimal point
       549822         6 decimal places from right to
      628368          left in the product.
     314184
     -----------
     3830.609874
```

PROBLEMS

1. Multiply 8.6542 by A.
2. Multiply 10.856 by B.
3. Multiply 24.678 by C.
4. Multiply 8.4967 by D.
5. Multiply 4.8976 by E.
6. Multiply 5.9654 by F.
7. Multiply 6.9876 by G and then subtract .87654.
8. Multiply 6.8763 by H and then add 8.6957.
9. Multiply the sum of 3.8756 and J, by the difference of 4.8643 and 2.7632.
10. Multiply the difference of 8.5438 and K by the sum of 5.9875 and 2.8737.

VARIABLES

Prob.	Sym.	No. 1	No. 2	No. 3	No. 4	No. 5	No. 6
1	A	.85375	5.8365	5.8495	.69857	7.3865	.48756
2	B	5.7532	.75985	8.763	.35674	.73745	7.3876
3	C	4.8565	5.9758	8.3957	4.9867	8.3865	.68756
4	D	.87495	.68996	.86075	.94765	.58764	6.8597
5	E	.09843	6.9847	.97846	.48967	5.8398	.97865
6	F	6.4623	3.8576	8.9476	.39874	8.0753	.58746
7	G	8.4965	.57849	3.9875	2.9578	2.7497	2.9175
8	H	4.9687	7.9687	.98576	6.9587	.64865	7.5987
9	J	7.8539	.68932	.75894	.68932	.85964	.57684
10	K	2.2	2.7	2.5	2.6	2.3	2.8

DIVISION OF DECIMALS

Division of decimals is a special application of long division. When a decimal is divided by another decimal, it is essential that the decimal point in the quotient be placed in the proper location. Misplacing the decimal point changes the value greatly. For each place that it is moved to the right, the value of the decimal is multiplied by 10; for each place it is moved to the left, the value is divided by 10. Multiplying both dividend and divisor by the same number does not alter the value of the quotient. To move the decimal point to the right in both dividend and divisor the same number of places is the same as to multiply both by 10, or some power of 10; hence the value of the quotient remains the same.

In division of decimals the divisor may be reduced to a whole number. In order to do this, move the decimal point to the right as many places as there are figures to the right of the decimal point. Next, move the decimal point in the dividend the same number of places to the right, counting from the original position. Eliminate the original decimal points in the dividend and divisor by drawing a cross through them. If there are fewer figures to the right of the decimal point in the dividend than there are to the right of the decimal point in the divisor, annex enough ciphers to the right of the dividend to take care of the new decimal point. Next, divide as in whole numbers. Write the first figure of the quotient directly above the last figure of the dividend used in the first step. Thereafter each figure annexed to the quotient should be written directly above the successive figures in the dividend. Place the decimal point in the quotient directly above the decimal point in the dividend and proceed as in the following examples.

Example a: Divide 58.759787 by .73867, *i.e.*,

.73867)58.759787.

Solution: Since there are five decimal places in the divisor, the decimal point should be moved five places to the right in both dividend and divisor, thus: 73867.)5875978.7. Find by inspection the number of times the divisor is contained into the first group of figures. Place this partial quotient directly above the last figure used in the dividend as shown in the illustrative problem. Place the next partial quotient directly above the next figure used in the dividend, and so on until five figures have been obtained.

```
                79.548
73867.)5875978.7
       517069 ↖ Last figure used
        705288   in first step.
        664803
         404857
         369335
          355220
          295468
          597520
          590936
            6584
```

Example b: Divide .03959 by 8.9752, *i.e.*,

8.9752).03959.

Solution:

```
              .00441
89752./0395.900  ← Hundredths figure.
       359 008   ← Tenths figure.
        36 8920
        35 9008
           99120
           89752
            9368
```

Proceed as in Example *a*, but if the divisor is not contained into the dividend by using the tenths figure, a cipher is placed in the tenths place in the quotient. If the divisor is not contained in the dividend by using the hundredths figure, a second cipher is placed in the quotient in the hundredths place. Keep on adding ciphers until the divisor is contained into the dividend.

From the previous examples the student will notice by placing the partial quotient directly above the last figure used in the dividend, that if the last figure used in the dividend is in tens, the partial quotient is in tens; if the last figure used is in units, the partial quotient is in units; if the last figure used is in tenths, the partial quotient is in tenths; etc.

PROBLEMS

1. Divide the product of 7.9854 and 6.5437 by B.
2. Divide 6.8647 by C. **3.** Divide 7.9754 by D.
4. Divide 6.984 by 23.765 and then multiply by E.
5. Multiply 5.8746 by .26376 and then divide by F.
6. Divide .87654 by G and then multiply by 7.9867.
7. Divide 5.9876 by H. **8.** Divide J by .076543.
9. Divide .008765 by K and then add .76534.

VARIABLES

Prob.	Sym.	No. 1	No. 2	No. 3	No. 4	No. 5	No. 6
1	B	5.7487	3.8765	7.9754	8.6432	5.9732	2.9648
2	C	.76542	.58352	.95275	.48275	.92746	.92648
3	D	.06542	.06327	.07625	.09642	.08426	.08532
4	E	4.9876	5.9264	6.9375	4.9863	7.9543	6.3965
5	F	.98534	.95432	7.9532	.97532	.96427	.96425
6	G	6.9476	8.6543	7.9642	3.9742	8.9542	8.7533
7	H	.46538	.96536	.67486	.98743	.96438	.94672
8	J	8.4852	7.5837	5.4375	8.6548	4.8769	6.4653
9	K	.08764	.07378	.08754	.08765	.05328	.0987

10. Multiply .09867 by 3.7652 and then divide by L.

11. Divide the product of M and 2.3649 by the quotient of 4.9867 divided by .76548.

12. Multiply 7.5876 by N and then divide by 5.9837.

13. Divide 4.8956 by P. **14.** Divide .008765 by R.

15. Divide 5.8646 by S.

16. Divide .48769 by T and then multiply by .056849.

Change the following fractions to decimal numbers:

17. $\frac{17}{U}$. **18.** $\frac{V}{189}$. **19.** $\frac{262}{W}$.

VARIABLES

Prob.	Sym.	No. 1	No. 2	No. 3	No. 4	No. 5	No. 6
10	L	8.8654	6.0984	8.4653	5.9836	5.3869	4.9724
11	M	4.3856	3.9875	3.6582	3.2857	2.9876	2.6895
12	N	.68954	.62893	.56894	.49873	.47239	.42189
13	P	5.8796	5.9873	6.3897	6.7893	7.1389	7.5693
14	R	.03985	.04763	.05784	.06895	.07289	.08396
15	S	4.6879	4.1389	3.8976	3.6895	3.1896	2.9876
16	T	.06895	.07329	.08395	.08962	.09137	.09654
17	U	21	23	25	27	29	31
18	V	2123	2133	2143	2153	2163	2173
19	W	567	589	603	625	647	669

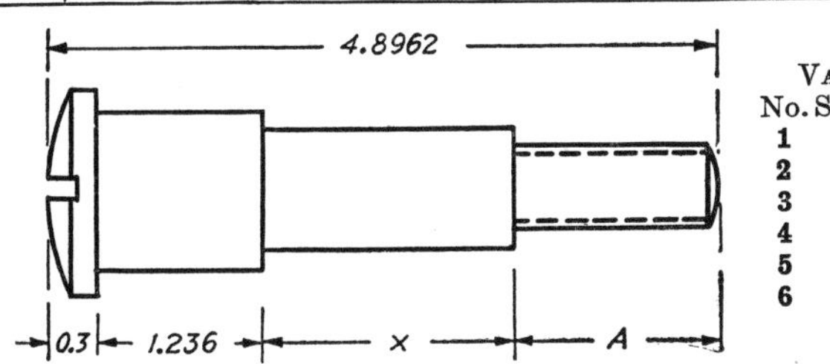

VARIABLE

No.	Sym.	Value
1	A	.8792
2	A	.8967
3	A	.9063
4	A	.9278
5	A	.9487
6	A	.9672

20. Determine the distance x.

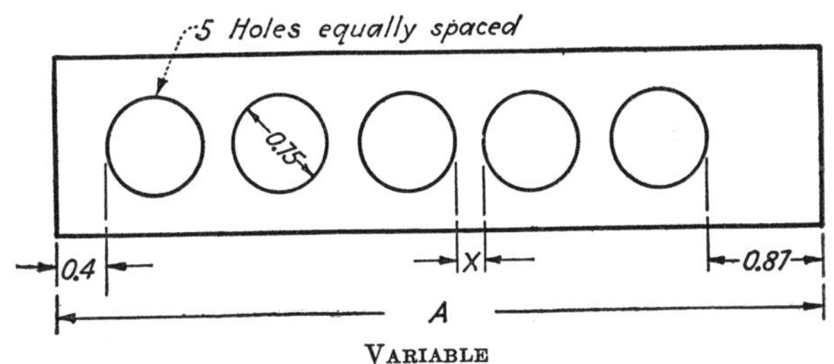

VARIABLE

1. A = 10.783 **2.** A = 11.642 **3.** A = 11.875
4. A = 11.964 **5.** A = 12.137 **6.** A = 12.379

21. Determine the distance x.

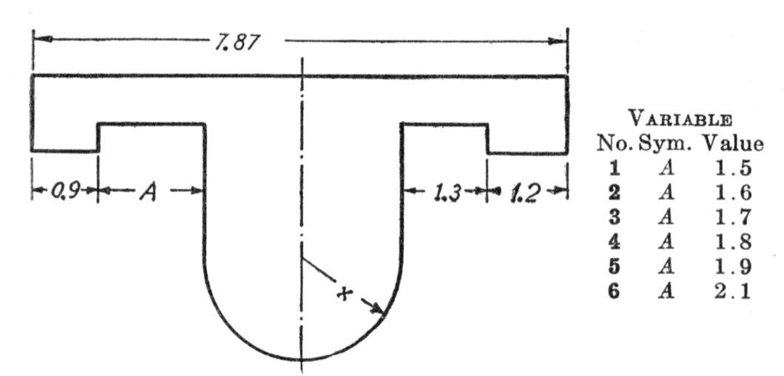

Variable		
No.	Sym.	Value
1	A	1.5
2	A	1.6
3	A	1.7
4	A	1.8
5	A	1.9
6	A	2.1

22. Determine the radius x.

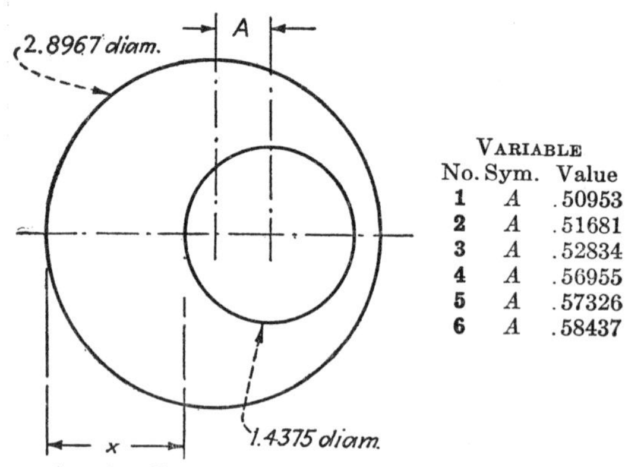

Variable		
No.	Sym.	Value
1	A	.50953
2	A	.51681
3	A	.52834
4	A	.56955
5	A	.57326
6	A	.58437

23. Determine the distance x.

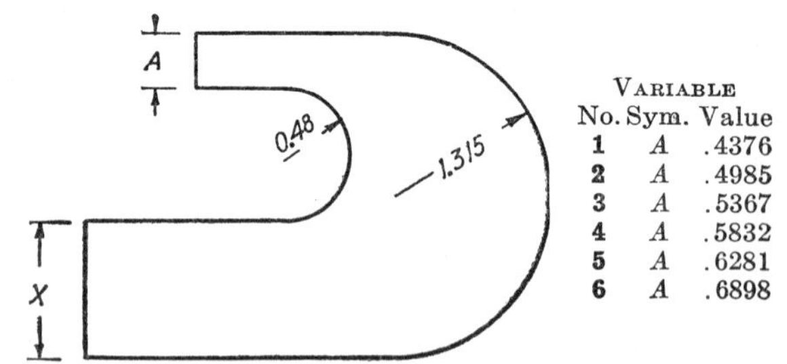

Variable		
No.	Sym.	Value
1	A	.4376
2	A	.4985
3	A	.5367
4	A	.5832
5	A	.6281
6	A	.6898

24. Determine the distance x.

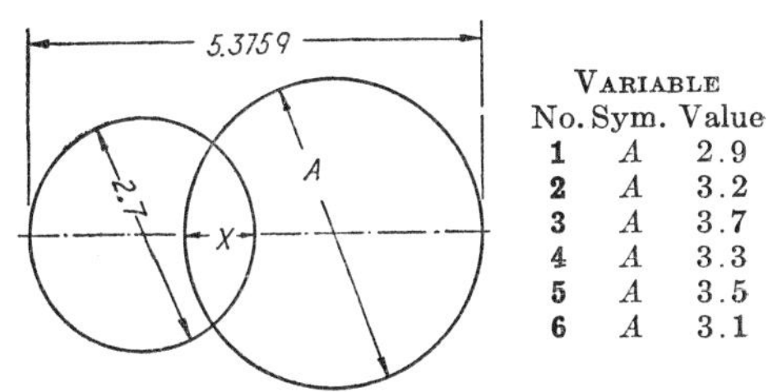

VARIABLE		
No.	Sym.	Value
1	*A*	2.9
2	*A*	3.2
3	*A*	3.7
4	*A*	3.3
5	*A*	3.5
6	*A*	3.1

25. Determine the distance x.

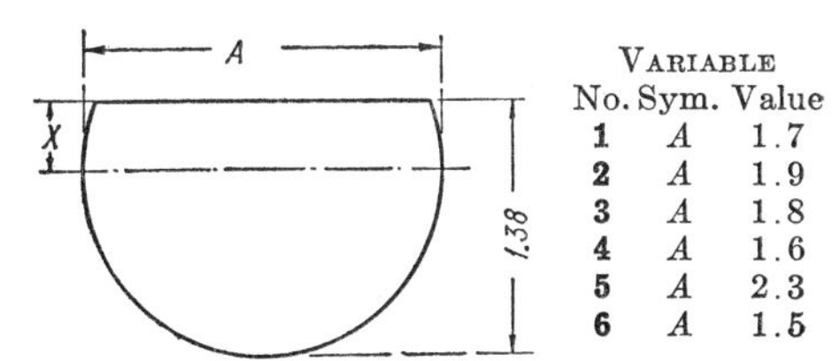

VARIABLE		
No.	Sym.	Value
1	*A*	1.7
2	*A*	1.9
3	*A*	1.8
4	*A*	1.6
5	*A*	2.3
6	*A*	1.5

26. Determine the distance x.

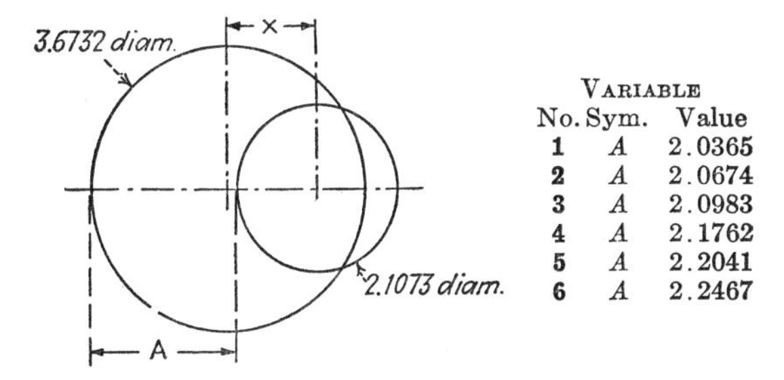

VARIABLE		
No.	Sym.	Value
1	*A*	2.0365
2	*A*	2.0674
3	*A*	2.0983
4	*A*	2.1762
5	*A*	2.2041
6	*A*	2.2467

27. Determine the distance x.

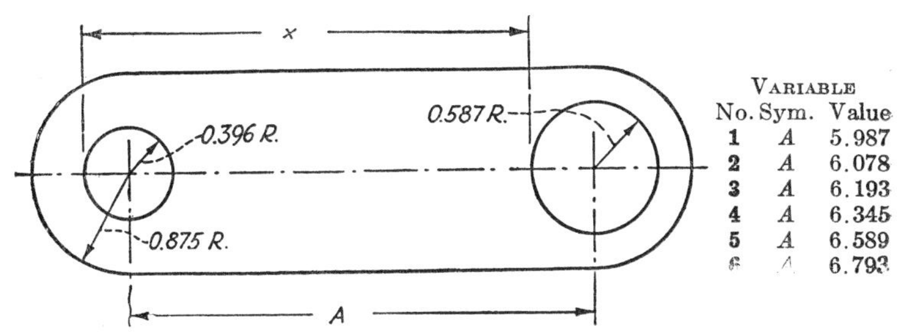

VARIABLE		
No.	Sym.	Value
1	*A*	5.987
2	*A*	6.078
3	*A*	6.193
4	*A*	6.345
5	*A*	6.589
6	*A*	6.793

28. Determine the distance x.

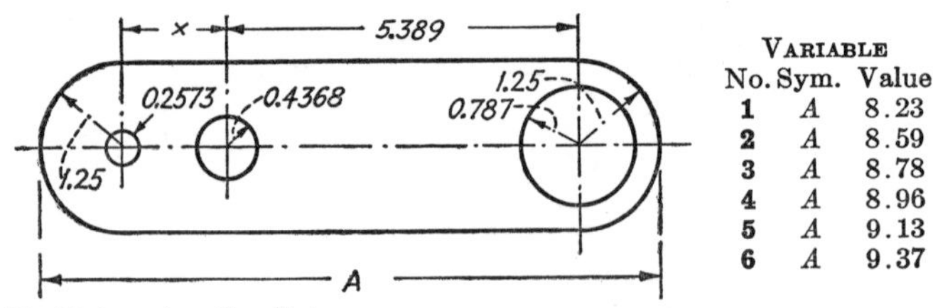

Variable		
No.	Sym.	Value
1	*A*	8.23
2	*A*	8.59
3	*A*	8.78
4	*A*	8.96
5	*A*	9.13
6	*A*	9.37

29. Determine the distance x.

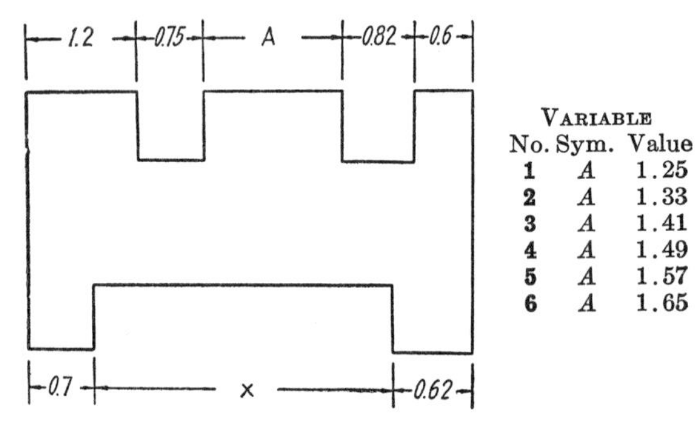

Variable		
No.	Sym.	Value
1	*A*	1.25
2	*A*	1.33
3	*A*	1.41
4	*A*	1.49
5	*A*	1.57
6	*A*	1.65

30. Determine the distance x.

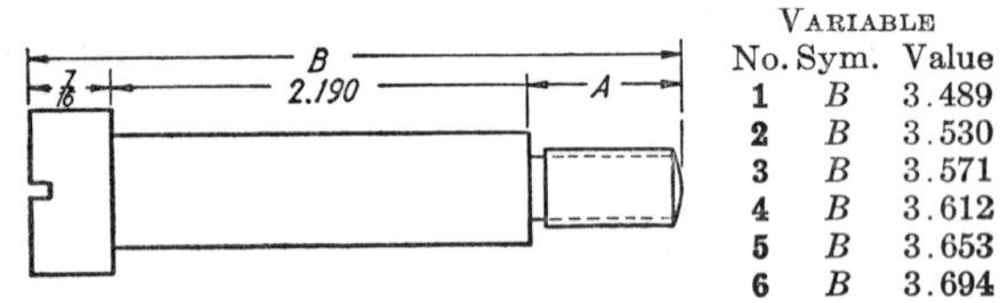

Variable		
No.	Sym.	Value
1	*B*	3.489
2	*B*	3.530
3	*B*	3.571
4	*B*	3.612
5	*B*	3.653
6	*B*	3.694

31. Determine the distance A.

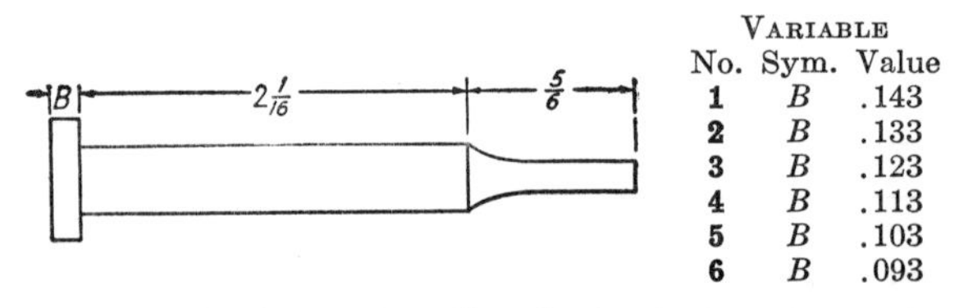

Variable		
No.	Sym.	Value
1	*B*	.143
2	*B*	.133
3	*B*	.123
4	*B*	.113
5	*B*	.103
6	*B*	.093

32. Allowing $\frac{1}{8}$ in. for cutting off each pin, (*a*) how many pins can be made from a 36 in. bar and (*b*) how much material will be left?

CHAPTER IV

MICROMETERS, VERNIERS, AND BEVEL PROTRACTORS

MICROMETERS

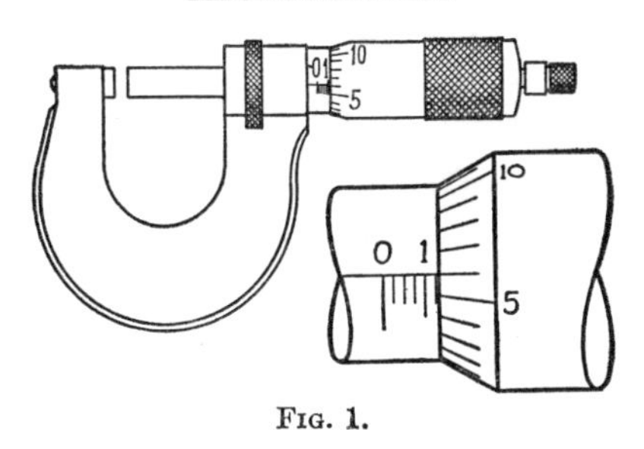

FIG. 1.

The scale on the barrel of the micrometer is divided into tenths of an inch and each tenth is subdivided into four parts, each part representing twenty-five thousandths of an inch. The screw of the micrometer has 40 threads per inch, so each revolution of the thimble will open the micrometer .025 in. The tapered end of the thimble is graduated into 25 equal divisions. Hence, rotating the thimble one division opens the micrometer .001 in.

To read a micrometer, observe the number of tenths and subdivisions up to the last line exposed to view by the thimble and express this value in thousandths of an inch. Add to this the thousandths given by the reading of the thimble which is just opposite the longitudinal line of reference which runs through the main scale.

In the foregoing figure, the exposed portion of the barrel shows $.1 + .025 = .125$. The reading on the thimble is .006 (to the nearest thousandth). Hence, the reading of the instrument is $.125 + .006 = .131$.

PROBLEMS

Determine the readings for the following settings of a micrometer:

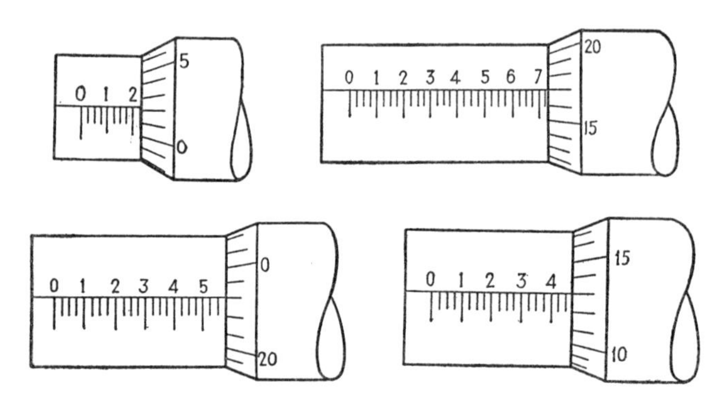

VERNIERS

The vernier caliper has two principal parts, the vernier proper and the main scale. The vernier proper is usually rather short with a definite number of divisions, while the scale may be of great length with a larger number of divisions. In the English-system vernier, the major divisions of the scale are inches; these are subdivided into tenths of an inch; each tenth is subdivided into four or five parts, each part being twenty-five or twenty thousandths, respectively.

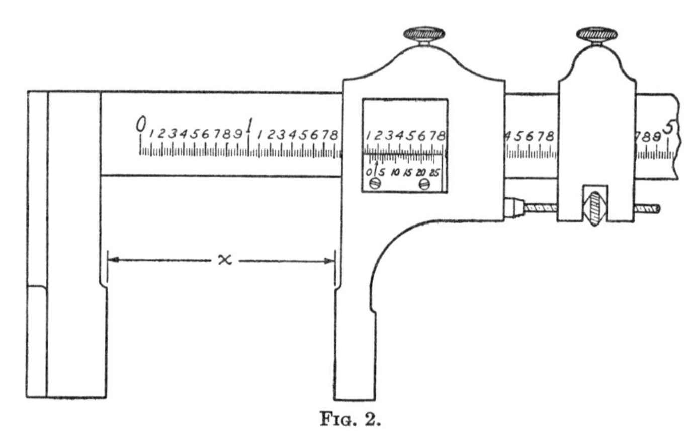

Fig. 2.

The vernier proper of a Starrett vernier caliper is .6 in. in length and is divided into 25 divisions. The scale has 24 divisions within the same length; therefore the two end lines of the vernier proper can be made to coincide with two lines on the scale. Each division of the vernier proper is equal to $\frac{.600}{25}$ or .024 in. and each division of the scale is $\frac{.600}{24}$ or .025 in.

If the two end lines of the vernier coincide with two lines of the scale, the distance from line 1 of the vernier proper to line 1 on the scale is the difference of .025 and .024 in., or .001 in. This value (.001 in.), which is the smallest value that can be read with the instrument, is called the least count. Then if the vernier is moved so that line one of the vernier proper coincides with line one of the scale, the vernier will have moved .001 in.; in like manner if line 5 of the vernier proper coincides with line 5 on the scale, the vernier will have moved .005 in.; if line 15 of the vernier proper coincides with line 15 of the scale, the vernier will have moved .015 in., etc.

When the jaws of the vernier are closed, the zero line of the vernier should coincide with (make a continuous line with) the zero line of the main scale.

The distance between the jaws of the vernier caliper is the same as the distance between the zero line of the vernier and the zero line of the main scale. To determine this, first read the inches and fractional part of an inch up to the line on the main scale, which is to the immediate left of the zero line of the vernier proper, and to this add the number of thousandths as given by the number of the vernier line which coincides with a line of the main scale.

In the foregoing figure, the zero line of the vernier is to the right of the first small division past the 2.1-in. mark, which means that the reading is more than $2.1 + .025 = 2.125$. The arrow indicates that the third line of the vernier proper coincides with a line of the main scale. Hence, the true reading of the instrument is $2.125 + .003 = 2.128$.

For some machines there are special verniers. The following problem will show how any vernier may be read.

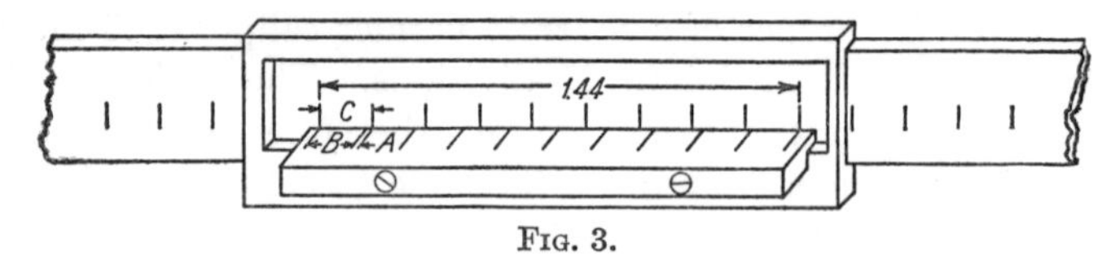

FIG. 3.

Illustrative Problem: The smallest main scale division is $C = \frac{1.44}{9} = .160$. The smallest vernier proper division is $B = \frac{1.44}{10} = .144$. Thus A, which is the distance between the first line on the vernier and the first line on the main scale, $= .160 - .144 = .016$. Hence, if the first line of the vernier coincides with the first line of the main scale, the vernier must have moved .016. If the fifth line of the vernier coincides with the fifth line of the main scale, the vernier proper has moved $5 \times .016 = .08$, etc.

PROBLEMS

If the zero line on the vernier proper is just beyond the line indicated by the arrow on the scale, and the Nth line of the vernier proper makes a continuous line with a line on the scale, what is the reading of the vernier? The number attached to each arrow indicates the different settings.

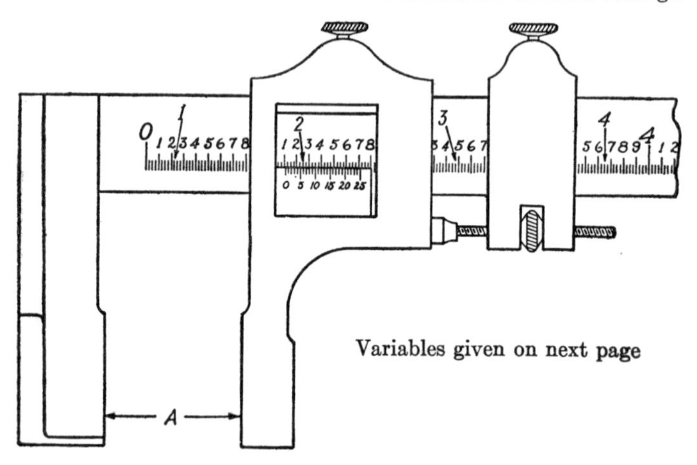

Variables given on next page

1. Determine A in setting 1.

2. Determine A in setting 2.

3. Determine A in setting 3.

4. Determine A in setting 4.

The number attached to each arrow indicates the different settings.

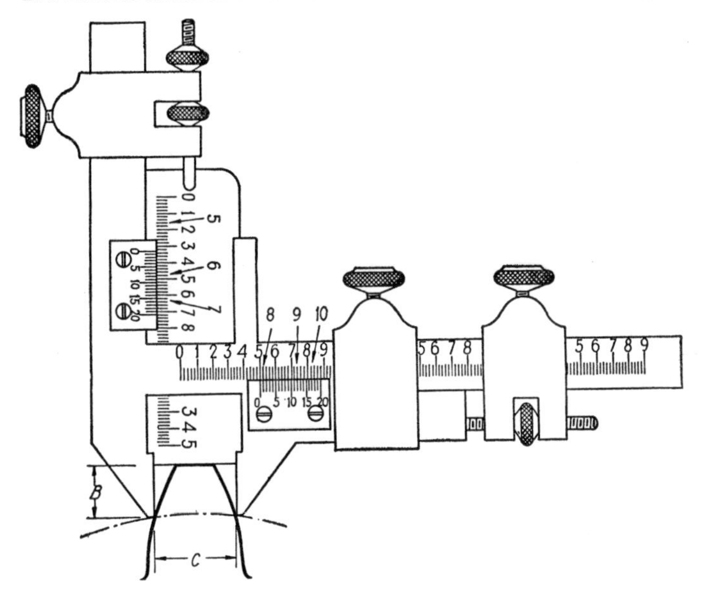

5. Determine B in setting 5.
6. Determine B in setting 6.
7. Determine B in setting 7.
8. Determine C in setting 8.
9. Determine C in setting 9.
10. Determine C in setting 10.

VARIABLES

Prob.	Sym.	No. 1	No. 2	No. 3	No. 4	No. 5	No. 6
1	N	4	6	8	10	12	14
2	N	23	5	4	3	2	9
3	N	14	16	18	20	22	24
4	N	3	5	7	9	11	13
5	N	17	20	18	16	14	2
6	N	11	9	7	5	3	20
7	N	19	15	2	4	6	8
8	N	7	9	11	13	15	17
9	N	10	12	14	16	18	20
10	N	19	20	18	3	5	7

BEVEL PROTRACTORS

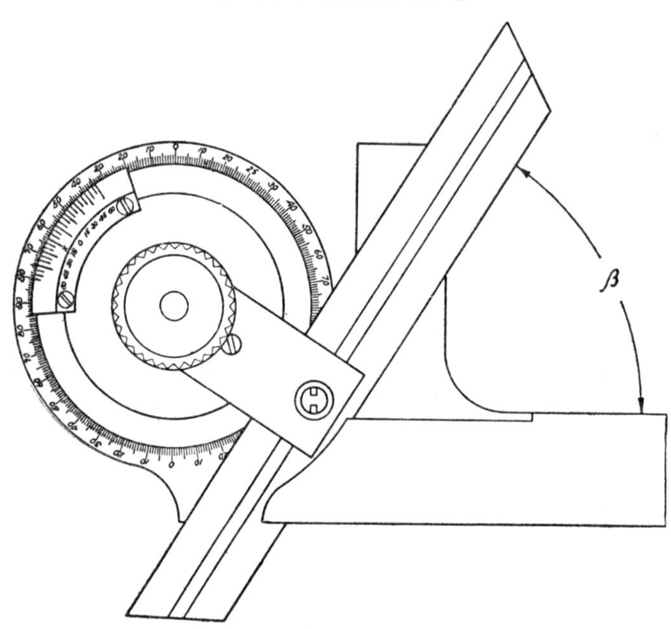

FIG. 4.

The bevel protractor used in shops is usually of the vernier type. The main scale is a stationary circular scale which is divided into degrees (see Fig. 4). The vernier proper is attached to a circular disk which rotates inside the main circular scale. The Brown and Sharpe instrument has a double vernier proper, each half of which is divided into 12 divisions.

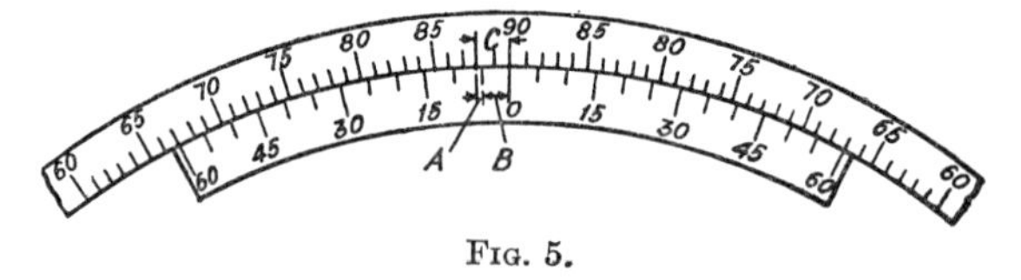

FIG. 5.

The principle of reading a vernier attached to a protractor is the same as that of a vernier attached to a scale. The only difference is that instead of the vernier being within a certain number of inches it is within a certain number of degrees. Reduce these degrees to minutes and proceed as before. There are 60 minutes (60′) in one degree. First of all observe that the vernier is within 23 degrees (23°). Reduce 23° to minutes, which gives 23 × 60 or 1380′. Next note that the vernier is divided into 12 equal parts, each part being equal to one-twelfth of 1380′ or 115′ indicated by B (Fig. 5). Since there are 60′ in 1°, then in 2° there are 120′ indicated by C (Fig. 5). $C - B = A$ or $120' - 115' = 5'$ which is the curved distance the vernier moves from one line to the next consecutive line on the protractor scale.

Hence, if the fourth line of the vernier proper coincides with a line of the main scale, four times 5′ or 20′ must be added to the reading in degrees on the main scale as determined by the position of the zero line of the vernier scale. As shown by Fig. 5, the main scale is graduated in degrees in both directions from zero. If the zero of the vernier proper is to the left of the zero of the main scale, the left half of the double vernier proper is used, and *vice versa.* Thus, in the illustration (Fig. 4) the reading is $57° + 3 \times 5' = 57°\ 15'$.

PROBLEMS

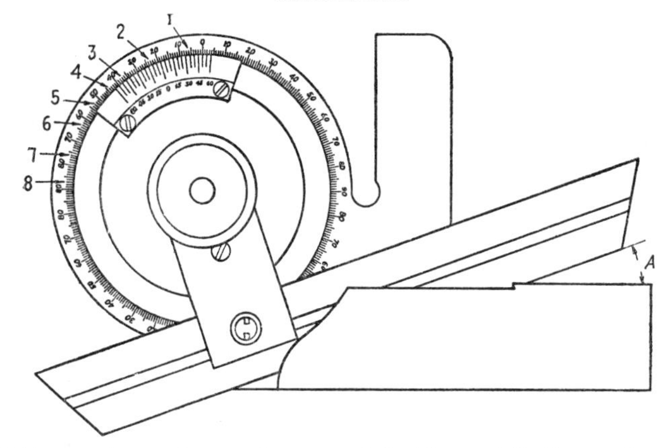

If the zero line on the vernier proper is just beyond the line on the outer dial indicated by the arrow, and the Lth line of the vernier proper makes a continuous line with a graduation on the outer dial, what is the value of angle A?

1. Determine angle A in setting 1.
2. Determine angle A in setting 2.
3. Determine angle A in setting 3.
4. Determine angle A in setting 4.
5. Determine angle A in setting 5.
6. Determine angle A in setting 6.
7. Determine angle A in setting 7.
8. Determine angle A in setting 8.

The answers to the following four problems are to be expressed in degrees and minutes.

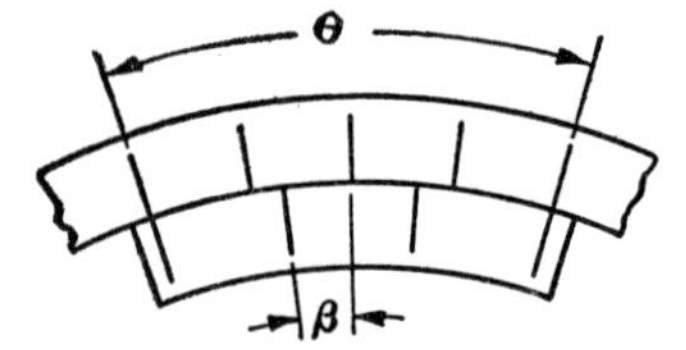

9. Determine the angle β.

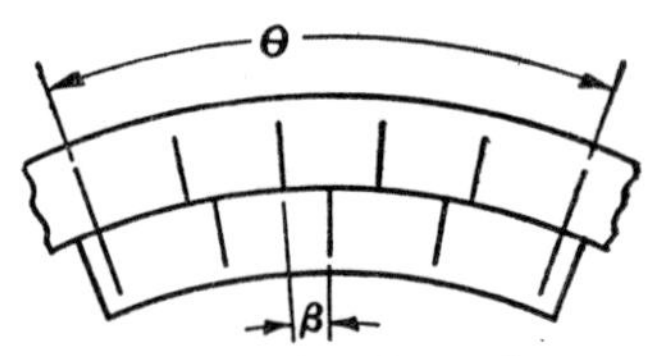

10. Determine the angle β.

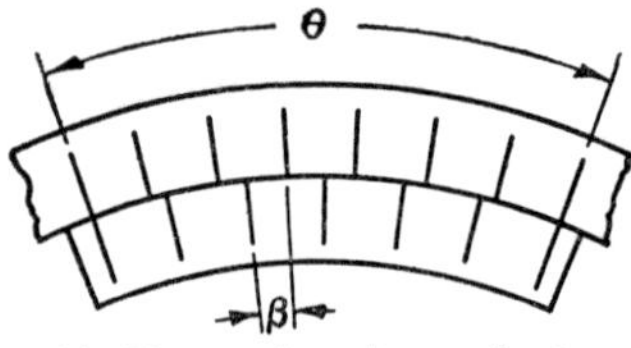

11. Determine the angle β.

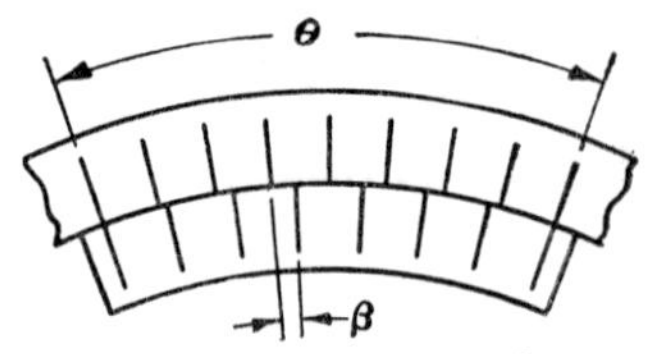

12. Determine the angle β.

VARIABLES

Prob.	Sym.	No. 1	No. 2	No. 3	No. 4	No. 5	No. 6
1	L	2	3	4	5	6	7
2	L	10	9	8	7	5	6
3	L	4	6	8	10	9	7
4	L	7	9	11	3	1	10
5	L	8	6	4	2	7	5
6	L	5	4	3	2	1	9
7	L	5	6	7	8	9	10
8	L	1	2	3	4	5	6
9	θ	20°	21°	22°	23°	24°	25°
10	θ	16°	17°	18°	19°	20°	21°
11	θ	19°	20°	21°	22°	23°	24°
12	θ	22°	23°	24°	25°	26°	27°

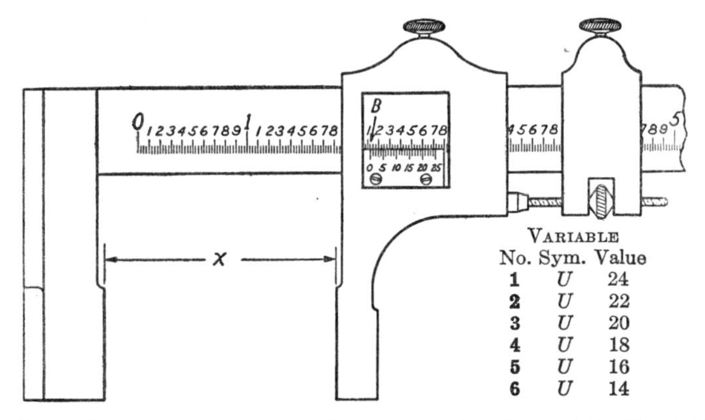

VARIABLE

No.	Sym.	Value
1	U	24
2	U	22
3	U	20
4	U	18
5	U	16
6	U	14

13. If the zero line on the vernier proper is just beyond the line indicated by the arrow (B), and the Uth line on the vernier proper makes a continuous line with a line on the scale, determine the distance x.

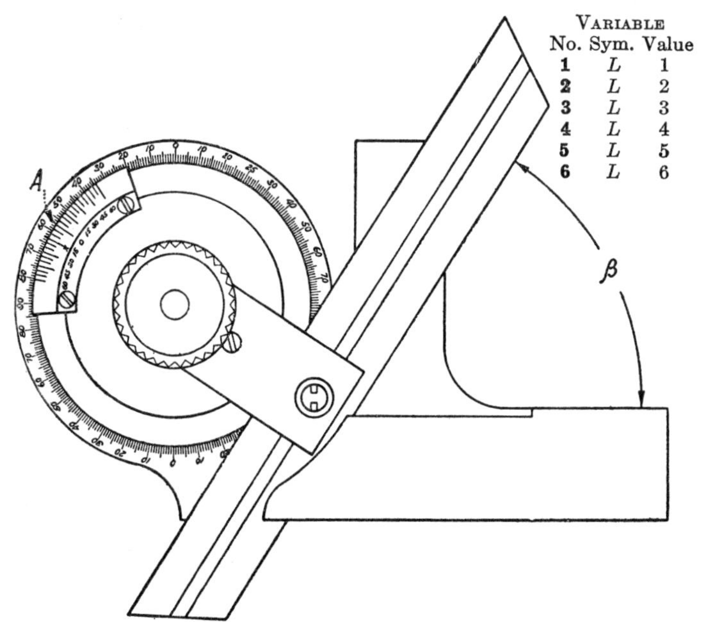

VARIABLE

No.	Sym.	Value
1	L	1
2	L	2
3	L	3
4	L	4
5	L	5
6	L	6

14. If the zero line of the vernier proper is just beyond the line indicated by the arrow (A), and the Lth line on the vernier proper makes a continuous line with a line on the protractor scale, determine the angle β.

CHAPTER V

ALGEBRA

Algebra is like arithmetic in that the principal operations used are addition, subtraction, multiplication, and division.

Algebra differs from arithmetic in that the quantities involved are often represented by letters instead of by numbers.

Letters are frequently used to represent unknown quantities in practical problems as a means of leading to a statement that two quantities are equal. Such a statement of equality is called an equation. One of the main functions of algebra is the solving of such equations, thus leading to the determination of the unknown quantity.

In this chapter those processes of algebra particularly used in the solution of shop problems will be discussed.

USE OF POSITIVE AND NEGATIVE NUMBERS

Numbers are commonly used to represent the magnitude of quantities. Thus the temperature on a certain summer day may be 70°F. (meaning 70 above 0°F.), a man may possess $5, or a certain place may have an altitude of +500 ft. (meaning 500 ft. above sea level).

Suppose it is now desired to represent a temperature of 6° below zero, the fact that a man owes $5, or the altitude of a place which is 200 ft. below sea level. These quantities are best represented by −6°F., −$5, and −200 ft., respectively. Thus negative numbers are used to represent magnitudes in the opposite sense to those which are arbitrarily chosen as positive. Usually the sign of operation is omitted in the case of positive numbers. Thus: 11 means +11.

The numerical value of a number regardless of its sign is called its absolute magnitude. The absolute magnitudes of −4, +8, and −3 are 4, 8, and 3, respectively.

ADDITION AND SUBTRACTION OF POSITIVE AND NEGATIVE NUMBERS

To add two positive numbers, add their absolute magnitudes and prefix the plus sign.

Example: $+7 + 22 = +29$.

To add two negative numbers, add their absolute magnitudes and prefix the minus sign.

Example: $-8 - 34 = -42$.

To add a positive number and a negative number, obtain the difference of their absolute magnitudes and prefix the sign of the number having the greater magnitude.

Examples: $-27 + 19 = -8, 44 - 18 = 26, 37 - 52 = -15$.

MULTIPLICATION OF POSITIVE AND NEGATIVE NUMBERS

The product of two numbers having like signs is positive and the product of two numbers having unlike signs is negative.

Examples: $5 \times 8 = 40$, $-20(-3) = 60$, $-12 \times 8 = -96$, $4(-13) = -52$. Note that the parentheses often take the place of a multiplication sign (see page 42).

DIVISION OF POSITIVE AND NEGATIVE NUMBERS

In division one number, called the divisor, is contained in another number, called the dividend, a certain number of times. This latter number is called the quotient. Thus:

$$\frac{48 \text{ (dividend)}}{8 \text{ (divisor)}} = 6 \text{ (quotient)}.$$

From this it follows that the dividend = divisor $\times$ quotient.

Since the law regarding signs in multiplication must apply to this, it follows that the law of signs for division is:

If the signs of the dividend and divisor are alike, the quotient is positive; if they are unlike, the quotient is negative.

Examples: $\frac{28}{4} = 7, \frac{56}{-4} = -14, \frac{-72}{9} = -8, \frac{-96}{-6} = 16.$

When the plus (+) and minus (−) signs occur with multiplication ($\times$) and division ($\div$) signs in an expression, the multiplication and division operations must be performed first, and then the addition and subtraction operations may be performed in the order in which they are written.

Examples:

$$12 \times 8 - 6 + 4 \times 12 = 96 - 6 + 48 = 138.$$
$$102 \div 6 - 6 \times 2 + 3 = 17 - 12 + 3 = 8.$$

It is a good policy to add all the plus quantities first, then add all of the minus quantities, and finally subtract the quantity having the lesser magnitude from the quantity having the greater magnitude and prefix the sign of the quantity having the greater magnitude to the result.

Example a: $12 - 6 + 4 - 2 + 9 - 18 + 5 = ?$

Solution: The sum of the plus quantities is $12 + 4 + 9 + 5 = 30$. The sum of the minus quantities is $-6 - 2 - 18 = -26$. The subtraction of the lesser magnitude from the greater is $30 - 26 = 4$.

Example b: $8 - 4 - 16 + 3 - 7 + 2 = ?$

Solution: The sum of the plus quantities is $8 + 3 + 2 = 13$. The sum of the minus quantities is $-4 - 16 - 7 = -27$. The subtraction of the lesser magnitude from the greater is $13 - 27 = -14$.

When several numbers are multiplied together, the product will be the same regardless of the order in which the multiplications are performed. Thus: $6 \times 35 \times 48$; $48 \times 6 \times 35$; $6 \times 48 \times 35$; $35 \times 48 \times 6$, etc., all have the same product.

PROBLEMS

1. Represent the following by the use of positive and negative numbers: a bank balance of \$95, a debt of \$20, a temperature of 10° below zero, a temperature of 72° above zero, the altitude of a place which is 800 ft. above sea level, the altitude of a place which is 100 ft. below sea level.

2. On a certain winter day, 20 tons of coal were burned in a factory and A tons were delivered. What is the net gain in the coal pile for the day?

3. At 6 A.M. a thermometer reads 60°F. The temperature rises B° during the next 3 hr.; rises 10° more during the next 3 hr.; falls 2° during the next 3 hr.; falls 12° during the next 3 hr. Compute the temperature at (*a*) 9 A.M., (*b*) noon; (*c*) 3 P.M., and (*d*) 6 P.M.

4. $4 - 8 + C - 17 + 4 = ?$

5. Multiply: (*a*) $+4$ and $+9$, (*b*) -5 and $+6$, (*c*) 8 and -7, (*d*) -6 and -13, (*e*) 21 and -8.

6. Divide: (*a*) 72 by 8, (*b*) 96 by -12, (*c*) -144 by 16, (*d*) -182 by -13, (*e*) 63 by -7.

Perform the indicated operations:

7. $8(-7) + D - 10 + 60.$

8. $E - 8(-3) = ?$

9. $-20 \times 6 + F - 28 + \frac{-40}{-4} = ?$

10. $G + \frac{35}{-7} = ?$

11. $\frac{30}{-5} + 6 - H + 7 \times 6 = ?$

12. $18 - J + 6 \times 8 = ?$

13. $8 - K + 12 \div 3 + 4 \times 9 = ?$

14. $27 - L + 3 = ?$

15. $16 \div 2 + 4 - 7 - M = ?$

16. $-9 + N - 24 \div 6 = ?$

17. $4 \times 8 - 9 - P = ?$

18. $12 - Q + 2 - 15 = ?$

19. $15 - R + 7 - 12 = ?$

20. $58 - 38 \div 2 - S = ?$

21. $23 + 7 - 48 \div 6 - 3 \times 7 + T - 26 + 8 = ?$

22. $55 \div 5 + 6 - 56 \div 8 + 2 \times 7 - U + 9 \times 3 = ?$

23. $12 + 7 - 20 - 36 - 52 - V + 2 \times 9 = ?$

24. $75 - 32 - 46 - W + 7 + 8 + 9 = ?$

25. $84 \div 2 - 4 \times 6 - 12 \div 2 - A + 3 - 8 = ?$

VARIABLES

Prob.	Sym.	No. 1	No. 2	No. 3	No. 4	No. 5	No. 6
2	*A*	40	42	44	46	48	50
3	*B*	16	18	20	22	24	26
4	*C*	10	12	14	16	18	20
7	*D*	3	4	5	6	7	8
8	*E*	5	7	9	11	13	15
9	*F*	70	74	78	82	86	90
10	*G*	7	9	11	13	15	17
11	*H*	14	16	18	20	22	24
12	*J*	18	19	20	21	22	23
13	*K*	2	4	6	8	10	12
14	*L*	35	37	39	41	43	44
15	*M*	5	7	9	11	13	15
16	*N*	18	20	22	24	26	28
17	*P*	41	43	45	47	49	51
18	*Q*	5	6	7	8	9	10
19	*R*	10	11	12	13	14	15
20	*S*	5	7	9	11	13	15
21	*T*	2	3	4	5	6	7
22	*U*	50	52	54	56	58	60
23	*V*	25	27	29	31	33	35
24	*W*	30	32	34	36	38	40
25	*A*	5	8	11	14	17	20

PARENTHESES AND GROUPING SYMBOLS

In a series of operations it frequently becomes necessary to use grouping symbols, such as parentheses (), brackets [], braces { }, or a vinculum ——. These symbols indicate that certain addition and subtraction operations should precede multiplication and division. They also indicate that the operations within should be carried out completely before the remaining operations are made. After these have been completed, the grouping symbols may be removed. If there are more than one pair of grouping symbols in an expression, the innermost pair should be removed first.

Examples: $7 + (6 - 2) = 7 + 4 = 11;$
$6 \times (8 - 5) = 6 \times 3 = 18.$

In an expression where grouping symbols are immediately preceded or followed by a number or quantity with the signs of operation omitted, multiplication is understood.

Examples: $8 + 6(4 - 1) = 8 + 18 = 26;$
$(6 + 2)(9 - 5) = 8 \times 4 = 32.$

Parentheses or other grouping symbols are often used in connection with subtraction and multiplication of negative quantities. Thus: plus 4 less negative 7 is written: $4 - (-7)$; plus 4 times negative 7 is written $4(-7)$.

To remove parentheses (or other grouping symbols) which are preceded by negative signs, the signs of all terms inside the grouping symbols must be changed (from plus to minus and minus to plus).

Examples: $4 - (-7) = 4 + 7 = 11;$
$8 - (7 - 4) = 8 - 3 = 5.$

Parentheses (or other grouping symbols) which are preceded by a plus sign may be removed without changing the signs of the terms within the grouping symbols.

Examples: $3 + (-8) = 3 - 8 = -5;$
$7 + (4 - 19) = 7 + (-15) = 7 - 15 = -8.$

When one set of grouping symbols is included within another set, remove the innermost set first.

Examples: $3[40 + (7 + 5)(8 - 2)] = 3[40 + 12 \times 6] =$
$3[40 + 72] = 3 \times 112 = 336;$

$$7 - 3\left[\frac{16 + 4}{5(8 - 6)} + 4\right] = 7 - 3\left[\frac{20}{5 \times 2} + 4\right] =$$
$$7 - 3[2 + 4] = 7 - 3 \times 6 = 7 - 18 = -11.$$

When several terms connected by + or − signs contain a common quantity, this common quantity may be placed in front of a parenthesis (or other grouping symbol) which encloses the results of dividing each of the several terms by the common quantity (called the common factor).

Example: In the expression $8x + 12$ the quantity 4 may be "factored out" giving $4(2x + 3)$. This is easily seen to be the reverse procedure of removing parentheses.

PROBLEMS

1. $M - (8 - 3) + 7 = ?$
2. $(7 - 2)(8 + N) = ?$
3. $L + 16 - 2[30 - (8 - 6)(2 + 7)] + 12 \div 4 = ?$
4. $K - \left[\frac{17 - 2}{5(6 - 4)}\right]5 + \frac{36 + 4}{8} = ?$
5. $P + 4(8 + 6)(9 - 2) \div 7 + 5 \times 8 = ?$
6. $Q - (8 + 2)(6 + 3) - 5 + 16 + 2 \times 8 + 72 = ?$
7. $5 - (-8) + R + (-14) = ?$
8. $[15 + 2(6 + 3)][3(A - 7) - 8] = ?$
9. $(8 - 3)(9 + B) - 6(8 + 7) - 21 = ?$
10. $6 \times 2 + 7(6 + 7) - 8(3 - 5)(C + 3) - 3 = ?$

VARIABLES

Prob.	Sym.	No. 1	No. 2	No. 3	No. 4	No. 5	No. 6
1	M	12	23	34	45	56	67
2	N	7.5	8.6	9.7	10.8	11.9	12.3
3	L	41	52	63	74	85	96
4	K	2	4	6	8	10	12
5	P	7	11	16	19	21	23
6	Q	56	67	78	89	95	99
7	R	6	8	10	12	14	16
8	A	2	3	4	5	6	7
9	B	8	7	6	5	4	3
10	C	3	5	7	9	11	13

11. $12 \div 4 + 18 \div 3 - D(8 - 2)(6 + 4) = ?$
12. $5[8 - 4(E - 4)][(6 - 3) + 2(7 + 2)] + 10 = ?$
13. $F + 12 \div 4 - 18 \div 6 - 24 \div 4 = ?$
14. $(12 - 4)(G - 5) + (8 - 12)(5 - 10) + 3 = ?$
15. $3 + 2 - H - 8 \div 4 - 3 \times 5 - 15 \div 5 + 7 = ?$
16. $3\{J + 2[6 - 3(6 - 3) + 3] - 6 \div 2\} + 10 = ?$
17. $2\{8 - 3[(4 + 5) - 3(6 \div 2)] + 5\}[S - 4(6 + 2)] = ?$
18. $5(6 + 5) - (T + 3)(7 - 5) + 3(8 \div 4)(5 \times 3 - 7) = ?$

VARIABLES

Prob.	Sym.	No. 1	No. 2	No. 3	No. 4	No. 5	No. 6
11	D	7	6	5	4	3	2
12	E	8	10	12	14	16	18
13	F	15	17	19	21	23	25
14	G	5	6	7	8	9	10
15	H	7	9	11	13	15	17
16	J	6	8	10	12	14	16
17	S	5	7	9	11	13	15
18	T	12	11	10	9	8	7

ALGEBRAIC SYMBOLS AND SIMPLE EQUATIONS

Frequently when the numerical value of a quantity is unknown, it is represented by a letter called an algebraic symbol. If enough data are given, the numerical value represented by the algebraic symbol can be obtained.

A **factor** of an expression is any one of the numbers or letters or groups which multiplied together give the expression.

Examples: The factors of 12 are 3 and 4. The factors of $5ac$ are 5, a, and c. The factors of $8x + 12$ are 4 and $2x + 3$. (See the example preceding the problems on page 43.)

An **equation** is a statement of equality between numbers or numbers and algebraic symbols.

The part of the equation which is to the left of the equality sign is called the left member (or left side) and the part to the right of the equality sign is called the right member (or right side).

Examples of equations:

$$12 = 6 \times 2; \quad 13 + 5 = 18;$$
$$2x + 9 = 15; \qquad y - 7 = 4y + 5.$$

Equations involving algebraic symbols to the first power only (the same symbol may occur more than once but only to the first power) is called a **simple equation** or linear equation.

Examples: $2x + 4 = 10$; $4x + 2 = 14x$;
$3x + 4y + 6 = 2y + 4.$

An **exponent** is the small number written at the right and a little above another number (or quantity) called the **base number** (or quantity). The exponent indicates the power of the base number (or quantity), *i.e.*, it expresses the number of times the base number (or quantity) is taken as a factor.

The **power** of a number is the result obtained by using the base number the specified number of times as a factor.

Example: $3^4 = 3 \times 3 \times 3 \times 3 = 81$. In the exponential expression 3^4, 3 is the base number, 4 is the exponent denoting that 3 is taken as a factor 4 times, and the result 81 is the fourth power of 3.

$a^2 = a \times a$. In this case, the base quantity is a, the exponent is 2 denoting that a is taken as a factor twice, and a^2 is the second power or "square" of a.

ADDITION AND SUBTRACTION OF EXPRESSIONS INVOLVING ALGEBRAIC SYMBOLS

In adding or subtracting several quantities, some of which involve algebraic symbols, only those terms which involve the same symbols and power can be combined. Thus:

$$10x + 14 - 7y^2 - 11a + 2x - 4 + 3y^2 - 4a + 8 =$$
$$10x + 2x - 7y^2 + 3y^2 - 11a - 4a + 14 - 4 + 8 =$$
$$12x - 4y^2 - 15a + 18.$$

PROBLEMS

Perform the indicated operations and combine similar terms.

1. $10x + 14 + 3y + 8 + 6x = ?$ *Ans.* $16x + 3y + 22$

2. $13 - (3x + 2) + (x - 8) = ?$

3. $4(2x + 8) - 9y - 3(4x + 7) = ?$

4. If $x = 2$, is $4x + 2 = 10$ a true equation?

5. If $x = 3$, is $4x + 2 = 10$ a true equation?

6. $5x^2 - 3y - 3x^2 + 5 - 4y + 7 - 3x = ?$

Ans. $2x^2 - 3x - 7y + 12$

7. $6(x - y) + 3(2x^2 - 4y) + 5 - x^2 + 7 + 5y = ?$

8. $4y^3 + 2y^2 - 5 + 3y^2 + 5x - 5y^3 = ?$

9. $5x - 2(3x - 4y) + 3x - 5 + y - 12 = ?$

10. $[7 + 3(x - 5y)][6 - 4(6 - 3)] + 3x - 2y = ?$

Ans. $-15x + 88y - 42$

11. $7x^2 + 3y^3 - 2x + 9 - 4(3x^2 - 4y^3) + 4x = ?$

12. $4(5x + 7) - 7(2y - x) + 8 - x - 4y = ?$

13. $8x^2 + 3x - 7 - 3x^2 - 5x + 6 - 3x = ?$

14. $12(2x - y) - 6(3 - 4x) + 5x - 6y - 15 = ?$

15. $7y^3 - 5y + 6y^2 - 15 - (6y - 3y^3 + 2y^2) = ?$

Ans. $10y^3 + 4y^2 - 11y - 15$

16. $4R^2 - 2R(3 - R) + 3R - 16 - (5R + 7) = ?$

17. $(7 - 5)(3 + 2)(6x - 1) - 4x + 17 - 3x + 8 = ?$

18. $3(2x^2 - 5) + 2(x - x^2) + 4(2x - 5x^2) + 19 - x = ?$

19. $11x - 5x^2 - 3y - 6 - 3x - 4x^2 + 18y = ?$

20. $(8 - 4)(6 + 3)x + 3(4x - 5) - (3 - 5)(6 - 9)x = ?$

Ans. $42x - 15$

21. $13(x + y) - 6(x - y) + x - 3y + 17 - 2(x + 7) = ?$

22. $7x^3 + 3 - 2x^2 - 4y + 6 - 5(x - 3) + 7 - 11x = ?$

23. $6y - 2x - 11 - (6 - 1)(8 + 2) + 6(x - 2y) = ?$

24. $3R - 2S + 5 - 2R - 7 + 3S + 3(2S - 4R) = ?$

AXIOMS REGARDING EQUATIONS

An **axiom** is a statement which because of its obvious truth is accepted without proof.

Algebraic Axiom I.—*The same number (or symbol) may be added to both members of an equation without changing its equality.*

Examples: If $8 + 6 = 14$, then $8 + 6 + 4 = 14 + 4$.
If $2x = 10$, then $2x + 6 = 10 + 6$.

Algebraic Axiom II.—*The same number (or symbol) may be subtracted from both members of an equation without changing its equality.*

Examples: If $9 + 5 = 14$, then $9 + 5 - 4 = 14 - 4$.
If $y + 12 = 20$, then $y + 12 - 8 = 20 - 8$.

Algebraic Axiom III.—*Each member of an equation may be multiplied by the same number (or symbol) without changing its equality.*

Examples: If $3m = 6$, then $2 \times 3m = 2 \times 6$ or $6m = 12$.
If $2x = 8 + 12$, then $4(2x) = 4(8 + 12)$.

Algebraic Axiom IV.—*Each member of an equation may be divided by the same number (or symbol) without changing its equality.*

Examples: If $6n = 12$, then $\frac{6n}{6} = \frac{12}{6}$ or $n = 2$.

If $4y + 8 = 20$, then $\frac{4y + 8}{4} = \frac{20}{4}$ or $y + 2 = 5$.

Algebraic Axiom V.—*The square root of each member of an equation may be extracted without changing its equality.*

Examples: If $4 + 5 = 9$, then $\sqrt{4 + 5} = \sqrt{9}$.
If $x^2 = 25$, then $\sqrt{x^2} = \sqrt{25}$.

Note: For meaning of the radical sign $\sqrt{\ }$ see page 67.

SOLUTION OF SIMPLE EQUATIONS

By the solution of an equation is meant the obtaining of the numerical value of the algebraic symbol which occurs in the equation.

Simple equations may be solved by the use of one or more of the foregoing axioms.

Example a: $2x = 10$.

Solution: Dividing both members by 2 gives $x = 5$.

Check: Substituting $x = 5$ in the original equation gives $10 = 10$.

Example b: $4x - 5 = 27$.

Solution: Adding 5 to both members of the equation gives: $4x - 5 + 5 = 27 + 5$ or $4x = 32$.

Dividing both members by 4, $x = 8$

Check: Substituting $x = 8$ in the original equation gives $32 - 5 = 27$.

Example c: $5x + 2 = 32$.

Solution: Subtracting 2 from both members,

$$5x + 2 - 2 = 32 - 2 = 30.$$

Dividing both members by 5, $x = 6$

Check: Substituting $x = 6$ in the original equation gives $30 + 2 = 32$.

Transposition.—The effect of adding the same number to both members or subtracting the same number from both members is seen in Examples a and b to be equivalent to removing the number from one side of the equality sign and placing it on the opposite side with its sign changed. This is called transposition.

Example a: $2x + 3 = 11$.

Solution: Transposing the 3, $2x = 11 - 3 = 8$.

Dividing by 2, $x = 4$.

Example b: $6x - 7 = 11$

Solution: Transposing the -7, $6x = 11 + 7 = 18$.

Dividing by 6, $x = 3$.

SOLUTION OF PROBLEMS

Consider the following problem: Divide 35 into two parts, so that four times the lesser equals three times the greater.

Let x = lesser number.

Then $35 - x$ = greater number.

By the condition of the problem,

$$4x = 3(35 - x).$$

Removing the parentheses, $4x = 105 - 3x$.

Transposing the $-3x$, $4x + 3x = 105$.

Adding the like terms, $7x = 105$.

Solving for x (by dividing both sides by 7), $x = 15$ = lesser number. $35 - x = 35 - 15 = 20$ = greater number.

These two numbers are seen to satisfy the conditions of the problem, since $4 \times 15 = 60$ and $3 \times 20 = 60$.

From the solution of the above problem, the general procedure for the solution of problems involving simple equations is seen to be as follows:

1. Represent the unknown number (or one of the unknown numbers) by some letter (such as x).
2. If there is a second unknown number, represent it in terms of the same letter according to the conditions given in the problem.
3. Use the relations given by the statement to form an equation.

4. Solve the equation.

5. Check the results by showing that they fulfill the conditions of the problems.

Success in solving problems of the above type will depend upon obtaining from a careful reading of the problem the necessary relation for Steps 2 and 3.

PROBLEMS

Solve the following equations and check the result in each case.

1. $11x + 4 = 37.$ *Ans.* $x = 3$

2. $14 + 3y = -5y - 18.$

3. $-3n + 6 = 14 - 5n.$

4. $3(2x - 4) = -4x + 28.$

5. $4m + 8(3m - 9) = 6 - \frac{72}{-9}.$ *Ans.* $m = 3.0714$

6. $12 - 6x + 4 - 8x - 2(x - 8) = 0$

7. $7 - 3(4 + 5)(3R - 2) - 6R - 7(2R + 3) = 0$

8. $(3x - 6)(7 - 2) - 2x - 7 + 3x(8 - 3) = 0$

9. $19 - 2y - (6 - 4)(7y - 2) + 3y + 15 = 0$

10. $12m - 6 + 2m(8 - 5) - (3m - 6)(7 - 2) = 0$ *Ans.* $m = -8$

11. $7 + 3x(2 - 5) - 5x + 4(2 - 3x) = 0$

12. $10 + 3R - 7 - 6(9 - 3)(3 - 5R) + 9R \div 3 = 0$

13. $(15 - 6)2y + 3(4 - 7) - 2 + 3y - 3(2 - 4y) = 0$

14. $17 - 3x(7 - 2x)(8 + 3) + 5x - 3(2x - 5) + 5x = 0$

15. The sum of two numbers is 76 and their difference is 4. Find the numbers. *Ans.* 36 and 40

16. If 8 is added to three times a number, the result exceeds twice the number by 17. Find the number.

17. The age of A is three times that of B, but in 5 years A will be only twice as old as B. Find their present ages.

18. A boy is one-half as old as his father and one-fourth as old as his grandfather. The sum of the three ages is 105. How old is each?

19. The sides of a square have been increased and decreased 8 and 6 in., respectively, without changing its area. Determine the length of the side of the square.

20. A cistern can be filled by three pipes operating separately in 10, 15, and 25 min. respectively. In what time can it be filled if all the pipes operate simultaneously? *Ans.* 4.8387 min.

RATIO AND PROPORTION

The **ratio** of one quantity to another is the first divided by the second. The ratio of a to b is $a \div b$, or $\frac{a}{b}$. The ratio of 7 to 3 is $7 \div 3$ or $\frac{7}{3}$.

An **inverse ratio** is the reciprocal of a given ratio and is, hence, equal to the ratio inverted. The inverse ratio of $\frac{a}{b}$ is $\frac{b}{a}$.

A **proportion** is an equality of two ratios. For example, $\frac{a}{b} = \frac{c}{d}$. A proportion is often written $a:b::c:d$, and in either form is read a is to b as c is to d. Another example of a proportion is $\frac{7}{3} = \frac{14}{6}$ or $7:3::14:6$.

In any proportion, the four terms are numbered in the order in which they occur. Thus in the proportion $a:b::c:d$ $\left(\frac{a}{b} = \frac{c}{d}\right)$, a is the first term, b the second, c the third, and d the fourth.

The first and fourth terms of a proportion are called the **extremes,** and the second and third are called the **means.** Thus in the proportion $\frac{a}{b} = \frac{c}{d}(a:b::c:d)$, a and d are the extremes and b and c are the means. In a ratio like $\frac{x}{y} = \frac{y}{z}$ $(x:y::y:z)$, y is called the **mean proportional** between x and z.

FUNDAMENTAL THEOREMS OF PROPORTION

Seven fundamental theorems of proportion, which will be referred to later in some of the geometry proofs, will now be stated and derived and numbered with Roman numerals for future reference.

I. In any proportion, the product of the extremes is equal to the product of the means.

Given: $\frac{a}{b} = \frac{c}{d}$. *To prove:* $ad = bc$.

Multiply both sides by bd (Axiom III), $\frac{a \times bd}{b} = \frac{c \times bd}{d}$.

Canceling the b terms on the left and the d terms on the right (see page 9):

$$ad = bc.$$

Numerical Illustration: If $\frac{4}{2} = \frac{2}{1}$, then $4 \times 1 = 2 \times 2$.

II. The mean proportional between two quantities is equal to the square root of their product.

If $\frac{x}{y} = \frac{y}{z}$, then by I, $xz = y \times y$, or $xz = y^2$.

Extracting the square root of both sides (Axiom V):

$$y = \sqrt{xz}.$$

III. If four quantities are in proportion, they are in proportion by **alternation**; *i.e.*, the first term is to the third as the second term is to the fourth.

Given: $\frac{a}{b} = \frac{c}{d}$. *To prove:* $\frac{a}{c} = \frac{b}{d}$.

By I, $ad = bc$.

Divide both sides by cd (Axiom IV), $\frac{ad}{cd} = \frac{bc}{cd}$.

Canceling the d's on the left and the c's on the right:

$$\frac{a}{c} = \frac{b}{d}.$$

IV. If four quantities are in proportion, they are in proportion by **inversion**; *i.e.*, the second term is to the first as the fourth is to the third.

Given: $\frac{a}{b} = \frac{c}{d}$. *To prove:* $\frac{b}{a} = \frac{d}{c}$.

By I, $bc = ad$.

Dividing both sides by ac and canceling (Axiom IV):

$$\frac{b\cancel{c} = \cancel{a}d}{a\cancel{c} = \cancel{a}c}$$

or

$$\frac{b}{a} = \frac{d}{c}.$$

V. If four quantities are in proportion, they are in proportion by **composition**; *i.e.*, the sum of the first two terms is to the second term as the sum of the last two terms is to the last term.

Given: $\frac{a}{b} = \frac{c}{d}$. *To prove:* $\frac{a+b}{b} = \frac{c+d}{d}$.

Adding one to each side of the given proportion (Axiom I):

$$\frac{a}{b} + 1 = \frac{c}{d} + 1.$$

This may be written

$$\frac{a}{b} + \frac{b}{b} = \frac{c}{d} + \frac{d}{d}$$

or

$$\frac{a + b}{b} = \frac{c + d}{d}.$$

It also may be shown that $\frac{a + b}{a} = \frac{c + d}{c}$ (by first inverting $\frac{a}{b} = \frac{c}{d}$ and then applying V).

VI. If four quantities are in proportion, they are in proportion by **division**; *i.e.*, the difference of the first two is to the second as the difference of the last two is to the last.

Given: $\frac{a}{b} = \frac{c}{d}$. *To prove:* $\frac{a - b}{b} = \frac{c - d}{d}$.

Subtracting one from each side of the given proportion (Axiom II):

$$\frac{a}{b} - 1 = \frac{c}{d} - 1,$$

which may be written

$$\frac{a}{b} - \frac{b}{b} = \frac{c}{d} - \frac{d}{d}$$

or

$$\frac{a - b}{b} = \frac{c - d}{d}.$$

VII. In a series of equal ratios, the sum of all the numerators is to the sum of all the denominators as any one numerator is to its denominator.

Given: $\frac{a}{b} = \frac{c}{d} = \frac{e}{f}$. *To prove:* $\frac{a + c + e}{b + d + f} = \frac{a}{b}$.

Let the equal ratios all equal r.

Then since $r = \frac{a}{b} = \frac{c}{d} = \frac{e}{f}$,

$$a = br,\ c = dr,\ e = fr \text{ (Axiom III).}$$

$\therefore a + c + e = br + dr + fr$ because equals added to equals give equals.

Factoring out the common term r on the right side (see page 43),

$$a + c + e = r(b + d + f).$$

Dividing both sides by $b + d + f$ (Axiom IV),

$$\frac{a + c + e}{b + d + f} = r.$$

But $$r = \frac{a}{b}.$$

$$\therefore \frac{a + c + e}{b + d + f} = \frac{a}{b}.$$

DIRECT AND INVERSE PROPORTION

Problems in ratio and proportion are proportional either directly or indirectly (often written inversely). They are directly proportional when an increase in one denomination will produce an increase in the other; thus, if 5 drills cost $4, then 7 drills must cost more. This greater cost in dollars is represented by x. By comparing quantities of like denominations is meant comparing men to men, hours to hours, bushels to bushels, dollars to dollars, etc.

In any proportion, the greater quantity of one denomination is to the lesser quantity of the same denomination as the greater quantity of a second denomination is to the lesser quantity of this second denomination.

Thus:

drills dollars
greater : lesser : : greater : lesser

$$\frac{7 \text{ drills}}{5 \text{ drills}} = \frac{x \text{ dollars}}{4 \text{ dollars}}.$$

$$5x = 7 \times 4 \qquad \text{by I}$$

$$\therefore x = \frac{28}{5} = \$5\frac{3}{5} = \$5.60.$$

A ratio is indirectly (or inversely) proportional when an increase in one quantity will produce a decrease in the other. For example, if 5 men can do a piece of work in 8 hr., 10 men can do it in fewer hours, which will be represented by x hr.

Since an increase in the number of men produces a decrease in the number of hours required, the proportion is indirect.

As in the previous problem, the greater quantity of one denomination is to the lesser quantity of the same denomination as the greater quantity of a second denomination is to the lesser quantity of this second denomination. Thus in this problem:

$$\begin{array}{c} \text{men} \qquad\qquad \text{hours} \\ \text{greater}:\text{lesser}::\text{greater}:\text{lesser} \end{array}$$

$$\frac{10 \text{ men}}{5 \text{ men}} = \frac{8 \text{ hr.}}{x \text{ hr.}}$$

$$10x = 5 \times 8 \qquad \text{by I.}$$

$$\therefore x = \frac{40}{10} = 4 \text{ hr.}$$

PROBLEMS

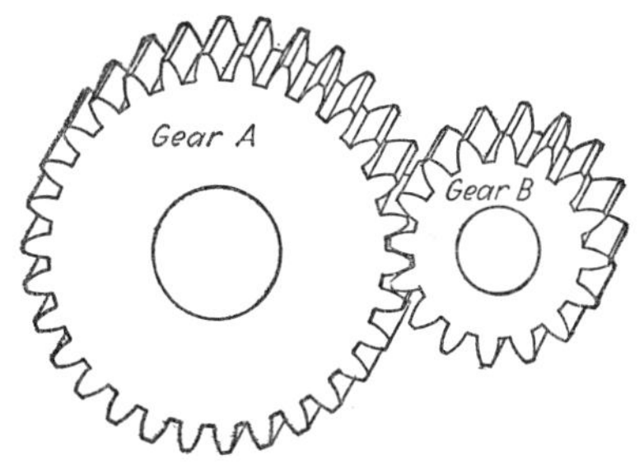

1. The numbers of teeth in gears A and B are 42 and N, respectively. What is the ratio of the numbers of teeth in the gears A and B?

2. The ratio of the numbers of teeth in the gears A and B is $\frac{5}{3}$. What number of teeth must B have if A has M?

3. If the numbers of teeth in the gears A and B are 73 and R, respectively, and if B makes 35.7 revolutions, how many revolutions will A make?

4. If 11 reamers cost T dollars, what would 7 reamers cost?

5. What is the ratio of 7.4859 and S? Answer to be a decimal number.

VARIABLES

Prob.	Sym.	No. 1	No. 2	No. 3	No. 4	No. 5	No. 6
1	N	26	27	28	29	30	31
2	M	25	30	35	40	45	50
3	R	45	48	52	56	59	62
4	T	12.50	13.25	14.0	14.75	15.5	16.25
5	S	8.1342	8.2641	9.2635	9.7654	10.308	10.865

6. If a rod of steel F in. long weighs 35.738 lb., what would a rod of the same steel weigh if it were 21.5 in. long?

7. What is the ratio of G and 5.9732? Answer to be a decimal number.

8. What common fraction must K be multiplied by to obtain 35?

9. If from a steel rod 23.7 ft. long L tapered pins can be machined, how many tapered pins can be machined from a steel rod 13.7 ft. long?

10. What is the ratio of $\frac{3}{7}$ and H? Answer to be a fraction.

11. What fraction must $\frac{4}{5}$ be multiplied by to obtain D?

12. If C pieces of work can be machined in 7 hr., how many hours will it take to machine 9 pieces of work?

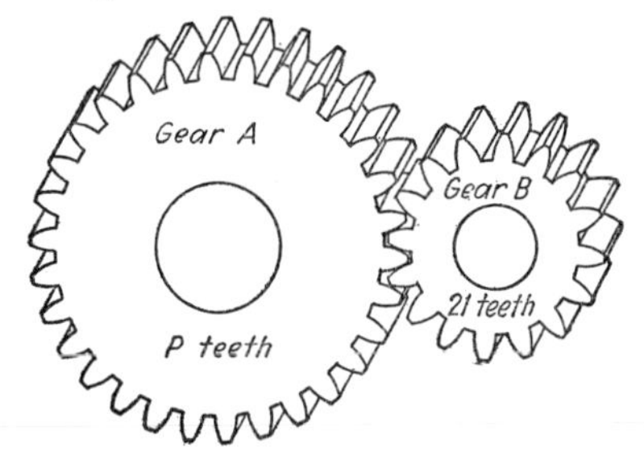

13. When gear A makes 7.5 revolutions, how many revolutions will gear B make?

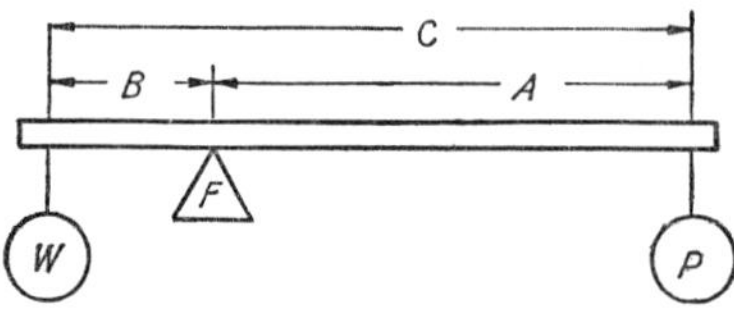

P = applied force = variable. W = weight. F = fulcrum.
A = 17 in. B = 5 in.

14. Determine the weight W.

VARIABLES

Prob.	Sym.	No. 1	No. 2	No. 3	No. 4	No. 5	No. 6
6	F	12.5	14.3	15.6	17.4	18.8	19.5
7	G	1.106	1.241	1.385	1.824	2.244	2.361
8	K	42	46	51	57	62	65
9	L	25	27	29	31	33	35
10	H	12	15	18	21	25	33
11	D	$\frac{7}{8}$	$\frac{5}{6}$	$\frac{9}{11}$	$\frac{2}{3}$	$\frac{1}{2}$	$\frac{3}{8}$
12	C	55	58	68	75	82	93
13	P	41	44	47	51	53	56
14	P	21	32	53	64	75	96

COMPOUND RATIO AND PROPORTION

A **compound ratio** is the product of simple ratios.

If a quantity is multiplied by a ratio greater than unity, the quantity is increased: thus

$$7 \times \frac{17}{14} = 8.5.$$

If a quantity is multiplied by a ratio less than unity, the quantity is decreased; thus

$$7 \times \frac{11}{14} = 5.5.$$

In many problems the quantity to be determined is affected by more than one condition. The effect of each of these conditions may be expressed by a simple ratio, and the effect of all of the conditions acting simultaneously may be obtained by multiplying these simple ratios together.

If $\begin{pmatrix}\text{decreasing}\\ \text{increasing}\end{pmatrix}$ one of the conditions causes an increase in the quantity to be determined, the two are said to be in $\begin{pmatrix}\text{indirect}\\ \text{direct}\end{pmatrix}$ proportion, and the quantity to be determined is obtained by multiplying the original value of that quantity by a ratio greater than unity.

If $\begin{pmatrix}\text{decreasing}\\ \text{increasing}\end{pmatrix}$ one of the conditions causes a decrease in the quantity to be determined, the two are said to be in $\begin{pmatrix}\text{direct}\\ \text{indirect}\end{pmatrix}$ proportion, and the quantity to be determined is obtained by multiplying the original value of that quantity by a ratio less than unity.

Example: If 9 men working 8 hr. per day can unload 24 carloads of castings in 3 days, how many days would be required for 5 men working 10 hr. per day to unload 18 carloads of castings?

Solution: Start with the quantity of the same denomination as the question, which in this case is 3 days. Then compare

the men, 9 in the first case and 5 in the second. In arranging the ratio, the number of days required is the only thing that should be considered. Thus 5 men will require more days than 9 men, therefore, the ratio should be $\frac{9}{5}$. Next, the hours per day should be considered. Since working 10 hr. per day in the second case will require fewer days, the ratio should be $\frac{8}{10}$. Finally, the number of carloads must be considered. 18 carloads will require fewer days than 24 carloads. Therefore, this ratio should be $\frac{18}{24}$. The solution of the problem then becomes

$$\not{3} \times \frac{9}{5} \times \frac{\not{8}}{\underset{5}{\not{10}}} \times \frac{\overset{9}{\not{18}}}{\underset{\not{3}}{\not{24}}} = \frac{81}{25} \text{ or } 3.24 \text{ days.}$$

PROBLEMS

1. If 9 iron bars 7 ft. long, 3 in. broad, and 1.2 in. thick, weigh B lb., what will be the weight of 5 bars of the same material 10 ft. long, 4 in. broad, and 2.3 in. thick?

2. If C men working 9 hr. per day, 7 days per week, can machine 236 castings in 23 weeks, how many weeks will it take 11 men working 6 hr. per day, 5 days per week, to machine 729 castings?

3. A pulley D in. in diameter runs at a speed of 125 r.p.m. and drives another pulley at a speed of 425 r.p.m. What is the diameter of the other pulley?

4. If 7 men do E pieces of work in 8 hr., how long will it take 9 men to do 55 pieces of the same work?

5. The diameter of two pulleys connected by a belt are, respectively, 35 and G in., and the smaller makes 217 r.p.m. Find the number of r.p.m. of the larger pulley.

6. If a piece of steel 16 ft. long, 5 ft. wide, and 2.3 ft. thick weighs R lb. what will be the weight of a block of the same kind of steel 9 ft. long, 2 ft. wide, and 1.2 ft. thick?

7. A grinder spindle is to be driven from a main shaft line making 167 r.p.m. It is found necessary to employ two countershafts. The pulley on the main shaft line is S in. in diameter and drives a pulley 7 in. in diameter. On the same shaft with the 7-in. pulley is a 12-in. pulley which in turn drives a 9-in. pulley, and on the same shaft with the 9-in. pulley is a 16-in. pulley which in turn drives a 5-in. pulley on the grinder spindle. What is the speed of the grinder spindle?

8. If 12 men working 10 hr. per day, 6 days per week, can dig a trench 4 ft. wide, 7 ft. deep, and Q ft. long, in 3 weeks, how many men working 8 hr. per day, 7 days per week, will be required to dig a trench 2.5 ft. wide, 7 ft. deep, and 10,587 ft. long in 4 weeks?

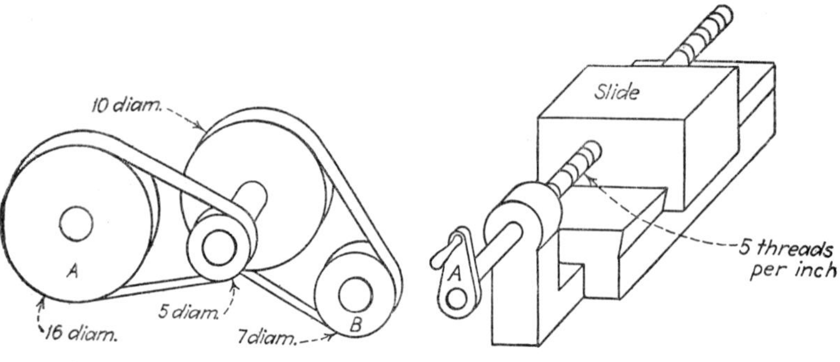

9. When shaft A makes T revolutions, how many revolutions will B make?

10. When A makes H revolutions, how far will the slide move?

VARIABLES

Prob.	Sym.	No. 1	No. 2	No. 3	No. 4	No. 5	No. 6
1	B	912	953	993	1,034	1,074	1,115
2	C	81	108	135	162	189	217
3	D	2.9411	3.235	3.5294	3.8235	4.1176	4.4117
4	E	13	12	11	10	9	8
5	G	130.20	142.6	155	167.4	179.8	192.2
6	R	241	252.39	259.43	275.16	287.6	295.35
7	S	17.9	19.6	21.4	23.6	25	26.8
8	Q	50,639	48,706	44,775	40,901	39,024	35,845
9	T	164.57	187.42	214.85	242.28	269.71	297.14
10	H	.442	.464	.506	.528	.550	.592

PERCENTAGE

Per cent means hundredths. For example, 5 per cent mean five hundredths (which may be written .05 or $\frac{5}{100}$). The symbol for per cent is %, which is written after the number. Thus, 9 per cent is written 9%. A quantity is always $\frac{100}{100}$ or 100% of itself.

Example a. Find 24% of 85.

Since per cent means hundredths, we must find .24 of 85 which is $.24 \times 85 = 20.4$.

Example b. 15 is what per cent of 65? That is, 15 is how many hundredths of 65?

$$\frac{15}{65} = .23076, \text{ which is } 23.076\%.$$

This problem may also be considered as a simple proportion.

Thus: Since 65 is 100%, $\frac{15}{65} = \frac{x\%}{100\%}$

$$x = 100\% \times \frac{15}{65} = 23.076\%.$$

PROBLEMS

1. N is what per cent of 73? **2.** 21 is what per cent of L?

3. If in a certain machine G of the energy supplied to the machine is lost in friction, what is the per cent of efficiency?

4. 13 is $M\%$ of what number? **5.** H is 3% of what number?

6. The usual allowance for shrinkage when casting iron pipes is $\frac{1}{8}$ in. per foot. What is the per cent of the allowance?

7. The indicated horsepower of an engine is F, the actual effective horsepower is 12.3. The actual horsepower is what per cent of the indicated horsepower?

8. What is the net price per barrel of oil, the list price of which is C dollars, subject to a discount of $15\frac{1}{4}\%$ and 5% off for cash?

9. Find the cost of an article that is listed at S dollars, 35% and 7% off for cash.

10. The clearance between a punch and die for a certain metal is 7% of the thickness of the stock. For stock of thickness T, determine the value of the clearance.

11. The thickness R of a tooth on a certain rotary cutter should measure $\frac{6}{7}$ in. The company manufacturing this cutter agreed to take a loss of $1 for each per cent that the specified dimension is undersize. The dimension R was found to be $\frac{13}{16}$ in., and accordingly the manufacturing company received N dollars. What was the list price?

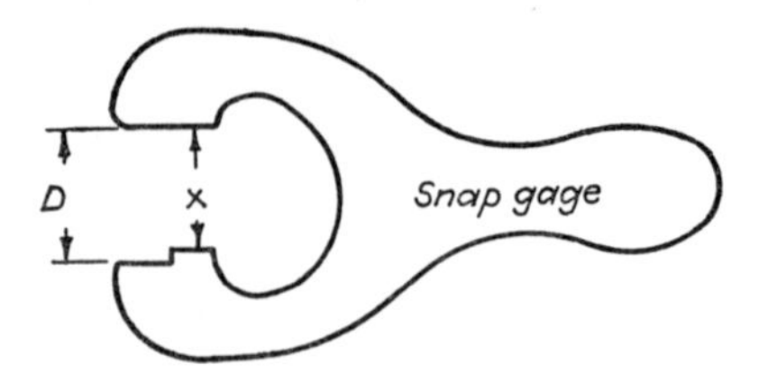

12. A snap gage is to be made for checking a shaft which has an allowance of .06% undersize. If the correct diameter is D, determine the distance x on the gage.

13. One of the best bearing metals is a Babbitt having the following composition:

Tin	84%
Antimony	10%
Copper	5%
Lead	1%

Find the number of pounds of each of the constituents in a bearing weighing A lb.

14. Certain alloys having low melting points are used in overhead safety sprinkler systems. Of such alloys, Wood's metal has about the lowest melting point (154°F.) and has the following composition:

Bismuth	50%
Lead	25%
Tin	$12\frac{1}{2}$%
Cadmium	$12\frac{1}{2}$%

Calculate the weight of each constituent in a mass of this alloy which weighs B lb.

VARIABLES

Prob.	Sym.	No. 1	No. 2	No. 3	No. 4	No. 5	No. 6
1	N	6	8	11	15	17	19
2	L	118	135	154	178	193	209
3	G	$\frac{1}{3}$	$\frac{1}{5}$	$\frac{2}{7}$	$\frac{3}{8}$	$\frac{2}{9}$	$\frac{1}{4}$
4	M	4	5	6	7	8	9
5	H	15	17	19	21	23	27
6		Complete	Complete	Complete	Complete	Complete	Complete
7	F	13.2	13.7	14.1	14.6	14.9	15.1
8	C	5.5	5.75	6.25	6.75	7.4	7.75
9	S	18.44	22.64	26.39	29.18	39.75	49.55
10	T	.093	.102	.112	.117	.125	.156
11	N	28.9	35.4	38.5	41.25	45.75	47.4
12	D	.431	.562	.623	.684	.775	.916
13	A	9.875	9.75	9.625	9.5	9.375	9.25
14	B	639	619	599	579	559	539

Illustrative Problem: How much lead must be added to 485 lb. of tinsmith's solder (59% tin, 41% lead) to change it to plumber's solder which is 35% tin and 65% lead?

Solution: In 100 lb. of the tinsmith's solder, there are 59 lb. of tin and 41 lb. of lead. In the plumber's solder the percentage is to be 35% tin and 65% lead. Hence

$$\frac{x \text{ lb. of lead}}{59 \text{ lb. of tin}} = \frac{65\% \text{ of lead}}{35\% \text{ of tin}}$$

or

$$\frac{x}{59} = \frac{65}{35}.$$

$$x = 59 \times \frac{65}{35} = 109.57 \text{ lb. of lead.}$$

Since the tinsmith's solder already contains 41 lb. of lead per 100 lb. of solder, the amount of lead to be added per 100 lb. of tinsmith's solder is $109.57 - 41 = 68.57$ lb.

Since 485 lb. is 4.85 times as great as 100 lb., the total amount of lead to be added is $4.85 \times 68.57 = 332.56$ lb.

15. How many pounds of lead must be added to change E lb. of a batch of solder which is 42% tin and 58% lead to a new batch of solder which is 30% tin and 70% lead?

16. In making cloth for upholstering a certain car, colored fibers were mixed in the following proportions: yellow 20%, green 30%, red 15%, black 10%, and blue 25%. It was found that a better color mixture would be obtained if the following proportions were used: yellow 30%, green 25%, red 10%, black 10%, and blue 25%. In order to add the least number of pounds, how many pounds of each color must be added to J lb. of the original mixture to produce the desired mixture?

VARIABLES

Prob.	Sym.	No. 1	No. 2	No. 3	No. 4	No. 5	No. 6
15	E	1476	1567	1689	1742	1821	1935
16	J	485	435	395	362	337	296

TAPER PER FOOT

There are two general forms of tapers as shown below:

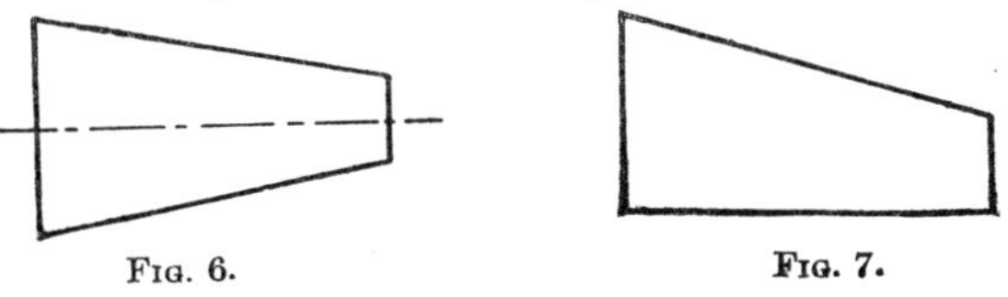

FIG. 6. FIG. 7.

Tapers of the form shown in Fig. 6 are used for tapered plugs and gages, tapered spindles and bearings, and taper fits as in the case of a hub on an axle, etc.

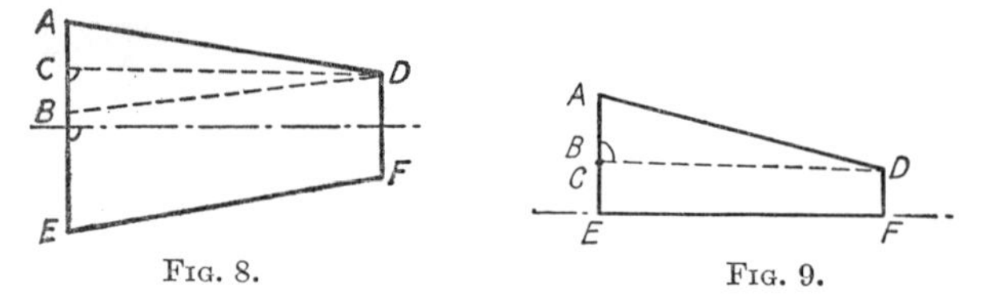

FIG. 8. FIG. 9.

Tapers of the form shown in Fig. 7 are used for the gibs of milling and grinding machines, rams of shaper heads, etc.

In Figs. 8 and 9, the taper is AB for the length CD. BD is drawn parallel to EF and CD is drawn parallel to the axis. By **taper per foot** is meant the distance corresponding to AB when CD is 1 ft.

The problem of determining taper per foot is obviously one of simple proportion.

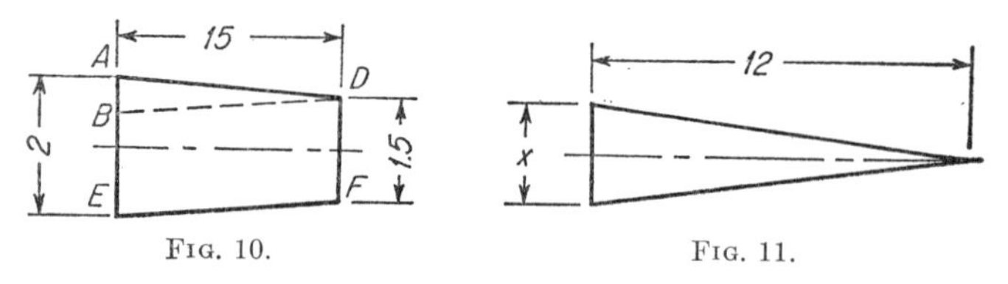

FIG. 10. FIG. 11.

Illustrative Example a: Determine the taper per foot for Fig. 10.

Solution: Draw BD parallel to EF, which shows the taper to be .5 in. for a length of 15 in. Draw another figure, which may be called the master figure, having the same shape and a length of 12 in. (see Fig. 11). Then x is the taper per foot.

$$\frac{x}{.5} = \frac{12}{15} \text{ or } x = .5 \times \frac{12}{15} = .4.$$

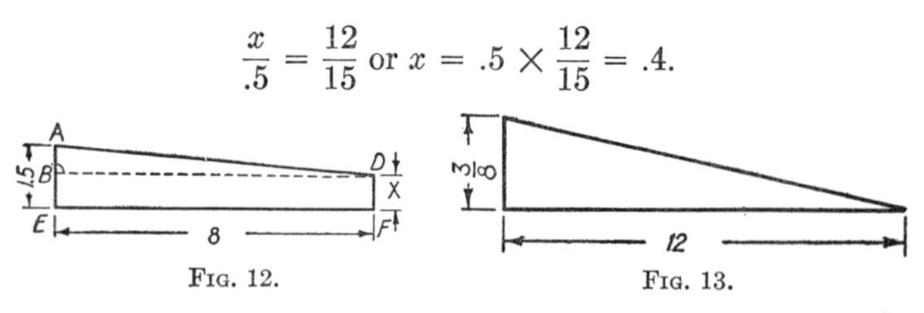

FIG. 12. FIG. 13.

Illustrative Example b: Determine the distance x in Fig. 12. Taper per foot $= \frac{3}{8}$. (*Solution on next page.*)

Solution for preceding problem: Draw BD parallel to EF, and construct a triangle similar to ABD having a length of 12 in. (Fig. 13).

$$\frac{AB}{\frac{3}{8}} = \frac{8}{12} \text{ or } AB = \frac{3}{8} \times \frac{8}{12} = \frac{1}{4}.$$

Hence $$x = 1.5 - .25 = 1.25 \text{ in.}$$

PROBLEMS

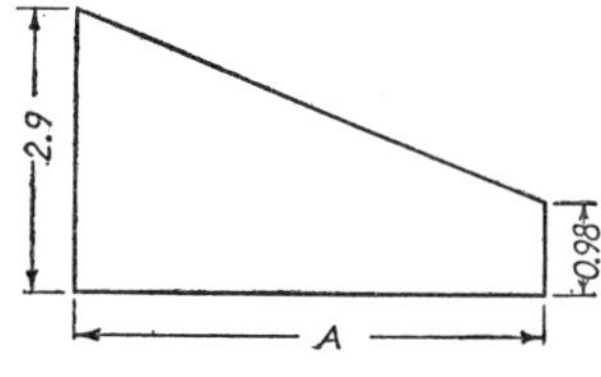

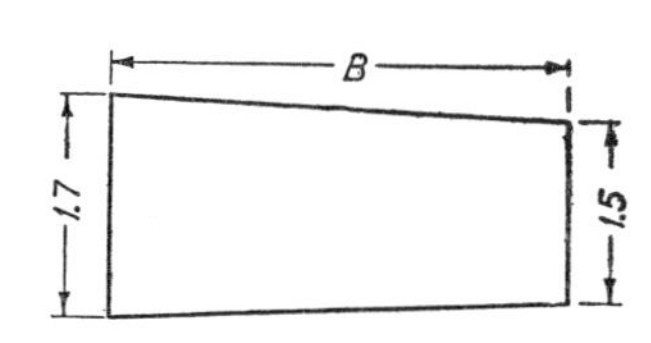

1. Determine the taper per foot. **2.** Determine the taper per foot.

VARIABLES

Prob.	Sym.	No. 1	No. 2	No. 3	No. 4	No. 5	No. 6
1	A	5.5	5.7	5.9	6.1	6.3	6.5
2	B	3.25	3.6	3.95	4.45	4.78	5.32

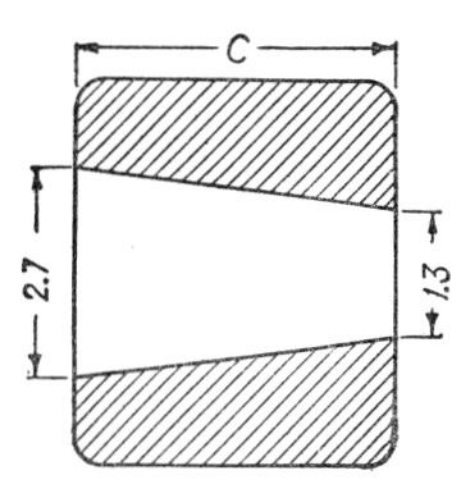

3. Determine the taper per foot.

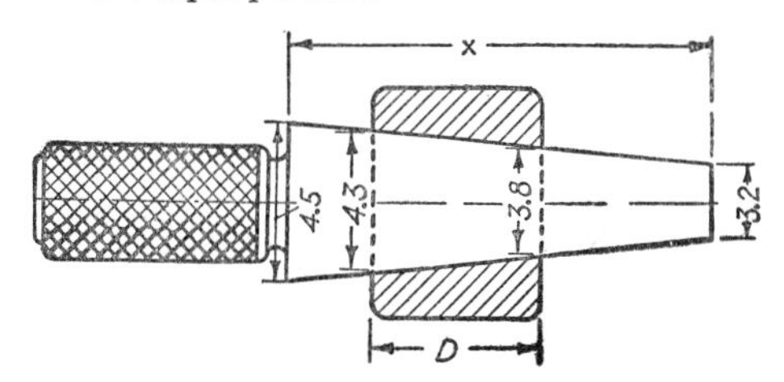

4. Determine the distance x.

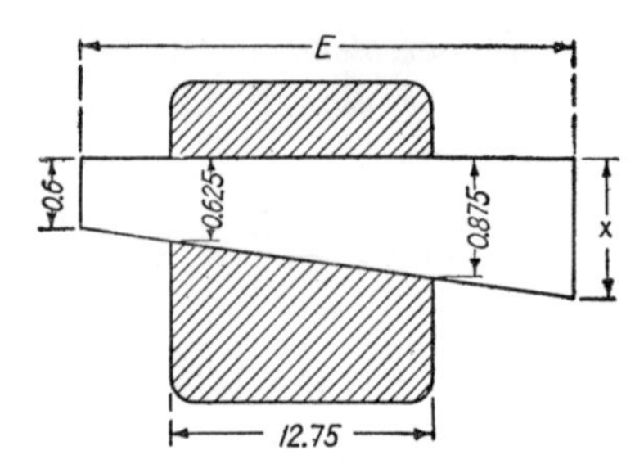

5. Determine the distance x.

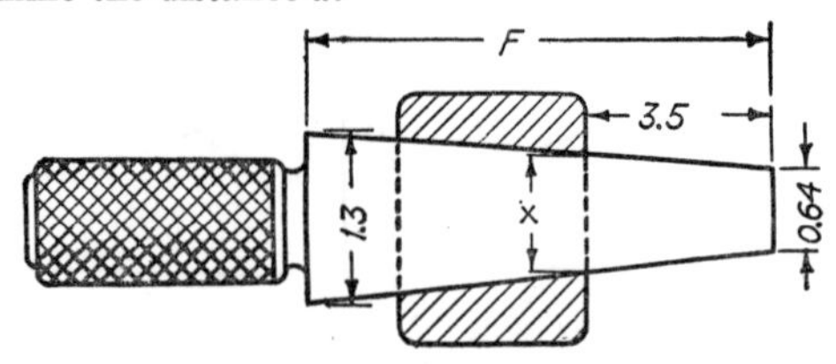

6. Determine the distance x.

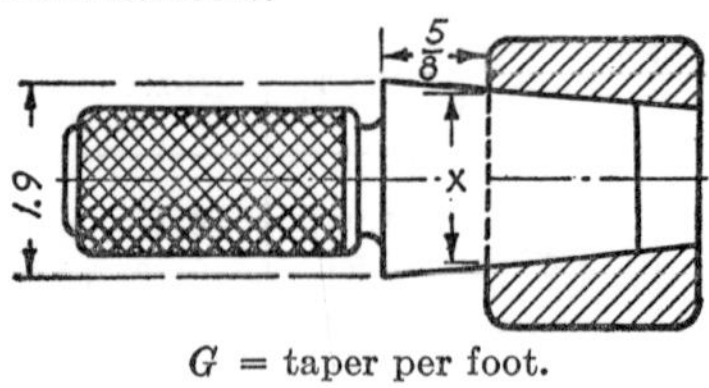

G = taper per foot.

7. Determine the diameter x.

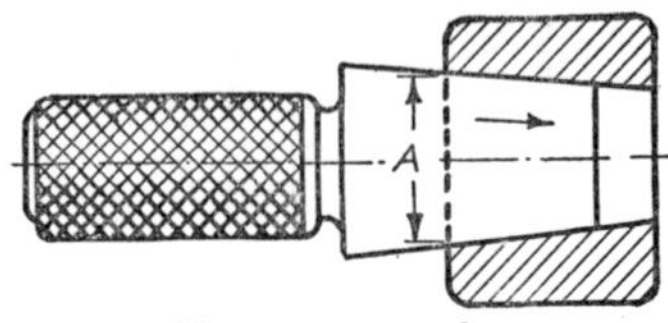

H = taper per foot.

8. If A were made .017 larger, how far would the taper plug advance?

VARIABLES

Prob.	Sym.	No. 1	No. 2	No. 3	No. 4	No. 5	No. 6
3	C	14.3	16.2	16.6	16.9	17.4	17.8
4	D	4.8	5.2	5.6	5.9	6.2	6.5
5	E	15	15.2	15.4	15.6	15.8	16
6	F	9.2	9.5	9.9	10.3	10.5	10.9
7	G	.125	.1875	.25	.3125	.375	.4375
8	H	.4375	.375	.3125	.25	.1875	.125

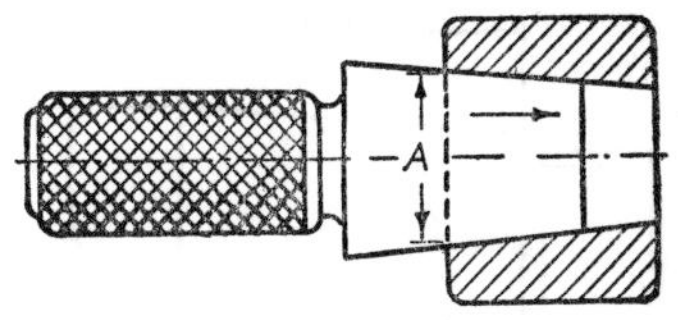

J = taper per foot.

9. How much must the diameter A be increased in order that the taper plug will advance .007?

VARIABLES

Prob.	Sym.	No. 1	No. 2	No. 3	No. 4	No. 5	No. 6
9	J	2.51	2	3	4	3.5	4.56

ADDITIONAL PROBLEMS ON RATIO AND PROPORTION, PERCENTAGE, AND TAPER PER FOOT

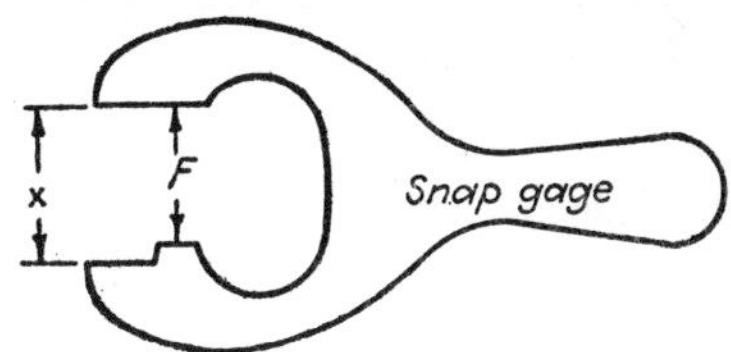

1. The above snap gage for checking shafts has an allowance of .06% oversize. Determine the distance x.

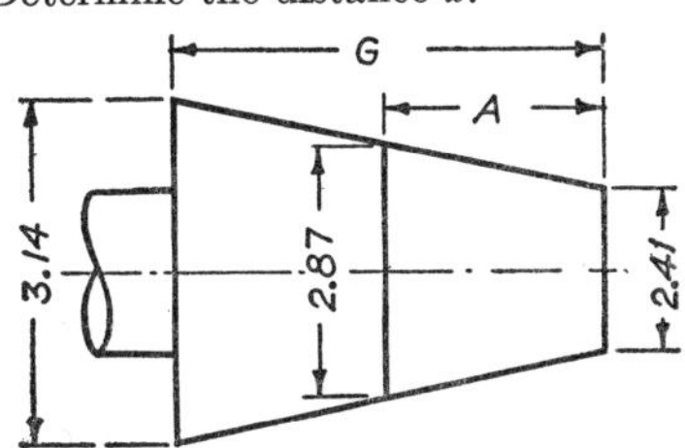

2. Determine the distance A.

3. H is what per cent of 91?

4. J is 7.8% of what number?

5. How much is 4.6% of K?

6. How much is 6.7% of L?

7. What decimal would express the ratio between 19.2 and S?

8. If 10 men can grind T castings in 21 days by working 9 hr. a day, how many days will it require 13 men working 8 hr. a day to grind 8913 castings?

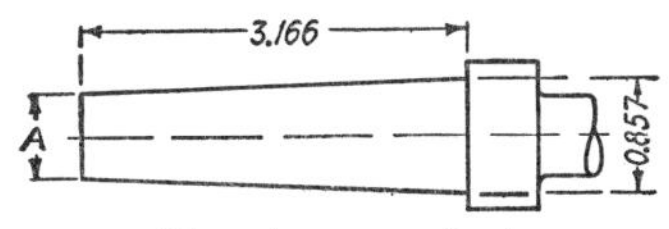

M = taper per foot.

9. Determine the distance A.

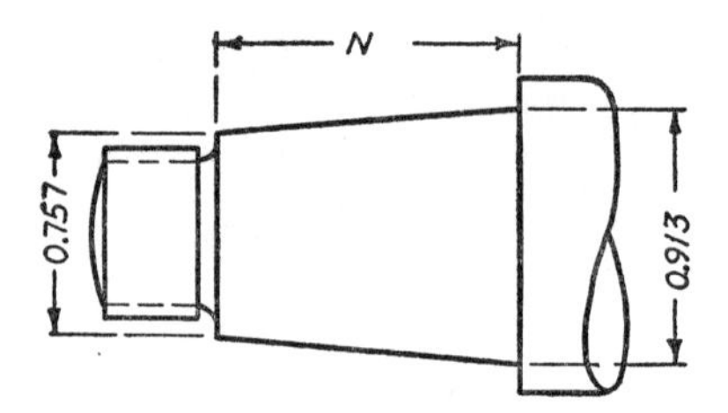

10. Determine the taper per foot.

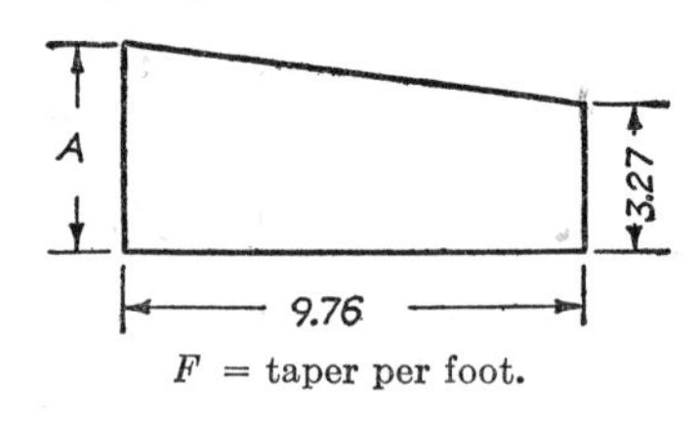

F = taper per foot.

11. Determine the value of A.

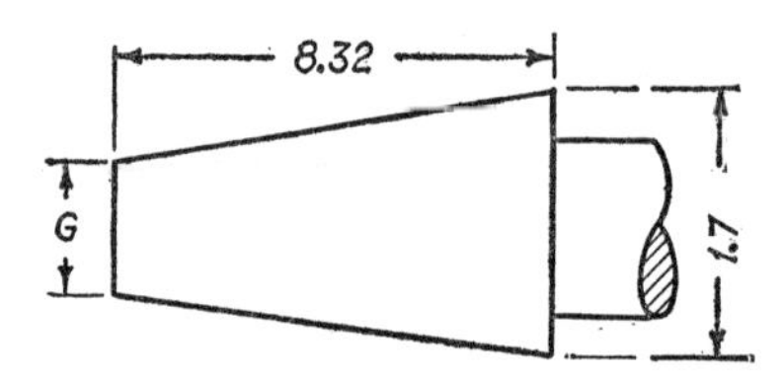

12. Determine the taper per foot in the figure above.

VARIABLES

Prob.	Sym.	No. 1	No. 2	No. 3	No. 4	No. 5	No. 6
1	*F*	.505	.525	.545	.565	.585	.605
2	*G*	7.25	6.75	6.25	5.75	5.25	4.75
3	*H*	13	16	19	22	25	28
4	*J*	89.2	87.2	85.2	83.2	81.2	79.2
5	*K*	459.29	479.39	499.49	519.59	539.69	559.79
6	*L*	196.4	189.6	183.4	179.5	174.3	168.6
7	*S*	40.8	39.6	38.4	37.2	35.8	34.6
8	*T*	6625	6675	6725	6775	6825	6875
9	*M*	.406	.431	.456	.481	.496	.521
10	*N*	3.331	3.352	3.373	3.384	3.395	3.406
11	*F*	.211	.221	.241	2.61	.281	.301
12	*G*	.783	.763	.743	.723	.703	.683

13. If 18 pipes, each delivering 5 gal. per minute, fill a tank in 3 hr. and 10 min., how long will it take 11 pipes each delivering H gal. per minute to fill a tank three times as large as the first?

14. If 7 men can build a wall in J days, how long would it take 13 men to build a wall of the same size?

15. What per cent of his time does a man rest when he sleeps K hr. out of every 24?

16. A man had $1200. He gave 30% of this to his son, and L% of the remainder to his daughter. How much did the daughter receive?

17. If it requires 4500 tiles, 8 in. long by 4 in. wide, to pave a court-yard 40 ft. long by 32 ft. wide, how many tiles, 10 in. square, will be needed to pave a hall M ft. long and 28 ft. wide?

18. If 9 gages cost $27.63, how much would N of the same kind of gages cost?

19. Determine the cost of an article listed at P dollars with 27% and 5% off for cash.

20. If 6 men working 9 hr. a day can build Q rods of fence in 5 days, how many rods of fence can 11 men build by working 7 hr. a day for 13 days?

VARIABLES

Prob.	Sym.	No. 1	No. 2	No. 3	No. 4	No. 5	No. 6
13	H	2.75	3	3.25	3.5	3.75	4
14	J	13	12	11	10	9	8
15	K	8.5	9	9.5	10	10.5	11
16	L	32	30	28	26	24	22
17	M	85	87	89	91	93	95
18	N	22	21	20	19	18	17
19	P	56.75	59.75	62.75	65.75	58.75	71.75
20	Q	849	859	869	879	889	899

SQUARE ROOT

The **square root** of a number or quantity is one of the two equal factors which, when multiplied together, will produce the given number or quantity.

The **radical sign** ($\sqrt{}$) preceding a number, or group of numbers, indicates that the square root of the number, or group of numbers, is to be found.

The **square** of a number is the number multiplied by itself. *Example:* The square of 4 is 4×4.

The length of the **vinculum** (———) attached to the radical sign indicates the extent of the figures to be considered when finding the square root.

The number from which the square root is to be extracted is called the **radicand.**

The algebraic method for obtaining square root can be readily understood from the accompanying diagram.

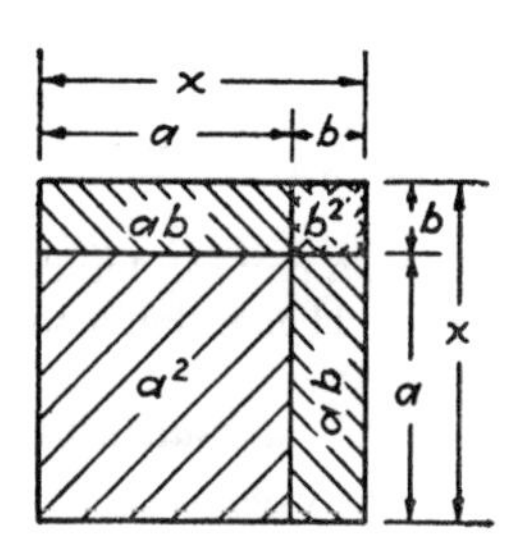

If the square root of a quantity (x^2) is desired, that quantity may be represented by the area of the largest square in the figure above.

This largest square (x^2) is seen to be made up of a large square (a^2), a small square (b^2), and two rectangles (each equal to ab). Thus:

$$x^2 = (a + b)^2 = a^2 + b^2 + ab + ab$$

or

$$x^2 = (a + b)^2 = a^2 + 2ab + b^2.$$

Hence the square of $a + b$ is seen to be the square of the first term plus two times the product of the first and second terms plus the square of the second term.

Similarly the square of $a - b$ is $(a - b)^2 = a^2 - 2ab + b^2$.

From the above relation, $x^2 = (a + b)^2 = a^2 + 2ab + b^2$, it follows that $\sqrt{a^2 + 2ab + b^2}$ is $a + b$ or x (by Axiom V).

The details of obtaining $a + b$ as the square root of the quantity $a^2 + 2ab + b^2$ will not be given in this text but a full explanation for obtaining the square root of a number which is based on this algebraic process is given as follows:

Rules for Extracting Square Root.

1. *Separate the radicand into groups consisting of two figures. This must be done by starting at the decimal point and counting to both the left and right. Indicate the groups by a prime symbol. Should the last group to the right of the decimal point consist of a single figure, a cipher should be added to complete the group.*

2. *Find the largest number which when squared will be contained by the first group. Write this number as the first figure of the root directly above the first group.*

3. *Subtract the square of the first figure of the root from the first group of figures and annex to this the second group to form the new partial radicand.*

4. *Form the trial divisor,*[1] *by multiplying the root by* 2, *and add a small cipher. This small cipher represents the next figure of the root and should be replaced by that figure after it has been obtained. Write this trial divisor to the left of the partial radicand.*

5. *Find how many times the partial radicand contains the trial divisor and write this figure over the second group as a second figure of the root, and also write it immediately above the small cipher to complete the exact divisor.*

6. *Multiply this exact divisor by the last figure of the root. Write this product under the partial radicand and subtract.* (*Note: If the product is larger than the partial radicand, a smaller number must be used for the last digit of the exact divisor* (*see Example a*). *Annex to this remainder the next group of figures to form the new partial radicand.*

7. *Continue to apply Rules* 4, 5, *and* 6 *until sufficient figures are obtained in the root.*

8. *Place the decimal point in the root directly above the decimal point in the radicand.*

Example a: Extract the square root of 762.879

[1] This is called the trial divisor because its last figure is not complete. The last figure will always be the same as the next figure in the root.

Solution: Follow the procedure outlined:

$$
\begin{array}{rr}
 & 2\ \ 7.\ 6\ \ 2 \\
 & \sqrt{7'62.87'90'} \\
 & 4 \\
4_0^7\,/ & 3\ 62 \\
 & 3\ 29 \\
54_0^6\,/ & 33\ 87 \\
 & 32\ 76 \\
552_0^2\,/ & 1\ 11\ 90 \\
 & 1\ 10\ 44 \\
 & 1\ 46
\end{array}
$$

1. Note that a cipher is added to complete the last group to the right of the decimal point.

2. The largest number which when squared will be contained by the first group at the left (7) is 2, which is placed directly above the 7 and the vinculum.

3. Subtracting the square of 2 from 7 gives 3, and bringing down the next group gives 362 as the new partial radicand.

4. Multiplying the first figure of the root (2) by 2 gives 4. The trial divisor consists of this 4 with a small cipher to the right of it, *i.e.*, 4_0. This small cipher represents the next number in the root and should be replaced by that number after it has been determined.

5. The partial radicand (362) contains the trial divisor (40) nine times. However, when the small cipher is replaced by 9 and the exact divisor (49) is multiplied by 9, the result is considerably larger than the radicand. Similarly, 8 is found to be too large, and 7 must be used as the second figure of the root. Place the 7 directly above the small cipher, making the exact divisor 47. (*Note:* If the partial radicand contains the trial divisor exactly without a remainder, this divisor will usually be too large.) Practice will enable the student easily to estimate the proper number for completing the exact divisor.

6. Multiplying the exact divisor 47 by 7, the latter being placed directly above the second group, gives 329, which is subtracted from 362, leaving 33. The next group (87) is brought down giving 3387 as the new partial radicand.

7. 2×27, followed by a small cipher, 54_0, is the new trial divisor. 3387 contains the trial divisor about 6 times, so 6 is the next number of the root and is placed directly over the third group and also replaces the small cipher of the trial divisor 54_0, giving 546 as the exact divisor. Multiplying 546 by 6 gives 3276 which is subtracted from 3387, leaving 111, which with the next group 90 gives 11,190 as the next partial radicand, etc.

8. The decimal point for the root is placed directly above the decimal point in the radicand of the original number, giving 27.62 as the square root of 762.879.

Example b: Extract the square root of 16.1456.

Solution: Proceed according to the rules already given. In carrying out step 5, it is obvious that 14 will not contain 8_0, so the small cipher after the 8 must be replaced by 0 and another small cipher added after this new trial divisor. Thus: 80_0. Bringing down the next group (56) gives 1456 as the new radicand. 1456 contains 800 only once so the exact divisor is 801, etc.

$$
\begin{array}{rr}
 & 4.\ 0\ \ 1\ \ 8 \\
 & \sqrt{16.14'56'00'} \\
 & 16 \\
8^{01}_{00}/ & 14\ 56 \\
 & 8\ 01 \\
802^{8}_{0}/ & 6\ 55\ 00 \\
 & 6\ 42\ 24 \\
 & 12\ 76
\end{array}
$$

Example c: Extract the square root of .007431.

Solution: When the number whose root to be obtained is less than .01 (as in this case) so that the first group is 00, the first number of the root is zero.

In a problem of this nature there will be as many zeros following the decimal point in the root as there are groups of zeros following the decimal point in the radicand.

$$
\begin{array}{rr}
 & .\ 0\ \ 8\ \ 6 \\
 & \sqrt{.00'74'31'} \\
 & 64 \\
16^{6}_{0}/ & 10\ 31 \\
 & 9\ 96 \\
 & 35
\end{array}
$$

Whenever it is necessary to extract the square root of a fraction, it is simpler first to divide the numerator by the denominator and then extract the square root of the quotient. Thus:

$$\sqrt{\frac{8.36485}{11.415}} = \sqrt{.732794} = .856.$$

Checking Square Root by Excess of Nines

To check the result of the square root by the excess of nines, multiply the excess of nines in the root by itself and add the excess of nines of this product to the excess of nines in the remainder. This sum, if the square root has been correctly obtained, will equal the excess of nines in the radicand.

PROBLEMS

Extract the square root of:

1. *A.* **2.** *B.* **3.** *C.* **4.** *D.* **5.** *E.* **6.** *F.* **7.** *G.* **8.** *H.*
9. *J.* **10.** *K.* **11.** *L.* **12.** *M.* **13.** *N.* **14.** *P.* **15.** *R.* **16.** *S.*
17. *N.* **18.** *P.* **19.** *R.* **20.** *S.* **21.** *A.* **22.** *B.* **23.** *C.* **24.** *D.*

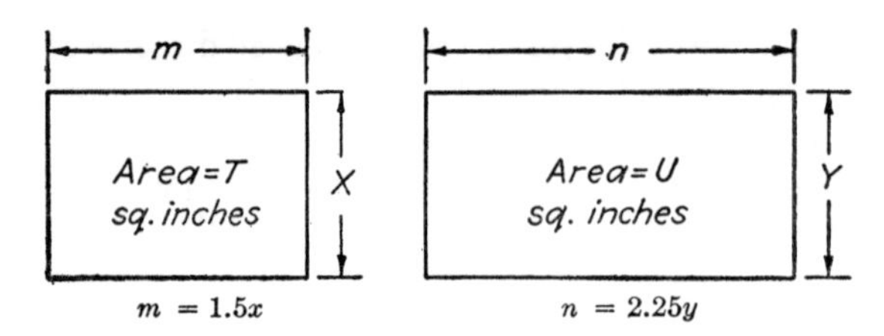

$m = 1.5x$ $n = 2.25y$

25. Determine the distance x. **26.** Determine the distance y.

VARIABLES

Prob.	Sym.	No. 1	No. 2	No. 3	No. 4	No. 5	No. 6
1	*A*	48361	50149	82743	39133	67876	91435
2	*B*	196043	276135	342753	876345	991873	146373
3	*C*	26 ÷ 37	50 ÷ 65	37 ÷ 82	65 ÷ 82	26 ÷ 50	50 ÷ 82
4	*D*	78543	12371	23945	76198	67342	81148
5	*E*	38.141	22.345	32.176	85.131	96.127	33.236
6	*F*	.00235	.00721	.00684	.00875	.00937	.00751
7	*G*	9.1436	9.2639	9.1832	9.1935	9.7623	9.9891
8	*H*	784.136	724.733	234.132	625.142	321.123	438.143
9	*J*	.00432	.00178	.00812	.00625	.00562	00473
10	*K*	.43215	.67823	.14538	.78912	.49625	.86535
11	*L*	.04623	.07948	.091191	.08608	.07843	.04535
12	*M*	45.631	32.1812	25.2253	16.3484	17.3515	18.3546
13	*N*	.000076	.000067	.000057	.000045	.000032	.000027
14	*P*	.59121	.75261	.34375	.20857	.91546	.46172
15	*R*	16.464	25.789	37.078	49.750	64.952	82.023
16	*S*	532.31	1230.9	1856.8	2709.6	3972.6	5196.1
17	*N*	51.387	53.251	55.377	57.451	59.459	59.673
18	*P*	4.5632	5.6387	5.9693	6.1636	6.3637	6.5756
19	*R*	.16387	.18936	.20137	.20443	.22445	.24934
20	*S*	$\frac{439}{818}$	$\frac{457}{829}$	$\frac{489}{841}$	$\frac{516}{618}$	$\frac{537}{839}$	$\frac{551}{887}$
21	*A*	226.35	238.42	247.55	262.91	277.25	289.33
22	*B*	5.962	6.143	6.896	7.385	7.956	8.236
23	*C*	9.894	10.278	10.536	10.897	11.286	11.957
24	*D*	19.289	19.546	19.972	20.132	20.495	20.867
25	*T*	4.364	4.512	4.937	5.139	5.426	5.634
26	*U*	7.193	7.483	7.729	7.916	8.287	8.542

MEANING OF FORMULAS AND METHOD OF SUBSTITUTION IN FORMULAS

A formula is a rule expressed in letters or symbols. The letters or symbols used in a formula simply represent given figures which are to be substituted in their respective places when the formula is evaluated. The multiplication sign in a formula is generally omitted. When a number, letter, or symbol immediately precedes or follows another letter or symbol without any operation symbol between them, it is understood that multiplication should be performed.

The evaluation of an expression is the process of determining its value by substituting definite numbers for the letters and then performing the operations as indicated.

The formula $A = \pi r^2$, where A stands for the area of a circle, r for the radius of that circle, and π for the constant 3.1416, tells us that the area of any circle may be obtained by squaring the radius of that circle and multiplying that result by π.

Evaluation: Compute the area of a circle having a radius of 10 in.

$$A = \pi r^2 = 3.1416\ (10)^2 = 3.1416 \times 100 = 314.16 \text{ sq. in.}$$

As a second example of evaluating a formula by substituting numbers for letters in a formula, compute the volume of the frustum of a right circular cone by using the formula:

$$\text{Volume} = .2618\ H(D^2 + d^2 + Dd).$$

Where $H = 2.5$, $D = 5.1$, and $d = 3.4$,

$$\begin{aligned} \text{Volume} &= .2618 \times 2.5(5.1^2 + 3.4^2 + 5.1 \times 3.4) \\ &= .6545(26.01 + 11.56 + 17.34) = .6545(54.91) \\ &= 35.94 \text{ cu. in.} \end{aligned}$$

Usually in a formula the one quantity desired is placed on the left side of the equation, and all the other quantities involved are on the right side. If the formula is in the form of a fraction without plus or minus signs between terms in either the numerator or the denominator, the quantity desired is directly proportional to all quantities in the numerator and indirectly (inversely) proportional to all quantities in the denominator.

PROBLEMS

Determine the left member of the following formulas:

1. $A = \frac{h}{2}(b + c)$. $h = 13$. $b = 7$. $c =$ variable.
2. $h = r + \frac{1}{2}\sqrt{4r^2 - c^2}$. $c = 6.5$. $r =$ variable.

3. $d = \frac{t}{3}\sqrt{rs}$. $t = 7$. $r = 4.2$. s = variable.

4. $A = B + C - P$. $C = 7.3$. $P = 2.6$. B = variable.

5. $P = \frac{N + n}{2C}$. $N = 38$. $C = 5.7$. n = variable.

6. $S = \frac{1.157}{P} - A$. $A = .035$. P = variable.

7. $C = \frac{N - n}{2P}$. $n = 21$. $P = 8$. N = variable.

8. $W = V + r - R$. $V = 20$. $r = 3$. R = variable.

9. $B = \frac{2P}{SP - N}$. $S = 9$. $N = 20$. P = variable.

10. $t = T\frac{C - F}{C}$. $C = 4.5$. $F = 2.8$. T = variable.

11. $A = \frac{BC}{D} + C$. $C = 6.8$. $D = 2.5$. B = variable.

12. $M = AB^2 + D$. $A = 3.9$. $D = 4.3$. B = variable.

13. $D = .3183NP$. $P = .437$. N = variable.

14. $C = \frac{3.1416d}{L}$. $d = 4.875$. L = variable.

15. $A = 2\sqrt{2S(D - 2S)}$. $S = 3.6$. D = variable.

16. $D = \frac{2CN}{N + n}$. $C = 5.5$. $n = 20$. N = variable.

VARIABLES

Prob.	Sym.	No. 1	No. 2	No. 3	No. 4	No. 5	No. 6
1	c	11.2	12.3	13.4	14.5	15.6	16.7
2	r	3.5	4.3	5.4	6.5	7.6	8.7
3	s	3	4	5	6	7	8
4	B	2.3	3.4	4.5	5.6	6.7	7.8
5	n	15	16	17	18	19	20
6	P	8	9	10	11	12	13
7	N	50	52	54	56	58	60
8	R	4	5	6	7	8	9
9	P	3	4	5	6	7	8
10	T	23	25	27	29	31	33
11	B	4.6	5.3	6.8	7.4	8.7	9.2
12	B	3.7	4.6	5.5	6.4	7.3	8.2
13	N	21	22	23	24	25	26
14	L	3.2	4.4	5.3	6.7	7.6	8.9
15	D	12.8	14.7	16.5	18.3	19.6	20.4
16	N	25	28	31	34	37	40

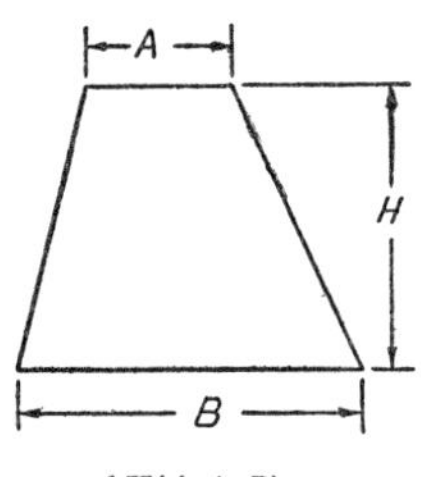

Area $= \frac{1}{2}H(A + B)$
$A = 3.5$
$B = 5.7$

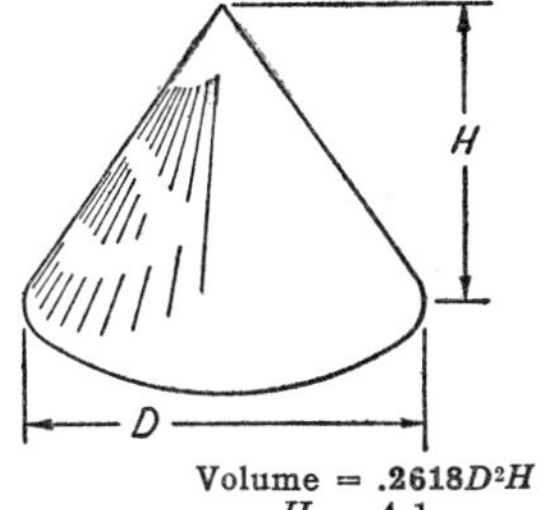

Volume $= .2618D^2H$
$H = 4.1$

17. Determine the area.

18. Determine the volume.

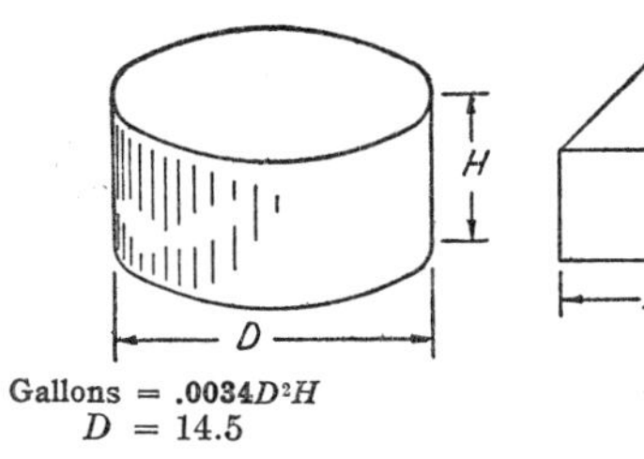

Gallons $= .0034D^2H$
$D = 14.5$

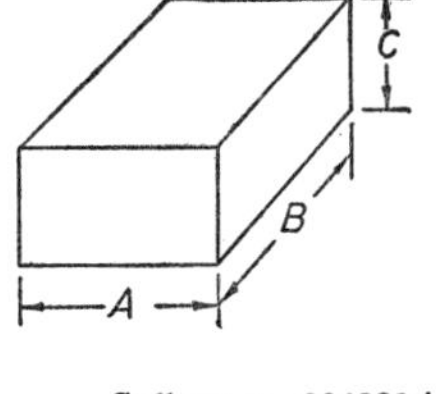

Gallons $= .004329ABC$
$A = 12.6$
$C = 5.8$

19. Determine number of gallons.

20. Determine number of gallons.

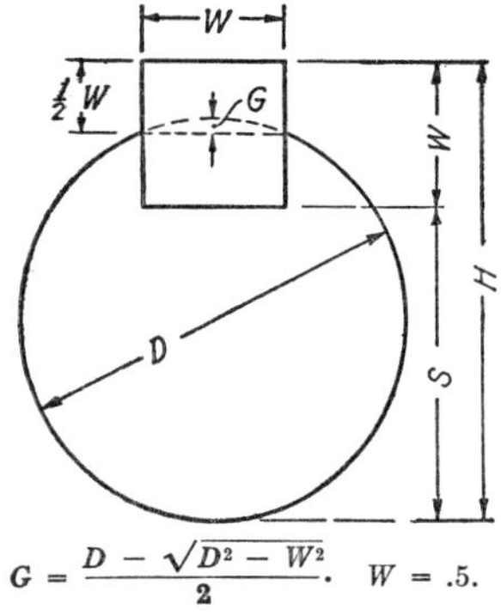

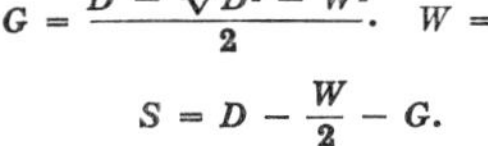

$$G = \frac{D - \sqrt{D^2 - W^2}}{2}. \quad W = .5.$$

$$S = D - \frac{W}{2} - G.$$

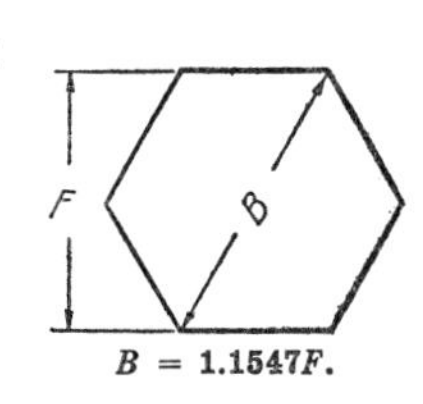

$B = 1.1547F.$

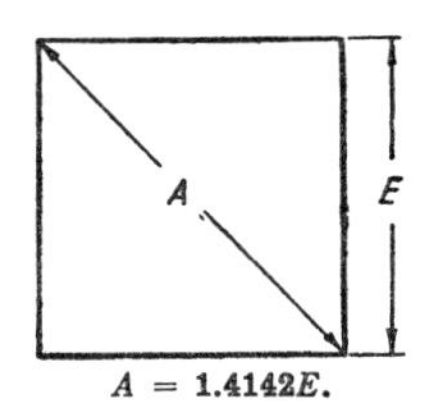

$A = 1.4142E.$

21. Determine S.
22. Determine G.

23. Determine B.
24. Determine B.

25. Determine A.
26. Determine A.

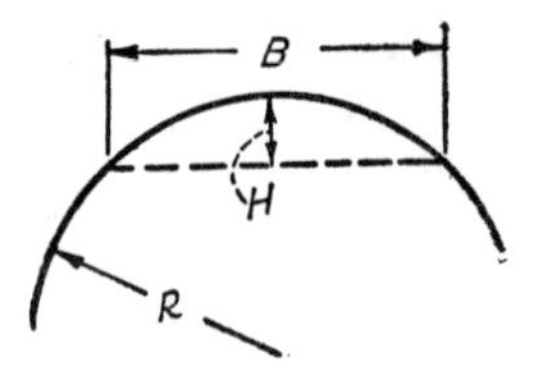

$B = 3.5. \quad R = \frac{B^2 + 4H^2}{8H}.$

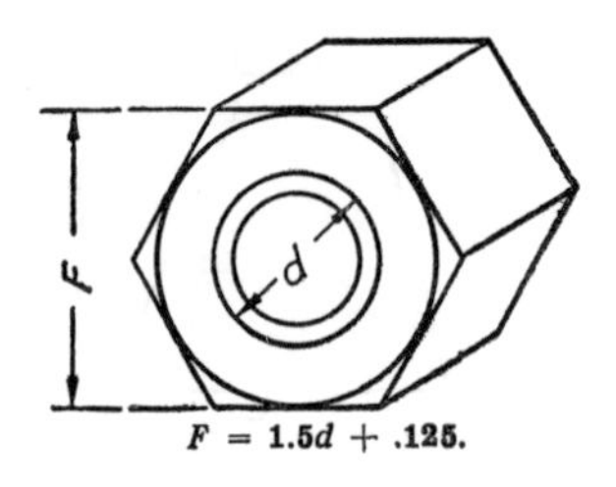

$F = 1.5d + .125.$

27. Determine the radius R.

28. Determine the radius R.

29. Determine the distance F.

30. Determine the distance F.

Sharp V Thread

P = pitch = distance between two successive peaks = .625.
$B = 1.5W - .866P.$

31. Determine the distance B.

32. Determine the distance B.

VARIABLES

Prob.	Sym.	No. 1	No. 2	No. 3	No. 4	No. 5	No. 6
17	H	7.8	7.9	8.1	8.2	8.3	8.4
18	D	4.6	4.7	4.8	4.9	5.1	5.2
19	H	7.3	7.5	7.7	7.9	8.1	8.3
20	B	18.7	18.9	19.1	19.3	19.5	19.7
21	D	1.325	1.437	1.525	1.781	1.785	1.937
22	D	1.325	1.437	1.525	1.781	1.785	1.937
23	F	1.25	1.375	1.625	1.775	1.875	2.225
24	F	2.3	2.5	2.7	2.9	3.2	3.5
25	E	1.375	1.4375	1.775	1.876	1.937	1.875
26	E	1.799	1.909	2.019	2.129	2.239	2.349
27	H	.75	.875	.9375	1.0625	1.125	1.25
28	H	1.057	.997	.937	.877	.817	.757
29	d	.525	.613	.687	.775	.844	.912
30	d	.436	.576	.642	.723	.875	.967
31	W	.375	.403	.437	.468	.505	.544
32	W	.623	.675	.714	.778	.826	.868

American National Thread

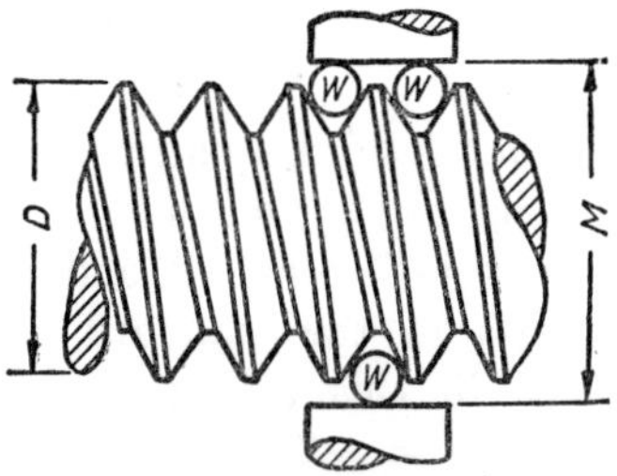

P = pitch = distance between two successive peaks = .5625. W = .375.
$M = D - 1.5155P + 3W.$

33. Determine the value of M.

34. Determine the value of M.

35. $n = \frac{2CP}{CP + T}$. $C = 3.7$. $P = 8$. T = variable.

36. $D = \frac{2\pi R}{L}$. $R = 6.3$. $\pi = 3.1416$. L = variable.

VARIABLES

Prob.	Sym.	No. 1	No. 2	No. 3	No. 4	No. 5	No. 6
33	D	3.177	3.288	3.399	3.4	3.511	3.622
34	D	1.654	1.797	1.824	1.963	2.186	2.25
35	T	3	4	5	6	7	8
36	L	5.6	6.7	7.2	8.3	9.5	10.4

SOLUTION OF QUADRATIC EQUATIONS

A quadratic equation is an equation involving an unknown quantity to the second power. The unknown quantity may also occur in the equation to the first power. Thus the following are quadratic equations:

(1) $x^2 - 9x + 20 = 0$. (2) $2y^2 = 50$.
(3) $7x^2 = 2x + 4$. (4) $4.16z^2 + 2.37z = 20.98$.

The standard form of the quadratic equation is

$$ax^2 + bx + c = 0$$

where x is the single unknown and a, b, and c represent numbers. Note that all terms are on the left side of the equation.

In the first equation above, x is the unknown and $a = 1$, $b = -9$, and $c = 20$. To put the second equation in the standard form, the 50 must be transposed giving $2y^2 - 50 = 0$. Then it is seen that the unknown is y and $a = 2$, $b = 0$, and $c = -50$.

The standard form of the third equation is $7x^2 - 2x - 4 = 0$ (obtained by transposing the $2x$ and the 4). In this equation, the unknown is x, and $a = 7$, $b = -2$, and $c = -4$.

The fourth equation in the standard form is $4.16z^2 + 2.37z - 20.98 = 0$, so in this case z is the unknown and $a = 4.16$, $b = 2.37$, and $c = -20.98$.

The general solution of the standard form of the quadratic equation is obtained by a method involving the completion of a square and the extraction of the square root. The details of this process will not be given in this text, but the resultant formula will be stated, and the method of obtaining the value of the unknown by use of this formula will be explained.

For any quadratic equation of the standard form

$$ax^2 + bx + c = 0,$$

the solution is

$$x = \frac{-b \pm \sqrt{b^2 - 4ac}}{2a}$$

Applying this formula to equation (1), $x^2 - 9x + 20 = 0$,

$$x = \frac{-(-9) \pm \sqrt{(-9)^2 - 4 \times 1 \times 20}}{2 \times 1}$$

$$= \frac{+9 \pm \sqrt{81 - 80}}{2} = \frac{+9 \pm 1}{2} = \frac{10}{2} \text{ or } \frac{8}{2} = 5 \text{ or } 4.$$

Note that there are two solutions, *i.e.*, two values of x which will satisfy the equation. In general the number of solutions of an unknown in an equation is equal to the highest power to which the unknown occurs in the equation.

The values of the unknown obtained should always be substituted in the equation as a check. Thus in the above problem substituting $x = 5$ in the original equation gives

$$5^2 - 9 \times 5 + 20 = 0 \text{ or } 25 - 45 + 20 = 0 \text{ or } 0 = 0,$$

which proves that $x = 5$ is a correct solution. Similarly checking the value $x = 4$,

$$4^2 - 9 \times 4 + 20 = 0 \text{ or } 16 - 36 + 20 = 0 \text{ or } 0 = 0.$$

The simple method for obtaining y in equation (2) is to divide both members by 2 giving $y^2 = 25$ and to extract the square root giving $y = \pm 5$. To show that the general formula solution will give the same results,

$$2y^2 - 50 = 0$$

$$y = \frac{-b \pm \sqrt{b^2 - 4ac}}{2a} = \frac{-0 \pm \sqrt{0^2 - 4 \times 2 \times (-50)}}{2 \times 2}$$

$$= \frac{\pm\sqrt{400}}{4} = \frac{\pm 20}{4} = \pm 5.$$

Substituting $y = +5$ in the original equation gives $2 \cdot (5)^2 = 50$ or $50 = 50$ and substituting $y = -5$ gives $2 \cdot (-5)^2 = 50$ or $50 = 50$.

Equation (3) in the standard form is $7x^2 - 2x - 4 = 0$.

Applying the formula:

$$x = \frac{-b \pm \sqrt{b^2 - 4ac}}{2a} = \frac{-(-2) \pm \sqrt{(-2)^2 - 4 \times 7 \times (-4)}}{2 \times 7}$$

$$= \frac{+2 \pm \sqrt{4 + 112}}{14} = \frac{2 \pm \sqrt{116}}{14} = \frac{2 \pm 10.7703}{14}$$

$= .9122$ or $-.6265$. Check for $x = .9122$.

$$7(.9122)^2 = 2 \times .9122 + 4 \text{ or } 5.8247 = 5.8244.$$

The slight discrepancy in the check is due to the fact that the last number in .9122 is not exactly 2 (it is nearer 2 than 3 or 1). The student should check the value $x = -.6265$.

Equation (4) in the standard form is $4.16z^2 + 2.37z - 20.98 = 0$.

Applying the formula,

$$z = \frac{-2.37 \pm \sqrt{(2.37)^2 - 4 \times 4.16 \times (-20.98)}}{2 \times 4.16}$$

$$= \frac{-2.37 \pm \sqrt{5.6169 + 349.107}}{8.32} = \frac{-2.37 \pm 18.8341}{8.32}$$

$$= \frac{-2.37 + 18.8341}{8.32} = 1.9788$$

or

$$z = \frac{-2.37 - 18.8341}{8.36} = -2.5363$$

The student should check both values of z.

As an example of how a quadratic equation may originate, consider the following geometrical problem.

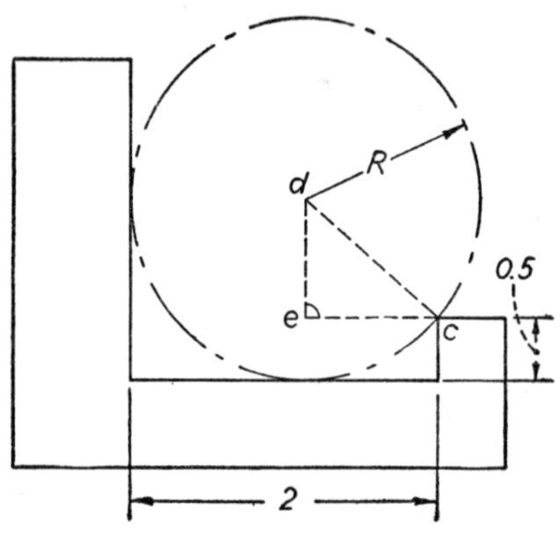

Compute the radius of a cylindrical plug which will touch the gage at the point C and be tangent to the other two surfaces.

$R^2 = \overline{cd}^2 = \overline{ce}^2 + \overline{ed}^2$ (see geometric proposition 31 on page 159).

$$ce = 2 - R \text{ and } ed = R - .5.$$

Hence $R^2 = (2 - R)^2 + (R - .5)^2$,

$$R^2 = 4 - 4R + R^2 + R^2 - R + .25 \qquad \text{(see page 68)}.$$

Collecting and putting in the standard form

$$R^2 - 5R + 4.25 = 0$$

Hence the above quadratic equation has been formed from the conditions given in the problem. The actual value of the radius may be obtained by solving this equation by means of the general quadratic formula as follows:

$$R = \frac{-b \pm \sqrt{b^2 - 4ac}}{2a} = \frac{-(-5) \pm \sqrt{(-5)^2 - 4 \times 1 \times 4.25}}{2 \times 1}$$

$$= \frac{5 \pm \sqrt{25 - 17}}{2} = \frac{5 \pm \sqrt{8}}{2} = \frac{5 \pm 2.8283}{2} = \frac{7.8283}{2} \text{ and}$$

$$\frac{2.1717}{2} = 3.9142 \text{ and } 1.0859\text{, respectively.}$$

Of the two possible values given by the solution on page 80, only the value less than 2 (*i.e.*, 1.0859) is seen to fit the condition of the problem.

PROBLEMS

Determine the value of the unknown symbol:

1. $x^2 - 6x + 8 = 0.$ *Ans.* $x = +2, +4.$

2. $2y^2 - 6 = 9y - 7.$

3. $5R^2 + 5R - 12 = 0.$

4. $x^2 + 7x - 12 = 0.$

5. $2y^2 - y = 4 + 3y.$ *Ans.* $y = 2.732, -.732.$

6. $3R^2 - 12R = 7R^2 + 6.$

7. $y^2 + 11y - 15 = 0.$

8. $2x^2 - 8x - 5 = 0.$

9. $7y^2 + 2 = y^2 - 3y + 7.$

10. $5x^2 = 9x^2 + 2 + 7x.$ *Ans.* $x = -.35961, -1.3904.$

11. $6R^2 - 7R - 3 = 0.$

12. $x^2 - 12x + 27 = 0.$

13. $3x^2 + 12x - 36 = 0.$

14. $34x - x^2 - 225 = 0.$

15. $16x^2 - 16x + 3 = 0.$ *Ans.* $x = .25, .75.$

16. $3R^2 - 10R + 3 = 0.$

17. $2y^2 - 12y + 10 = 0.$

18. $5x^2 - 3x - 2 = 0.$

19. $9R^2 - 24R + 16 = 0.$

20. $y^2 - 4 = 4y - 7.$ *Ans.* $y = 1, 3.$

21. $R^2 - 2R - 3 = 0.$

22. $6x^2 - 5x - 1 = 0.$

23. $y^2 - 14y - 51 = 0.$

24. $R^2 - 6R + 8 = 0.$

25. $5x^2 - 4x - 1 = 0.$ *Ans.* $x = 1, -.2.$

26. $2y - y^2 = 4y - 3.$

27. $R^2 + R - 20 = 0.$

28. $x^2 - x - 12 = 0.$

29. $5y^2 - 2 = 7y + 6.$

30. $3R^2 + 5R = 7.$ *Ans.* $R = .90672, -2.5734.$

31. $2x^2 - 3x - 4 = 0.$

32. $3x - 7 = -7x^2 + 4.$

33. $9R^2 + 8R - 6 = 0.$

34. $y^2 - 2y - 1 = 0.$

35. $6x^2 - 4x - 5 = 0.$ *Ans.* $x = 1.3051, -.63849.$

36. $4R^2 + 9R - 6 = 0.$

37. $y^2 - 3y - 7 = 0.$

38. $8x = 3x^2 - 5.$

39. $2R^2 + 7R + 4 = 0.$

40. $x^2 - 9x + 7 = 0.$ *Ans.* $x = .85995, 8.1400.$

The following eight geometrical algebraic problems are to be solved by the aid of the general quadratic formula. Each problem has a variable A, and for convenience the value of A should be inserted in the problem before forming a solution.

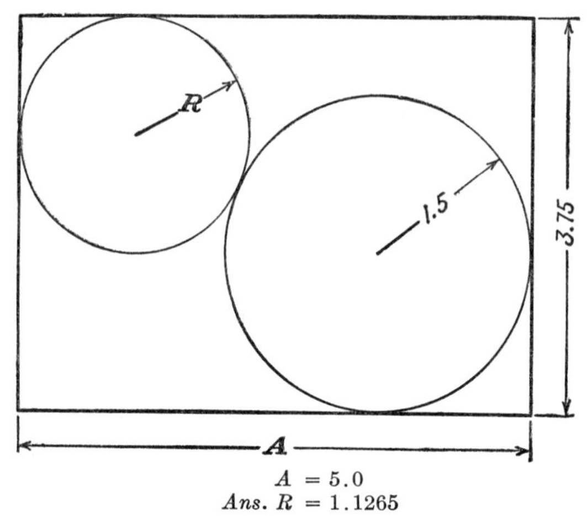

VARIABLE		
No.	Sym.	Value
1	A	4.25
2	A	4.37
3	A	4.50
4	A	4.62
5	A	4.87
6	A	4.93

$A = 5.0$
Ans. $R = 1.1265$

1. Determine the radius R.

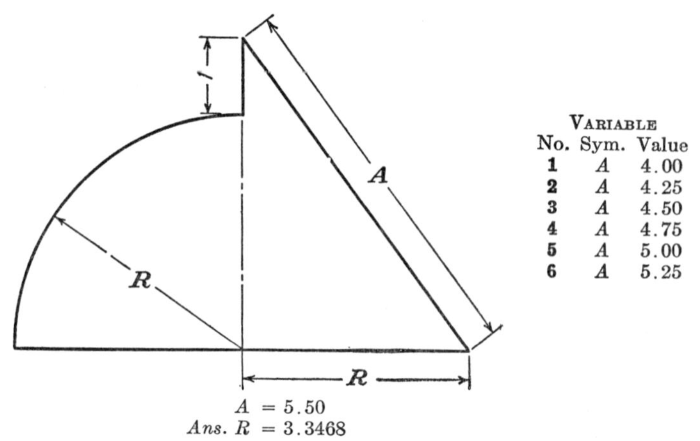

VARIABLE		
No.	Sym.	Value
1	A	4.00
2	A	4.25
3	A	4.50
4	A	4.75
5	A	5.00
6	A	5.25

$A = 5.50$
Ans. $R = 3.3468$

2. Determine the radius R.

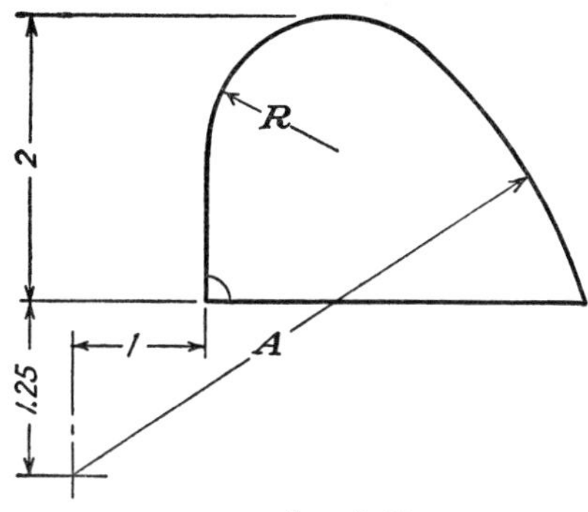

VARIABLE		
No.	Sym.	Value
1	A	3.76
2	A	3.80
3	A	3.84
4	A	3.88
5	A	3.92
6	A	3.96

$A = 4.00$
Ans. $R = .98865$

3. Determine the radius R.

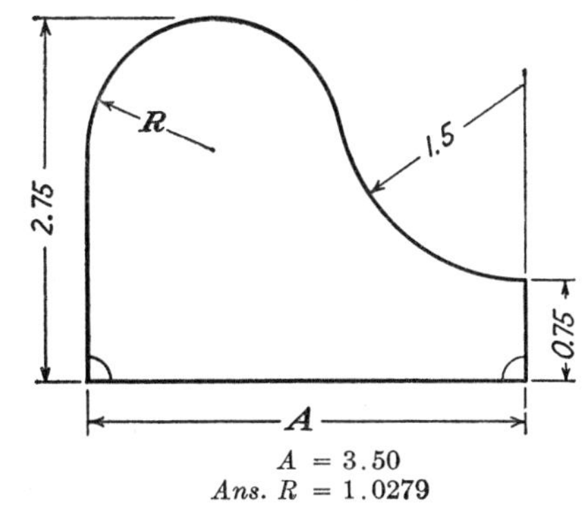

VARIABLE		
No.	Sym.	Value
1	A	3.20
2	A	3.25
3	A	3.30
4	A	3.35
5	A	3.40
6	A	3.45

$A = 3.50$
Ans. $R = 1.0279$

4. Determine the radius R.

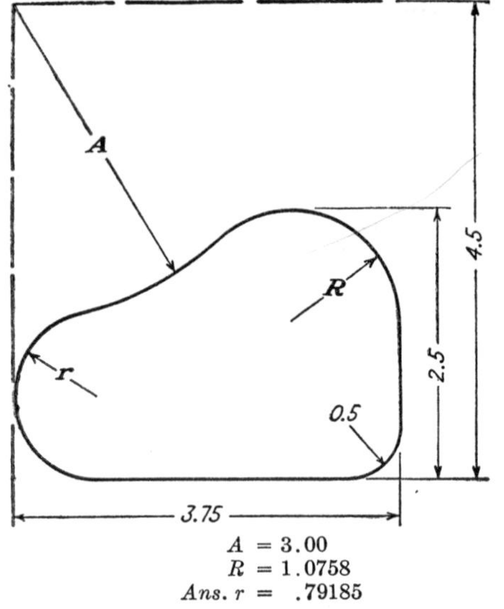

Variable		
No.	Sym.	Value
1	A	3.75
2	A	3.31
3	A	3.25
4	A	3.18
5	A	3.12
6	A	3.06

$A = 3.00$
$R = 1.0758$
Ans. $r = .79185$

5. Determine the radius R.

6. Determine the radius r.

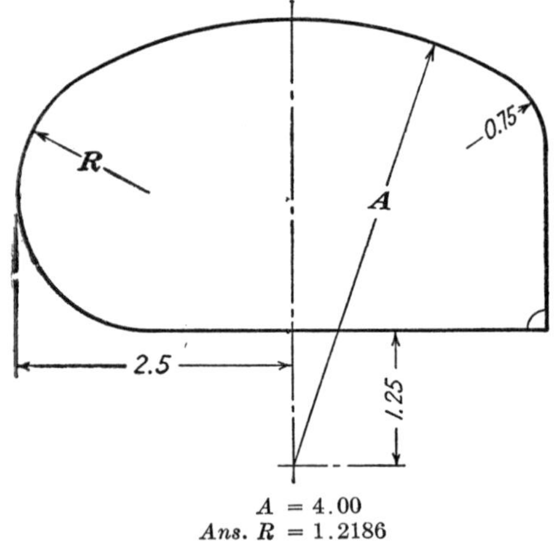

Variable		
No.	Sym.	Value
1	A	3.81
2	A	3.84
3	A	3.87
4	A	3.90
5	A	3.93
6	A	3.96

$A = 4.00$
Ans. $R = 1.2186$

7. Determine the radius R.

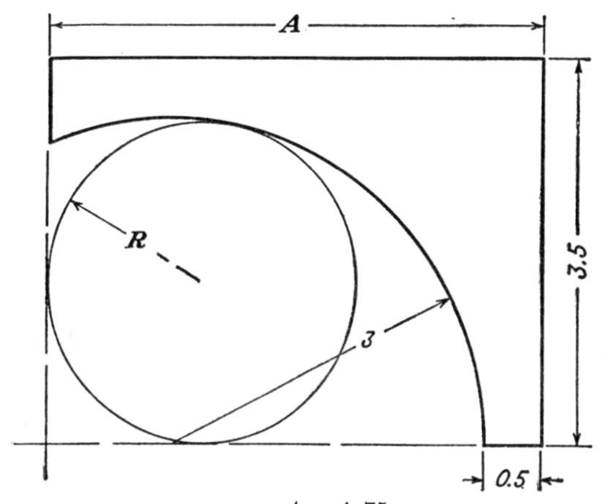

Variable		
No.	Sym.	Value
1	A	4.37
2	A	4.43
3	A	4.50
4	A	4.56
5	A	4.62
6	A	4.68

$A = 4.75$
Ans. $R = 1.4903$

8. Determine the radius R.

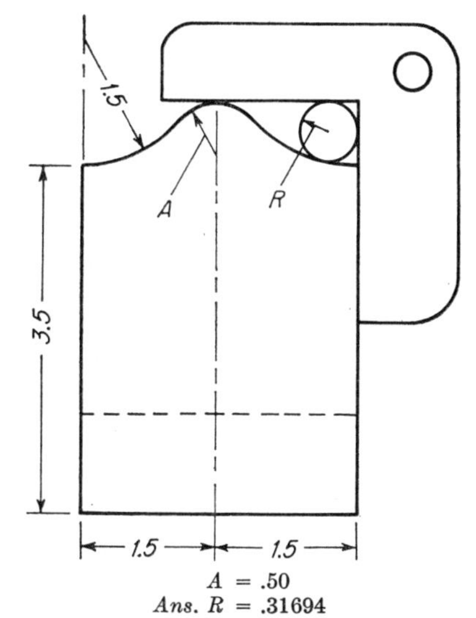

Variable		
No.	Sym.	Value
1	A	.38
2	A	.40
3	A	.42
4	A	.44
5	A	.46
6	A	.48

$A = .50$
Ans. $R = .31694$

9. Determine the radius R.

CHAPTER VI

GEAR RATIOS AND LEAD SCREWS

SPUR GEAR TERMINOLOGY

Gears are used to transmit power from one rotating shaft to another by means of intermeshing teeth. This process prevents the loss of power due to slippage that occurs when power is transmitted by two disks in contact. The teeth are constructed on the circumferences of disks with alternately formed lugs and recesses as indicated in the drawing below.

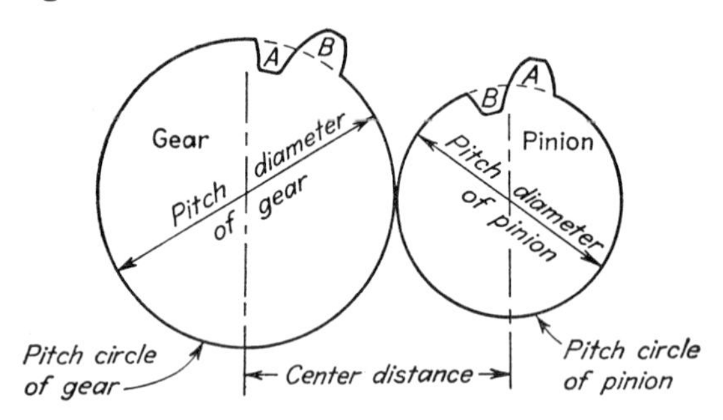

The circumferences of these disks are called the **pitch circles** of the gears, and the diameters are called the **pitch diameters.** The distance between the centers of the two pitch circles is called the **center distance.**

The diametral pitch of a gear is equal to the number of teeth within π in. of the pitch circumference, which is the same as the number of teeth within π in. along a straight line. The diametral pitch states the size of the tooth in the same manner that the number of threads per inch of a screw states the size of the thread. For example, a 7-diametral pitch gear is one that has 7 teeth within π in. of the pitch circumference.

The circular pitch is the distance from the center of one tooth to the center of the next measured along the pitch circumference. It follows that the circular pitch is equal to π divided by the diametral pitch. Circular pitch is usually given in multiples of sixteenths of an inch expressed decimally, thus: .1875, .250, etc. . . . In a pair of gears, the one having the greater number of teeth is called the **gear,** and the one having the lesser number is called the **pinion.**

GEAR AND PINION RATIOS

When two shafts are connected by a pair of gears, the numbers of revolutions of the shafts are governed by the numbers of teeth in the gears. The point A, indicated by the arrow, in Fig. 14 is a fixed point on the intersection of the center line and the pitch circumferences of gear and pinion. For brevity the point A is called the **common point of contact.**

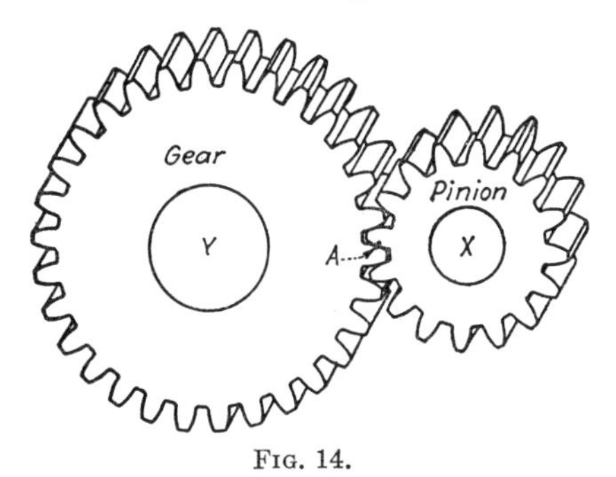

FIG. 14.

In Fig. 14, consider that the gear has 35 teeth and the pinion has 19 teeth. If the gear is revolved 1 revolution, 35 of its teeth will pass the point A. If it makes $\frac{5}{7}$ of a revolution, $\frac{5}{7}$ of 35 (or 25) of its teeth will pass the point A. If it makes $\frac{1}{35}$ of a revolution, $\frac{1}{35}$ of 35 (or 1) of its teeth will pass the point A. Thus when one tooth passes the point A, the gear makes $\frac{1}{35}$ of a revolution. Similarly, if 10 teeth pass the point A, the gear makes $\frac{10}{35}$ of a revolution, etc. From this it is evident that the number of teeth that pass the point A divided by the number of teeth in the gear is equal to the number of revolutions that the gear makes.

In Fig. 14, since 1 revolution of the gear causes 35 of its teeth to pass the point A, and since the teeth of the gear and pinion are intermeshed, it will also cause 35 teeth of the pinion to pass the point A. The number of revolutions made by the pinion is equal to the number of teeth passing point A (35) divided by the number of teeth in the pinion (19) which is $\frac{35}{19}$. It is thus seen that the pinion makes the greater number of revolutions, and the gear the lesser number. Therefore the gear is placed on the shaft which makes the lesser number of revolutions, and the pinion is placed on the shaft that makes the greater number of revolutions.

By the foregoing reasoning it was found that the pinion makes $\frac{35}{19}$ of a revolution while the gear makes 1 revolution. The ratio of the number of revolutions of the pinion to the gear is $\frac{35}{19}$ to 1 (or $\frac{35}{19}$). However, the ratio of the number of teeth in the gear to the number of teeth in the pinion is also 35 to 19 (or $\frac{35}{19}$). Hence the following relation:

$$\frac{\textbf{Number of revolutions of pinion shaft}}{\textbf{Number of revolutions of gear shaft}} = \frac{\textbf{number of teeth in the gear}}{\textbf{number of teeth in the pinion}}$$

which means that the number of revolutions vary inversely as the number of teeth.

This inverse relation means that the number of revolutions of one shaft may be considered as the number of teeth in the gear on the other shaft and *vice versa.*

Example: Let the shaft X make 41 revolutions and the shaft Y 21 revolutions. In order to have this relation, the gear on shaft Y can have 41 teeth and the pinion on shaft X can have 21 teeth.

Check: The number of teeth of the pinion passing the point A in 41 revolutions is 41×21. Since there are 41 teeth in the gear, the number of revolutions of the gear will be

$$\frac{\text{Number of teeth which pass a point } A}{\text{Number of teeth in gear being revolved}} = \frac{41 \times 21}{41} \text{ or } 21.$$

Increasing or Reducing Gear Teeth (Ratio Unaltered)

The number of teeth in a pair of mating gears can be raised or lowered without changing the ratio of the revolutions of the shafts. This can be done by multiplying or dividing the numbers of teeth in both mating gears by the same number, in the same manner in which the terms of a fraction are raised or lowered.

Example a: In order to maintain a revolution ratio of 19 to 24, the gear teeth ratio should be 24 to 19. In case a 19 tooth gear is not available, a pair of gears having 2×24 and 2×19, or 48 and 38, teeth may be used. Still another possible set would be 3×24 and 3×19, or 72 and 57 teeth, etc.

Example b: If the revolution ratio of the shafts is 171 to 69, the ratio of the gear teeth will be 69 to 171. A more practical set would be $\frac{69}{3}$ and $\frac{171}{3}$, or 23 and 57, teeth.

SPUR-GEAR AND RACK RATIOS

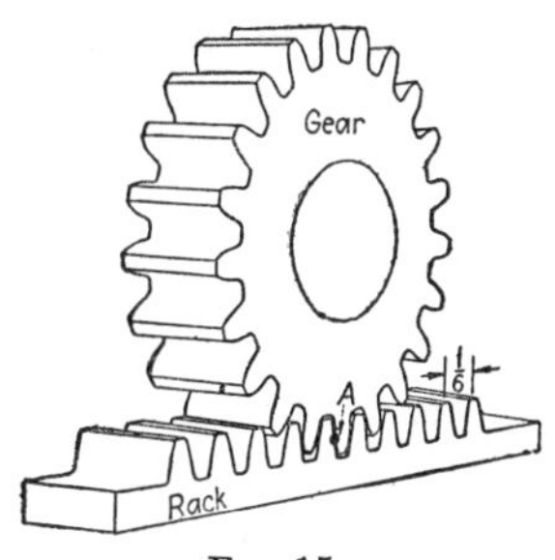

Fig. 15.

Consider that a gear having 17 teeth is in mesh with a rack whose pitch is $\frac{1}{6}$ in. One revolution of the gear will cause 17 teeth of the rack to pass the point of contact A (Fig. 15), thus moving the rack forward seventeen times $\frac{1}{6}$ in. or $2\frac{5}{6}$ in. If the gear makes $\frac{4}{5}$ of a revolution, the rack moves $\frac{4}{5}$ as far or $\frac{4}{5} \times 2\frac{5}{6}$ or $\frac{34}{15}$ or $2\frac{4}{15}$ in.

The reverse of the foregoing is to consider the rack to move forward a certain distance and to compute the corresponding number of revolutions of the gear. Thus, if the rack moves $\frac{5}{6}$ in., the number of teeth of the rack that will pass the point A is $\frac{5}{6}$ divided by $\frac{1}{6}$ or 5. Since five teeth of the gear are thus caused to pass A, the gear will make $\frac{5}{17}$ of a revolution.

Again, if the rack moves forward $1\frac{3}{8}$ in., the number of teeth passing A is $1\frac{3}{8}$ divided by $\frac{1}{6}$ or $8\frac{1}{4}$. Since $8\frac{1}{4}$ teeth of the gear have passed the point A, the gear will make $\frac{8\frac{1}{4}}{17} = \frac{33}{68}$ of a revolution.

BEVEL-GEAR RATIOS

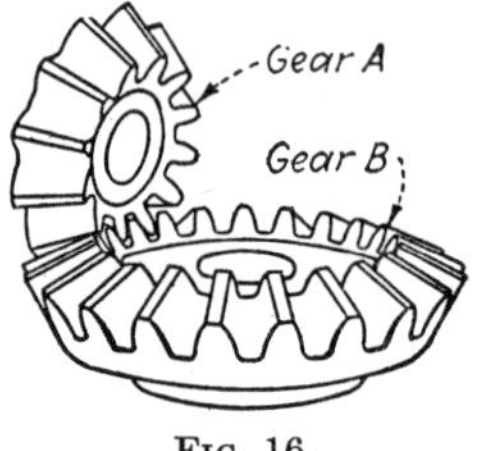

FIG. 16.

In bevel gears, the inverse relation between the numbers of revolutions and the numbers of teeth holds as in the case of spur gears, and thus we may again consider the number of revolutions of either shaft as the number of teeth of the gear on the opposite shaft and *vice versa.* Consider that the bevel gears A and B of Fig. 16 have 16 and 70 teeth, respectively. If gear A makes 1 revolution, gear B will make $\frac{16}{70}$ of a revolution. Conversely, if gear B makes 1 revolution, gear A will make $\frac{70}{16}$ revolutions. If gear B makes 8 revolutions, gear A will make 8 times $\frac{70}{16} = 35$ revolutions.

WORM AND WORM-WHEEL RATIOS

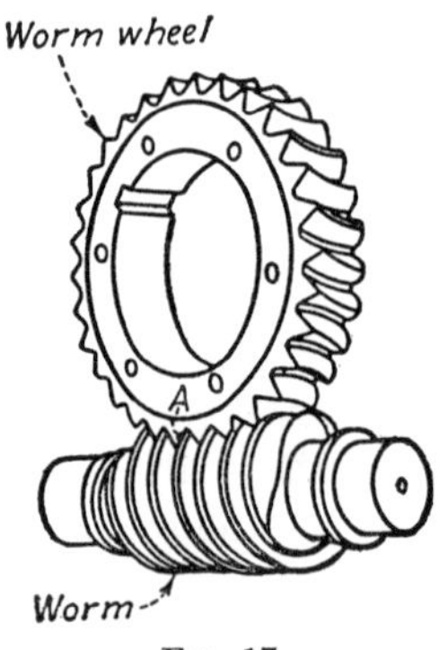

FIG. 17.

In Fig. 17, if the worm is single threaded, one revolution of the worm will cause one tooth of the worm wheel to pass the point of contact A; if the worm is double threaded, one revolution of the worm will cause two teeth of the worm wheel to pass point A; etc.

Consider that the worm wheel of Fig. 17 has 32 teeth and worm is triple threaded. When the worm makes one revolution, it will cause 3 teeth of the worm wheel to pass point A, or the worm wheel will make $\frac{3}{32}$ of a revolution.

If the worm wheel makes 1 revolution, the worm will make as many revolutions as the number of teeth in the worm wheel contains the number of threads in the worm. If the worm wheel of Fig. 17 makes $\frac{5}{8}$ revolution, the worm will make $\frac{\frac{5}{8} \times 32}{3}$ or $6\frac{2}{3}$ revolutions.

LEAD SCREW AND SLIDE

The **lead** of a screw is the distance a thread advances along a line parallel to the axis in one complete revolution.

The **pitch** is the distance from the center of one thread to the center of the next measuring along a line parallel to the axis.

The distance that the slide advances in one revolution of the screw is equal to the lead. On a single-threaded screw, the lead is equal to the pitch; on a double-threaded screw, the lead is equal to twice the pitch; etc.

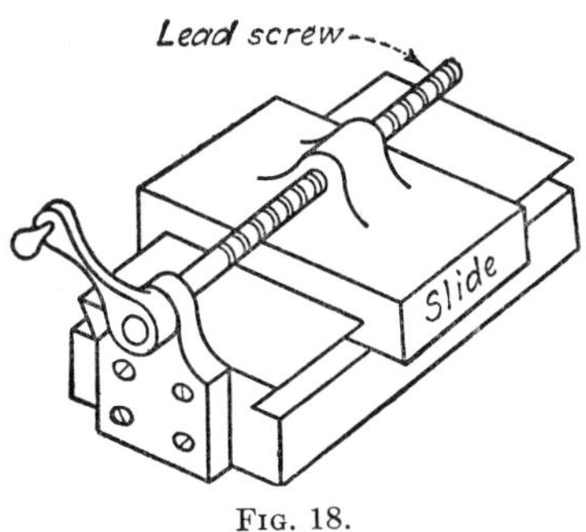

FIG. 18.

The screw of Fig. 18 is a single-threaded screw and has a pitch of $\frac{1}{7}$ in. Hence the lead is $\frac{1}{7}$ in. and the slide will move forward $\frac{1}{7}$ in. for each revolution of the screw. If the screw makes $3\frac{1}{2}$ revolutions, the slide will move forward $3\frac{1}{2}$ times $\frac{1}{7}$ or $\frac{1}{2}$ in.

If the slide moves $\frac{5}{8}$ in., the screw will make as many revolutions as $\frac{5}{8}$ contains $\frac{1}{7}$, or $4\frac{3}{8}$ revolutions.

If the screw of Fig. 18 were a triple-threaded screw having a pitch of $\frac{1}{7}$ in., for each revolution of the screw the slide would move forward three times $\frac{1}{7}$ or $\frac{3}{7}$ in. For $5\frac{1}{2}$ revolutions of the screw, the slide will move $3 \times \frac{1}{7} \times 5\frac{1}{2}$, or $2\frac{5}{14}$ inches.

Conversely, if the slide moved $\frac{3}{4}$ in., the screw would make as many revolutions as $\frac{3}{4}$ contains $3 \times \frac{1}{7}$ or $1\frac{3}{4}$ revolutions.

PROBLEMS

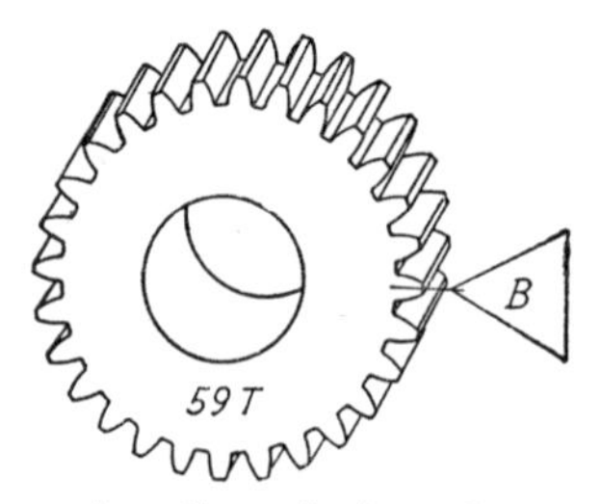

1. When the gear makes G revolutions, how many teeth pass the pointer B?

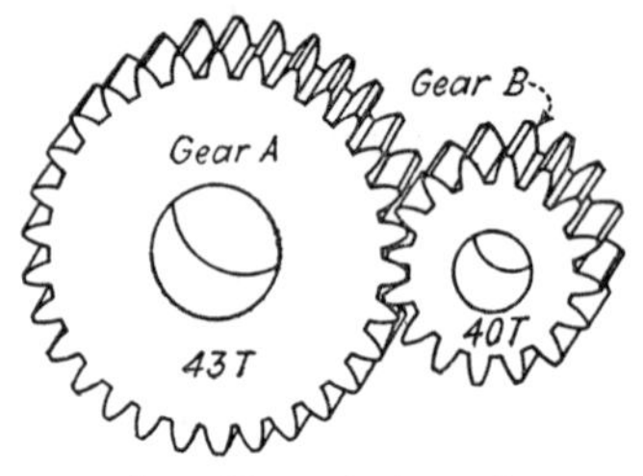

2. When the B gear makes H revolutions, how many revolutions will the gear A make?

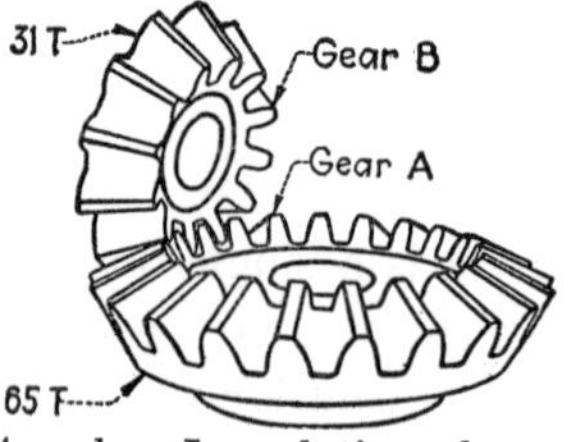

3. When the gear A makes J revolutions, how many revolutions will the gear B make?

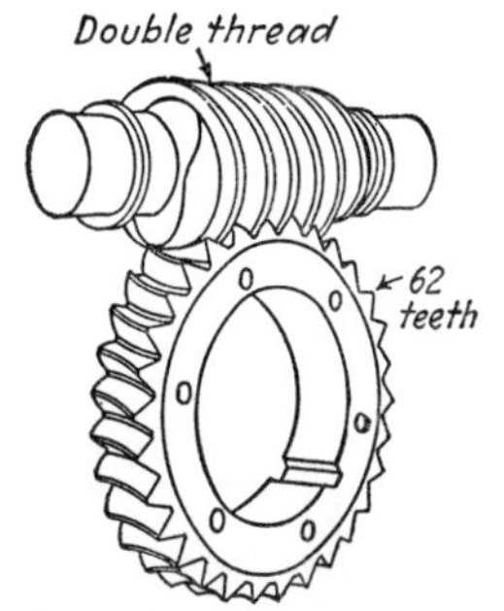

4. How many revolutions of the worm will cause the worm wheel to make T revolutions?

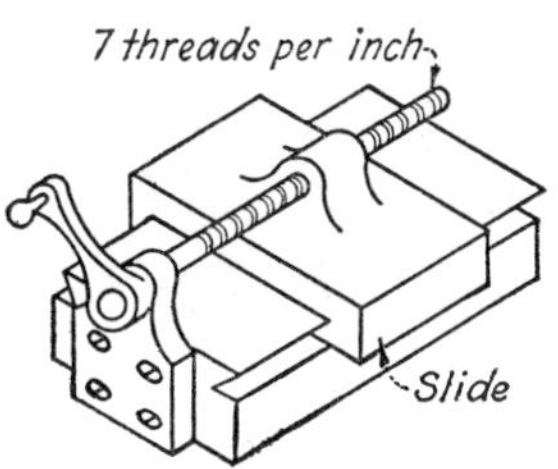

5. How many revolutions must the crank make to cause the slide to move S in.? The fixture above has single thread screw.

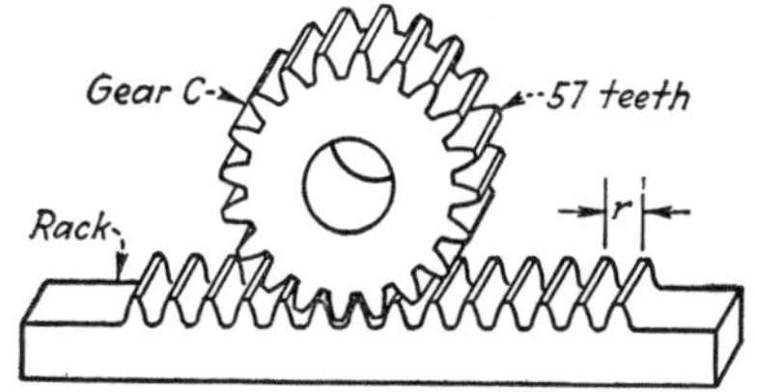

6. When the rack moves 7.532 in., how many revolutions will gear C make?

VARIABLES

Prob.	Sym.	No. 1	No. 2	No. 3	No. 4	No. 5	No. 6
1	G	7.6841	10.305	11.742	13.612	14.735	14.838
2	H	3.405	4.218	5.312	6.687	7.781	8.887
3	J	11.2	12.6	14.5	16.3	17.8	19.5
4	T	1.21	1.78	2.26	2.88	3.42	4.19
5	S	2.115	2.465	2.835	3.223	3.614	4.008
6	r	.375	.5	.4375	.5625	.625	.6875

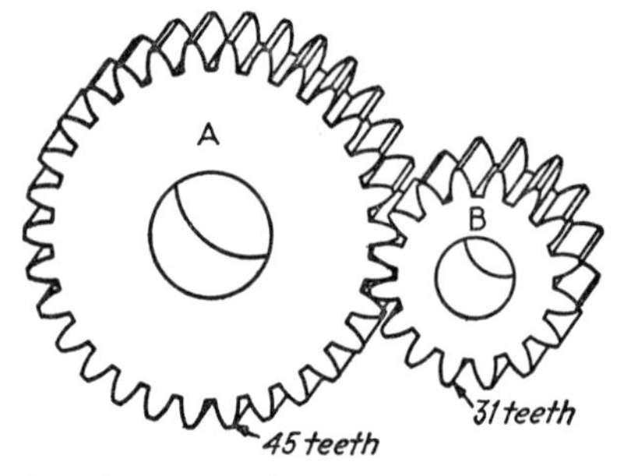

7. When gear A makes U revolutions, how many revolutions will the gear B make?

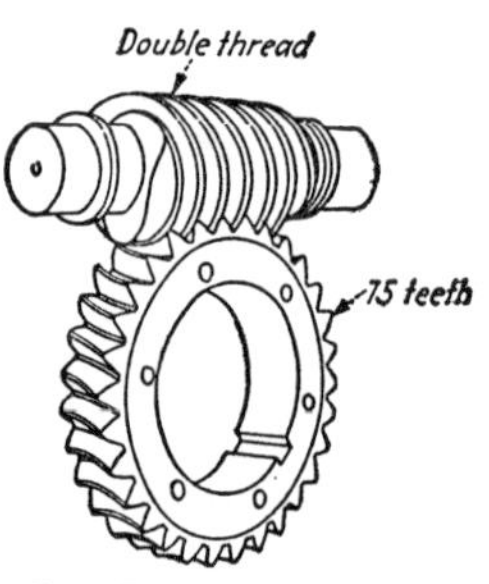

8. When the worm makes L revolutions, how many revolutions will the worm wheel make?

VARIABLES

Prob.	Sym.	No. 1	No. 2	No. 3	No. 4	No. 5	No. 6
7	U	4.32	5.46	6.18	6.92	7.76	8.56
8	L	21.1	24.2	27.3	31.4	34.5	38.6

IDLER GEARS

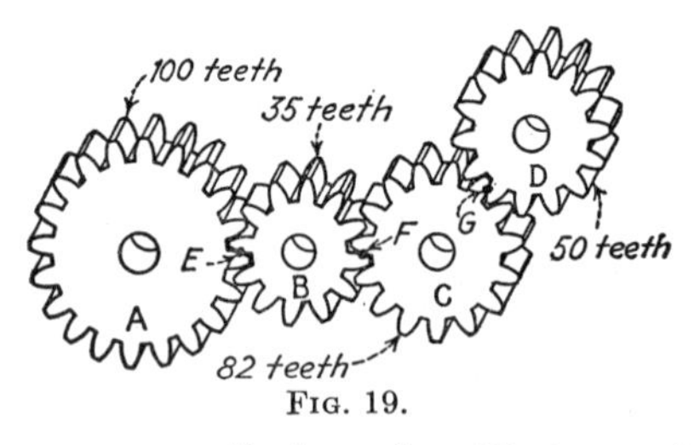

FIG. 19.

Idler gears are gears that mesh with two other gears and are used to transmit power from one shaft to another when the shafts are too far apart to be connected directly by a pair of gears. Another effect of an idler gear is to change the direction of rotation of the shaft. Idler gears have no effect on the shaft ratio. Thus, in the train of gears *A*, *B*, *C*, and *D* of Fig. 19, when *A* makes 1 revolution, *D* will make as many revolutions as the number of teeth in *A* contains the number of teeth in *D*; *i.e.*, $\frac{100}{50}$ or 2 revolutions. Since *B* and *C* are idler gears, the number of teeth in *B* and *C* are disregarded. The reason for this is as follows:

When *A* makes 1 revolution, 100 teeth will pass the contact point *E*; this will cause 100 teeth to pass the point *F*; this, in turn, will cause 100 teeth to pass *G*, and hence *D* will make $\frac{100}{50}$ or 2 revolutions.

TRAIN OF GEARS

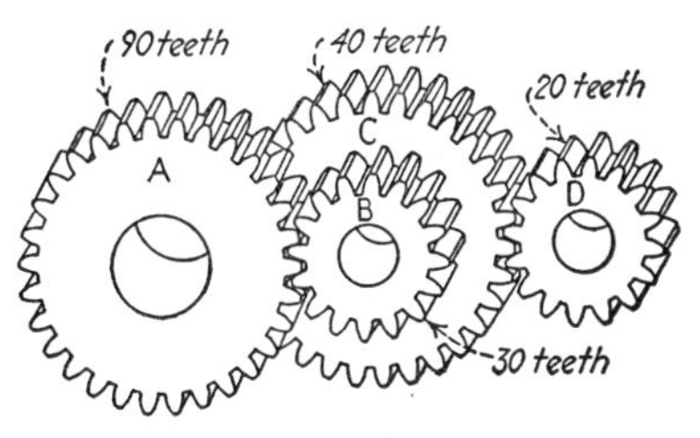

Fig. 20.

In order to determine the number of revolutions that *D* makes when *A* makes one revolution in the train of gears of Fig. 20, obtain the two ratios of the two sets of single gears separately and multiply them together.

Example: When *A* makes 1 revolution, *B* makes $\frac{90}{30}$ or 3 revolutions; when *C* makes 1 revolution, *D* will make $\frac{40}{20}$ or 2 revolutions. Since gears *B* and *C* are keyed to the same shaft, they will make the same number of revolutions. Since *B* and therefore *C* makes $\frac{90}{30}$ revolutions (for 1 revolution of *A*) *D* will make $\frac{90}{30} \times \frac{40}{20}$ or 6 revolutions. Conversely, when *D* makes 1 revolution, *A* will make $\frac{20}{40} \times \frac{30}{90}$ or $\frac{1}{6}$ revolution. When *D* makes $\frac{5}{8}$ of a revolution, *A* will make $\frac{20}{40} \times \frac{30}{90} \times \frac{5}{8}$, or $\frac{5}{48}$ of a revolution.

COMBINATION OF GEARS AND LEAD SCREW

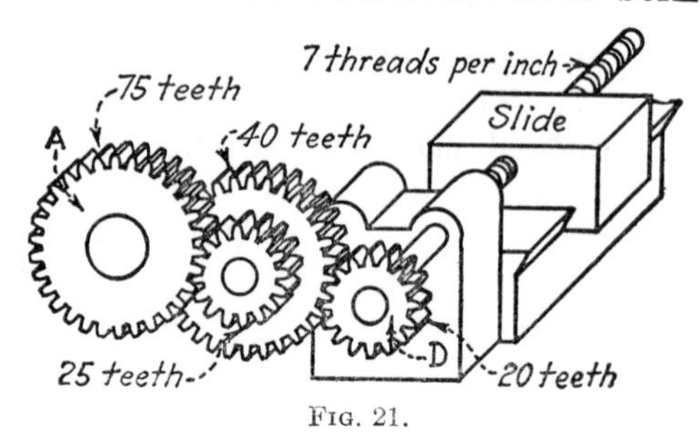

FIG. 21.

Figure 21 shows a train of gears connected with a lead screw and slide. The lead screw is a single threaded screw with seven threads per inch. Consider the problem of determining the distance the slide moves when gear A makes 1 revolution. When A makes 1 revolution, D will make $\frac{75}{25} \times \frac{40}{20}$ or 6 revolutions. When D makes 1 revolution, the slide will move a distance equal to the lead which is $\frac{1}{7}$ in. Since 1 revolution of A causes D to make $\frac{75}{25} \times \frac{40}{20}$ revolutions, and since each revolution of D moves the slide $\frac{1}{7}$ in., the slide will move $\frac{75}{25} \times \frac{40}{20} \times \frac{1}{7}$ or $\frac{6}{7}$ in.

Conversely, if the slide moves $\frac{3}{4}$ in., determine the number of revolutions A makes. Since $\frac{1}{7}$ in. represents 1 revolution of the screw, the gear D, which is attached to the screw, will make as many revolutions as $\frac{3}{4}$ contains $\frac{1}{7}$, or $5\frac{1}{4}$ revolutions. When D makes 1 revolution, A makes $\frac{20}{40} \times \frac{25}{75}$ or $\frac{1}{6}$ revolution. Since D makes $5\frac{1}{4}$ revolutions, A makes $\frac{20}{40} \times \frac{25}{75} \times 5\frac{1}{4}$ or $\frac{7}{8}$ of a revolution.

COMBINATION OF SPUR GEARS AND RACK

In Fig. 22, determine the distance the rack moves when C makes 1 revolution. When A makes 1 revolution, the rack moves a distance equal to the pitch circumference of gear A. The distance from the center of one tooth to the center of the next (on the rack) is equal to the circular pitch which in this case is $\frac{1}{4}$ in. Since there are 20 teeth in gear A, the pitch circumference is equal to $20 \times \frac{1}{4}$ or 5 in., which is the distance the rack moves when A makes 1 revolution. When C makes 1 revolution, B makes $\frac{25}{35}$ revolution. Gears A and B are keyed to the same shaft and therefore make the same number of revolutions. Therefore, when C makes 1 revolution, the rack will move a distance equal to $\frac{25}{35} \times 20 \times \frac{1}{4}$ or $3\frac{4}{7}$ in.

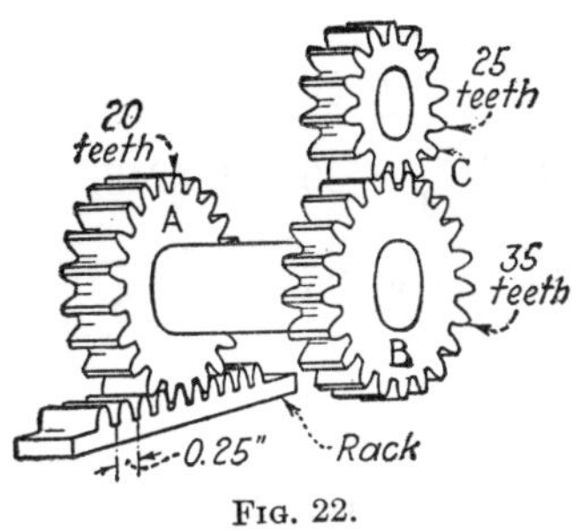

FIG. 22.

Conversely, when the rack moves $\frac{7}{8}$ in., determine the number of revolutions of C. Since the rack moves $\frac{1}{4}$ in. for each tooth, there will be as many teeth passing the point of contact as $\frac{7}{8}$ contains $\frac{1}{4}$, or $3\frac{1}{2}$ teeth. Gear A makes $\dfrac{3\frac{1}{2}}{20}$ or $\frac{7}{40}$ revolution. Since A makes $\frac{7}{40}$ revolution, C makes $\frac{7}{40} \times \frac{35}{25}$ or $\frac{49}{200}$ revolution.

COMBINATION OF RACK AND SPUR GEAR, AND WORM AND WORM WHEEL

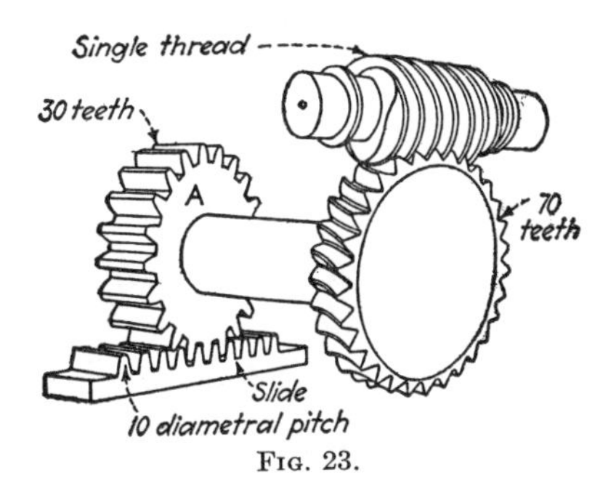

FIG. 23.

In Fig. 23, determine the distance that the rack moves when the worm makes 1 revolution. The rack is of 10 diametral pitch and the worm is single threaded. When the worm makes 1 revolution, the worm wheel makes $\frac{1}{70}$ of a revolution. Gear A is keyed to the same shaft as the worm wheel and therefore makes the same number of revolutions. Since 1 revolution of A causes the rack to move a distance equal to the pitch circumference of the gear A, $\frac{1}{70}$ of a revolution causes the rack to move $\frac{1}{70}$ as far. The pitch circumference of the gear is equal to $\frac{\pi}{10} \times 30$ or 9.4248. Thus for 1 revolution of the worm the rack will move $\frac{1}{70} \times \frac{\pi}{10} \times 30$ or .13464 in.

PROBLEMS

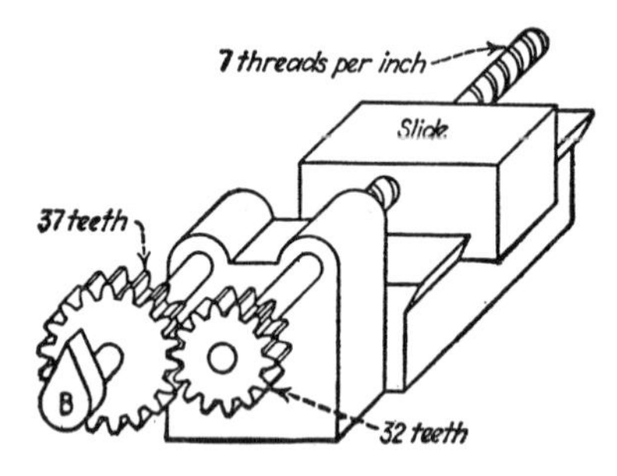

1. When the slide is caused to move G in., how many revolutions will B make?

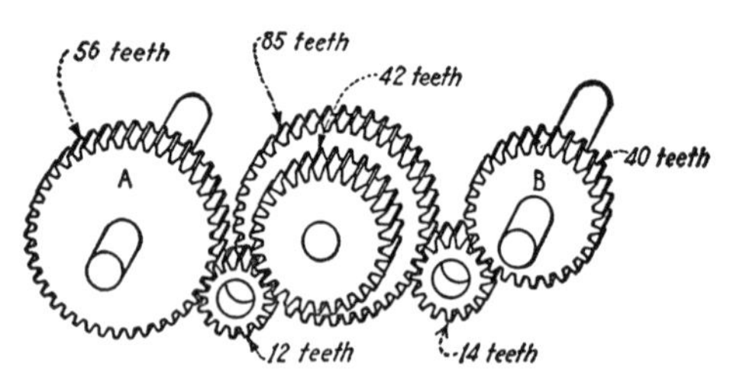

2. When B makes J revolutions, how many revolutions will A make?
3. How many idler gears are there in this problem?

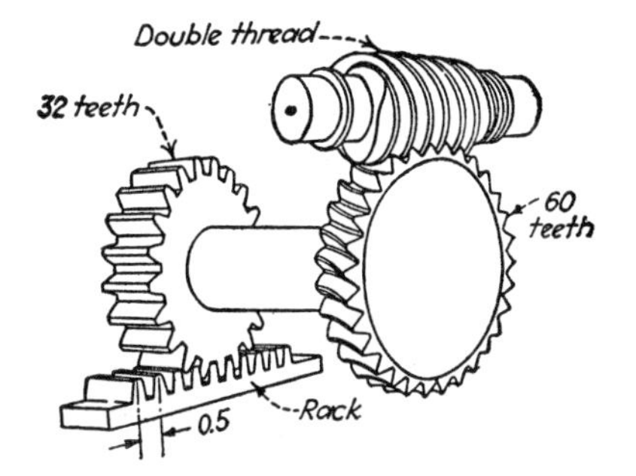

4. How many revolutions must the worm make in order that the rack will move H in.?

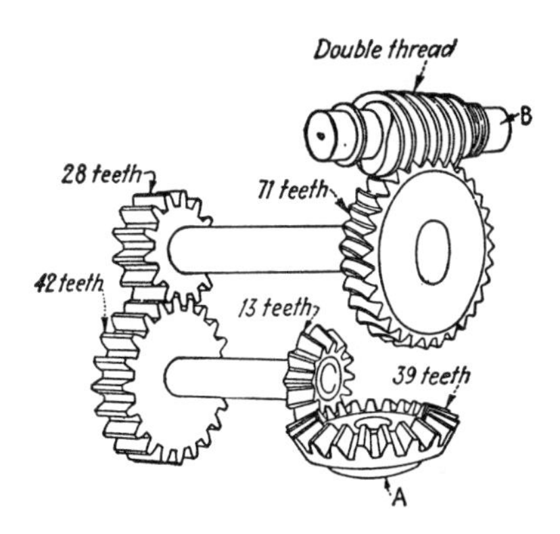

5. When A is caused to make T revolutions, how many revolutions will B make?

6. When B makes R revolutions, how many revolutions will A make?

VARIABLES

Prob.	Sym.	No. 1	No. 2	No. 3	No. 4	No. 5	No. 6
1	G	1.328	1.514	1.815	2.315	2.468	2.735
2	J	10.3	10.8	11.4	11.9	12.6	13.4
3	J	10.3	10.8	11.4	11.9	12.6	13.4
4	H	2.21	2.46	2.67	2.88	3.05	3.32
5	T	12.5	13.2	13.8	14.5	15.3	15.9
6	R	25.6	27.4	29.3	31.4	33.6	34.5

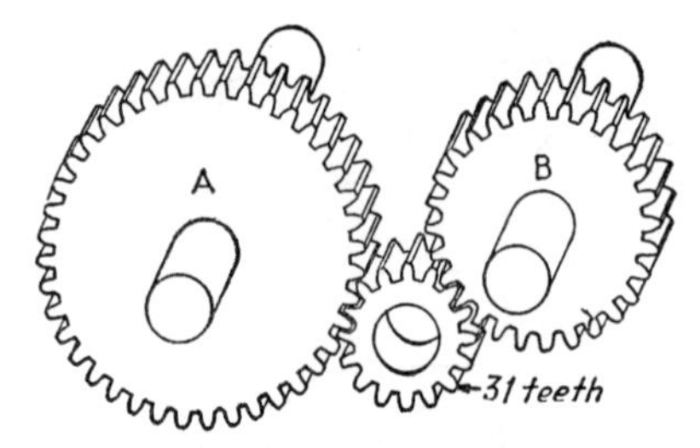

7. When A makes 5 revolutions, B must make 7. If gear A has S teeth, what number of teeth must B have?

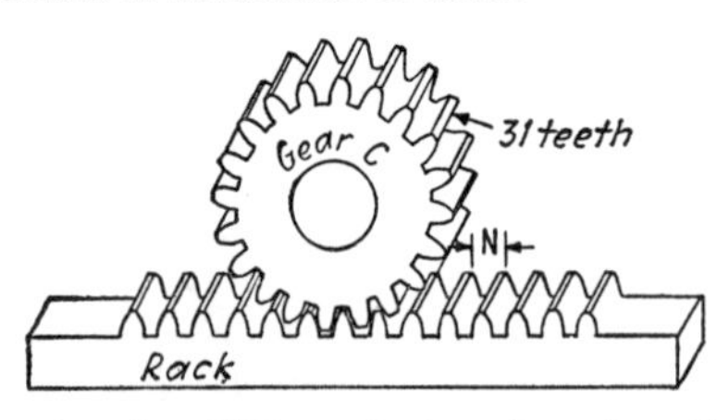

8. When gear C makes 5.73 revolutions, how far will the rack move?

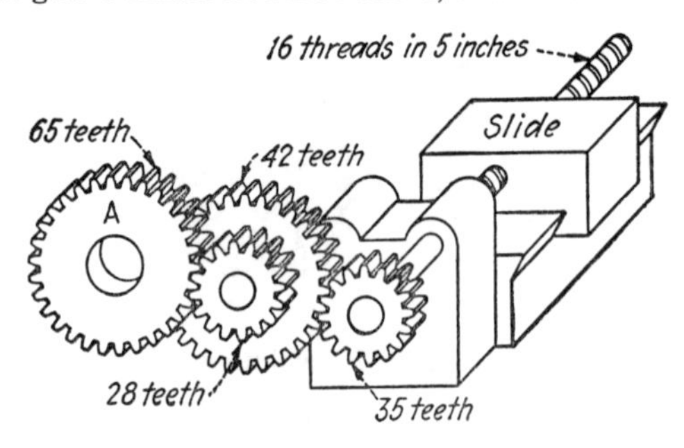

9. When the slide is caused to move P in., how many revolutions will A make?

10. When A makes L revolutions, how far will the slide move?

VARIABLES

Prob.	Sym.	No. 1.	No. 2.	No. 3.	No. 4.	No. 5.	No. 6.
7	S	42	49	56	63	70	35
8	N	.25	.3125	.375	.4375	.5	.5625
9	P	2.68	2.86	2.97	3.12	3.24	3.27
10	L	10.1	12.2	13.3	15.4	17.5	18.6

CHAPTER VII

PLAIN AND DIFFERENTIAL INDEXING

PLAIN INDEXING

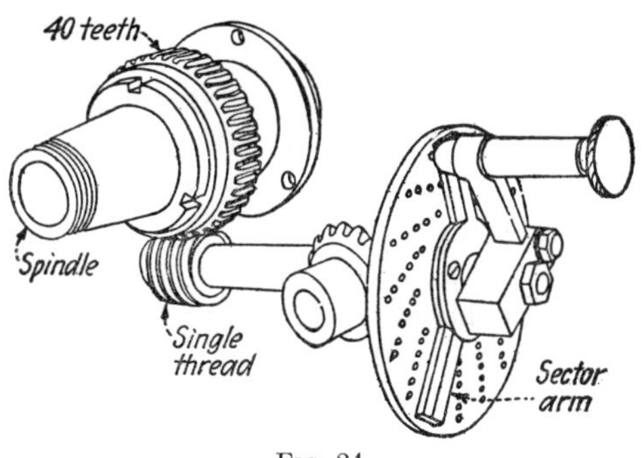

Fig. 24.

The principal working parts of a dividing head are the worm, worm wheel, spindle, crank, sector arm, and index plate, as shown in Fig. 24. The worm and crank are attached to the same shaft, therefore making the same number of revolutions. The worm wheel is attached to the spindle. The index plate is placed on the same shaft with the worm and can be either stationary or movable. For plain indexing the index plate is held stationary with a pin located at the back of the index plate. For differential indexing the index plate is allowed to rotate, either in the same direction as the crank or in the opposite direction, depending upon the nature of the problem. When the number of divisions are such that they cannot be made with plain indexing, they are generally determined by differential indexing. The sector arm is used to indicate the fraction of a turn of the crank required by the given problem and can be adjusted to give any fractional part of a turn.

There are three index plates furnished with a Brown and Sharpe dividing head. The numbers of holes in the different circles contained by these three index plates are **15-16-17-18-19-20-21-23-27-29-31-33-37-39-41-43-47-49.**

For plain indexing, assume that the worm and worm wheel are in the ratio of 40 to 1. When the spindle makes 1 revolution, the crank makes 40 revolutions. Suppose that it is required to cut 12 teeth on a gear. This gear is mounted on a chuck on the spindle of the dividing head which means that the spindle must make $\frac{1}{12}$ revolution when going from one tooth to the next. Hence the crank makes one-twelfth of 40, or $3\frac{1}{3}$ revolutions (meaning that the crank makes three complete turns and one-third of a turn). Mount, on the same shaft with the crank, an index plate in which the number of holes in a circle is evenly divisible by the denominator of the fractional part of a revolution that the crank must make. In this case, any of the possible circles given above, which are evenly divisible by 3, may be used, *i.e.*, 15, 18, 21, 27, 33, and 39. If a circle having 33 holes is used to obtain one-third of a turn, the two-part sector arm must be expanded to one-third of 33 or 11 circular divisions where a circular division is the fractional part of a turn of the crank between the centers of two successive holes. In this case a circular division represents one thirty-third of a turn.

Rules for Plain Indexing

Rule 1.—*Form a fraction having for its numerator the number of turns the crank makes when the spindle makes one revolution, and for its denominator the required number of divisions to be indexed. This gives the whole and fractional part of a revolution of the crank for one division of the work.*

Rule 2.—*Reduce this fraction to a mixed number, or to its lowest terms. When the fraction reduces to a mixed number, the whole number indicates the complete turns that the crank must make, and the fraction (reduced) the fractional part of a turn.*

Rule 3.—*Choose an index plate having a number of holes evenly divisible by the denominator of the fraction obtained by Rule 2. To obtain the number of circular divisions that the sector arm must expand, multiply the number just obtained by the fraction of Rule 2.*

Angular Indexing

One revolution of the crank revolves the spindle $\frac{1}{40}$ revolution, or 9°. When the crank is turned 1 hole in a 9-hole circle, the spindle revolves 1°; when the crank is turned 1 hole in an 18-hole circle, the spindle revolves $\frac{1}{2}$°; when the crank is turned 1 hole in a 27-hole circle, the spindle revolves $\frac{1}{3}$°, etc.

PROBLEMS

The following problems refer to Fig. 24.

1. How many revolutions must the crank make in order that the spindle will make E revolutions?

2. When the spindle makes F revolutions, the crank makes how many?

3. How many revolutions must the crank make in order that the spindle will make G revolutions?

4. When the spindle makes H revolutions, the crank makes how many?

5. How many holes must the index circle have in order to index J divisions?

6. How many holes must the index circle have in order to index K divisions?

7. How many holes must the index circle have in order to index L divisions?

8. What number of circular divisions must the sector arm expand in order to index M divisions?

9. What number of circular divisions must the sector arm expand in order to index N divisions?

10. What number of circular divisions must the sector arm expand in order to index P divisions?

VARIABLES

Prob.	Sym.	No. 1	No. 2	No. 3	No. 4	No. 5	No. 6
1	E	$\frac{3}{8}$	$\frac{7}{16}$	$\frac{1}{2}$	$\frac{5}{8}$	$\frac{11}{16}$	$\frac{7}{8}$
2	F	3.5	4.5	5.5	6.5	7.5	8.5
3	G	$\frac{3}{13}$	$\frac{4}{13}$	$\frac{5}{13}$	$\frac{6}{13}$	$\frac{7}{13}$	$\frac{8}{13}$
4	H	$\frac{1}{37}$	$\frac{1}{39}$	$\frac{1}{41}$	$\frac{1}{43}$	$\frac{1}{47}$	$\frac{1}{49}$
5	J	11	12	13	14	15	16
6	K	21	22	23	24	25	26
7	L	32	33	34	35	36	37
8	M	41	42	43	44	45	46
9	N	82	84	85	86	88	90
10	P	50	52	54	55	56	58

DIFFERENTIAL INDEXING

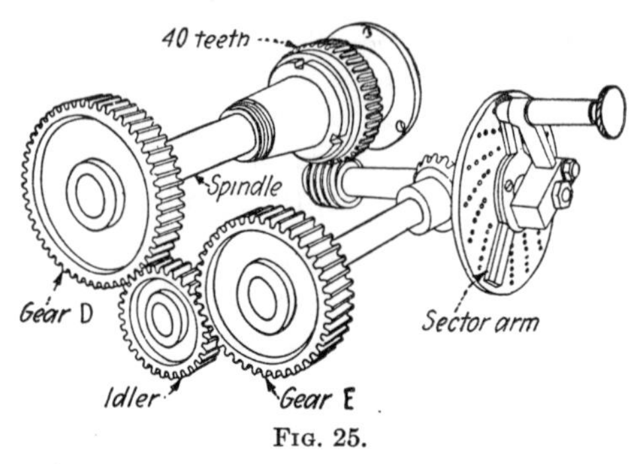

FIG. 25.

Differential indexing is used to index prime and fractional numbers that cannot be obtained by plain indexing.

The procedure for differential indexing can be best illustrated by the following two problems.

Example a: Assume that 73 divisions are to be indexed.

Solution: Applying Rule 1 of plain indexing, the fraction is $\frac{40}{73}$, the denominator of which cannot be reduced to any of the given numbers of holes in the index plates. Therefore, select some number close to 73 which will give a fraction that can be reduced within the range of plain indexing, say 70. If 70 is used, the crank will rotate $\frac{40}{70}$ or $\frac{4}{7}$ revolution each time it is operated. Since 73 divisions must be indexed, the crank is operated seventy-three times while the spindle is supposed to make 1 revolution. The number of revolutions that the crank makes when it is operated seventy-three times is $73 \times \frac{4}{7}$ or $41\frac{5}{7}$ revolutions, but the crank should make 40 revolutions while the spindle makes 1. Therefore, the crank has made $1\frac{5}{7}$ or $\frac{12}{7}$ revolutions too many. This means that the index plate must rotate $\frac{12}{7}$ revolutions in the direction opposite to that of the crank while the spindle makes 1 revolution. This is to be accomplished by means of gears *D* and *E* of Fig. 25, *D* being the gear on the spindle and *E* the gear on the shaft which actuates the gear attached to the index plate. The numbers of revolutions $\frac{12}{7}$ to 1 or 12 to 7 correspond to the numbers of teeth in the gears on the opposite shafts. Therefore, the numbers of teeth in the gears on the spindle and index plate are in the ratio of 12 to 7, or 36 to 21, respectively. In this case two idler gears are used in order that the index plate will rotate in an opposite direction to that of the crank.

Example b: Assume that 461 divisions are to be indexed.

Solution: Select some number very nearly equal to the number of divisions required, which is within the range of plain indexing, *e.g.*, 470. In order to cut 470 divisions, the crank must make $\frac{40}{470}$ or $\frac{4}{47}$ of a turn for each division. Since each time the crank is operated, it rotates $\frac{4}{47}$ of a turn, then when the crank is operated 461 times, it will make $461 \times \frac{4}{47}$, or $39\frac{11}{47}$ revolutions, which is $\frac{36}{47}$ revolution less than it should make. Now the index plate must rotate $\frac{36}{47}$ revolution in the same direction as the crank, while the spindle makes 1 revolution. For the same reason given above, the gears on the spindle and index plate are in the ratio of 36 to 47, respectively, or the gear on the spindle has 36 teeth and the gear on the index plate has 47 teeth. In this case one idler gear is used.

In each of the foregoing problems, it may be noticed that the gear ratio is equal to the product of the fractional part of a revolution that the crank makes for each division, and the difference between the number of divisions required and the number of divisions selected. This relation can be proved algebraically.

Rules for Differential Indexing

Rule 1.—*Select some number nearly equal to the number of divisions required, which, when divided by one of the factors of* 40 (2, 4, 5, 8, 10, 20, 40), *will give a number that can be indexed by plain indexing.*

Rule 2.—*Form a fraction with* 40 *as the numerator and the selected number as the denominator and reduce it to its lowest terms.*

Rule 3.—*Choose an index plate having a number of holes evenly divisible by the denominator of the fraction obtained by Rule* 2. *Multiply this number by the fraction of Rule* 2 *to give the number of circular divisions that the sector arm must expand.*

Rule 4.—*Multiply the fraction of Rule* 2 *by the difference between the number selected and the number of divisions required. This product is equal to the gear ratio, which, if necessary, may* *be raised to higher terms in order to get numbers which are within the usual range for gear teeth. The numerator and denominator of this fraction are equal to the number of teeth in the gear on the spindle and the number of teeth in the gear on the index plate, respectively.*

Rule 5.—*When the index plate must be rotated in the same direction as the crank only one idler gear is needed, and for the index plate to rotate in the opposite direction as the crank two idler gears are needed.*

PROBLEMS

The following problems refer to Fig. 25.

Notation

A = number of divisions in index circle that the crank must move
B = number of holes in circle to be used
D = number of teeth in gear on spindle
E = number of teeth in gear on worm

The numbers of teeth in gears must be within the limits of 20 and 100.

1. Number of divisions to be indexed equals R. Determine the values of A, B, D, and E.

2. Number of divisions to be indexed equals S. Determine the values of A, B, D, and E.

3. Number of divisions to be indexed equals T. Determine the values of A, B, D, and E.

4. Number of divisions to be indexed equals U. Determine the values of A, B, D, and E.

Variables

Prob.	Sym.	No. 1	No. 2	No. 3	No. 4	No. 5	No. 6
1	R	101	107	113	117	121	131
2	S	357	361	373	383	323	347
3	T	563	571	585	592	551	523
4	U	1841	1762	1993	1684	1805	1606

CHAPTER VIII

THE SLIDE RULE

The slide rule is an instrument that greatly simplifies the common mathematical processes of multiplication, division, proportion, squares, square roots, etc.

Practical shop problems occurring in toolrooms, die rooms, and machine repair departments, which must be solved by the mechanic, the draftsman, and the engineer, require accuracy to five significant figures. Slide-rule computations are reliable to only three significant figures, so for most actual machine-shop problems, the slide rule cannot be used. However, in carrying out practice problems, where the main object is to obtain practice on the geometrical phases of the problem and the numerical result is of secondary importance, the use of the slide rule in getting approximate answers will save the student hours of time. Furthermore, a slide-rule solution may be used as a rapid check on the ordinary method of multiplication, division, etc.

BRIEF THEORY OF THE SLIDE RULE

In Chap. V on algebra, the idea was exemplified that when quantities expressed with exponents are multiplied, the exponents are added. Thus: $(a^3)(a^2) = a^5$. When quantities are divided, the exponents are subtracted. Thus:

$$\frac{a^5}{a^3} = a^{5-3} = a^2.$$

In common logarithms the base number is 10, and the exponent is the degree of the power to which 10 must be raised to give the number. Thus: $10 = 10^1$, $100 = 10^2$, $1000 = 10^3$. A number between 10 and 100 will have an exponent between 1 and 2. Thus: $83 = 10^{1.91908}$, a number between 100 and 1000 will have an exponent between 2 and 3. Thus: $624 = 10^{2.79518}$. The integer part (1 in the case of 83 and 2 in the case of 624) is called the **characteristic** and is determined by inspection. The fractional part of the exponent is called the **mantissa** and has been carefully worked out for all numbers and is given in tables of common logarithms.

To multiply 2 by 3, the characteristics are seen to be 0 (any number between 1 and 10 has a characteristic 0). The mantissas are sought in a five-place "log" table and found to be .30103 and .47712, respectively. Thus $2 \times 3 = (10^{0.30103})(10^{0.47712}) = 10^{0.30103+0.47712} = 10^{0.77815}$, by adding exponents. Reversing the procedure for finding the mantissa of a number, the number having the mantissa of .77815 is found from the table to be 6. Thus $2 \times 3 = 6$. This seems a lot of work to obtain the result, but the amount of work and time is no greater in multiplying 347 by 728.

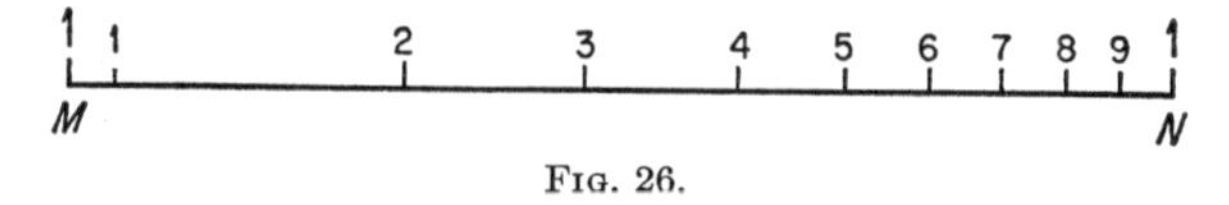

Fig. 26.

In the ordinary slide rules the scales C and D are laid off to represent the mantissas of numbers from 1 to 10.

To show how this is done consider that the line MN in Fig. 26, which is 3 in. long, is to be made into the C scale of a slide rule. The point M is labeled 1, since the log 1 is zero. Point N is also labeled 1, although we may better think of it as 10 for this discussion.

The mantissa of the log 2 is .30103. Hence the number to represent 2 must be .30103 of the distance from M to N (1 to 10). Thus $.30103 \times 3$ in. = .90309 in., which is the distance of 2 from M. Likewise, the mantissa of the log of 3 being .47712, the number 3 is located $.47712 \times 3$ in. = 1.43136 in. to the right of M. Similarly, the digits 4, 5, 6, 7, 8, and 9 are located.

Subdivisions on the scale are determined in the same manner. Thus the location of the small 1 between the left digit 1 and the digit 2 (which represents 1.1) is obtained by multiplying the mantissa of the log of 1.1 (which is .04139) by 3 in. (which gives .12417 in.).

The D scale is made identical to the C scale. Hence, when the left index (the 1) of scale C is placed above the large number 2 on the scale D and the indicator is moved to the large 3 on the C scale, the distance representing the mantissa of 2 is being added to the distance representing the mantissa of 3, thus giving on the D scale under the indicator the mantissa of 6.

To divide 6 by 3, we have

$$\frac{6}{3} = \frac{10^{0.77815}}{10^{0.47712}} = 10^{0.77815-0.47712} = 10^{0.30103}$$

The table of logarithms shows that the number having the exponent (mantissa) .30103 is 2.

To divide 6 by 3 using the slide rule, set the large 3 of the C scale directly over the large 6 of the D scale using the indicator (the hairline on slide) to line them up carefully. The answer 2 is found on the D scale directly under the left index of the C scale. The indicator should be moved to this index of the C scale in order to read the answer on the D scale as accurately as possible. The student should realize that in this process the exponent to which 10 must be raised to give 3 (the length on the C scale from left index to the 3) has been subtracted from the exponent to which 10 must be raised to give 6 (the length on the D scale from the left index to the 6), thus giving the exponent to which 10 must be raised to give 2 (the length on the D scale from the left index to the 2).

USE OF THE SLIDE RULE

Multiplication

Rule.—*To multiply two numbers, set the index (the figure 1) of the C scale directly above one of the numbers on the D scale and read the answer on the D scale under the other number on the C scale.*

Note: If setting the *left* index of the C scale over one number brings the other number beyond the range of the D scale, the *right* index of the C scale must be used.

The above procedure is summarized in formula (1), page 118.

Example a: Multiply 2×3.

Set the left index of the C scale directly over the large 2 of D scale (see Fig. 27) and read, under the large 3 of the C scale, the answer 6 on the D scale.

The student should note that the distance between successive integers on the C and D scales diminishes as the numbers increase. For that reason the space between 1 and 2 is first divided into 10 parts (divisions numbered) and each of these

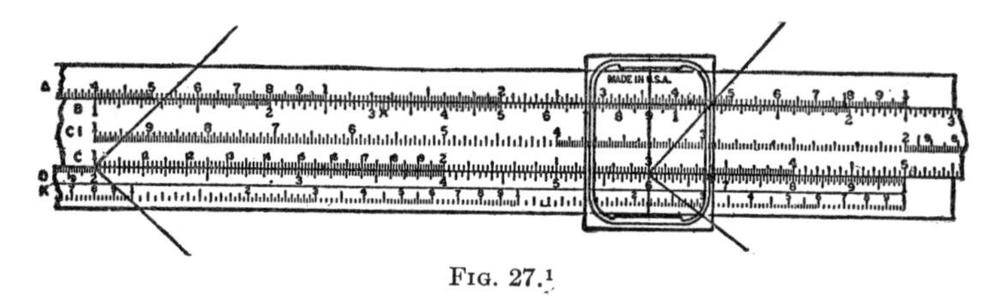

FIG. 27.[1]

parts is subdivided in 10 parts (divisions not numbered). Thus 1.5, which must be midway between 1 and 2, is at the graduation line labeled with the small 5 located between 1 and 2. Since there are 10 graduated divisions between 1.5 and 1.6, each large division represents 1 unit; hence 1 small division beyond this small 5 is 1.51, etc.

Between the large 2 and the large 3 there are also 10 main divisions (not numbered) each of which is divided into 5 subdivisions. The longest line (midway between the large 2 and the large 3) is 2.5. Since there are only 5 graduations between 2.5 and 2.6, each graduation represents 2 units. Hence the first small line beyond 2.5 is 2.52. Halfway along this small division, since each graduation represents 2 units, is one greater than the previous number, or the reading is 2.53, etc.

The space between 3 and 4 is divided in the same manner as the space between 2 and 3 except that the divisions are smaller.

The space between 4 and 5 (and 5 and 6, etc.) is divided into 10 main divisions, each of which is divided into 2 subdivisions. Thus the longest line (midway between 4 and 5) is 4.5. Since there are only 2 graduations between 4.5 and 4.6, each graduation represents 5 units, and the next (small) line beyond 4.5 represents 4.55. The student should learn to estimate readings on this part of the scale (from 4 to the right index, which

[1] The slide rule cuts are by courtesy of the Keuffel & Esser Co.

is 10) to three figures. Thus if the hairline of the indicator seems to be about two-fifths (slightly less than half) of the division beyond that corresponding to 4.55, the reading is estimated to be 4.57. If the hairline is at four-fifths of that division (nearly to the 4.60 line), the reading is estimated to be 4.59, etc.

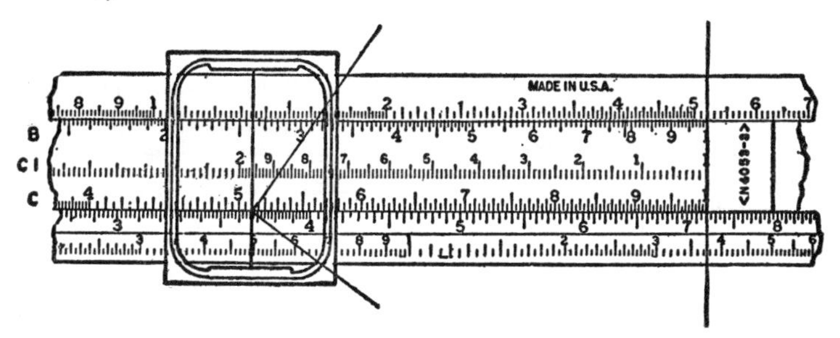

FIG. 28.

The student must also understand that the large 3 may represent 3, in which case 4 represents 4, 5 is 5, etc., or the large 3 may represent 30, in which case the 4 represents 40, the 5 is 50, etc. Similarly, the large 3 may represent 300, 3000, .3, .003, etc. The 1.51 already mentioned can also represent 15.1, 151, 1510, .151, .0151, etc.

Example b: Multiply 72 × 51.

Set the *right* index of the C scale directly over 72 (same place as 7.2) of the D scale as in Fig. 28 and under 51 (same as 5.1) of the C scale read the answer 3672 on the D scale. Actually the slide-rule reading would give only the first three figures 367, which would give an answer of 3670, but in this case it is noted that the product of the last two figures (1 × 2) is 2.

Determination of the Position of the Decimal Point.—To determine the location of the decimal point the student should mentally carry out the process using simple numbers which approximate the actual numbers. Thus in the previous problem, 72 × 51 carried out mentally is 70 × 50, which is equal to 3500. This shows that the answer is 3670 rather than 367 or 36.7.

Example c: Multiply 2.47 × 34.2.

Set the *left* index of C directly over 2.47 on the D scale

(2.47 is halfway between the lines representing 2.46 and 2.48) as in Fig. 29 and under 34.2 (same as 3.42) on the C scale read the answer 84.5 on the D scale. The answer is known to be 84.5 rather than 8.45 or 845, because using the approximate simple numbers 2×30 gives 60, which is nearer 84.5 than 8.45 or 845.

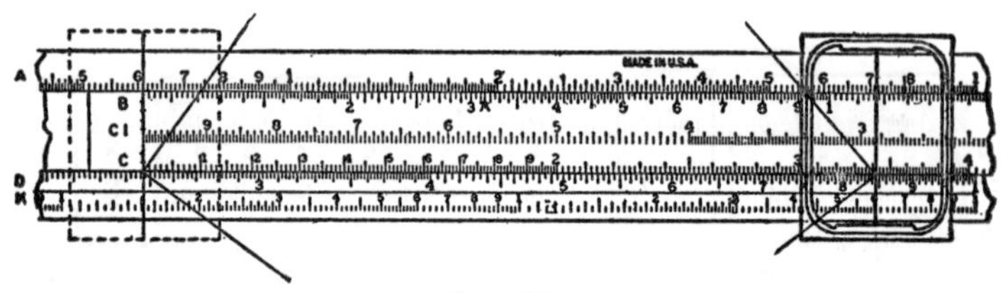

Fig. 29.

Actually 2.47×34.2 is 84.474, but the slide rule can be used only to the first three significant figures, which in this case would be 84.5.

Division

Rule.—*To divide one number (the dividend) by another (the divisor), set the divisor on the C scale directly above the dividend on the D scale, and under the index of the C scale, read the answer (the quotient) on the D scale.*

The above rule is summarized in Formula 2 on page 118.

Example a: Divide 6 by 3.

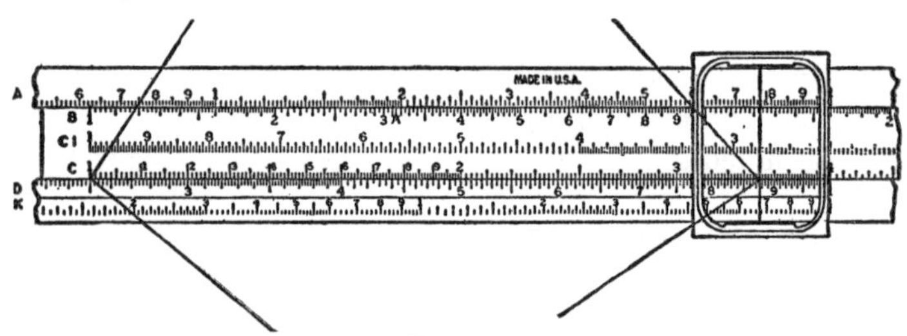

Fig. 30.

Since division is the reverse of multiplication, Fig. 27 can be used. The divisor 3 of the C scale is placed (with the aid of the indicator) directly above the dividend 6 of the D scale, and under the index of the C scale is the answer 2.

Example b: Divide 875 by 35.

Place the divisor 35 on the C scale directly above the dividend 875 on the D scale as in Fig. 30. Under the index of the C scale read the quotient 25 on the D scale. Using approximate numbers, $900 \div 30 = 30$, which shows that the answer is 25 rather than 2.5 or 250.

Multiplication and Division

In a problem involving several multiplications and divisions, first carry out a division, then a multiplication, then another division, then another multiplication, etc. It is not necessary to record the intermediate answers in such problems.

Squares and Square Roots

The A scale consists of two complete logarithmic scales each half as long as the logarithmic scales of C and D. The B scale is similar to the A scale, and multiplication and division can be carried out with the A and B scales. However, this is seldom done, as less accurate estimates can be made with shorter scales.

The principal use of the A scale is in obtaining squares and square roots when used in conjunction with the D scale.

Rule for Squares.—*Set the indicator line on any number on the D scale, and the square of that number will be found under the indicator line on the A scale.* See formula 4 on page 118.

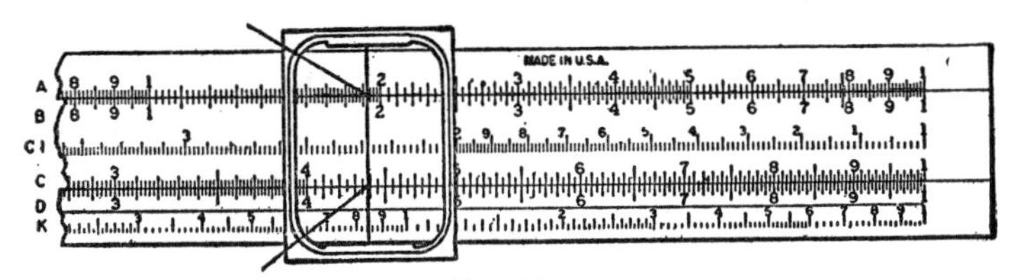

FIG. 31.

Example: Obtain the square of 43.8.

Set the indicator line on 43.8 on the D scale as in Fig. 31, and under the line on the A scale read the answer 1920. To determine the position of the decimal point, note that the square of the approximate simple number 40 is (by inspection) 1600, so that the answer had to be 1920 rather than 192 or 19200. Actually $(43.8)^2$ is 1918.44. However, the slide rule gives only three significant figures, so if there are to be four figures in the answer one zero must be added to the three numbers given by slide rule. The square of 438 would be 192000, since $(400)^2$ is 160000.

Rule for Square Roots.

a. To find the square root of a number having an **odd** *number of figures before the decimal point or, in the case of a decimal fraction, having an* **odd** *number of zeros immediately to the right of the decimal point, set the indicator line on the number on the* ***left*** *half of the A scale and read the square root on the D scale under the indicator line.*

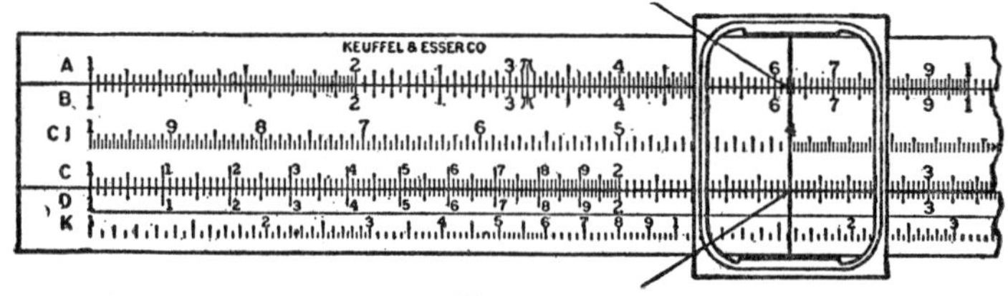

Fig. 32.

b. To find the square root of a number having an **even** *number of figures before the decimal point or, in the case of a decimal fraction, having an* **even**[1] *number of zeros immediately to the right of the decimal point, set the indicator line on the number on the* ***right*** *half of the A scale and read the square root on the D scale under the indicator line.*

See formula 5, on page 118.

Example a: Obtain the square root of 625.

Since there are three figures before the decimal point, Rule *a* applies. With the indicator line on 625 of the left A scale as in Fig. 32, the square root is found on the D scale under the indicator line to be 25.0. The answer is 25, rather than 2.5 or 250, since using approximate simple numbers gives $20 \times 20 = 400$.

Example b: Find the square root of 6250.

Since there are four figures before the decimal point, Rule *b* applies. Setting the indicator line on 6250 on the right A scale as in Fig. 33, the square root is found on the D scale

[1] A decimal fraction with no zeros such as .432 is equivalent to having an even number of zeros.

under the indicator line to be 79.1. The position of the decimal point is determined by noting that $80 \times 80 = 6400$.

Example c: Find the square root of .0506.

Since this is a decimal fraction with one zero immediately to the right of the decimal point, Rule *a* applies. Setting the indicator line on .0506 of the left A scale gives the square root

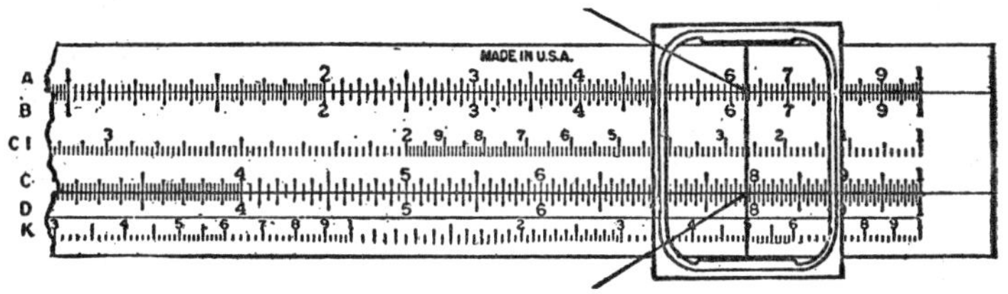

FIG. 33.

on the D scale to be .224. This is the correct position of the decimal point, since using approximate simple numbers gives $.2 \times .2 = .04$.

Proportion

Many problems may be solved by setting up a proportion. Thus if it is known that 8 castings weigh 23.2 lb., how much will 67 castings weigh? How much will 39 of these castings weigh?

$$\frac{23.2}{8} = \frac{x}{67}.$$

Set 8 on the C scale above 23.2 on the D scale, and under 67 on the C scale read the answer 194 on the D scale. Without moving the slide (the middle movable section of the slide rule), move the indicator to 39 on the C scale, and on the D scale under the indicator read the second answer 113.

Note that the above procedure is summarized in formula 3 on page 118.

Example: If 50 bolts from a bin weigh 113 lb., compute: the weight of 6000 such bolts; the weight of 10 gross (1440) such bolts.

$$\frac{113}{50} = \frac{x}{6000} = \frac{y}{1440}.$$

Set 50 on the C scale directly above 113 on the D scale.

Under 6000 on the C scale read on the D scale the answer for x (13560 lb.).

Since 1440 on the C scale is not above the D scale, the left index must be moved to the reading occupied by the right index (with the aid of the indicator). Then under 1440 on the C scale is the answer on the D scale for y (3254). Note that the last figure of the answer is very uncertain; however, the answer is obviously somewhat more than 3250 and less than 3260.

Problems Involving Sines, Tangents, Cosecants, Cotangents, Cosines, and Secants

This part of this chapter will be better understood after the student has studied the material given on trigonometric functions given in Chap. X.

If the slide is reversed, the S scale is adjacent to the A scale and the T scale to the D scale.

Rule for Obtaining Sine of an Angle.—*With the ends of the S and A scales coincident, place the indicator over the angle on the S scale and the sine of this angle is under the indicator on the A scale.* See formula 16 on page 119.

Note: (*a*) All sines read on the *right* half of the A scale have the decimal point just before the first figure. (*b*) All sines read on the *left* half of the A scale have a zero between the decimal point and the first significant figure.

Example a: Obtain the sine of 30°. Since 500 on the *right-hand* A scale is just above 30 on the S scale, sine 30° = .500.

Example b: Obtain the sine of 4°. Since 698 on the *left-hand* A scale is just above 4 on the S scale, sine 4° = .0698.

The procedure just discussed may be reversed to give an angle corresponding to a given value of the sine.

Example a: Obtain the angle when its sine is .0425.

Under 425 of the *left-hand* A scale, the angle is found to be 2° 26′.

Example b: Obtain the angle α when $\sin \alpha = .623$.

Under 623 of the *right-hand* A scale, the angle is found to be 38° 30′.

Rule for Obtaining Tangent of an Angle.—*With the ends of the T and D scales coincident, set the indicator over the angle* *on the T scale, and the tangent of this angle is under the indicator on the D scale.*

Example a: Obtain the tangent of 30°.

Under 30° on the T scale, .577 is seen to be the tangent.

The angle corresponding to a given value of the tangent may be obtained by reversing the procedure just given.

Example b: Obtain the angle α when $\tan \alpha = .352$.

Just over .352 on the D scale is the answer 19° 23′ on the T scale.

The use of the T and D scales gives directly the tangents of angles from 5° 43′ to 45°. However, by using the relations, $\tan \alpha = \frac{1}{\cot \alpha}$, page 204 and $\cot \alpha = \tan (90° - \alpha)$, page 204, the tangents of angles from 45° to 84° 17′ can also be found. For example,

$$\tan 62° = \frac{1}{\cot 62°} = \frac{1}{\tan (90° - 62°)} = \frac{1}{\tan 28°}.$$

For angles less than 5° 43′, $\tan \alpha = \sin \alpha$ nearly, so the tangent is obtained on the slide rule by looking up the sine.

If the cosine of an angle is needed, it may be obtained from the relation, $\cos \alpha = \sin (90° - \alpha)$, page 204. Thus $\cos 40° = \sin (90° - 40°) = \sin 50°$, which is given on the S scale to be .766.

Cotangents for angles up to 45° may be obtained from the relation $\cot \alpha = \frac{1}{\tan \alpha}$, and for angles over 45°, $\cot \alpha = \tan (90° - \alpha)$.

Cosecants may be handled on the slide rule from the relation $\csc \alpha = \frac{1}{\sin \alpha}$, page 204.

Secants are obtained from the S scale by using the relations $\sec \alpha = \frac{1}{\cos \alpha} = \frac{1}{\sin (90° - \alpha)}$, page 204.

In problems in trigonometry, frequently a number must be multiplied or divided by the sine of an angle or by the tangent, by the cosecant, by the cotangent, by the cosine, or by the secant.

Example a: Multiply 12 sin 28°.

Set the index (beginning mark) of the S scale under 12 of the A scale. The answer is on the A scale directly above 28 on the S scale and is found to be 5.63.

Example b: Compute 17 cot 25°.

Since $\cot 25 = \frac{1}{\tan 25°}$, the problem is $\frac{17}{\tan 25°}$. Set 25 on the T scale over 17 on the D scale and read the answer 36.5 on the D scale under the index of the T scale.

Example c: Compute 44 sec 48°.

Since $\sec 48° = \frac{1}{\cos 48°} = \frac{1}{\sin (90° - 48°)}$, the problem is $\frac{44}{\sin 42°}$. Set 42° on the S scale under 44 of the right half of the A scale and read the answer 65.8 on the A scale over the index of the T scale.

SUMMARY OF SLIDE-RULE MANIPULATIONS

The following slide-rule "formulas" are summaries of preceding operations with the addition of certain other combinations.

1. $X = a \times b$	Set 1 on C to a on D; at b on C read X on D.
2. $X = a \div b$	Set b on C to a on D; at 1 on C read X on D.
3. $X = a \times b \div c$	Set c on C to a on D; at b on C read X on D.
4. $X = a^2$	Over a on D, read X on A.
5. $X = \sqrt{a}$	Under a on A, read X on D.
6. $X = a \times b^2$	Set 1 on B to a on A; over b on on C, read X on A.
7. $X = a \div b^2$	Set b on C under a on A; at 1 on B read X on A.
8. $X = a^2 \div b$	Set b on B over a on D; at 1 on B read X on A.
9. $X = a^2 \times b^2$	Set 1 on C to a on D; over b on C read X on A.
10. $X = a^2 \div b^2$	Set b on C to a on D; over 1 on C read X on A.

11. $X = a^2 \times b \div c$ — Set c on B to b on A; over a on C read X on A.

12. $X = a \times b \div c^2$ — Set c on C under a on A; over b on B read X on A.

13. $X = a^2 \times b^2 \div c$ — Set c on B over a on D; over b on C read X on A.

14. $X = a^2 \times b \div c^2$ — Set c on C to a on D; at b on B read X on A.

15. $X = a^2 \times b^2 \div c^2$ — Set c on C to a on D; over b on C read X on A.

16. $X = \sin \alpha$ — Set index of S to index of A; over α on S read X on A.

17. $X = \tan \alpha$ — Set index of T to index of D; under α on T read X on D.

18. $\sin \alpha = b$ — Set index of S to index of A; under b on A read α on S.

19. $\tan \alpha = b$ — Set index of T to index of D; over b on D read α on T.

20. $X = b \sin \alpha$ — Set index of S to b on A; over α on S read X on A.

21. $X = b \tan \alpha$ — Set index of T to b on D; under α on T read X on D.

22. $X = b \csc \alpha$ — Set α on S to b on A; over index of S read X on A.

23. $X = b \cot \alpha$ — Set α on T to b on D; under index of T read X on D.

24. $\sin \alpha = a \div b$ — Set index of S to b on A; under a on A read α on S.

25. $\tan \alpha = a \div b$ — Set index of T to b on D; over a on D read α on T.

26. $\csc \alpha = a \div b$ — Set index of S to a on A; under b on A read α on S.

27. $\cot \alpha = a \div b$ — Set index of T to a on D; over a on D read α on T.

The last 12 formulas will be used by the student after he has studied Chap. X but are placed here to avoid splitting the discussion on the use of the slide rule.

PROBLEMS

Use a slide rule to solve:

1. $8.31 \times A = ?$ **2.** $.325 \times B = ?$ **3.** $\frac{59.3}{C} = ?$ **4.** $\frac{7.12}{D} = ?$

5. $\frac{76.4 \times E}{28.3} = ?$ **6.** $\frac{1.286 \times 94.5}{F} = ?$ **7.** $G^2 = ?$ **8.** $H^2 = ?$

9. $\sqrt{I} = ?$ **10.** $\sqrt{J} = ?$ **11.** $.918 \times K^2 = ?$ **12.** $L \times (.416)^2 = ?$

13. $\frac{M}{(13.6)^2} = ?$ **14.** $\frac{962}{N^2} = ?$ **15.** $\frac{P^2}{4.18} = ?$ **16.** $\frac{(13.7)^2}{Q} = ?$

17. $(32.1)^2 \times R^2 = ?$ **18.** $S^2 \times (4.93)^2 = ?$ **19.** $\frac{T^2}{(6.38)^2} = ?$

20. $\frac{(19.6)^2}{U^2} = ?$ **21.** $\frac{(2.54)^2 \times V}{19.8} = ?$ **22.** $\frac{W^2 \times 78.6}{624} = ?$

23. $\frac{7.31 \times A}{(62.4)^2} = ?$ **24.** $\frac{29.3 \times 82.4}{B^2} = ?$ **25.** $\frac{(22.7)^2 \times C^2}{13.2} = ?$

26. $\frac{(39.4)^2 \times (.132)^2}{D} = ?$ **27.** $\frac{(1.56)^2 \times E}{(14.6)^2} = ?$ **28.** $\frac{F^2 \times 16.5}{(19.4)^2} = ?$

29. $\frac{(21.8)^2 \times G^2}{(16.7)^2} = ?$ **30.** $\frac{(3.17)^2 \times (13.8)^2}{H^2} = ?$ **31.** $\sin J = ?$

32. $\sin K = ?$ **33.** $\tan L = ?$ **34.** $\tan M = ?$

35. $\sin \alpha = N, \alpha = ?$ **36.** $\sin \alpha = P, \alpha = ?$ **37.** $\tan \alpha = Q, \alpha = ?$

38. $\tan \alpha = R, \alpha = ?$ **39.** $24.6 \sin S = ?$ **40.** $T \sin 15° 37' = ?$

41. $642 \tan U = ?$ **42.** $V \tan 6° 53' = ?$ **43.** $68.3 \csc W = ?$

44. $A \csc 32° 14' = ?$ **45.** $93.6 \cot B = ?$ **46.** $C \cot 5° 55' = ?$

47. $\sin \alpha = \frac{D}{12.4}, \alpha = ?$ **48.** $\sin \alpha = \frac{61.2}{E}, \alpha = ?$ **49.** $\tan \alpha = \frac{F}{76.2}, \alpha = ?$

50. $\tan \alpha = \frac{2.84}{G}, \alpha = ?$ **51.** $\csc \alpha = \frac{H}{39.4}, \alpha = ?$

52. $\csc \alpha = \frac{82.9}{J}, \alpha = ?$ **53.** $\cot \alpha = \frac{K}{32.4}, \alpha = ?$

54. $\cot \alpha = \frac{78.6}{L}, \alpha = ?$

VARIABLES

Prob.	Sym.	No. 1	No. 2	No. 3	No. 4	No. 5	No. 6
1	A	5.62	5.82	6.02	6.22	6.42	6.62
2	B	132	134	136	138	142	144
3	C	7.26	7.28	7.32	7.34	7.36	7.38
4	D	14.28	14.32	14.34	14.36	14.38	14.42
5	E	24.7	26.7	28.7	30.7	31.7	32.7
6	F	.305	.307	.309	.311	.313	.315
7	G	3.11	3.31	3.51	3.71	3.91	4.11
8	H	31.1	33.1	35.1	37.1	39.1	41.1
9	I	53.9	56.0	58.1	60.2	62.3	64.4
10	J	539	560	581	602	623	644

Prob.	Sym.	No. 1	No. 2	No. 3	No. 4	No. 5	No. 6
11	*K*	11.2	11.4	11.6	11.8	12.2	12.4
12	*L*	25.9	26.4	26.9	27.4	27.9	28.4
13	*M*	42.8	43.9	45.0	46.1	47.2	48.3
14	*N*	7.12	7.17	7.22	7.27	7.32	7.37
15	*P*	7.81	7.92	8.03	8.14	8.25	8.36
16	*Q*	1.88	1.93	1.98	2.03	2.08	2.13
17	*R*	.198	.187	.176	.165	.154	.14[illegible]
18	*S*	6.38	6.49	6.60	6.71	6.32	6.93
19	*T*	9.02	9.13	9.24	9.35	9.46	9.57
20	*U*	40.2	39.2	38.2	37.2	36.2	35.2
21	*V*	5.65	5.76	5.87	5.98	6.09	6.20
22	*W*	5.58	5.62	5.64	5.66	5.68	5.70
23	*A*	22.6	23.6	24.6	25.6	26.6	27.6
24	*B*	37.7	36.7	35.7	34.7	33.7	32.7
25	*C*	20.8	20.6	20.4	20.2	19.8	19.6
26	*D*	88.6	89.6	90.6	91.6	92.6	93.6
27	*E*	396	407	418	429	440	451
28	*F*	31.7	31.2	30.7	30.2	29.7	29.2
29	*G*	27.9	28.3	28.9	29.3	29.8	30.4
30	*H*	40.2	40.7	41.2	41.7	42.2	42.7
31	*J*	3° 54′	3° 48′	3° 42′	3° 36′	3° 24′	3° 18
32	*K*	40° 10′	39° 40′	39° 10′	38° 40′	38° 10′	37° 40′
33	*L*	8° 28′	8° 17′	8° 6′	7° 55′	7° 4[illegible]	7° 33′
34	*M*	30° 57′	30° 52′	30° 46′	30° 41′	30° 33′	30° 25′
35	*N*	.0801	.0806	.0811	.0816	.0821	.0826
36	*P*	.694	.698	.702	.706	.710	.714
37	*Q*	.236	.233	.227	.224	.221	.218
38	*R*	.868	.878	.888	.898	.908	.918
39	*S*	48° 10′	48° 30′	48° 50′	49° 10′	49° 30′	49° 50′
40	*T*	45.4	44.4	43.4	42.4	41.4	40.4
41	*U*	44° 12′	44° 2′	43° 52′	43° 42′	43° 32′	43° 22′
42	*V*	3.66	3.76	3.86	3.96	4.06	4.16
43	*W*	2° 43′	2° 38′	2° 33′	2° 28′	2° 23′	2° 18′
44	*A*	11.9	12.4	12.9	13.4	13.9	14.4
45	*B*	41° 48′	41° 43′	41° 38′	41° 33′	41° 28′	41° 23′
46	*C*	30.4	29.9	29.4	28.9	28.4	27.9
47	*D*	1.84	1.89	1.94	1.99	2.04	2.09
48	*E*	93.3	92.3	91.3	90.3	89.3	88.3
49	*F*	8.74	8.64	8.54	8.44	8.34	8.24
50	*G*	4.01	4.11	4.21	4.31	4.41	4.51
51	*H*	76.6	77.6	78.6	79.6	80.6	81.6
52	*J*	44.4	45.4	46.4	47.4	48.4	49.4
53	*K*	198	218	238	258	278	298
54	*L*	70.2	69.2	68.2	67.2	66.2	65.2

CHAPTER IX

GEOMETRY

PLANE GEOMETRY

1. **Plane geometry** is a study of points, lines, triangles, quadrilaterals, circles, and other common figures. For this study we assume the truth of a certain number of fundamental statements called **axioms.**

2. From these axioms and certain proved statements, we reason the proofs of other statements. These proved statements are called **propositions** or **theorems.**

3. A statement, the truth of which is seen to be a direct consequence of a proposition or axiom, is called a **corollary** (abbreviated cor.).

AXIOMS

The following axioms will be referred to frequently:

Axiom I.—*Things which are equal to the same thing, or to equal things, are equal to each other.*

Axiom II.—*Any quantity may be substituted for its equal in a mathematical expression.*

Axiom III.—*If equals are added to equals, the sums are equal.*

Axiom IV.—*If equals are subtracted from equals, the remainders are equal.*

Axiom V.—*If equals are multiplied by equals, the products are equal.*

Axiom VI.—*If equals are divided by equals, the quotients are equal.*

Axiom VII.—*The whole is greater than any of its parts.*

Axiom VIII.—*The whole is equal to the sum of its parts.*

Axiom IX.—*Only one straight line can be drawn from one point to another. That is to say, two points determine a straight line.*

Corollary to Axiom IX.—*Two straight lines can intersect in only one point.*

For, if two straight lines could intersect in two points, we should have two straight lines drawn between the two points.

Axiom X.—*Through a given point only one line can be drawn parallel to a given line.*

Corollary to Axiom X.—*If two lines are each parallel to a third line, they are parallel to each other.*

For, if the two are not parallel, they would intersect, which would give two lines through the same point parallel to the original line, which is impossible by Axiom X.

DEFINITIONS

4. A **straight line** is the shortest line that can be drawn through two points. If any portion (or segment) of a straight line be placed with its extremities on another part of the straight line, the whole of the first part will lie along the second portion.

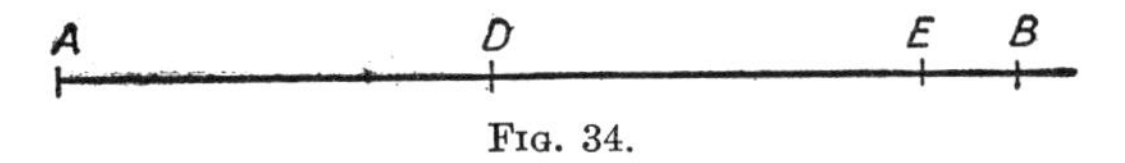

FIG. 34.

Thus, the line AB is the shortest line that can be drawn from A to B, and if AD is placed with its extremities at D and E, it coincides with DE throughout.

5. An **angle** is the figure formed by drawing two straight lines outward from a common point. The point is called the **vertex** of the angle, and the bounding straight lines are called the **sides** of the angle.

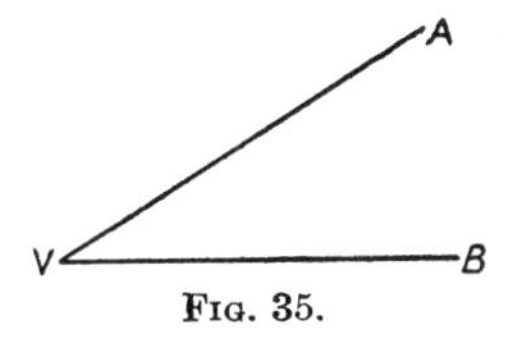

FIG. 35.

Thus the angle AVB (often written $\angle AVB$) has the vertex V and the sides VA and VB

6. If the two sides of an angle extend in opposite directions from the vertex, the angle is called a straight angle.

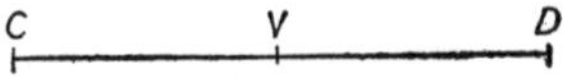

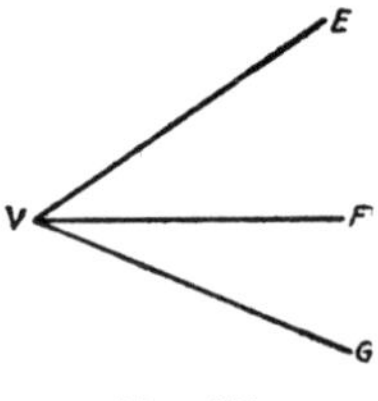

FIG. 36.

In Fig. 36, $\angle CVD$ is a straight angle.

7. Two angles are called **adjacent angles** if they have a common side. $\measuredangle EVF$ and FVG, having the common side VF are adjacent angles.

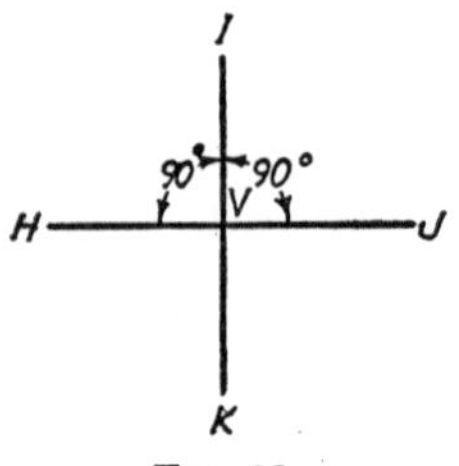

FIG. 37.

8. If two adjacent angles formed by the intersection of two straight lines are equal, each angle is a **right angle.** The equal adjacent angles HVI and IVJ are each right angles.

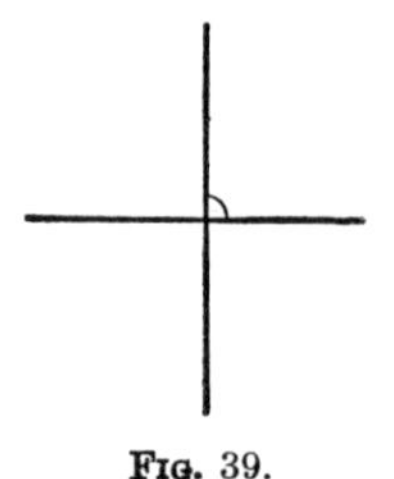

FIG. 38.

In the future, some of the right angles will be indicated by a small arc. Thus:

FIG. 39.

9. An angle is measured in degrees. A **degree** is $\frac{1}{360}$ part of a circle and is subdivided into 60 minutes, and a minute is subdivided into 60 seconds. Hence a **minute** is $\frac{1}{60}$ of a degree and a **second** is $\frac{1}{60}$ of a minute. The **symbols** used to indicate degrees, minutes, and seconds, placed at the upper right hand corner of a numeral, are as follows: 12° 15′ 45″, respectively.

10. Two angles are said to be **complementary** if their sum is equal to a right angle (90°). ∡s *LVM* and *MVN* are complementary. Complements of the same angle or of equal angles are equal (Axiom IV).

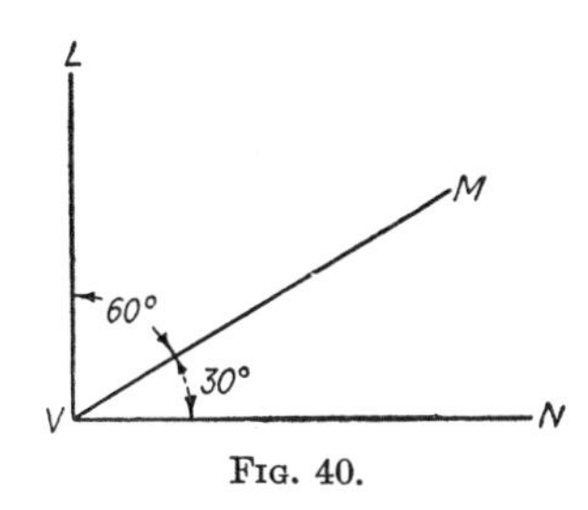

Fig. 40.

11. Two angles are said to be **supplementary** if their sum is equal to a straight angle (180°). Thus ∡s *OVP* and *PVQ* are supplementary. Supplements of the same angle or of equal angles are equal (Axiom IV).

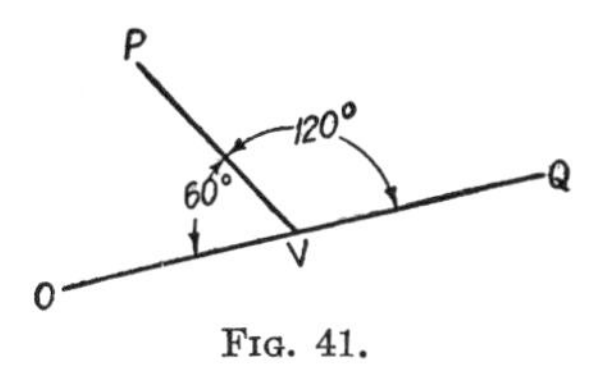

Fig. 41.

12. A **perpendicular** to a given line is a line which makes a right angle with the given line. In Figs. 38 and 40, *IK* is perpendicular to *HJ* (usually written *IK* ⊥ *HJ*), and *VN* ⊥ *VL*.

13. The point of intersection of the perpendicular with the given line is called the **foot of the perpendicular.** Point *V* is the foot of the perpendicular *NV* (Fig. 40).

14. Two straight lines are said to be **parallel** if they do not meet however far they are extended. Lines RS and TU are parallel (often written $RS \parallel TU$) (Fig. 42).

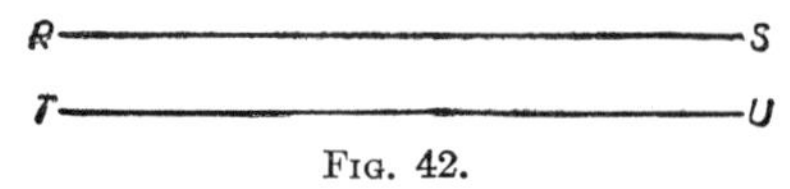

Fig. 42.

15. **A plane surface,** or **plane,** is a surface such that a straightedge will touch the surface at all points, no matter where the surface may be tested. The top of a table is a portion of a plane.

16. A **polygon** is a portion of a plane enclosed by three or more straight lines.

17. A **triangle** is a polygon of three sides.

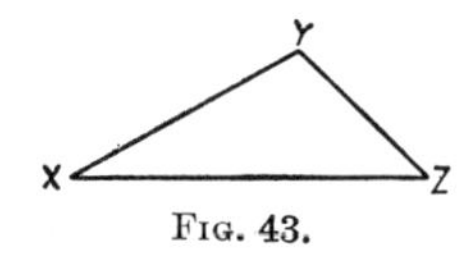

Fig. 43.

18. A **quadrilateral** is a polygon of four sides.

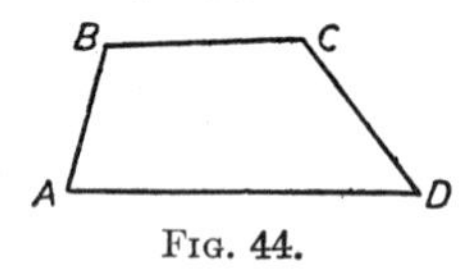

Fig. 44.

19. A **pentagon** is a polygon of five sides.

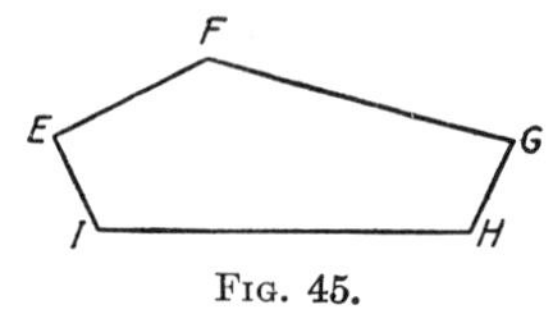

Fig. 45.

20. A **hexagon** is a polygon of six sides.

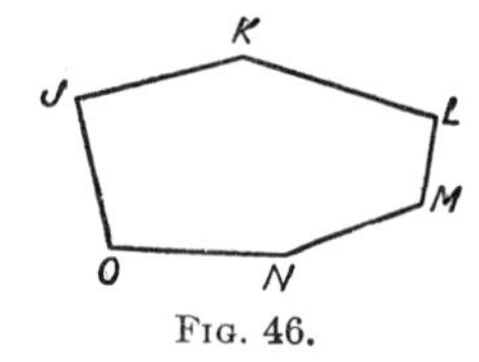

Fig. 46.

21. An **octagon** is a polygon of eight sides.

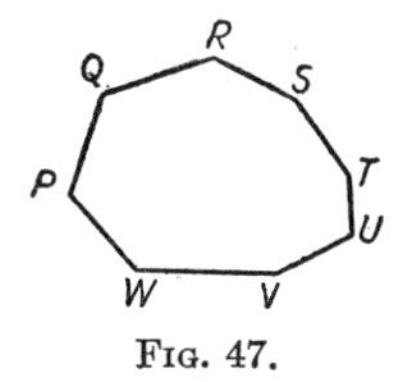

Fig. 47.

22. A **regular polygon** is one that is both equilateral and equiangular. Thus a square is a regular quadrilateral.

23. The **perimeter** of a polygon is the sum of the sides of the polygon.

24. A **parallelogram** is a quadrilateral having its opposite sides parallel. Thus $ABCD$ is a parallelogram if $AB \parallel DC$ and $BC \parallel AD$.

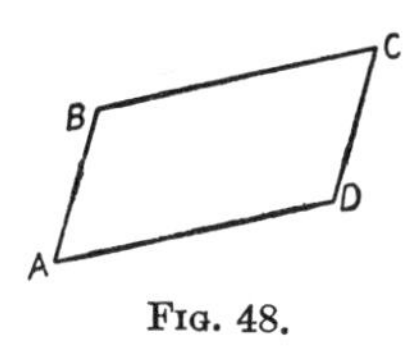

Fig. 48.

25. Two figures that exactly coincide when one figure is superposed on the other are said to be **congruent** (designated by the symbol $\cong$), and the corresponding parts, called **homologous** parts, are equal.

26. A **right-angled triangle** (or rt. $\triangle$) is a triangle one of whose angles is a right angle, as $\triangle EFG$, which has a right angle at F. The sides adjacent to the right angle are called the **legs** of the right triangle and the side opposite the right angle is called the **hypotenuse**.

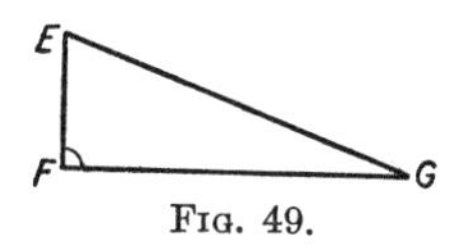

Fig. 49.

27. An **acute angle** is an angle less than 90°, as $\angle E$ or $\angle G$ (Fig. 49).

28. An **obtuse angle** is an angle greater than 90°, as $\angle HIJ$.

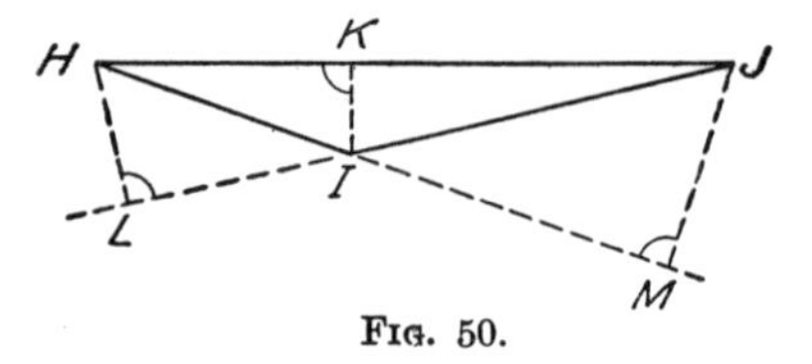

Fig. 50.

29. An **oblique triangle** is one not having any of its angles equal to a right angle as $\triangle HIJ$ (sometimes called an obtuse triangle) and $\triangle NOP$ (sometimes called an acute triangle).

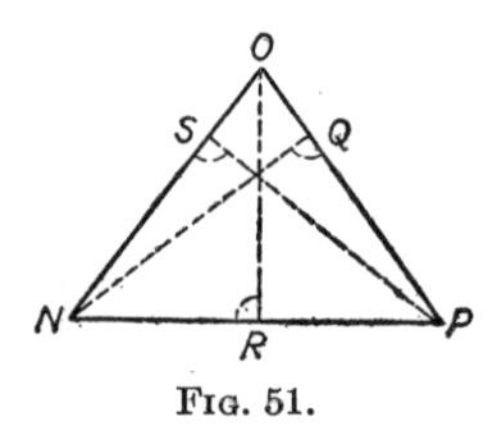

Fig. 51.

30. The three **altitudes** of an oblique triangle are the three perpendiculars from the three vertices to the opposite sides (extended if necessary) as IK, HL, and JM for $\triangle HIJ$ and NQ, OR, and PS for $\triangle NOP$.

31. An **isosceles triangle** is one having two of its sides equal, as $\triangle TUV$ (side TU = side VU).

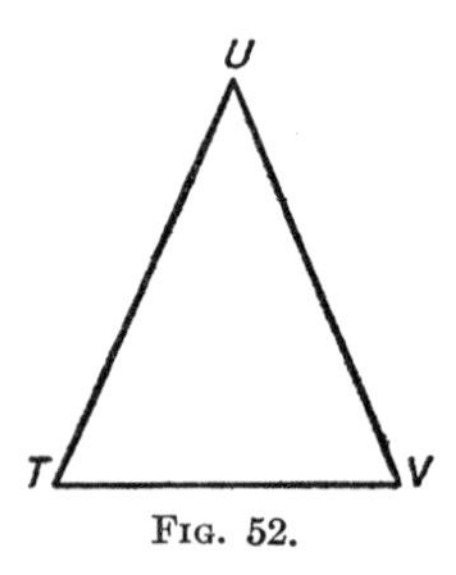

Fig. 52.

32. The **projection** of one line upon a second line is the segment of the second line included between the perpendiculars drawn to it from the extremities of the first line. Thus the

projection of AB on FG is HJ, and the projection of AB on CD is AE.

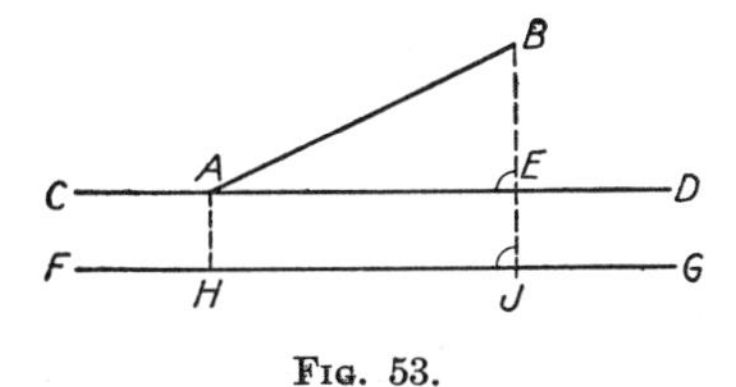

Fig. 53.

PROPOSITIONS

In the following work, axioms will be referred to as A-I, A-II, etc., definitions as D-1, D-2, etc., and propositions as P-1, P-2, etc. In the propositions to be proved, the given conditions will be referred to as the **hypothesis,** which will be abbreviated hyp.

PROPOSITION 1

If two straight lines intersect, the opposite or vertical angles are equal.

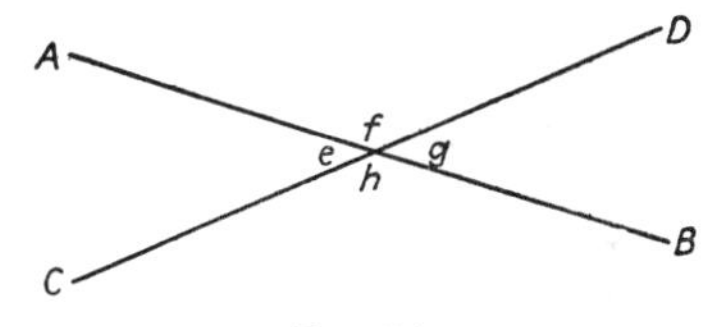

Fig. 54.

Given: The intersecting straight lines AB and CD, which form the two pairs of vertical angles, e and g and f and h.

To prove: $\angle e = \angle g$ and $\angle f = \angle h$.

$\angle f$ is the supplement of $\angle g$ (D-6 and D-11).

$\angle h$ is the supplement of $\angle g$ (D-6 and D-11).

$\therefore \angle f = \angle h$ (D-11).

Similarly it may be proved that $\angle e = \angle g$.

PROPOSITION 2

Two triangles are congruent if two sides and the included angle of the one are equal, respectively, to two sides and the included angle of the other.

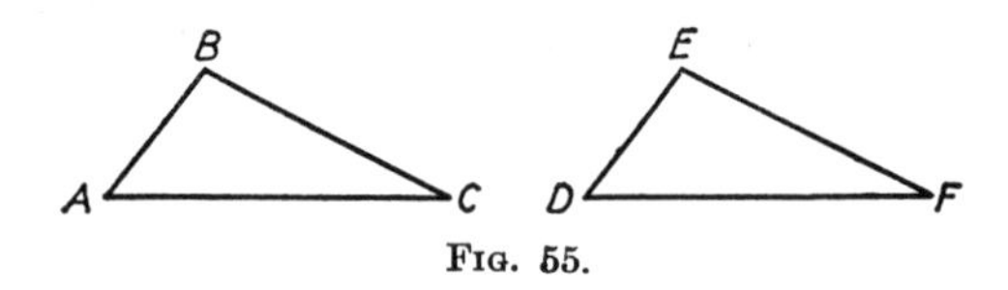

Fig. 55.

Given: $\triangle ABC$ and DEF having $AB = DE$, $BC = EF$, and $\angle ABC = \angle DEF$.

To prove: $\triangle ABC \cong \triangle DEF$. D-25.

Place $\triangle DEF$ on $\triangle ABC$ so that vertex E falls on vertex B, side EF along BC and side ED along BA ($\angle E = \angle B$ by hyp.).

Then F will fall on C ($EF = BC$ by hyp.)
and D will fall on A ($ED = BA$ by hyp.).

Hence line DF coincides with line AC (A-IX).

Thus the triangles can be made to coincide throughout and are therefore congruent. (D-25).

PROPOSITION 3

Two triangles are congruent if two angles and the included side of one are equal respectively to two angles and the included side of the other.

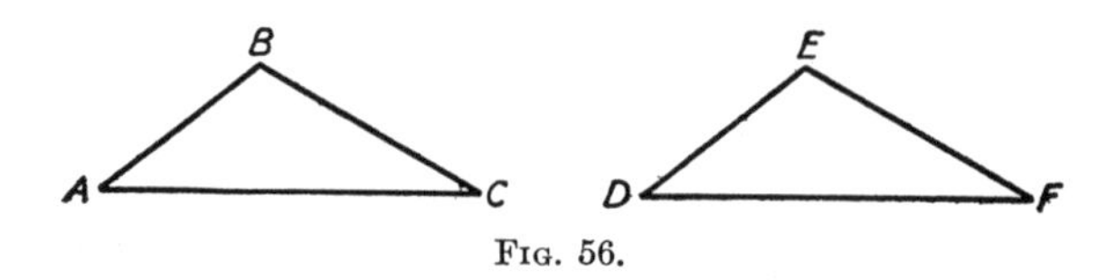

Fig. 56.

Given: $\triangle ABC$ and DEF having $\angle A = \angle D$, $\angle B = \angle E$, and $AB = DE$.

To prove: $\triangle ABC \cong \triangle DEF$.

Place $\triangle DEF$ on $\triangle ABC$ so that DE falls on AB
($DE = AB$ by hyp.).

Then side EF will fall along BC ($\angle E = \angle B$ by hyp.).
And side DF will fall along AC ($\angle D = \angle A$ by hyp.).
Hence point F will fall on point C (cor. to A-IX).

Thus the triangles can be made to coincide throughout and are therefore congruent. (D-25).

PROPOSITION 4

Only one perpendicular can be drawn from a given external point to a given line, and this perpendicular is the shortest line that can be drawn from the external point to the line.

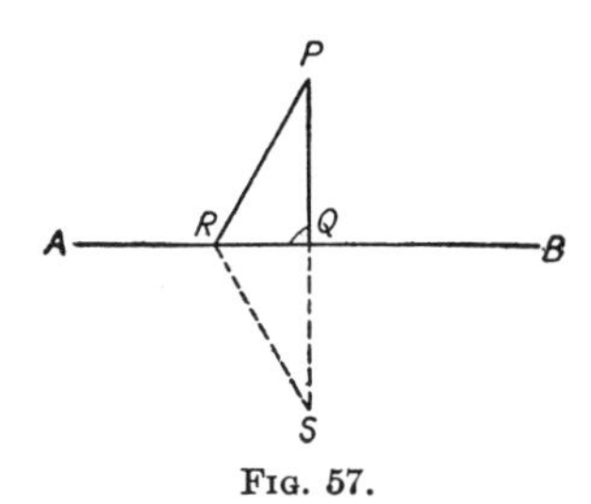

FIG. 57.

Given: Line AB and external point P, the $\perp$ PQ, and some other line PR from P to line AB.

To prove: 1. PR is not perpendicular to AB.

Produce PQ to S making $QS = PQ$.

Draw RS.

PQS is a straight line (by construction).

$\therefore$ PRS is not a straight line (A-IX).

And $\angle PRS$ is not a straight angle (D-6).

In $\triangle PQR$ and SQR

$PQ = QS$ (construction).

$RQ = RQ$ (common).

$\angle PQR = \angle SQR$ (both are rt. $\angle$s by D-12).

$\therefore$ $\triangle PQR$ and SQR are congruent (P-2).

Hence $\angle QRP = \angle QRS$ (D-25).

Since $\angle PRS$ is not a straight angle, $\angle QRP$, which is half of it, is not a right angle.

$\therefore$ PR is not $\perp$ AB (D-12).

To prove: 2. Line PQ is less than PR (often written $PQ < PR$).

$PQS < PRS$ (A-IX and D-4).

PQ is half of PQS (construction).

PR is half of PRS (since $PR = RS$ by D-25).

$\therefore$ $PQ < PR$ (A-VI).

PROPOSITION 5

Two lines in the same plane perpendicular to the same line are parallel.

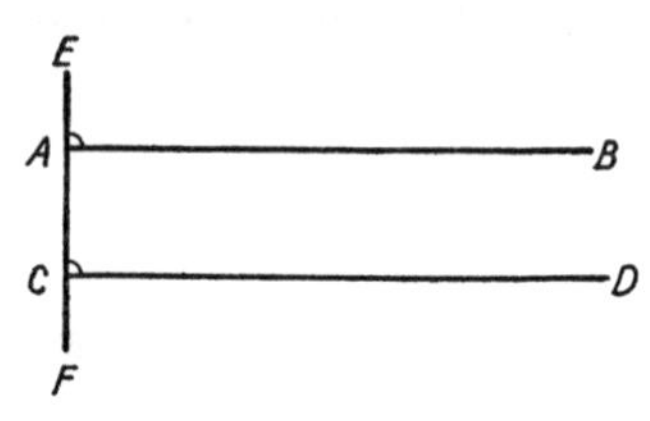

FIG. 58.

Given: Two lines AB and CD both perpendicular to EF.

To prove: $AB \parallel CD$.

If AB and CD, sufficiently extended, could meet, there would be two perpendiculars from that point to the same line. This is impossible (P-4).

Hence AB and CD cannot meet and therefore must be parallel (D-14).

PROPOSITION 6

If a line is perpendicular to one of two parallel lines, it is perpendicular to the other also.

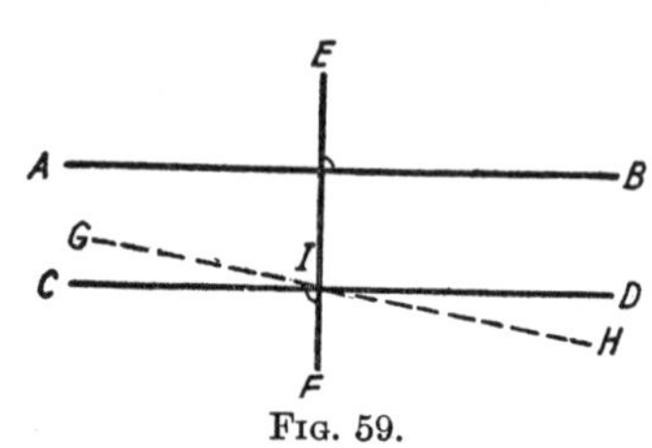

FIG. 59.

Given: Two parallel lines AB and CD, and $EF \perp AB$.

To prove: $EF \perp CD$.

If EF is not perpendicular to CD, let GH be a line through I that EF is perpendicular to.

Then $GH \parallel AB$ (P-5).

But $CD \parallel AB$ (hyp.).

$\therefore$ CD and GH must coincide (A-X).

Hence, since GH is perpendicular to EF, CD must be perpendicular to EF.

33. If two straight lines are cut by a third, the angles are named as follows:

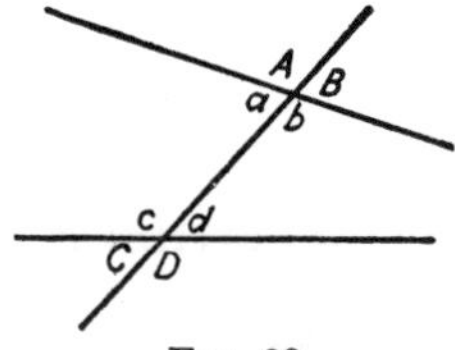

FIG. 60.

$\angle$s A, B, C, and D are **exterior** angles.

$\angle$s a, b, c, and d are **interior** angles.

$\angle$s A and D, and B and C, are pairs of **alternate-exterior** angles.

$\angle$s a and d, and b and c, are pairs of **alternate-interior** angles.

$\angle$s A and c, B and d, C and a, D and b are pairs of **exterior-interior** angles (often called corresponding angles).

PROPOSITION 7

If two parallel lines are cut by a third line, the alternate-interior angles are equal.

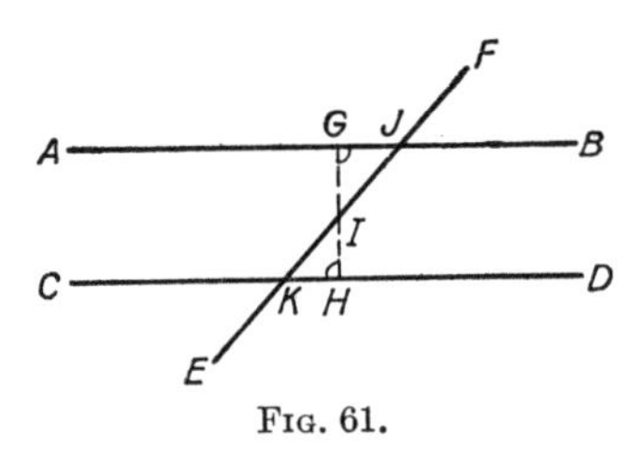

FIG. 61.

Given: Two parallel lines AB and CD cut by the third line EF.

To prove: $\angle GJI = \angle HKI$, and $\angle IJB = \angle IKC$.

Through I, midway between the two lines AB and CD and on the line EF, draw a line GH perpendicular to AB.

Then $GH \perp CD$ (P-6).

In the $\triangle IGJ$ and KHI,

$$\angle GIJ = \angle HIK \qquad \text{(P-1)},$$
$$\angle KHI = \angle JGI \qquad \text{(both rt. } \angle\text{s)}.$$
$$GI = IH \qquad (I \text{ taken as midway}).$$
$$\therefore \triangle IGJ \cong \triangle KHI \qquad \text{(P-3)}.$$
$$\text{Hence } \angle GJI = \angle HKI \qquad \text{(D-25)}.$$

Similarly, $\angle IJB$ may be proved equal to $\angle IKC$.

PROPOSITION 8

If two parallel lines are cut by a third line, the exterior-interior angles are equal.

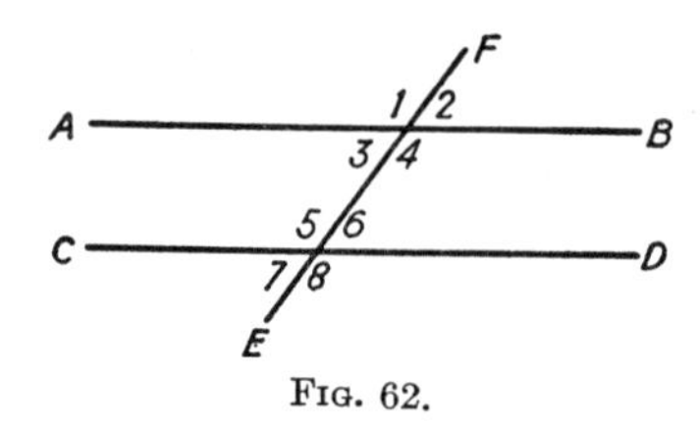

FIG. 62.

Given: Two parallel lines AB and CD cut by the third line EF.

To prove: $\angle 1 = \angle 5$, $\angle 2 = \angle 6$, $\angle 3 = \angle 7$, $\angle 4 = \angle 8$.

$$\angle 1 = \angle 4 \qquad \text{(P-1)}.$$
$$\angle 5 = \angle 4 \qquad \text{(P-7)}.$$
$$\therefore \angle 1 = \angle 5 \qquad \text{(A-I)}.$$

Similarly, the other pairs may be proved equal.

PROPOSITION 9

If two lines in the same plane are intersected by a third line, and the exterior-interior angles are equal, the two lines are parallel.

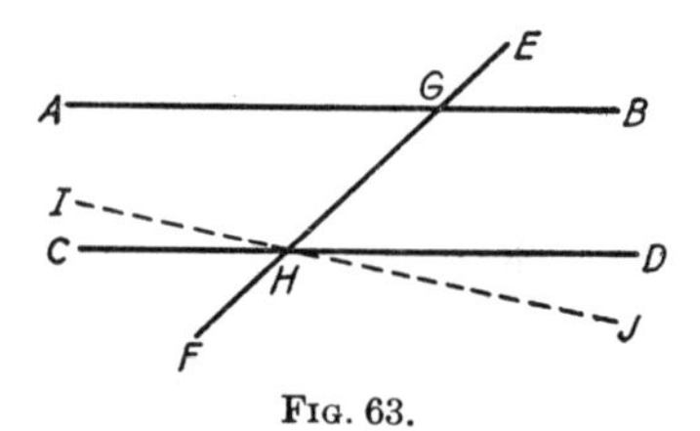

FIG. 63.

Given: Two lines AB and CD cut by the third line EF with $\angle EGB = \angle GHD$.

To prove: $CD \parallel AB$.

Assuming that CD is not parallel to AB, draw a line IJ through H parallel to AB.

Then $\angle EGB = \angle GHJ$ (P-8).

But $\angle EGB = \angle GHD$ (hyp.).

Hence $\angle GHJ = \angle GHD$ (A-I).

$\therefore$ IJ and CD must coincide since the vertices and other sides of the two equal angles coincide.

$\therefore$ $CD \parallel AB$ since CD coincides with IJ which was drawn parallel to AB.

PROPOSITION 10

The sum of the degrees of the three angles of any triangle is equal to 180°.

Given: $\triangle ABC$.

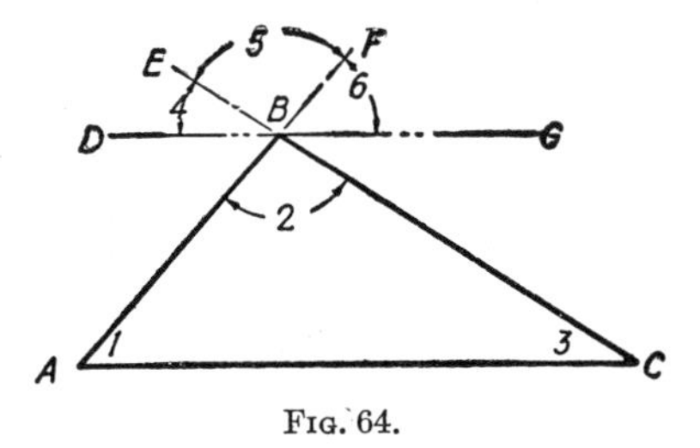

FIG. 64.

To prove: $\angle 1 + \angle 2 + \angle 3 = 180°$.

Extend AB to F, CB to E, and draw DG, through B, parallel to AC.

$$\angle 5 = \angle 2 \quad \text{(P-1)}.$$
$$\angle 4 = \angle 3 \quad \text{(P-8)}.$$
$$\angle 6 = \angle 1 \quad \text{(P-8)}.$$
$$\angle 4 + \angle 5 + \angle 6 = 180° \quad \text{(D-6)}.$$
$$\therefore \angle 1 + \angle 2 + \angle 3 = 180° \quad \text{(A-II)}.$$

Corollary to Proposition 10.—*The two acute angles of a right triangle are complementary.*

PROBLEMS

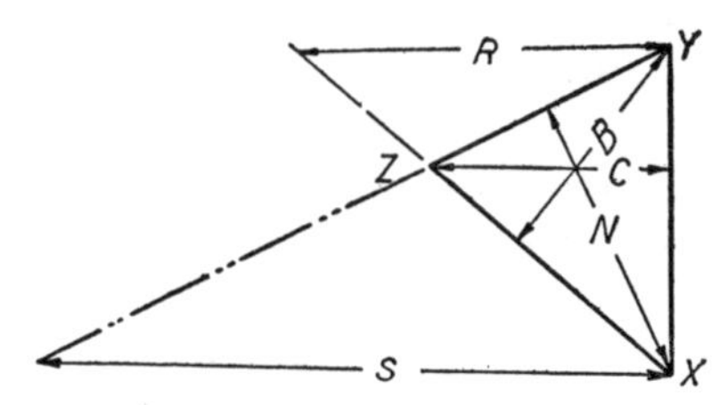

1. In the triangle XYZ, name the three altitudes. No Variable.

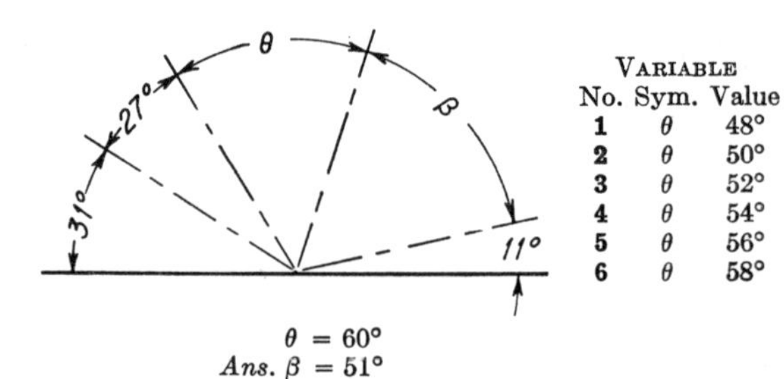

VARIABLE		
No.	Sym.	Value
1	θ	48°
2	θ	50°
3	θ	52°
4	θ	54°
5	θ	56°
6	θ	58°

$\theta = 60°$
Ans. $\beta = 51°$

2. Determine the angle β.

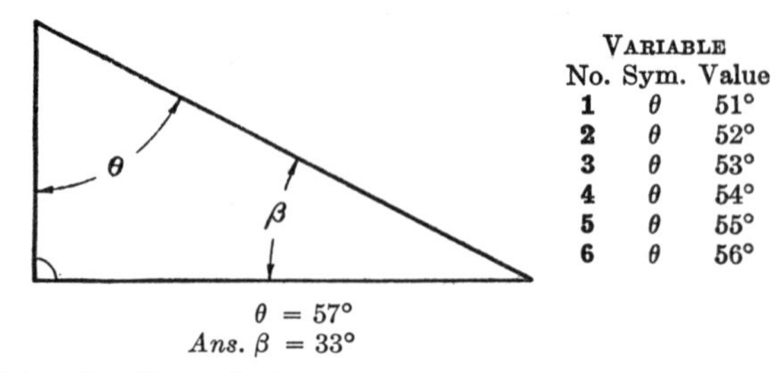

VARIABLE		
No.	Sym.	Value
1	θ	51°
2	θ	52°
3	θ	53°
4	θ	54°
5	θ	55°
6	θ	56°

$\theta = 57°$
Ans. $\beta = 33°$

3. Determine the angle β.

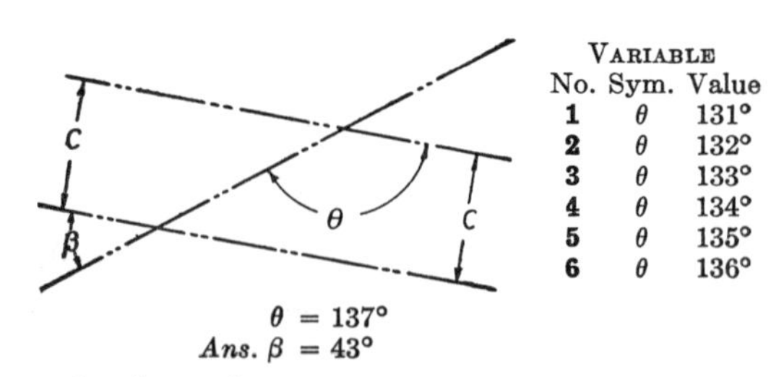

VARIABLE		
No.	Sym.	Value
1	θ	131°
2	θ	132°
3	θ	133°
4	θ	134°
5	θ	135°
6	θ	136°

$\theta = 137°$
Ans. $\beta = 43°$

4. Determine the angle β.

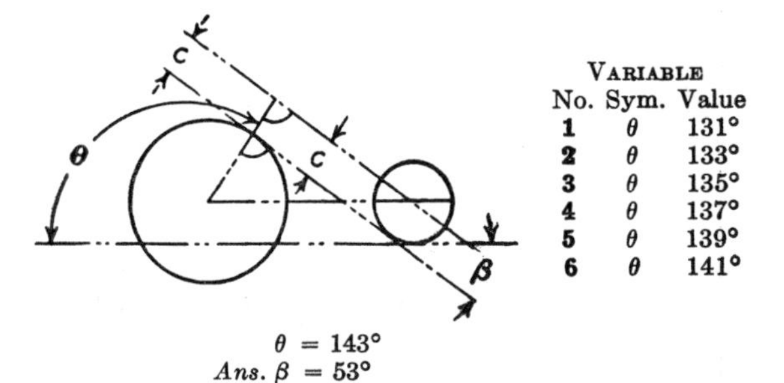

VARIABLE		
No.	Sym.	Value
1	θ	131°
2	θ	133°
3	θ	135°
4	θ	137°
5	θ	139°
6	θ	141°

$\theta = 143°$
Ans. $\beta = 53°$

5. Determine the angle β.

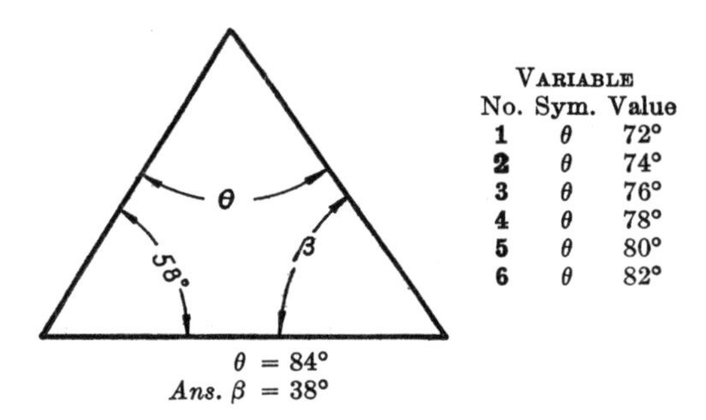

VARIABLE		
No.	Sym.	Value
1	θ	72°
2	θ	74°
3	θ	76°
4	θ	78°
5	θ	80°
6	θ	82°

$\theta = 84°$
Ans. $\beta = 38°$

6. Determine the angle β.

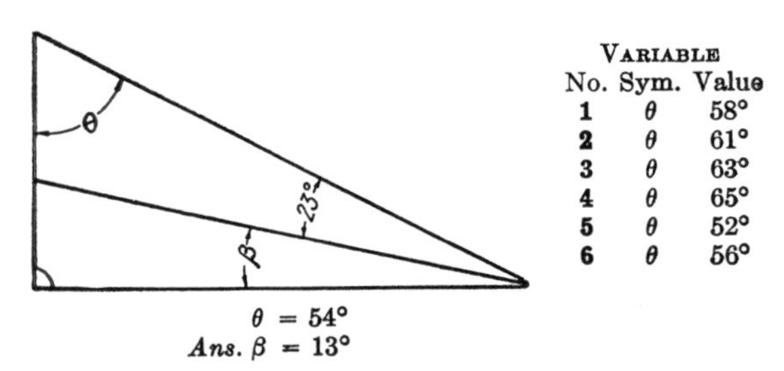

VARIABLE		
No.	Sym.	Value
1	θ	58°
2	θ	61°
3	θ	63°
4	θ	65°
5	θ	52°
6	θ	56°

$\theta = 54°$
Ans. $\beta = 13°$

7. Determine the angle β.

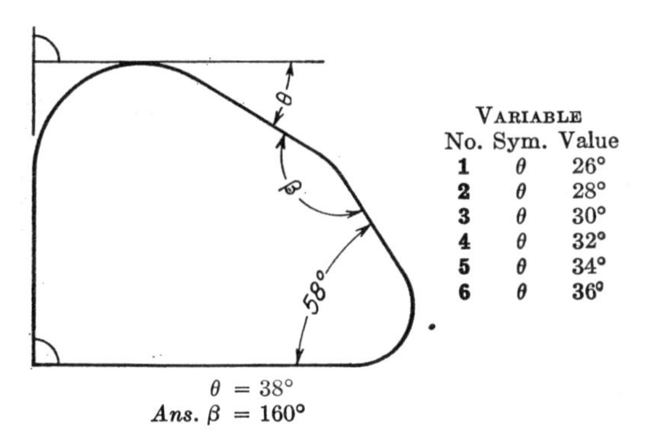

Variable		
No.	Sym.	Value
1	θ	26°
2	θ	28°
3	θ	30°
4	θ	32°
5	θ	34°
6	θ	36°

$\theta = 38°$
Ans. $\beta = 160°$

8. Determine the angle β.

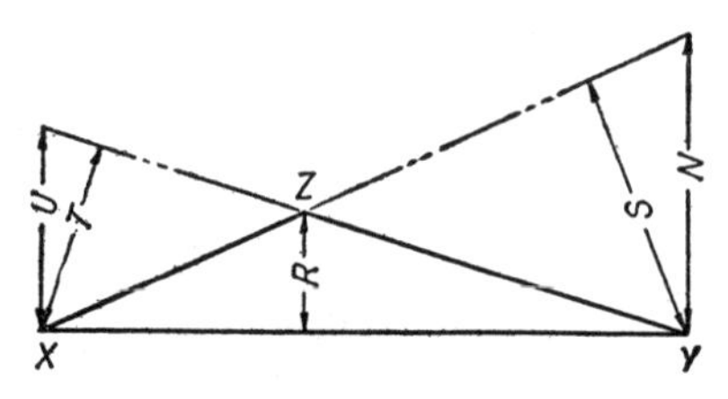

9. In the triangle XYZ, name the three altitudes. No Variable.

PROPOSITION 11

When two angles of a triangle are equal to two angles of another triangle, the third angles are equal.

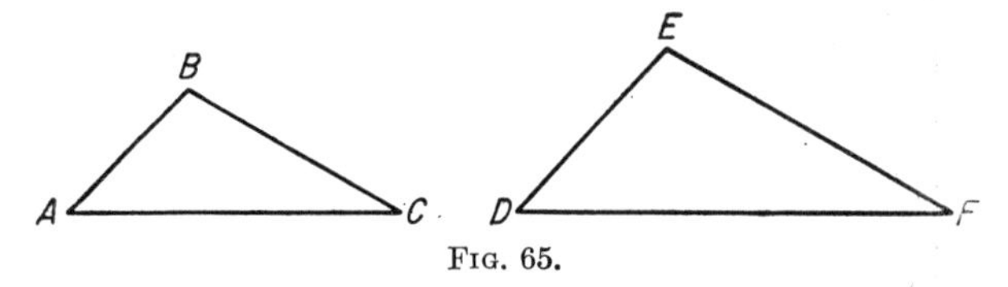

FIG. 65.

Given: Two $\triangle ABC$ and DEF with $\angle A = \angle D$ and $\angle C = \angle F$.
To prove: $\angle B = \angle E$.

$\angle A + \angle B + \angle C = 180°$ (P-10).
$\angle D + \angle E + \angle F = 180°$ (P-10).
$\angle A + \angle C = \angle D + \angle F$ (hyp. and A-III).
$\therefore \angle B = \angle E$ (A-IV).

PROPOSITION 12

The sum of the degrees of the interior angles of a polygon of N sides is (N − 2) times 180°.

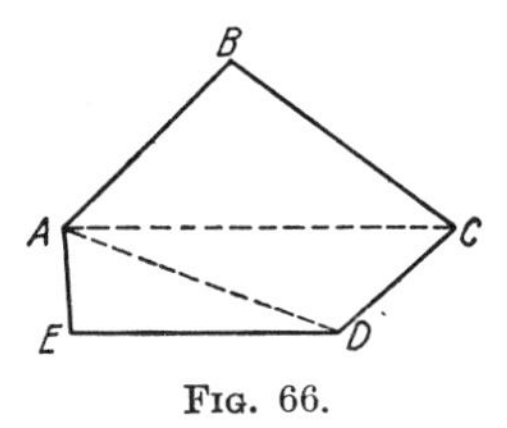

Fig. 66.

Given: Polygon $ABCDE$.

To prove: $\angle ABC + \angle BCD + \angle CDE + \angle DEA + \angle EAB = (N - 2)\ 180°$.

Draw diagonals AC and AD. This will divide the polygon into $(N - 2)$ triangles (one for each side except the adjacent sides AB and AE).

The sum of the degrees of the interior angles of these triangles is the sum of the degrees of the interior angles of the polygon.

The sum of the degrees in each triangle is 180° (P-10).

∴ the sum of the degrees of the interior angles of the polygon is $(N - 2)$ times 180°.

PROPOSITION 13

The exterior angle formed by prolonging one side of a triangle is equal to the sum of the two opposite interior angles.

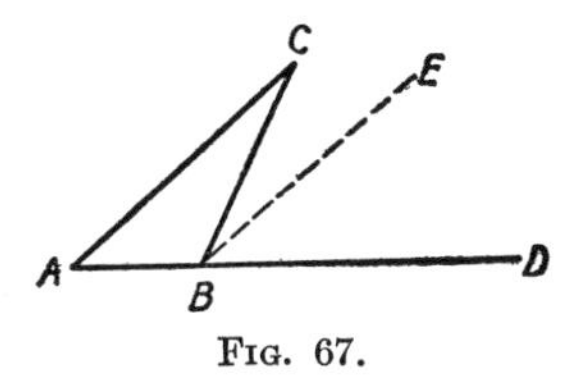

Fig. 67.

Given: $\triangle ABC$ with exterior $\angle CBD$.

To prove: $\angle CBD = \angle BAC + \angle BCA$.

Draw BE parallel to AC.

$\angle DBE = \angle BAC$ (P-8).

$\angle EBC = \angle BCA$ (P-7).

$\angle DBE + \angle EBC = \angle BAC + \angle BCA$ (A-III).
$\angle DBC = \angle DBE + \angle EBC$ (A-VIII).
$\angle DBC = \angle BAC + \angle BCA$ (A-I).

PROPOSITION 14

Two straight lines drawn from a point in a perpendicular to a given line, cutting on the given line equal segments from the foot of the perpendicular, are equal and make equal angles with the perpendicular.

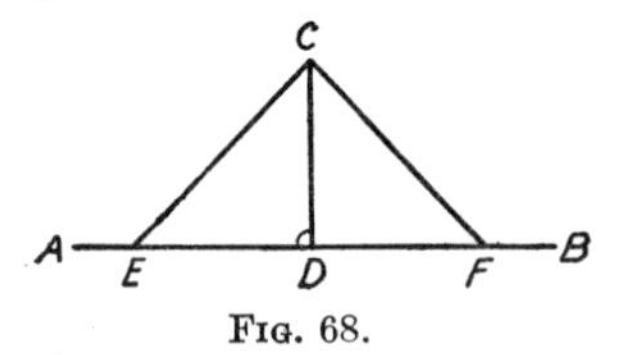

Fig. 68.

Given: CD perpendicular to line AB and oblique lines CE and CF cutting off equal segments ED and DF.

To prove: $CE = CF$ and $\angle ECD = \angle FCD$.

Fold over $\triangle CDE$ on CD as an axis until it falls on the plane to the right of CD.

ED will fall along DF ($\angle CDE = \angle CDF$, each being 90°).
Point E will fall on F ($DE = DF$ by hyp.).
$\therefore CE$ coincides with CF throughout (A-IX).
That is, $CE = CF$.
Also $\angle ECD = \angle FCD$ (vertices and sides coincide).

Corollary to Proposition 14.—*All points on the perpendicular bisector of a line are equidistant from the extremities of the line.*

In the foregoing figure, CD is the perpendicular bisector of EF, and EC has already been proved equal to CF.

PROPOSITION 15

Two angles are equal when their sides are parallel, right to right and left to left.

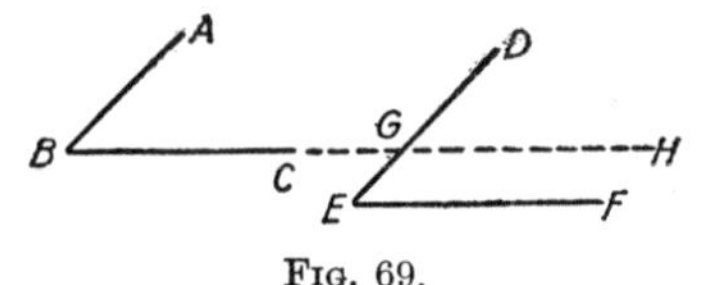

Fig. 69.

Given: ∠s B and E with $EF \parallel BC$ and $ED \parallel BA$.
To prove: $\angle E = \angle B$.
Let BC (extended if necessary) meet ED at G.
Then $\angle E = \angle DGH$ (P-8).
And $\angle B = \angle DGH$ (P-8).
$\therefore \angle E = \angle B$ (A-I).

PROBLEMS

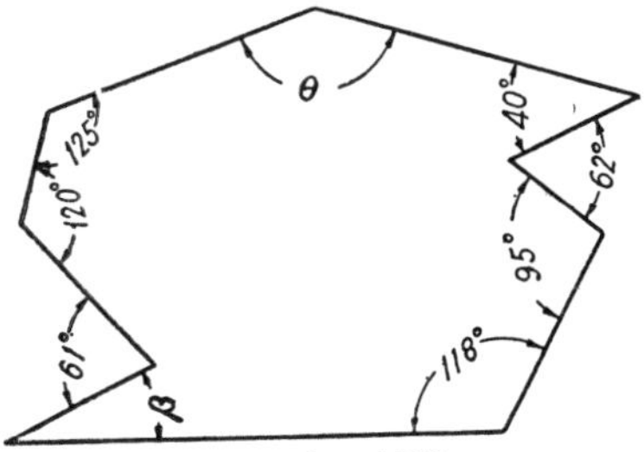

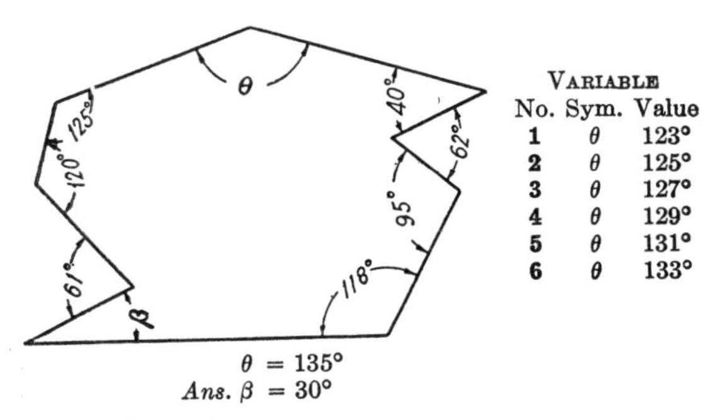

Variable		
No.	Sym.	Value
1	θ	123°
2	θ	125°
3	θ	127°
4	θ	129°
5	θ	131°
6	θ	133°

$\theta = 135°$
Ans. $\beta = 30°$

1. Determine the angle β.

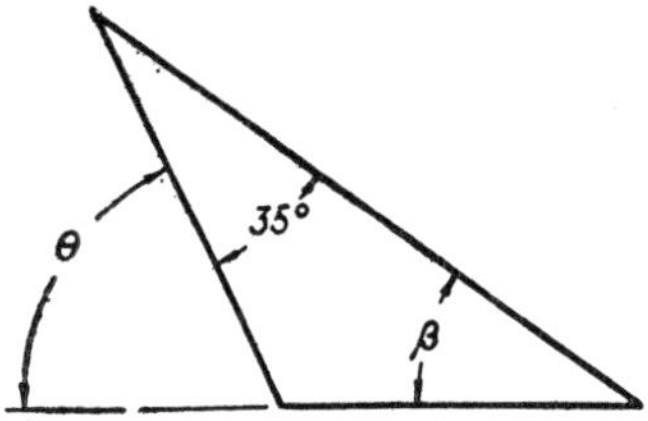

Variable		
No.	Sym.	Value
1	θ	75°
2	θ	76°
3	θ	77°
4	θ	78°
5	θ	79°
6	θ	80°

$\theta = 81°$
Ans. $\beta = 46°$

2. Determine the angle β.

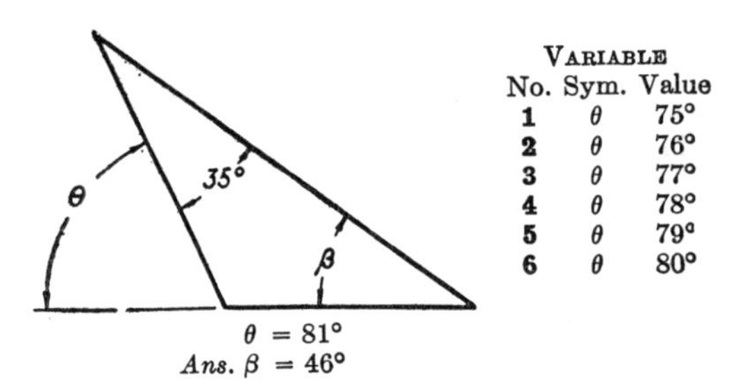

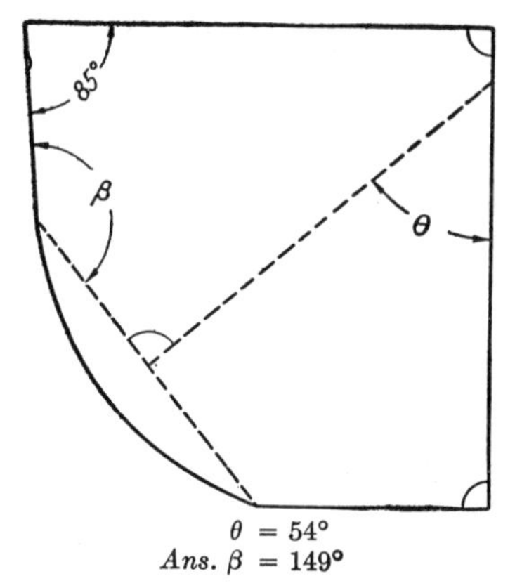

No.	Sym.	Value
1	θ	42°
2	θ	44°
3	θ	46°
4	θ	48°
5	θ	50°
6	θ	52°

$\theta = 54°$
Ans. $\beta = 149°$

3. Determine the angle β.

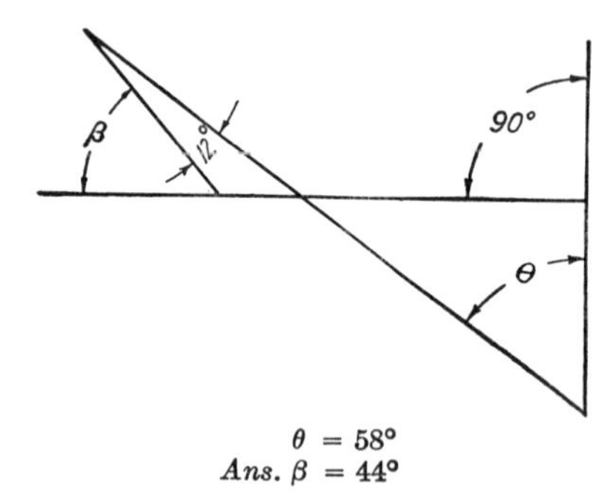

No.	Sym.	Value
1	θ	46°
2	θ	48°
3	θ	50°
4	θ	52°
5	θ	54°
6	θ	56°

$\theta = 58°$
Ans. $\beta = 44°$

4. Determine the angle β.

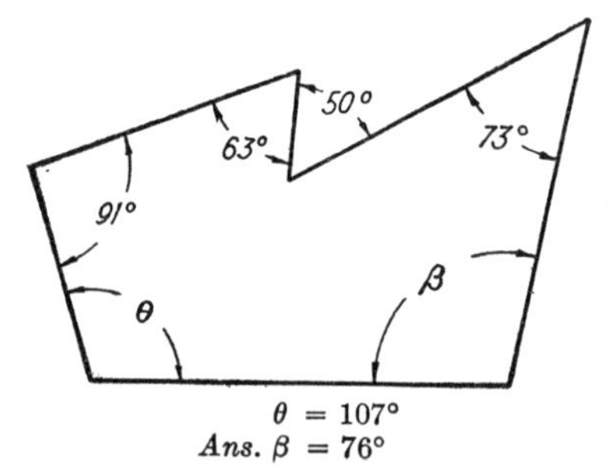

No.	Sym.	Value
1	θ	101°
2	θ	102°
3	θ	103°
4	θ	104°
5	θ	105°
6	θ	106°

$\theta = 107°$
Ans. $\beta = 76°$

5. Determine the angle β.

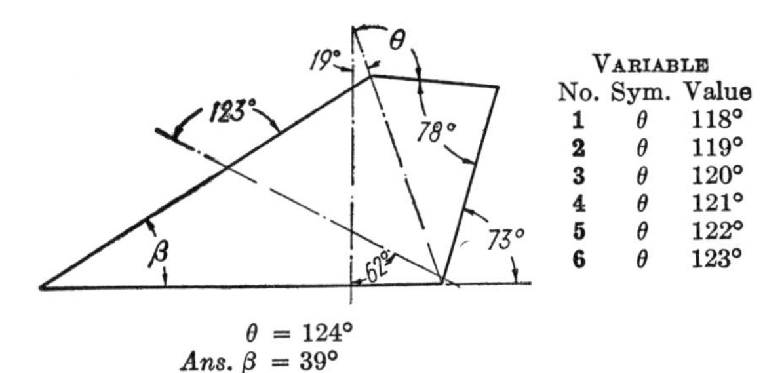

No.	Sym.	Value
1	θ	118°
2	θ	119°
3	θ	120°
4	θ	121°
5	θ	122°
6	θ	123°

$\theta = 124°$
Ans. $\beta = 39°$

6. Determine the angle β.

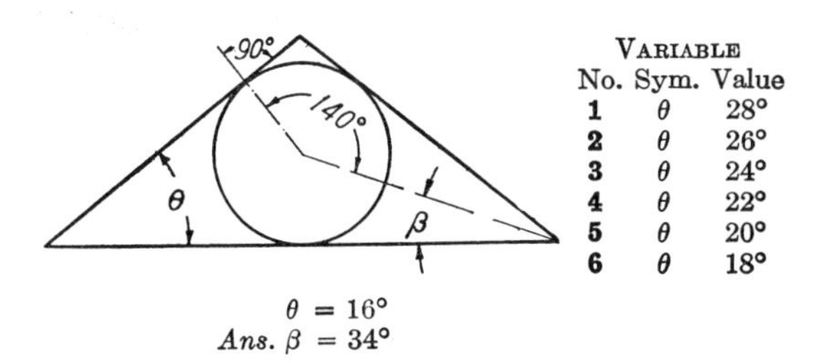

No.	Sym.	Value
1	θ	28°
2	θ	26°
3	θ	24°
4	θ	22°
5	θ	20°
6	θ	18°

$\theta = 16°$
Ans. $\beta = 34°$

7. Determine the angle β.

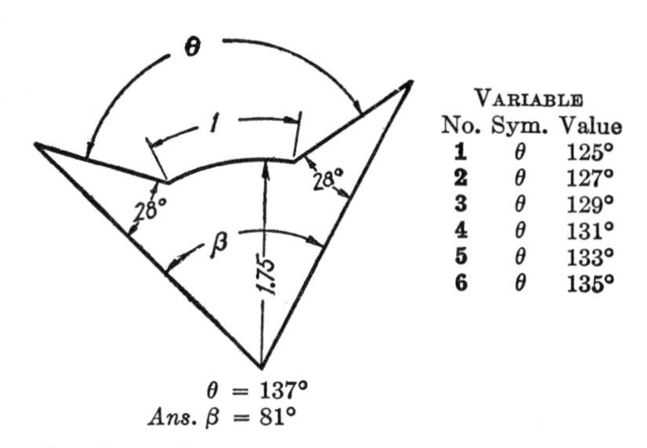

No.	Sym.	Value
1	θ	125°
2	θ	127°
3	θ	129°
4	θ	131°
5	θ	133°
6	θ	135°

$\theta = 137°$
Ans. $\beta = 81°$

8. Determine the angle β.

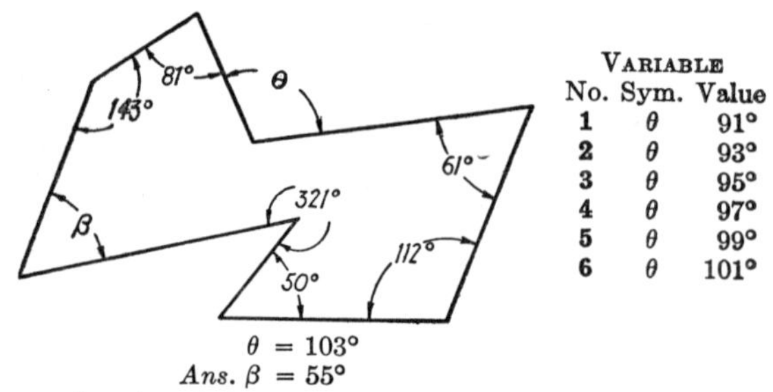

VARIABLE		
No.	Sym.	Value
1	θ	91°
2	θ	93°
3	θ	95°
4	θ	97°
5	θ	99°
6	θ	101°

$\theta = 103°$
Ans. $\beta = 55°$

9. Determine the angle β.

PROPOSITION 16

Two angles whose sides are perpendicular right to right and left to left are equal.

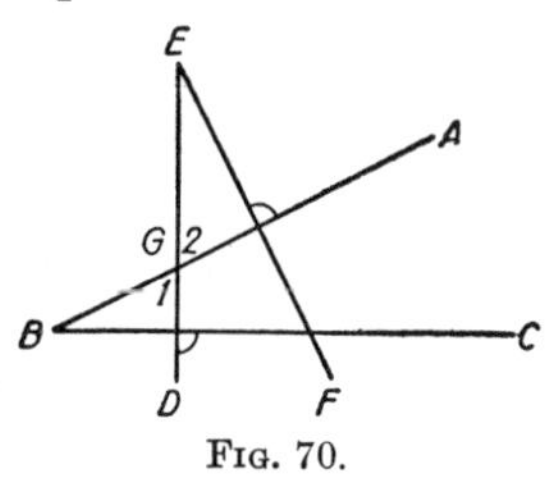

FIG. 70.

Given: $\angle$s ABC and DEF with right side ED perpendicular to right side BC and left side EF perpendicular to left side BA.

To prove: $\angle E = \angle B$.

$\angle B$ is the complement of $\angle 1$ (cor. to P-10).
$\angle E$ is the complement of $\angle 2$ (cor. to P-10).
$\angle 1 = \angle 2$ (P-1).
$\therefore \angle B = \angle E$ (D-10).

PROPOSITION 17

The angles at the base of an isosceles triangle are equal.

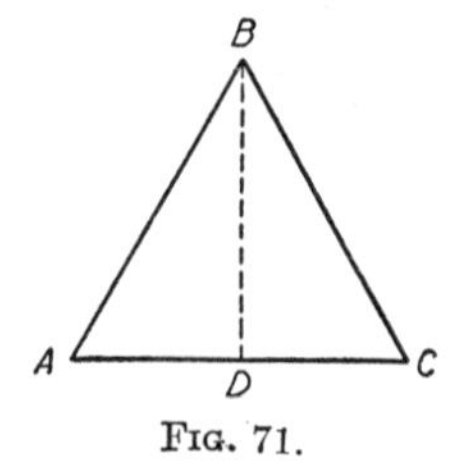

FIG. 71.

Given: Isosceles $\triangle ABC$ ($AB = BC$).

To prove: $\angle A = \angle C$.

Draw BD so as to bisect $\angle ABC$

In $\triangle ADB$ and CDB,

$\angle ABD = \angle CBD$ (by construction).

$AB = BC$ (hyp.).

$BD = BD$ (common).

$\therefore \triangle ADB \cong \triangle CDB$ (P-2).

and $\angle A = \angle C$ (D-25).

Corollary 1 to Proposition 17.—*The line from the vertex perpendicular to the base of an isosceles triangle bisects the base and the angle at the vertex.*

$\triangle ADB \cong \triangle CDB$ (Proof left to student).

$AD = DC$ (D-25).

and $\angle ABD = \angle CBD$ (D-25).

Corollary 2 to Proposition 17.—*If a triangle is equilateral, it is also equiangular.*

PROPOSITION 18

If equal lines are drawn from a point in a perpendicular to a given line, they cut off equal segments on that line from the foot of the perpendicular.

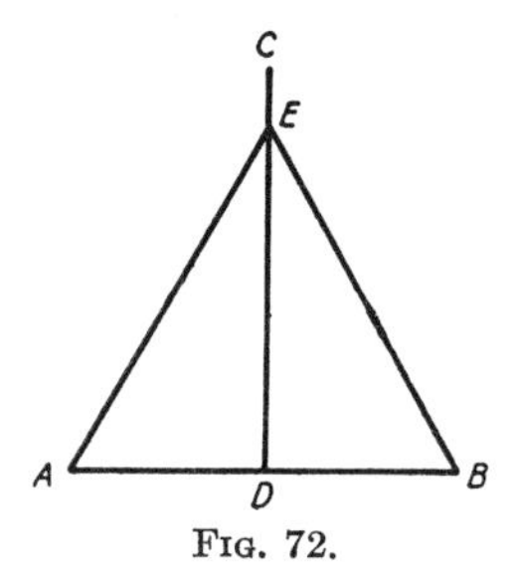

FIG. 72.

Given: $CD \perp AB$ and equal lines EA and EB drawn from point E to line AB.

To prove: $AD = DB$.

In $\triangle ADE$ and BDE,

$AE = BE$ (hyp.).

$ED = ED$ (common).

$\angle A = \angle B$ (P-17).

$\angle ADE = \angle BDE$ (both rt. $\angle$s).
$\therefore \angle AED = \angle DEB$ (P-11).
Hence $\triangle ADE \cong \triangle BDE$ (P-2).
$\therefore AD = DB$ (D-25).

PROPOSITION 19

If two angles of a triangle are equal, the sides opposite are equal (i.e., the triangle is isosceles).

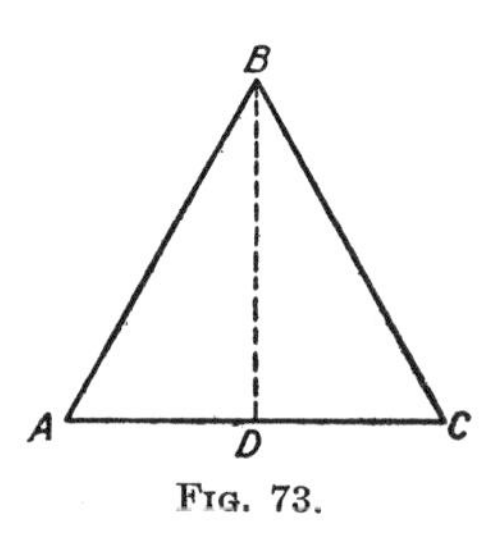

FIG. 73.

Given: $\angle A = \angle C$.
To prove: $AB = BC$.
Draw BD so as to bisect $\angle ABC$.

Since $\angle A = \angle C$ (hyp.),
and $\angle ABD = \angle DBC$ (construction),
$\angle ADB = \angle CDB$ (P-11).
$BD = BD$ (common).
$\therefore \triangle ADB \cong \triangle CDB$ (P-3).
Hence $AB = BC$ (D-25).

Corollary 1 to Proposition 19.—*The bisector of the vertical angle of an isosceles triangle is perpendicular to the base and bisects the base.*

$\angle ADB$ and $\angle CDB$ are both right angles (D-8).
$AD = DC$ (D-25).

Corollary 2 to Proposition 19.—*If a triangle is equiangular, it is also equilateral.*

PROPOSITION 20

Two triangles are congruent if the three sides of the one are equal respectively to the three sides of the other.

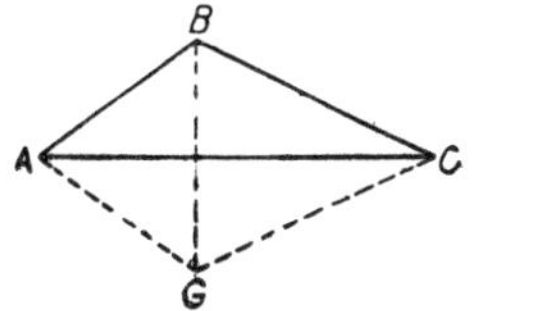

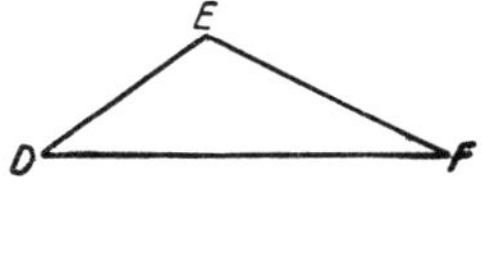

Fig. 74.

Given: $AB = DE$, $BC = EF$, and $AC = DF$.

To prove: $\triangle ABC \cong \triangle DEF$.

Place $\triangle DEF$ so that DF falls on its equal AC and point E falls on the opposite side from B, say at G. Draw BG.

In $\triangle BAG$, $\angle ABG = \angle AGB$ (P-17).

In $\triangle BCG$, $\angle GBC = \angle BGC$ (P-17).

$\angle ABG + \angle GBC = \angle AGB + \angle BGC$ (A-III).

That is, $\angle ABC = \angle AGC$ (A-VIII).

Hence, $\triangle ABC \cong \triangle AGC$ (P-2).

But $\triangle AGC \cong \triangle DEF$ (by construction).

$\therefore \triangle ABC \cong \triangle DEF$ (A-II).

PROPOSITION 21

Two right triangles are congruent if the hypotenuse and a leg of one are equal respectively to the hypotenuse and a leg of the other.

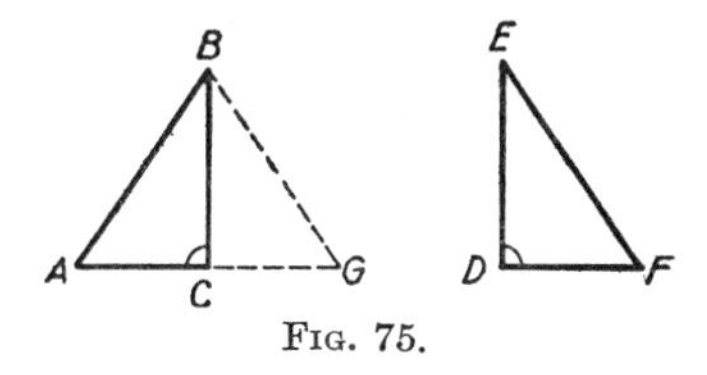

Fig. 75.

Given: Hypotenuse AB = hypotenuse EF and leg BC = leg ED.

To prove: $\triangle ABC \cong \triangle DEF$.

Place $\triangle DEF$ so that DE falls on its equal BC and point F falls on the opposite side from point A, say at G.

Side ACG is a straight line (D-6).
and $\triangle ABG$ is an isosceles triangle ($AB = EF$ by hyp.).
$AC = CG$ (cor. 1 to P-17).
$\therefore \triangle ABC \cong \triangle GCB$ (P-20).
But $\triangle GCB \cong \triangle DEF$ (by construction).
$\therefore \triangle ABC \cong \triangle DEF$ (A-II).

PROPOSITION 22

If two angles of a triangle are unequal, the sides opposite these angles are unequal and the longer side lies opposite the greater angle.

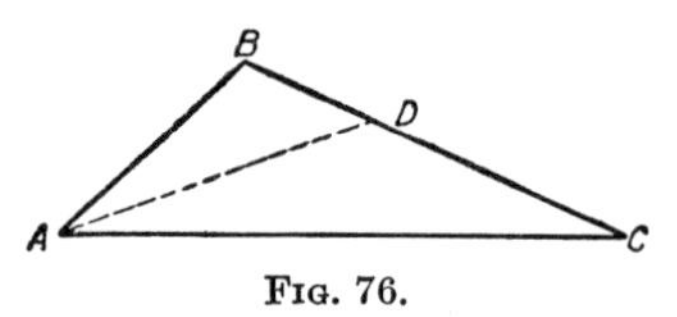

FIG. 76.

Given: $\triangle ABC$ with $\angle BAC$ greater than $\angle BCA$ (usually written $\angle BAC > \angle BCA$).

To prove: $BC > AB$.

Construct AD making $\angle DAC = \angle DCA$.

Then $AD = DC$ (P-19).

$AB < AD + DB$ (D-4).
$\therefore AB < DC + DB$ (A-II).
or $AB < BC$ (A-VIII).

That is, $BC > AB$.

PROPOSITION 23

The opposite sides of a parallelogram are equal.

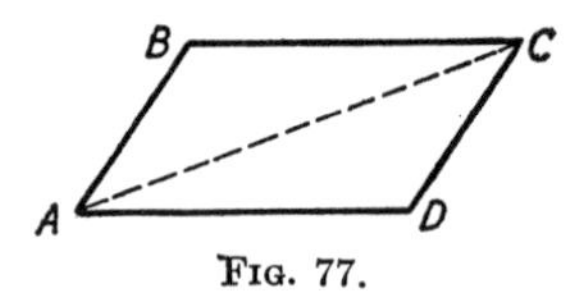

FIG. 77.

Given: Parallelogram $ABCD$ with opposite sides parallel, *i.e.*, $AB \parallel CD$ and $BC \parallel AD$.

To prove: $AB = CD$ and $BC = AD$.

Draw the diagonal AC.

In $\triangle$ ABC and ADC,

$\angle DAC = \angle BCA$, (P-7).

$\angle BAC = \angle DCA$ (P-7).

$AC = AC$ (common).

$\therefore \triangle ABC \cong \triangle ADC$ (P-3).

Hence $AB = CD$ and $BC = AD$ (D-25).

Corollary 1 to Proposition 23.—*Segments of parallel lines intercepted by parallel lines are equal.*

Corollary 2 to Proposition 23.—*A diagonal divides a parallelogram into two equal triangles.*

PROPOSITION 24

If three or more parallel lines intercept equal segments on one intersecting line (often called a transversal), they intercept equal segments on all intersecting lines.

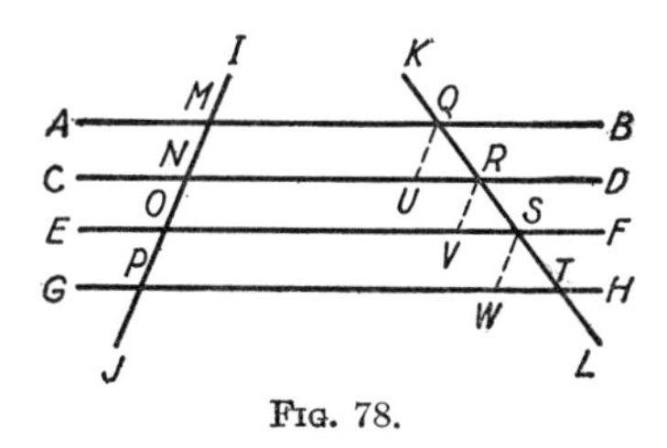

Fig. 78.

Given: Parallel lines AB, CD, EF, and GH cut by the intersecting lines IJ and KL with $MN = NO = OP$.

To prove: $QR = RS = ST$.

Through the points Q, R, and S, draw the lines QU, RV, and $SW \parallel IJ$. Then $QU \parallel RV \parallel SW$ (cor. to A-X).

$\angle QRU = \angle RSV = \angle STW$ (P-8).

$\angle RQU = \angle SRV = \angle TSW$ (P-8).

$\therefore \angle QUR = \angle RVS = \angle SWT$ (P-11).

$QU = RV = SW$ (cor. 1 to P-23).

Hence $\triangle QUR \cong \triangle RVS \cong \triangle SWT$ (P-3).

$\therefore QR = RS = ST$ (D-25).

PROPOSITION 25

If a line is drawn through two sides of a triangle parallel to the third side, it divides those sides proportionally.

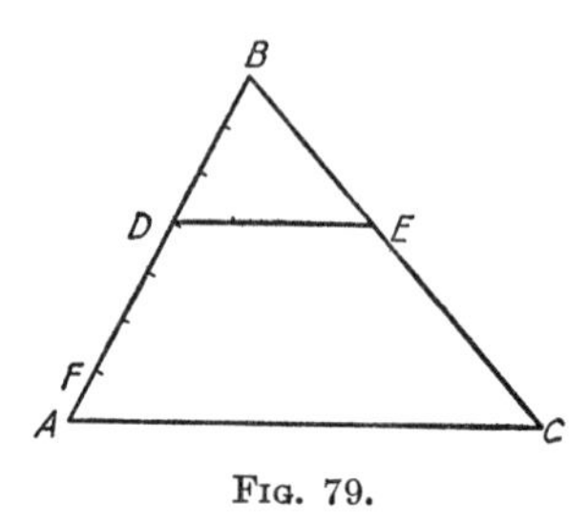

Fig. 79.

Given: $\triangle ABC$ with $DE \parallel AC$.

To prove: $\frac{AD}{DB} = \frac{CE}{EB}$.

Assume that AF is a small unit of length that fits an exact number of times into both AD and DB, say X ($X = 4$ in the figure) times in AD and Y ($Y = 3$ in the figure) times in DB.

Then $\frac{AD}{DB} = \frac{X}{Y}$.

Through the points of division of AD and DB draw lines parallel to AC. These lines will divide line BC into $X + Y$ equal parts, of which X will be in CE and Y in EB. (P-24).

Then $\frac{CE}{EB} = \frac{X}{Y}$.

Hence $\frac{AD}{DB} = \frac{CE}{EB}$ (A-I).

Note: In case no common unit can be found for the lengths AD and DB, the proposition may still be proved by using the method of limits.

Corollary 1 to Proposition 25.—*One side of a triangle is to either of its segments cut off by a line parallel to the base as the other side is to its corresponding segment.*

Since $\frac{AD}{DB} = \frac{CE}{EB}$, (P-25).

$\frac{AD + DB}{DB} = \frac{CE + EB}{EB}$ (Theorem V of Chap. V).

That is, $\frac{AB}{DB} = \frac{BC}{EB}$.

34. Two polygons are said to be similar if their homologous (corresponding) angles are equal and their homologous sides are proportional.

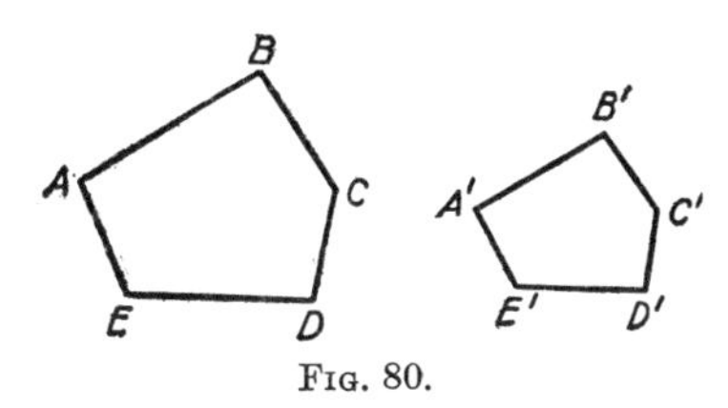

Fig. 80.

Thus polygons $ABCDE$ and $A'B'C'D'E'$ are similar (often written $ABCDE \sim A'B'C'D'E'$) if $\angle A = \angle A'$, $\angle B = \angle B'$, $\angle C = \angle C'$, $\angle D = \angle D'$, $\angle E = \angle E'$, and $\frac{AB}{A'B'} = \frac{BC}{B'C'} = \frac{CD}{C'D'} = \frac{DE}{D'E'} = \frac{EA}{E'A'}$.

PROPOSITION 26

If two triangles are mutually equiangular, their corresponding sides are proportional and hence the triangles are similar.

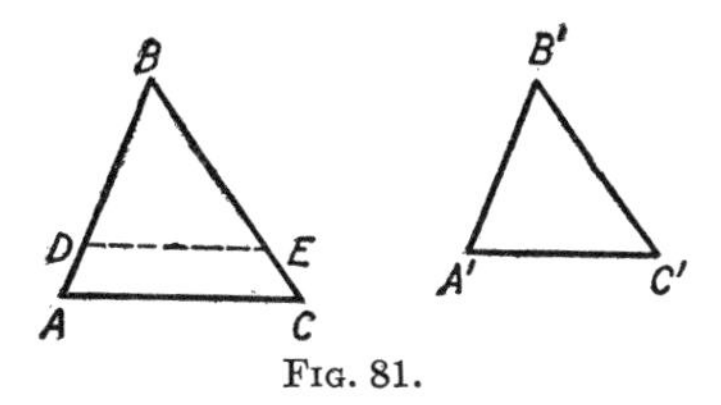

Fig. 81.

Given: $\triangle ABC$ and $A'B'C'$ with $\angle A = \angle A'$, $\angle B = \angle B'$ and $\angle C = \angle C'$.

To prove: $\frac{AB}{A'B'} = \frac{BC}{B'C'} = \frac{CA}{C'A'}$.

and $\triangle ABC \sim \triangle A'B'C'$.

Place $\angle A'B'C'$ on $\angle ABC$ so that $\angle B'$ coincides with $\angle B$ (vertex B' on vertex B and sides $B'A'$ and $B'C'$ falling along corresponding sides BA and BC).

A' will fall at some point D and C' at some point E. Thus the $\triangle A'B'C'$ takes the position BDE.

$$\angle BDE = \angle A \qquad \text{(hyp.)}.$$

$$\text{Hence } DE \parallel AC \qquad \text{(P-9)}.$$

$$\therefore \frac{AB}{DB} = \frac{BC}{BE} \qquad \text{(cor. 1 to P-25)}.$$

$$\text{That is, } \frac{AB}{A'B'} = \frac{BC}{B'C'} \qquad \text{(A-II)}.$$

Similarly by placing $\triangle A'B'C'$ on $\triangle ABC$ so that $\angle A'$ falls on $\angle A$, it may be shown that

$$\frac{AB}{A'B'} = \frac{AC}{A'C'}$$

$$\therefore \frac{BC}{B'C'} = \frac{AC}{A'C'} \qquad \text{(A-I)};$$

$$\text{and we have } \frac{AB}{A'B'} = \frac{BC}{B'C'} = \frac{AC}{A'C'}.$$

$$\therefore \triangle ABC \sim \triangle A'B'C' \qquad \text{(D-34)}.$$

Corollary 1 to Proposition 26.—*Two triangles are similar if two angles of one are equal to two angles of the other* (P-11 and P-26).

Corollary 2 to Proposition 26.—*Two right triangles are similar if they have an acute angle of one equal to an acute angle of the other.* (The right angles are also equal so Corollary 1 to Proposition 26 applies.)

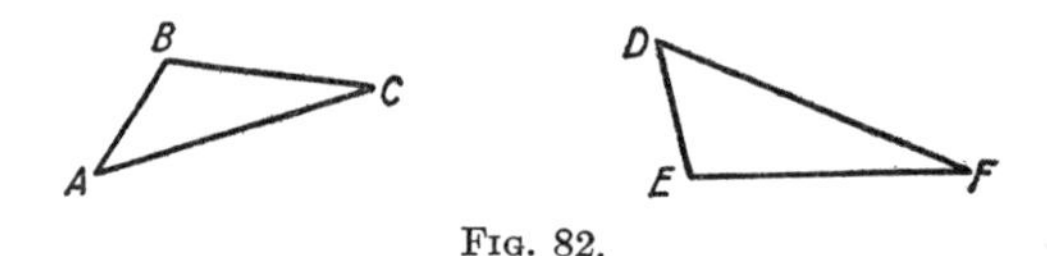

Fig. 82.

Note: It follows from Proposition 22 that the largest angle of a triangle is opposite its longest side, the next largest angle is opposite the next longest side, etc. Hence in two similar triangles as ABC and DEF, the angles that are equal are the angles opposite the sides corresponding in length. Thus $\angle C$, which is opposite AB, the shortest side of $\triangle ABC$, is equal to $\angle F$, which is opposite to DE, the shortest side of $\triangle DEF$, etc.

The student should be able to recognize corresponding angles at a glance by this method.

PROPOSITION 27

Two triangles are similar if their sides are respectively perpendicular.

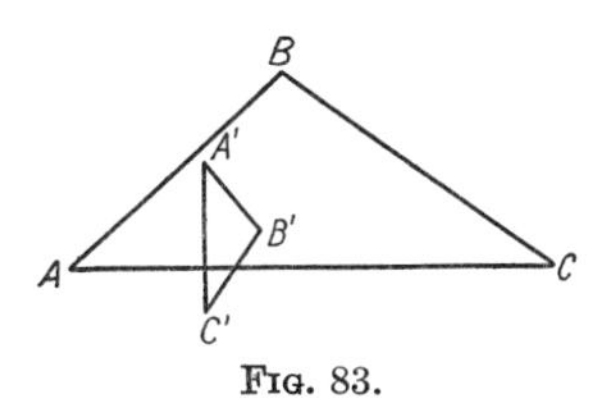

FIG. 83.

Given: $\triangle ABC$ and $A'B'C'$ with $A'B' \perp AB$, $B'C' \perp BC$, and $A'C' \perp AC$.

To prove: $\triangle ABC \sim \triangle A'B'C'$.

$\angle A = \angle A'$, $\angle B = \angle B'$, and $\angle C = \angle C'$ (P-16).

$\therefore \triangle ABC \sim \triangle A'B'C'$ (P-26).

PROPOSITION 28

Two triangles are similar if their sides are respectively parallel.

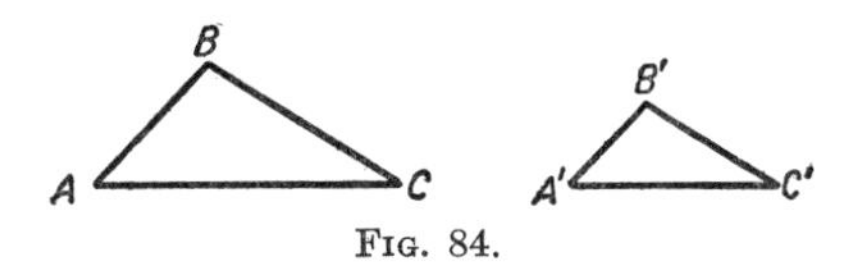

FIG. 84.

Given: $\triangle ABC$ and $A'B'C'$ with $A'B' \parallel AB$, $B'C' \parallel BC$, and $A'C' \parallel AC$.

To prove: $\triangle ABC \sim \triangle A'B'C'$.

$\angle A = \angle A'$, $\angle B = \angle B'$, $\angle C = \angle C'$ (P-15).

$\therefore \triangle ABC \sim \triangle A'B'C'$ (P-26).

PROPOSITION 29

If perpendiculars are drawn from two points on one side of an angle to the other side of the angle, the triangles formed are similar.

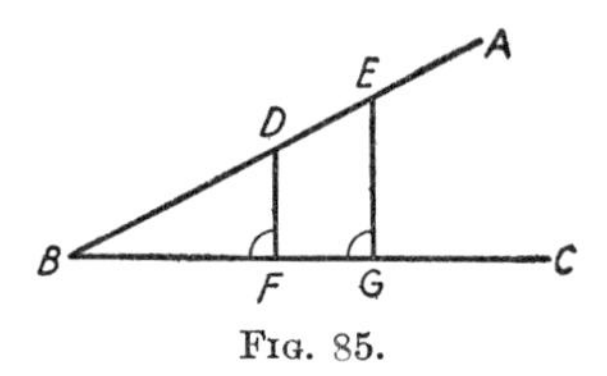

FIG. 85.

Given: $\angle ABC$ and the $\perp$s DF and EG drawn from the points D and E to the line BC.

To prove: $\triangle DBF \sim \triangle EBG$.

$\angle B$ is common and both triangles are right triangles (D-26).

$\therefore \triangle DBF \sim \triangle EBG$ (cor. 2 to P-26).

PROPOSITION 30

If, in a right triangle, a perpendicular is drawn from the vertex of the right angle to the hypotenuse:

a. The triangles formed on either side of the perpendicular are similar to the whole triangle and to each other.

b. The perpendicular is the mean proportional between the segments of the hypotenuse.

c. Each side adjacent to the right angle is a mean proportional between the hypotenuse and the segment of the hypotenuse adjacent to that side.

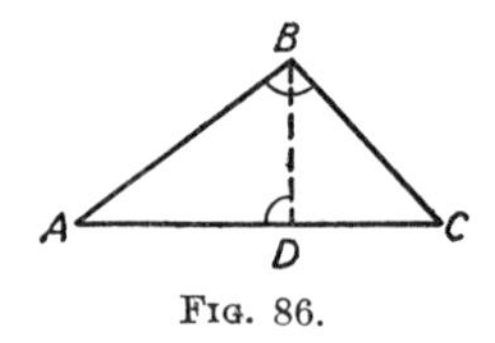

FIG. 86.

Given: $\triangle ABC$ with right angle at B and BD drawn from B perpendicular to the hypotenuse.

a. To prove: ⊿s ABC, ADB, and CDB, all similar.

Rt. $\triangle ADB \sim$ rt. $\triangle ABC$ since $\angle A$ is common (cor. 2 to P-26).

Rt. $\triangle CDB \sim$ rt. $\triangle ABC$ since $\angle C$ is common (cor. 2 to P-26).

$\triangle$s ADB and CDB, being both similar to $\triangle ABC$, have their angles equal to those of $\triangle ABC$ (D-34).

$\therefore$ Angles of $\triangle ADB$ = angles of $\triangle CDB$, respectively (A-I).

$\therefore \triangle ADB \sim \triangle CDB$ (P-26).

b. To prove: $\dfrac{AD}{BD} = \dfrac{BD}{DC}$

Since $\triangle ADB \sim \triangle BDC$ (P-30*a*),

$$\frac{AD}{BD} = \frac{BD}{DC} \qquad \text{(D-34).}$$

c. To prove: $\dfrac{AC}{AB} = \dfrac{AB}{AD}$ and $\dfrac{AC}{BC} = \dfrac{BC}{DC}$.

Since $\triangle ABC \sim \triangle ADB$ (P-30*a*),

$$\frac{AC}{AB} = \frac{AB}{AD} \qquad \text{(D-34).}$$

Since $\triangle ABC \sim \triangle CDB$ (P-30*a*),

$$\frac{AC}{BC} = \frac{BC}{DC} \qquad \text{(D-34).}$$

Note: The solutions of many problems depend on Proposition 30*a* and hence it is very important that the student recognize at once the equal angles of the similar triangles. The following statement will assist in recognizing the equal angles:

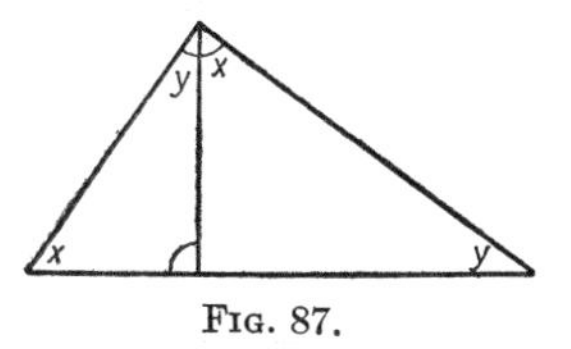

FIG. 87.

Corollary to Proposition 30.—*If a perpendicular is dropped from the vertex of the right angle to the hypotenuse, the angle opposite the perpendicular in one triangle is equal to the angle adjacent to this perpendicular in the other triangle.*

PROBLEMS

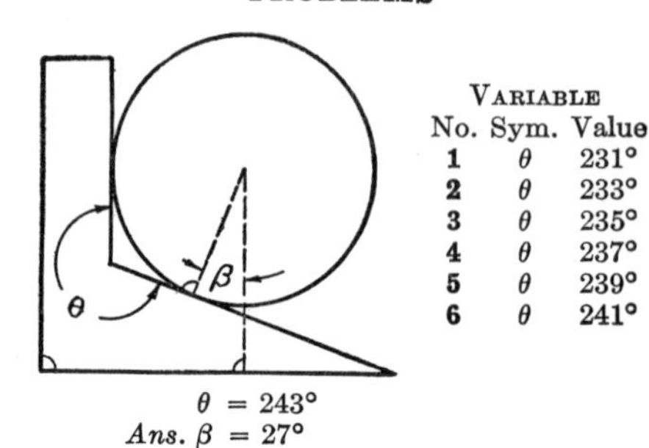

VARIABLE		
No.	Sym.	Value
1	θ	231°
2	θ	233°
3	θ	235°
4	θ	237°
5	θ	239°
6	θ	241°

$\theta = 243°$
Ans. $\beta = 27°$

1. Determine the angle β.

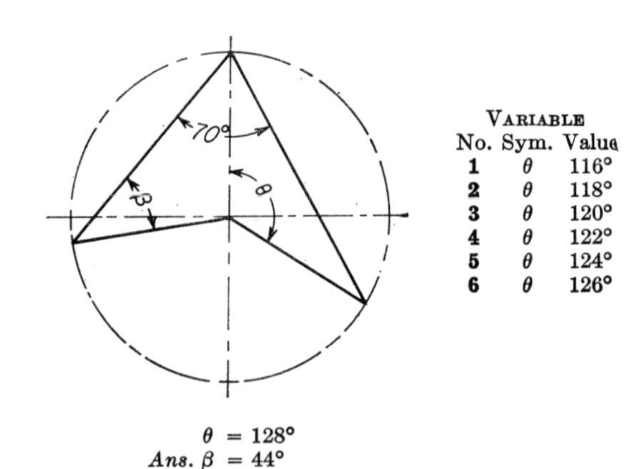

VARIABLE		
No.	Sym.	Value
1	θ	116°
2	θ	118°
3	θ	120°
4	θ	122°
5	θ	124°
6	θ	126°

$\theta = 128°$
Ans. $\beta = 44°$

2. Determine the angle β.

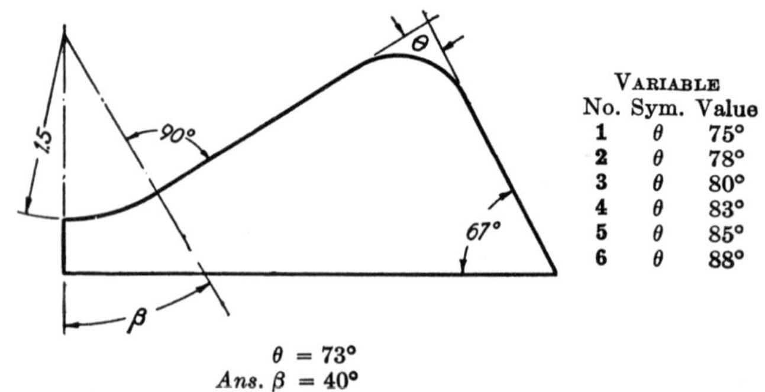

VARIABLE		
No.	Sym.	Value
1	θ	75°
2	θ	78°
3	θ	80°
4	θ	83°
5	θ	85°
6	θ	88°

$\theta = 73°$
Ans. $\beta = 40°$

3. Determine the angle β.

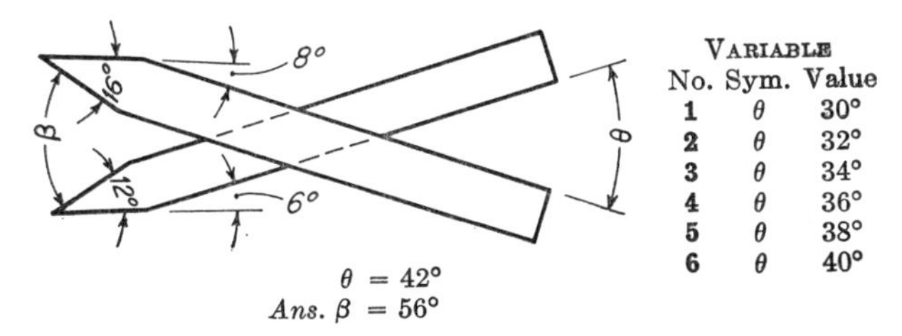

VARIABLE		
No.	Sym.	Value
1	θ	30°
2	θ	32°
3	θ	34°
4	θ	36°
5	θ	38°
6	θ	40°

$\theta = 42°$
Ans. $\beta = 56°$

4. Determine the angle β.

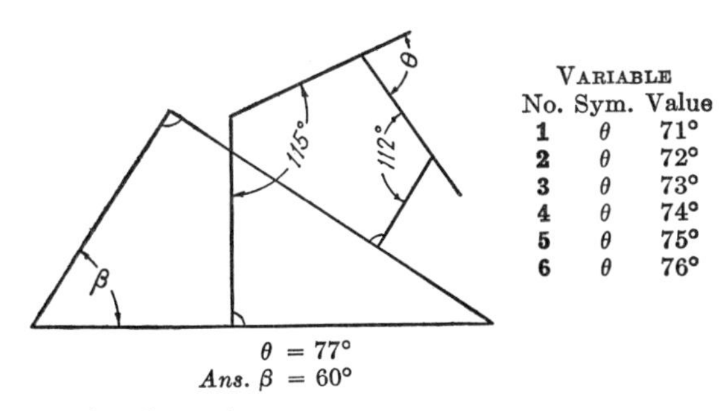

VARIABLE		
No.	Sym.	Value
1	θ	71°
2	θ	72°
3	θ	73°
4	θ	74°
5	θ	75°
6	θ	76°

$\theta = 77°$
Ans. $\beta = 60°$

5. Determine the angle β.

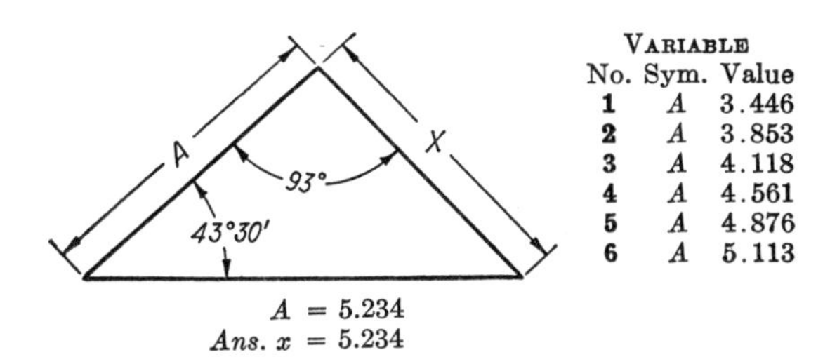

VARIABLE		
No.	Sym.	Value
1	A	3.446
2	A	3.853
3	A	4.118
4	A	4.561
5	A	4.876
6	A	5.113

$A = 5.234$
Ans. $x = 5.234$

6. Determine the distance x.

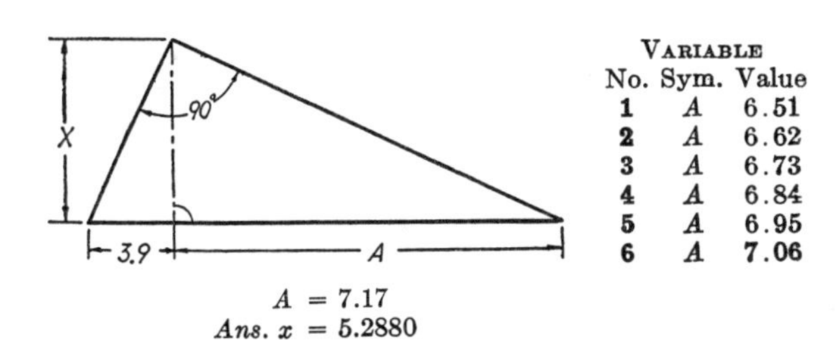

VARIABLE		
No.	Sym.	Value
1	A	6.51
2	A	6.62
3	A	6.73
4	A	6.84
5	A	6.95
6	A	7.06

$A = 7.17$
Ans. $x = 5.2880$

7. Determine the distance x.

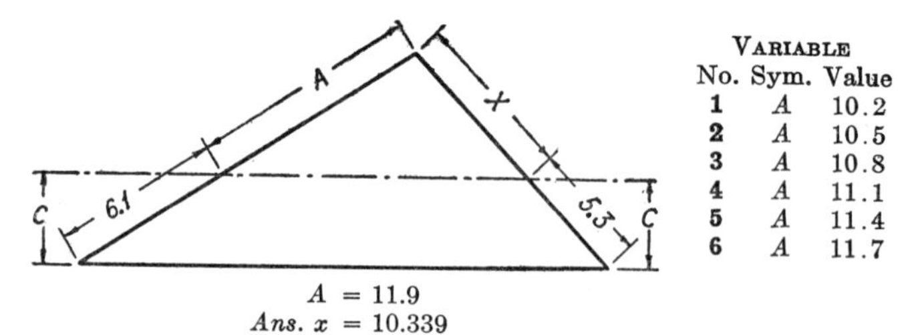

Variable		
No.	Sym.	Value
1	A	10.2
2	A	10.5
3	A	10.8
4	A	11.1
5	A	11.4
6	A	11.7

$A = 11.9$
Ans. $x = 10.339$

8. Determine the distance x.

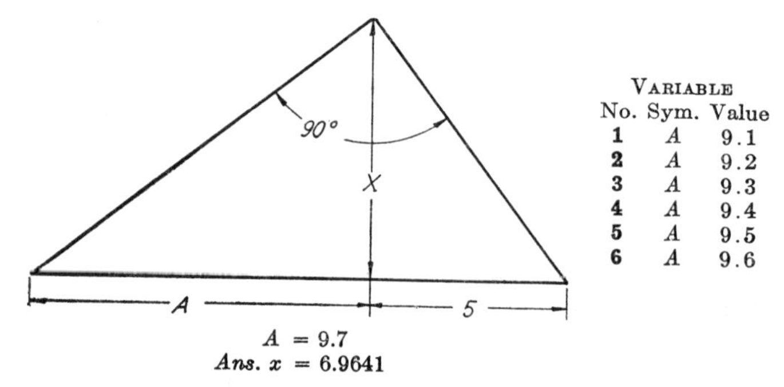

Variable		
No.	Sym.	Value
1	A	9.1
2	A	9.2
3	A	9.3
4	A	9.4
5	A	9.5
6	A	9.6

$A = 9.7$
Ans. $x = 6.9641$

9. Determine the distance x.

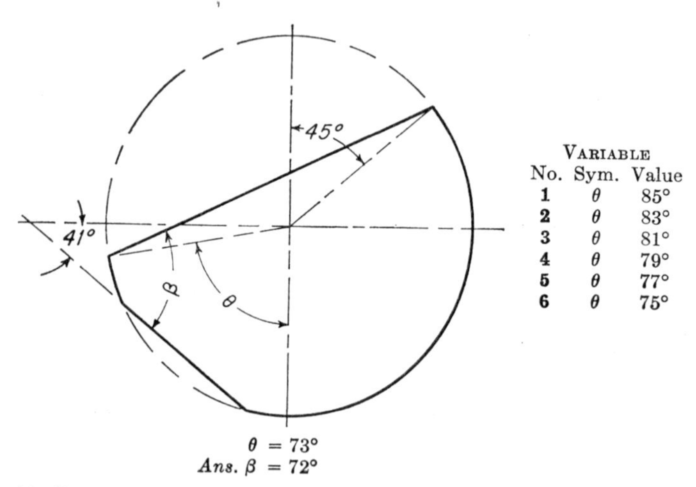

Variable		
No.	Sym.	Value
1	θ	85°
2	θ	83°
3	θ	81°
4	θ	79°
5	θ	77°
6	θ	75°

$\theta = 73°$
Ans. $\beta = 72°$

10. Determine the angle β.

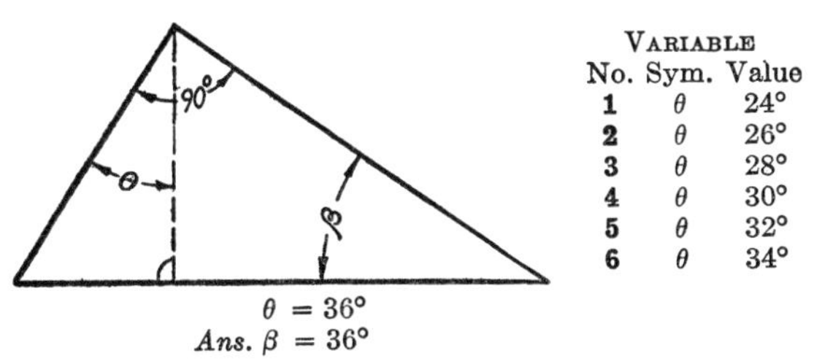

VARIABLE		
No.	Sym.	Value
1	θ	24°
2	θ	26°
3	θ	28°
4	θ	30°
5	θ	32°
6	θ	34°

$\theta = 36°$

Ans. $\beta = 36°$

11. Determine the angle β.

PROPOSITION 31

In any right triangle the square of the hypotenuse is equal to the sum of the squares of the two other sides.

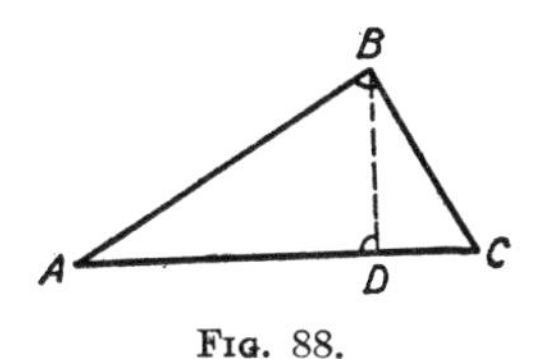

FIG. 88.

Given: $\triangle ABC$ with right angle at B.

To prove: $\overline{AC}^2 = \overline{AB}^2 + \overline{BC}^2$.

Draw line BD from B perpendicular to base AC.

$$\frac{AC}{AB} = \frac{AB}{AD} \qquad \text{(P-30}c\text{)}.$$

$\therefore \overline{AB}^2 = AC \times AD$ (Theorem 1 of Chap. V).

$$\frac{AC}{BC} = \frac{BC}{DC} \qquad \text{(P-30}c\text{)}.$$

$\therefore \overline{BC}^2 = AC \times DC$ (Theorem I of Chap. V).

$\overline{AB}^2 + \overline{BC}^2 = AC \times AD + AC \times DC$ (A-III).

$\overline{AB}^2 + \overline{BC}^2 = AC(AD + DC)$ (factoring out common term AC).

$\overline{AB}^2 + \overline{BC}^2 = AC(AC)$ (A-VIII and A-II).

$\overline{AB}^2 + \overline{BC}^2 = \overline{AC}^2$.

As an illustration of this theorem, consider a right triangle having one leg AB 4 units long and the other leg BC 3 units long. The hypotenuse of this right triangle will be found

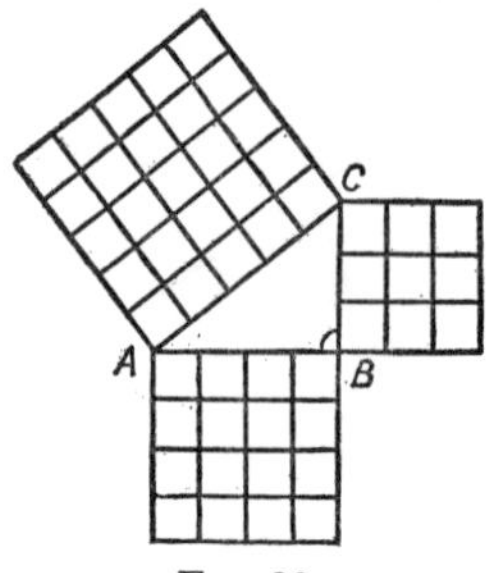

FIG. 89.

to be 5 units long. As seen from the figure the square on the hypotenuse contains 25 square units and those on the legs 16 and 9 square units, respectively.

$$25 = 16 + 9.$$

PROBLEMS

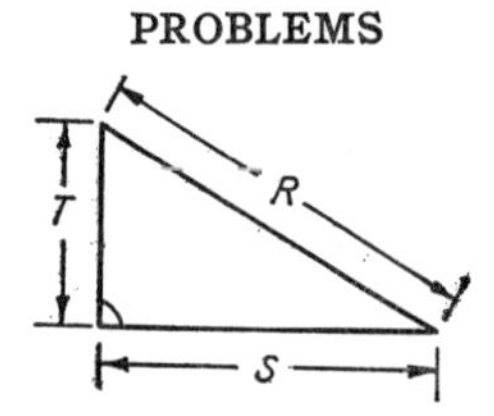

Prob.	R	S	T
1	A	?	5.76
2	B	7.53	?
3	?	12.95	C
4	20.53	?	D
5	?	E	5.876
6	17.32	F	?
7	?	8.95	G

Substitute the given values for the letters in the diagram above and solve for the unknown side. A, B, C, etc., are the variables.

VARIABLES

Prob.	Sym.	No. 1	No. 2	No. 3	No. 4	No. 5	No. 6
1	A	8.425	8.752	7.754	6.793	9.913	9.375
2	B	10.52	9.625	8.461	9.453	11.25	10.88
3	C	7.755	8.252	9.748	9.275	8.644	11.58
4	D	16.28	15.93	15.25	14.48	12.85	13.75
5	E	9.252	8.925	3.975	10.28	11.45	10.75
6	F	13.75	12.96	12.83	14.55	15.25	11.82
7	G	2.875	3.812	4.125	4.775	5.237	5.375

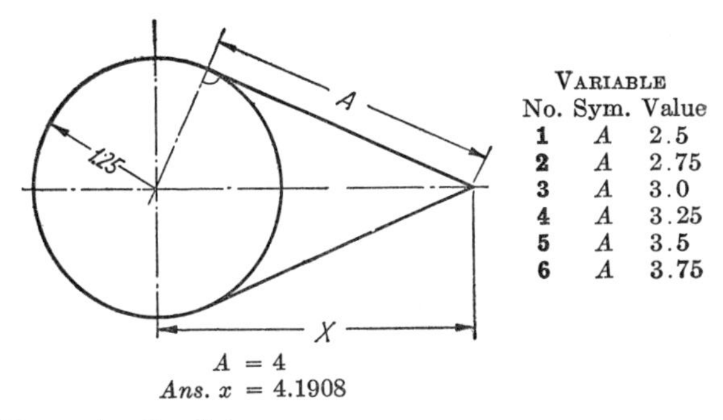

VARIABLE		
No.	Sym.	Value
1	A	2.5
2	A	2.75
3	A	3.0
4	A	3.25
5	A	3.5
6	A	3.75

$A = 4$
Ans. $x = 4.1908$

8. Determine the distance x.

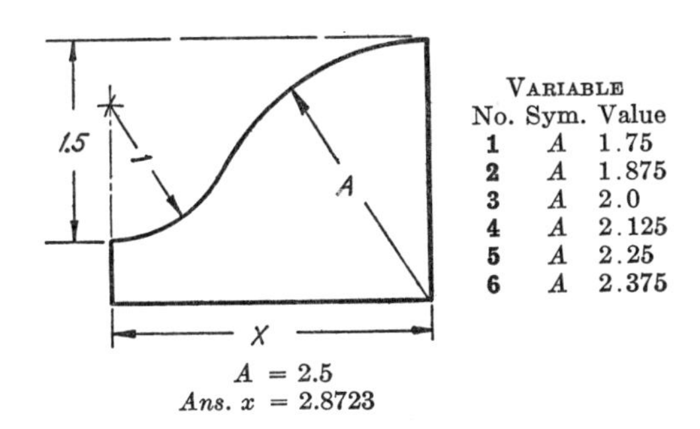

VARIABLE		
No.	Sym.	Value
1	A	1.75
2	A	1.875
3	A	2.0
4	A	2.125
5	A	2.25
6	A	2.375

$A = 2.5$
Ans. $x = 2.8723$

9. Determine the distance x.

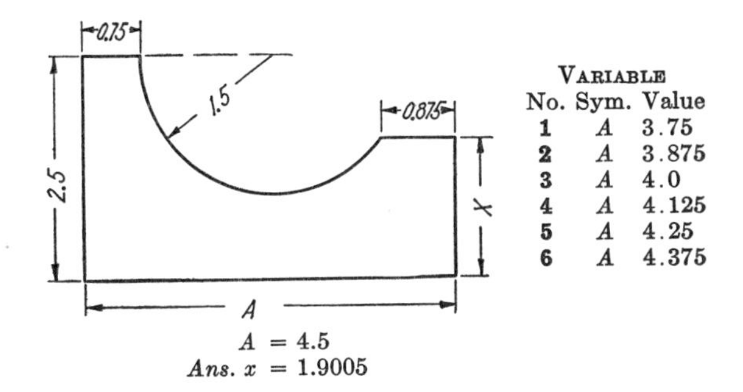

VARIABLE		
No.	Sym.	Value
1	A	3.75
2	A	3.875
3	A	4.0
4	A	4.125
5	A	4.25
6	A	4.375

$A = 4.5$
Ans. $x = 1.9005$

10. Determine the distance x.

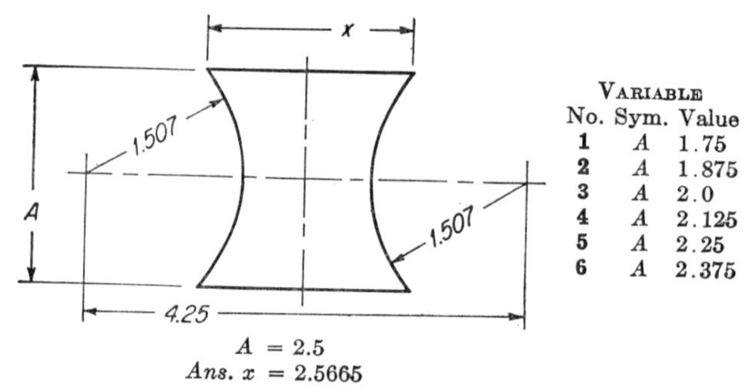

VARIABLE		
No.	Sym.	Value
1	A	1.75
2	A	1.875
3	A	2.0
4	A	2.125
5	A	2.25
6	A	2.375

$A = 2.5$
Ans. $x = 2.5665$

11. Determine the distance x.

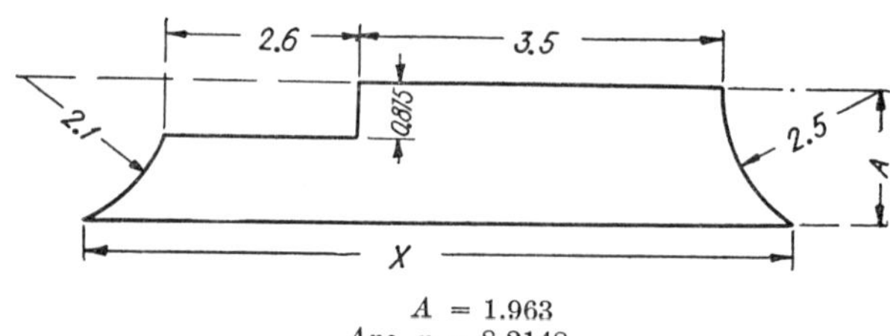

$A = 1.963$
Ans. $x = 8.2148$

VARIABLE

1. $A = 1.041$ **2.** $A = 1.122$ **3.** $A = 1.253$
4. $A = 1.624$ **5.** $A = 1.755$ **6.** $A = 1.886$

12. Determine the distance x.

PROPOSITION 32

The projection of a side of a triangle upon the base is equal to the square of this side plus the square of the base minus the square of the third side, divided by two times the base.

Note: For students who prefer the graphic method, the figures are used to lead directly to the required result.

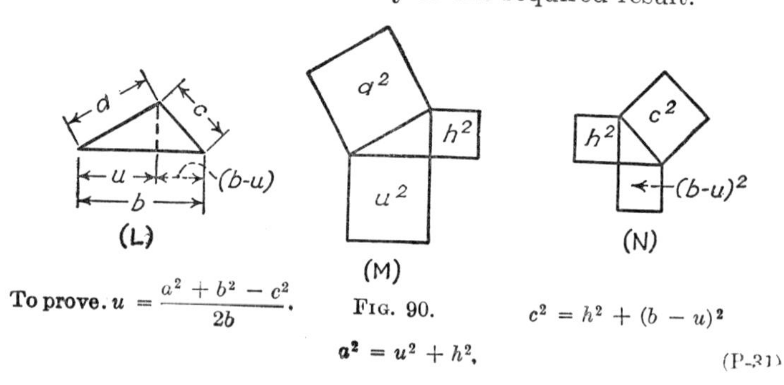

To prove. $u = \dfrac{a^2 + b^2 - c^2}{2b}$. FIG. 90. $c^2 = h^2 + (b - u)^2$

$a^2 = u^2 + h^2$.

(P-31)

The expression $a^2 + b^2 - c^2$ may be evaluated diagrammatically as follows: The order of the expression $a^2 + b^2 - c^2$ may be arranged thus: $b^2 + a^2 - c^2$. b^2 is the area of a square erected upon the base of the original triangle and is shown in figure R. $a^2 - c^2$ as shown in figures M and N is equivalent to u^2 minus $(b - u)^2$ since the h^2 in M minus the h^2 in N is zero.

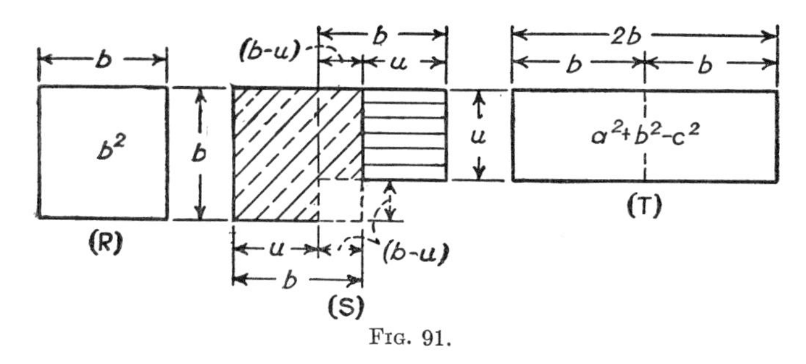

Fig. 91.

Figure S shows the combined areas expressed by b^2 plus u^2 minus $(b - u)^2$ or $(b^2 + a^2 - c^2)$. This area may be rearranged as shown in figure T into two rectangular pieces each of which has a length b and a width u. The area of T which is $a^2 + b^2 - c^2$ is thus seen to be equal to $2bu$.

Hence $2bu = a^2 + b^2 - c^2$

and $$u = \frac{a^2 + b^2 - c^2}{2b}.$$

For students who prefer the algebraic method, the following may be used:

$$\begin{aligned} a^2 + b^2 - c^2 &= u^2 + \cancel{h^2} + b^2 - \cancel{h^2} - (b - u)^2 \\ &= \cancel{u^2} + \cancel{b^2} - \cancel{b^2} + 2bu - \cancel{u^2} \text{ [expanding } (b - u)^2] \\ &= 2bu \end{aligned}$$

or $$u = \frac{a^2 + b^2 - c^2}{2b}.$$

Similarly, for an obtuse triangle as shown in Fig. 92, the following expression may be worked out for the projection v:

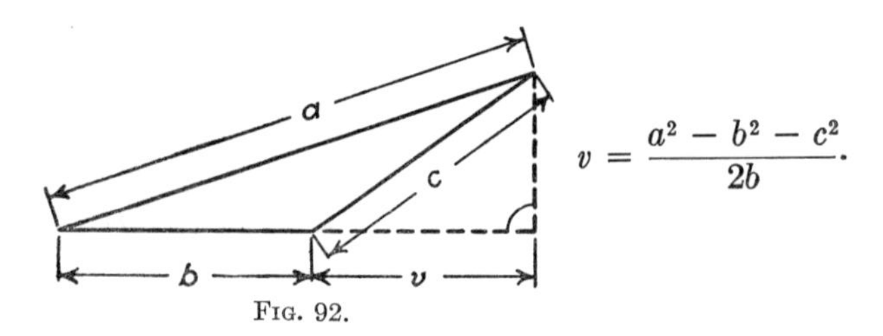

$$v = \frac{a^2 - b^2 - c^2}{2b}.$$

FIG. 92.

These two formulas for u and v will be used in trigonometry and will be referred to as the projection formulas.

PROBLEMS

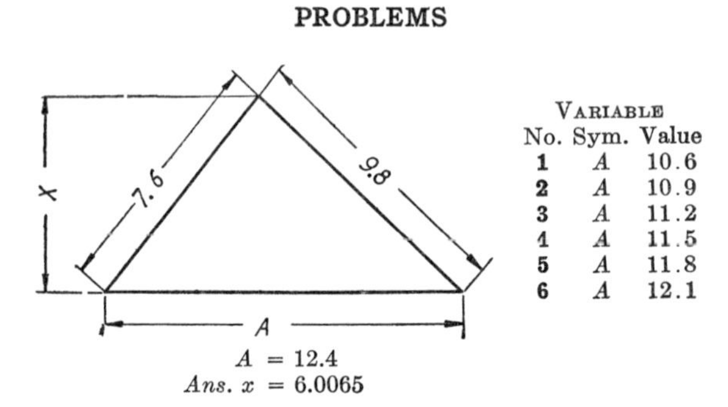

VARIABLE		
No.	Sym.	Value
1	A	10.6
2	A	10.9
3	A	11.2
4	A	11.5
5	A	11.8
6	A	12.1

$A = 12.4$
Ans. $x = 6.0065$

1. Determine the altitude x.

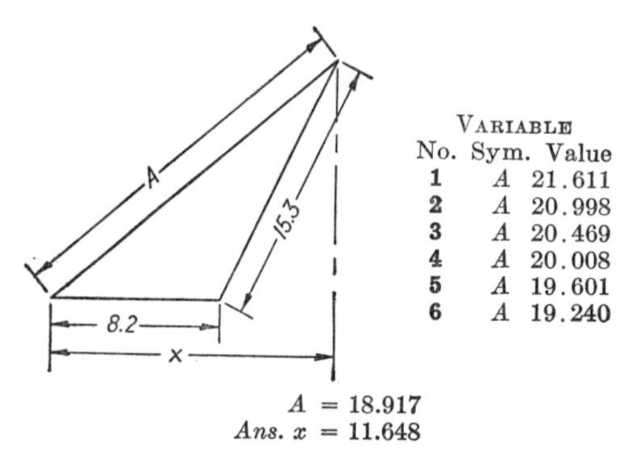

VARIABLE		
No.	Sym.	Value
1	A	21.611
2	A	20.998
3	A	20.469
4	A	20.008
5	A	19.601
6	A	19.240

$A = 18.917$
Ans. $x = 11.648$

2. Determine the distance x.

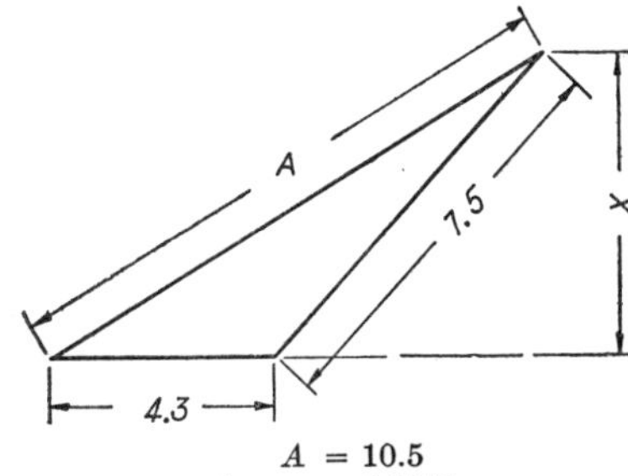

VARIABLE		
No.	Sym.	Value
1	A	8.8
2	A	9.1
3	A	9.3
4	A	9.5
5	A	9.8
6	A	10.3

$A = 10.5$
Ans. $x = 6.2611$

3. Determine the distance x.

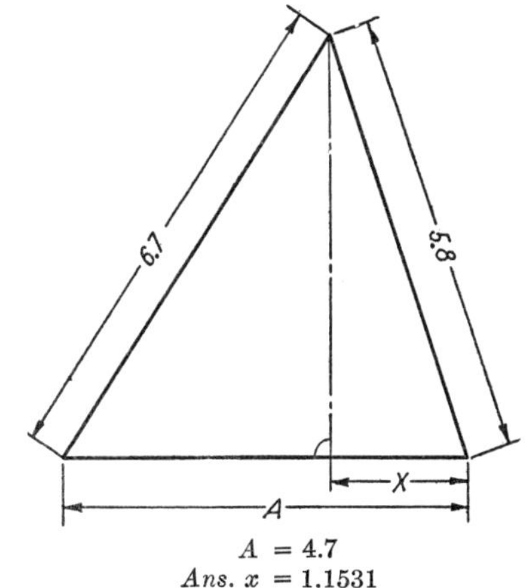

VARIABLE		
No.	Sym.	Value
1	A	4.1
2	A	4.0
3	A	3.9
4	A	3.8
5	A	3.7
6	A	3.6

$A = 4.7$
Ans. $x = 1.1531$

4. Determine the distance x.

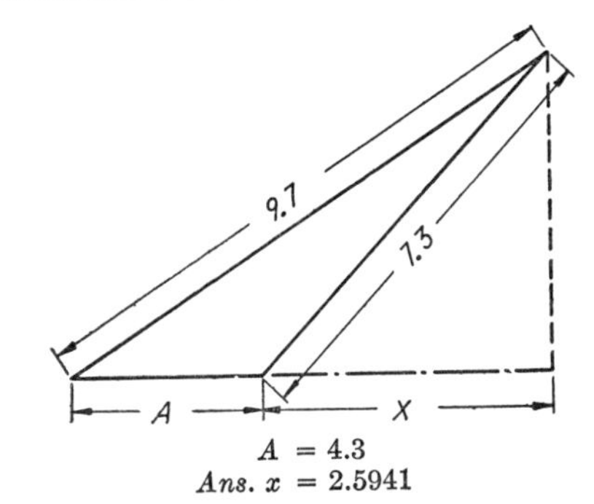

VARIABLE		
No.	Sym.	Value
1	A	5.5
2	A	5.3
3	A	5.1
4	A	4.9
5	A	4.7
6	A	4.5

$A = 4.3$
Ans. $x = 2.5941$

5. Determine the distance x.

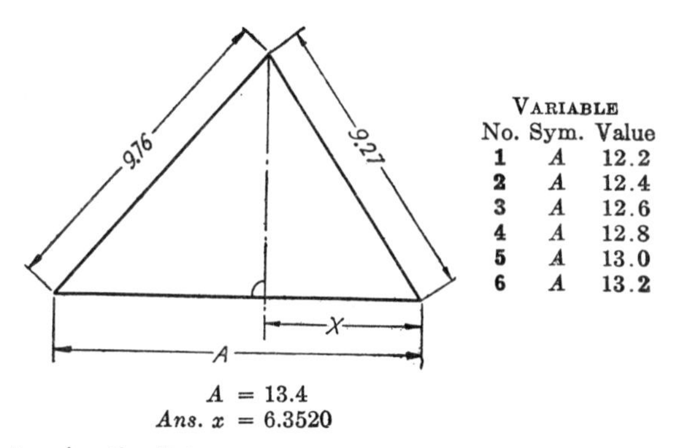

Variable		
No.	Sym.	Value
1	A	12.2
2	A	12.4
3	A	12.6
4	A	12.8
5	A	13.0
6	A	13.2

$A = 13.4$
Ans. $x = 6.3520$

6. Determine the distance x.

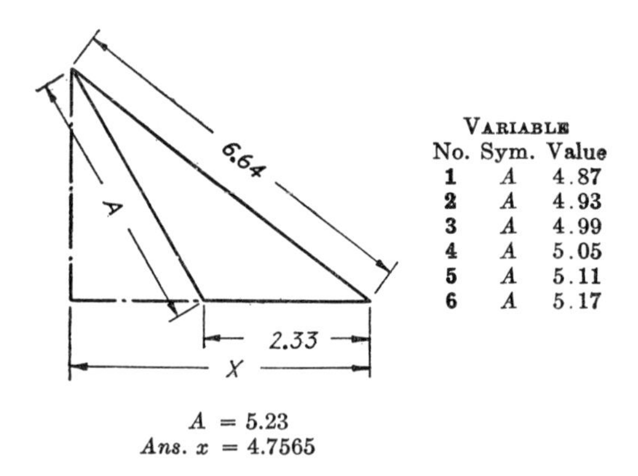

Variable		
No.	Sym.	Value
1	A	4.87
2	A	4.93
3	A	4.99
4	A	5.05
5	A	5.11
6	A	5.17

$A = 5.23$
Ans. $x = 4.7565$

7. Determine the distance x.

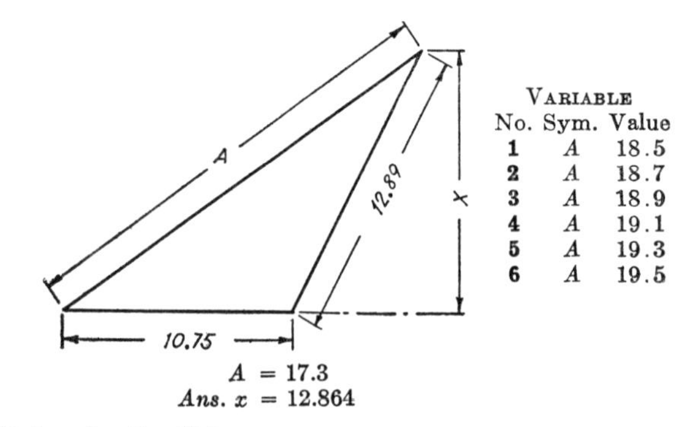

Variable		
No.	Sym.	Value
1	A	18.5
2	A	18.7
3	A	18.9
4	A	19.1
5	A	19.3
6	A	19.5

$A = 17.3$
Ans. $x = 12.864$

8. Determine the distance x.

CIRCLES

Definitions

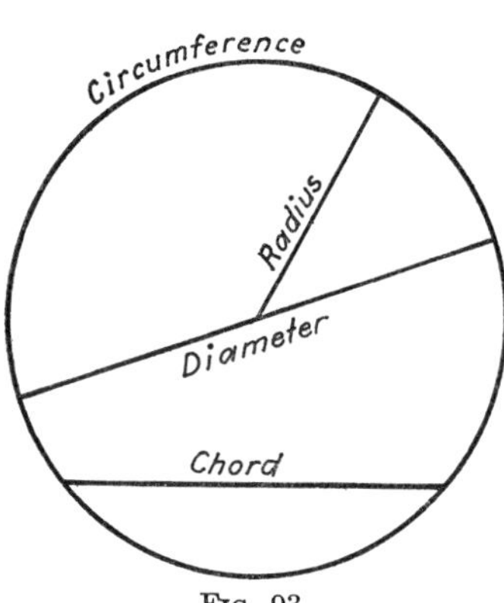

Fig. 93.

35. A **circle** is a plane figure bounded by a line called the **circumference,** all points of which are equidistant from a point within called the center.

36. A **radius** of a circle is a straight line from the center to a point on the circumference.

37. A **diameter** of a circle is a straight line through the center, having its ends in the circumference.

38. A **chord** of a circle is a straight line having its ends in the circumference.

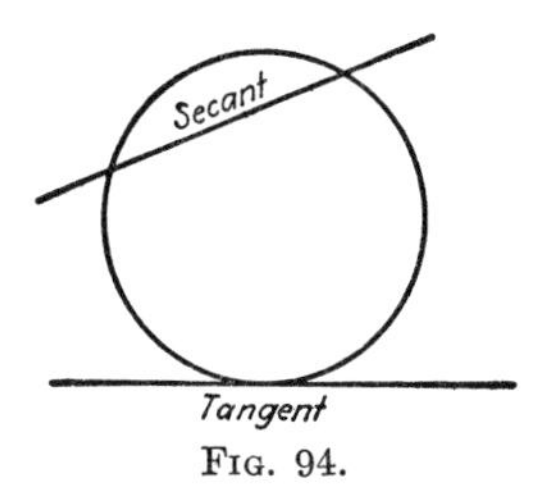

Fig. 94.

39. A **secant** is any straight line intersecting a circle.

40. A **tangent** to a circle is a straight line which touches the circumference in only one point.

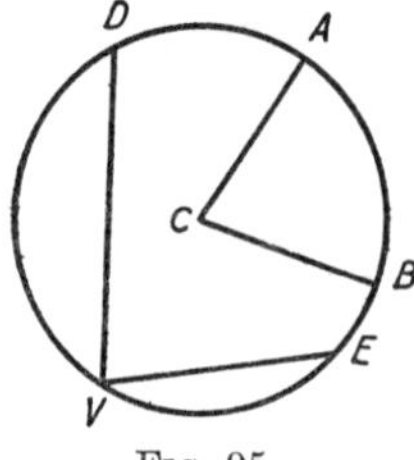

FIG. 95.

41. **A central angle** is an angle having its vertex at the center of the circle and radii of the circle for its sides, as $\angle ACB$.

42. An angle **inscribed in a circle** (usually called an *inscribed angle*) is an angle whose vertex lies on the circumference and whose sides are chords of the circle, as $\angle DVE$.

43. An angle is said to **intercept** the arc included within its sides. Thus $\angle ACB$ intercepts arc AB (often written $\widehat{AB}$) and $\angle DVE$ intercepts arc DAE ($\widehat{DAE}$). The arc is said to be subtended by the angle.

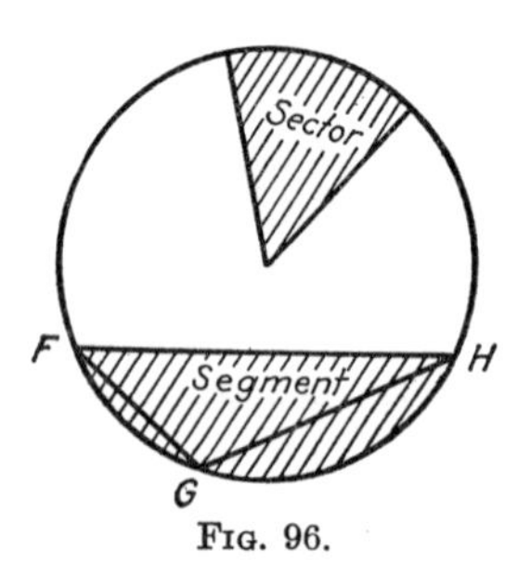

FIG. 96.

44. **A sector** of a circle is that portion bounded by two radii and the intercepted arc.

45. A **segment** of a circle is that portion bounded by a chord and the intercepted arc.

46. An angle is said to be **inscribed in a segment** if the vertex is on the circumference and the sides pass through the ends of the arc of the segment. Thus $\angle FGH$ is inscribed in the shaded segment.

47. A **regular circumscribed polygon** is a regular polygon having all of its sides tangent to a circle, as $EFGH$ (Fig. 97).

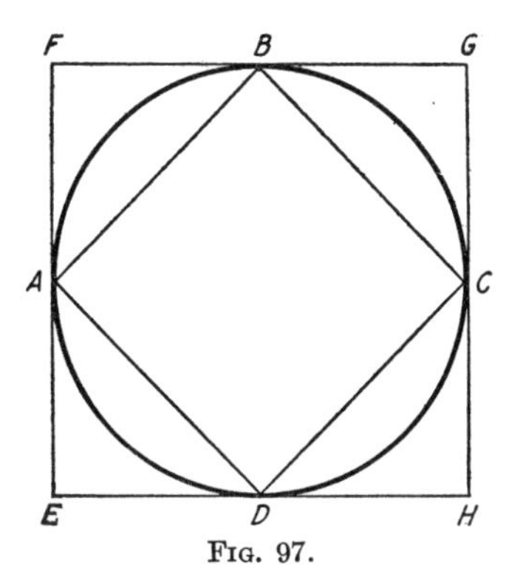

FIG. 97.

48. **A regular inscribed polygon** is a regular polygon having all of its vertices on the circumference of a circle, as *ABCD.*

49. The length of the circumference of a circle divided by its diameter is equal to the number 3.1416— which is called π (Pī), *i.e.*, the circumference is equal to the diameter multiplied by π.

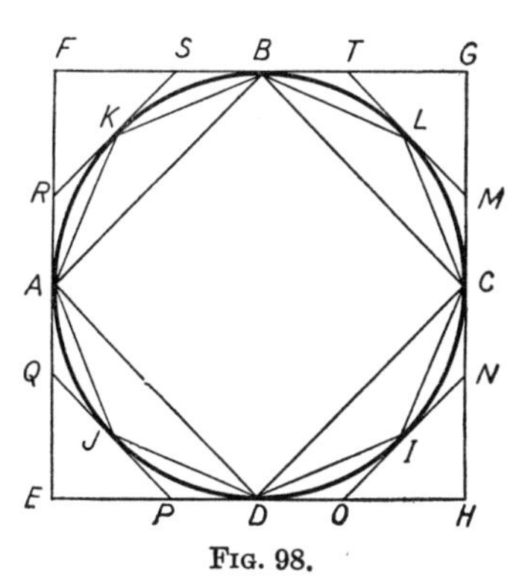

FIG. 98.

50. Informal discussion of the determination of π.

It can be seen from the accompanying figure that the perimeter of the inscribed regular polygon *ABCD* is less than the circumference of the circle in which it is inscribed, and that the perimeter of the circumscribed regular polygon *EFGH* is greater than the circumference of the circle around which it is circumscribed. It is also readily seen that, as the number of sides of the regular inscribed and circumscribed polygons is increased, their perimeters more nearly equal the circumference of the circle. Thus octagon $IDJAKBLCI >$ square $ABCD$ and octagon $MNOPQRSTM <$ square $EFGH$.

Hence, if the perimeter of an inscribed (or circumscribed) polygon of a large number of sides be divided by the diameter of the circle in which it is inscribed (or about which it is circumscribed) the quotient will closely approximate π as defined in D-49.

The following data will bring out this point:

Number of sides	Perimeter of inscribed regular polygon ÷ diameter	Perimeter of circumscribed regular polygon ÷ diameter	Difference in these perimeters
4	2.82843	4.00000	1.17157
8	3.06147	3.31371	0.25224
64	3.14033	3.14412	0.00379
512	3.14157	3.14163	0.00006

Thus π, which must be between the two ratios given in the second and third columns, is seen to be approximately 3.1416.

PROPOSITION 33

In the same circle or in equal circles, if two central angles are equal, they subtend equal arcs; conversely, if two arcs of the same circle, or of equal circles, are equal, they are subtended by equal central angles.

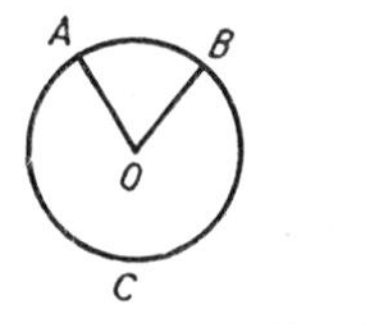

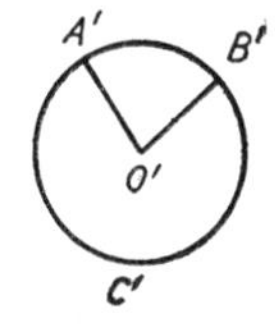

FIG. 99.

a. Given: Equal circles ABC and $A'B'C'$ with equal central angles AOB and $A'O'B'$.

To prove: $\widehat{AB} = \widehat{A'B'}$.

Place circle $A'B'C'$ on circle ABC with center O' falling on O and line $O'B'$ falling along line OB.

Point B' will fall on B

($O'B'$ and OB being radii of equal circles).

Line $O'A'$ will fall along line OA

($\angle A'O'B' = \angle AOB$ by hyp.).

Point A' will fall on A

($O'A' = OA$, being radii of equal circles).

Thus $\overset{\frown}{A'B'}$ is made to coincide with $\overset{\frown}{AB}$ and is equal to it.

b. Given: Equal circles with $\overset{\frown}{A'B'} = \overset{\frown}{AB}$.

To prove: $\angle A'O'B' = \angle AOB$.

Since both circles and the arcs are equal, circle $A'B'C'$ may be placed on circle ABC so that center O' falls on center O and $\overset{\frown}{A'B'}$ falls on its equal, $\overset{\frown}{AB}$.

Thus line $O'A'$ coincides with line OA (A-IX).

and line $O'B'$ coincides with line OB (A-IX).

$\therefore$ $\angle A'O'B'$ coincides with $\angle AOB$ and is equal to it.

PROPOSITION 34

In the same circle, or in equal circles, two central angles have the same ratio as their subtended arcs.

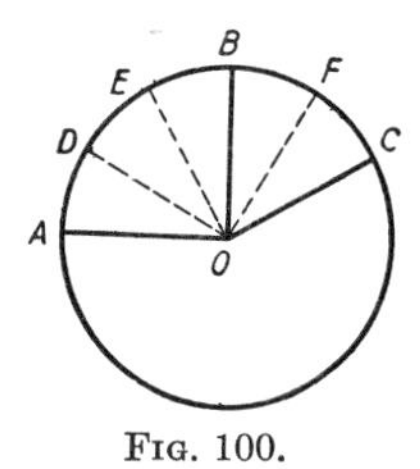

FIG. 100.

Given: Central $\measuredangle AOB$ and BOC.

To prove: $\dfrac{\angle AOB}{\angle BOC} = \dfrac{\overset{\frown}{AB}}{\overset{\frown}{BC}}$.

Assume that there is a certain small angle, such as $\angle AOD$, which is contained a whole number of times in both $\angle AOB$ and $\angle BOC$, say three times in $\angle AOB$ and twice in $\angle BOC$;

Then $\dfrac{\angle AOB}{\angle BOC} = \dfrac{3}{2}$

The radii drawn from O in applying the $\angle AOD$ to $\angle s\ AOB$ and BOC will divide the arcs AB and BC into 3 and 2 equal arcs, respectively (P-33a).

$$\therefore \frac{\widehat{AB}}{\widehat{BC}} = \frac{3}{2}.$$

$$\text{Hence } \frac{\angle AOB}{\angle BOC} = \frac{\widehat{AB}}{\widehat{BC}} \qquad \text{(A-I).}$$

Note: In case no common unit of angle can be found for $\angle s\ AOB$ and BOC, the proposition may still be proved by using the method of limits.

Corollary to Proposition 34.—The relation expressed in Proposition 34 is usually stated as follows: *A central angle is measured by its subtended arc.*

PROPOSITION 35

Construction

To circumscribe a circle about a given triangle.

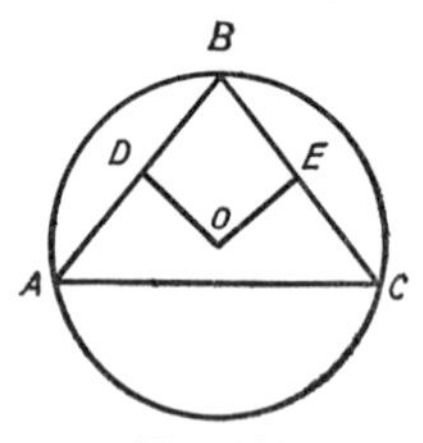

FIG. 101.

Given: $\triangle ABC$.

Required: To circumscribe a circle about $\triangle ABC$, *i.e.*, to draw a circle which will pass through the vertices A, B, and C.

At D, the mid-point of AB, erect a perpendicular to the line AB.

At E, the mid-point of BC, erect a perpendicular to the line BC.

Point O, the intersection of these two perpendiculars, is equidistant from points A and B (cor. to P-14).

That is, $AO = OB$.

Similarly, point O is equidistant from points B and C (cor. to P-14).

That is, $OB = OC$ and thus $OA = OB = OC$ (A-1).

A circle having O as a center and OA as a radius will pass through B and C (since $OA = OB = OC$) (D-35).

This is the only circle that can be drawn through the three points A, B, and C, for any other circle passing through these points must have its center on both of the perpendicular bisectors DO and OE. But there can be but one intersection of these two perpendiculars (cor. to A-IX).

Note: The foregoing relation may be stated as follows: *Through three given points, one, and only one, circle may be drawn; i.e., three points determine a circle.*

PROPOSITION 36

A straight line perpendicular to a radius at its extremity is tangent to the circle; conversely, the tangent at the extremity of a radius is perpendicular to the radius.

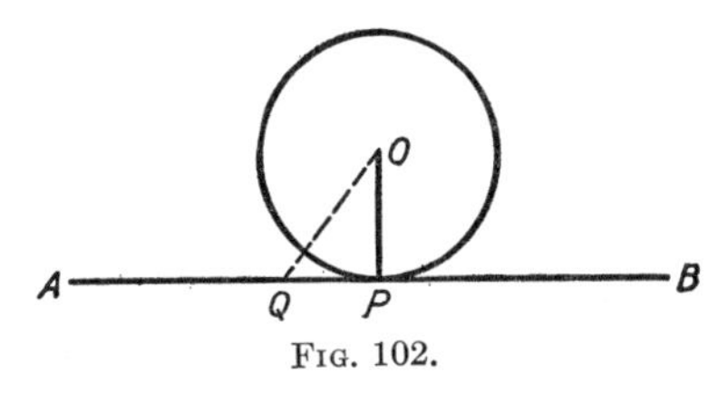

FIG. 102.

a. Given: $AB \perp OP$ (*i.e.*, $OP \perp AB$).

To prove: AB is tangent to the circle.

Draw any other line from O to the line AB, as OQ.

Then $OQ > OP$ (P-4).

$\therefore$ Q lies outside the circle (D-35).

Thus all points on AB except P lie outside the circle and hence AB is tangent to the circle (D-40).

b. Given: AB tangent to the circle at P.

To prove: $AB \perp OP$.

Since OP is the shortest line from O to the line AB (all points but the point of tangency lie outside the circle), $OP \perp AB$ or $AB \perp OP$ (P-4).

Corollary to Proposition 36. *A line perpendicular to a tangent at the point of contact passes through the center of the circle.*

PROPOSITION 37

Two lines drawn tangent to a circle from a given external point are equal and make equal angles with the line joining the point to the center of the circle.

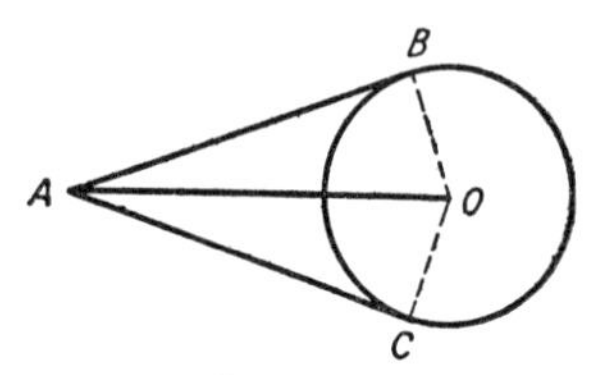

FIG. 103.

Given: Lines AB and AC drawn tangent to the circle.

To prove: $AB = AC$ and $\angle BAO = \angle CAO$.

Draw the radii OB and OC to the points of tangency.

$\angle ABO$ and $\angle ACO$ are both right angles (P-36).

$$OB = OC \qquad \text{(D-35).}$$
$$AO = AO \qquad \text{(common).}$$

$\therefore \triangle ABO \cong \triangle ACO$ (P-21).

Hence $AB = AC$ and $\angle BAO = \angle CAO$ (D-25).

Note: From the foregoing proposition, the following statement is true: *A line drawn from an external point through the center of a circle bisects the angle formed by the tangents drawn from that point.*

PROBLEMS

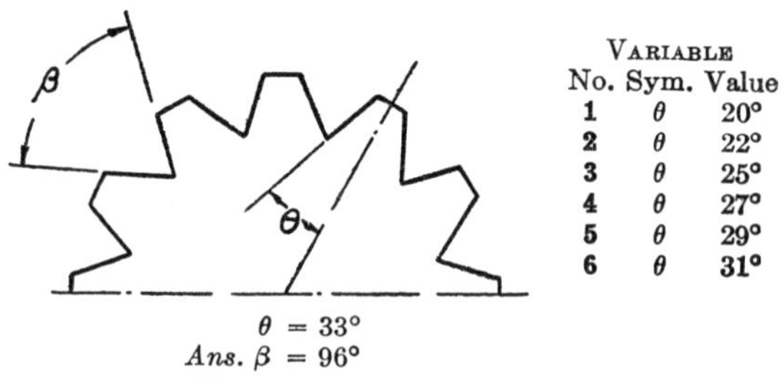

VARIABLE		
No.	Sym.	Value
1	θ	20°
2	θ	22°
3	θ	25°
4	θ	27°
5	θ	29°
6	θ	31°

$\theta = 33°$

Ans. $\beta = 96°$

The broach above has 12 teeth.

1. Determine the angle β.

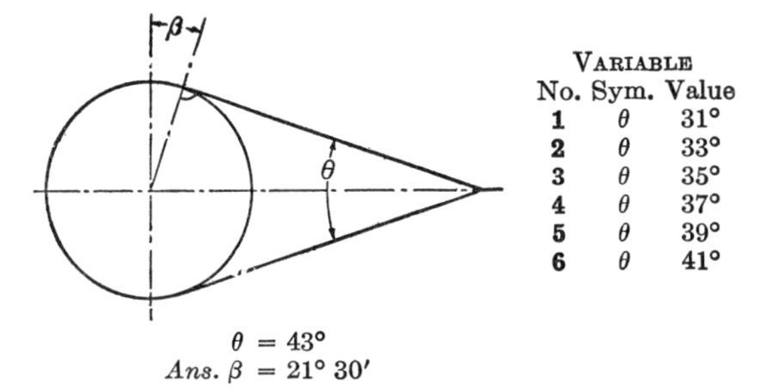

VARIABLE		
No.	Sym.	Value
1	θ	31°
2	θ	33°
3	θ	35°
4	θ	37°
5	θ	39°
6	θ	41°

$\theta = 43°$
Ans. $\beta = 21° \ 30'$

2. Determine the angle β.

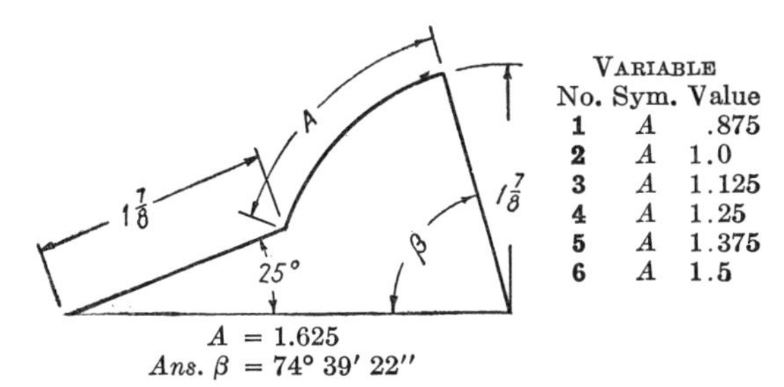

VARIABLE		
No.	Sym.	Value
1	A	.875
2	A	1.0
3	A	1.125
4	A	1.25
5	A	1.375
6	A	1.5

$A = 1.625$
Ans. $\beta = 74° \ 39' \ 22''$

3. Determine the angle β.

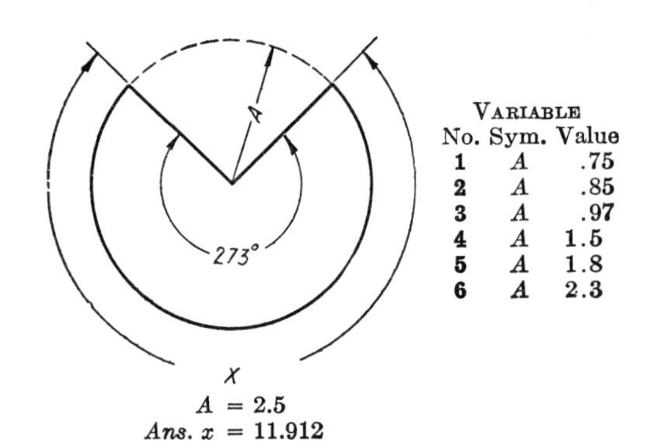

VARIABLE		
No.	Sym.	Value
1	A	.75
2	A	.85
3	A	.97
4	A	1.5
5	A	1.8
6	A	2.3

$A = 2.5$
Ans. $x = 11.912$

4. Determine the value of x.

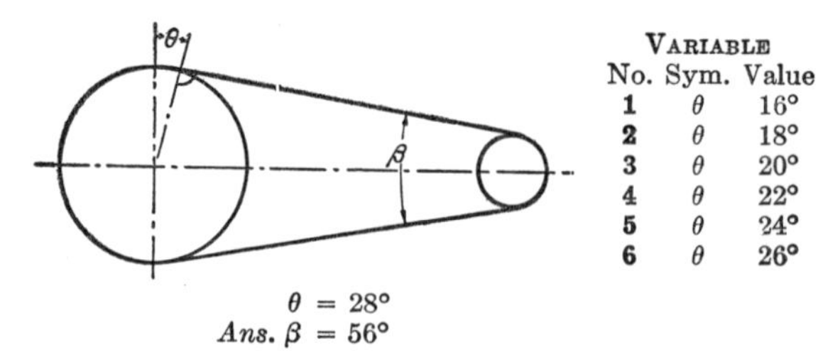

VARIABLE		
No.	Sym.	Value
1	θ	16°
2	θ	18°
3	θ	20°
4	θ	22°
5	θ	24°
6	θ	26°

$\theta = 28°$
Ans. $\beta = 56°$

5. Determine the angle β.

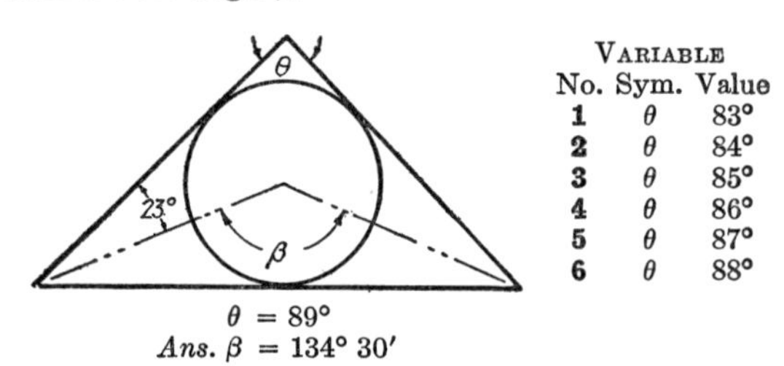

VARIABLE		
No.	Sym.	Value
1	θ	83°
2	θ	84°
3	θ	85°
4	θ	86°
5	θ	87°
6	θ	88°

$\theta = 89°$
Ans. $\beta = 134° \; 30'$

6. Determine the angle β.

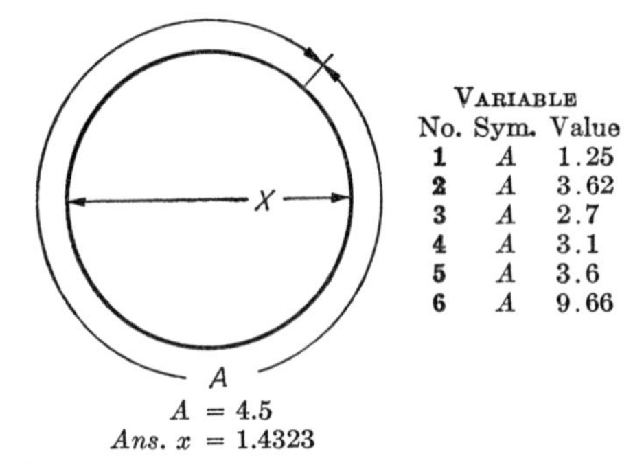

VARIABLE		
No.	Sym.	Value
1	A	1.25
2	A	3.62
3	A	2.7
4	A	3.1
5	A	3.6
6	A	9.66

$A = 4.5$
Ans. $x = 1.4323$

7. Determine the diameter x.

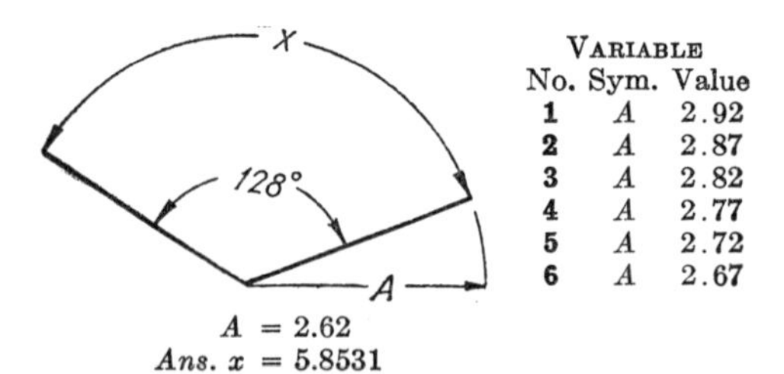

VARIABLE		
No.	Sym.	Value
1	A	2.92
2	A	2.87
3	A	2.82
4	A	2.77
5	A	2.72
6	A	2.67

$A = 2.62$
Ans. $x = 5.8531$

8. Determine the length of arc x.

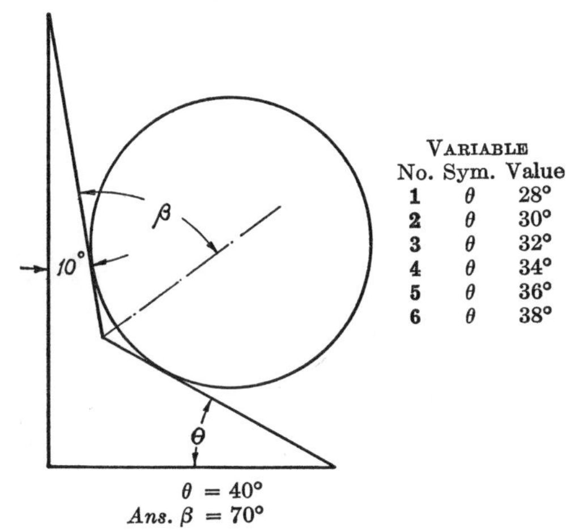

$\theta = 40°$
Ans. $\beta = 70°$

No.	Sym.	Value
1	θ	28°
2	θ	30°
3	θ	32°
4	θ	34°
5	θ	36°
6	θ	38°

9. Determine the angle β.

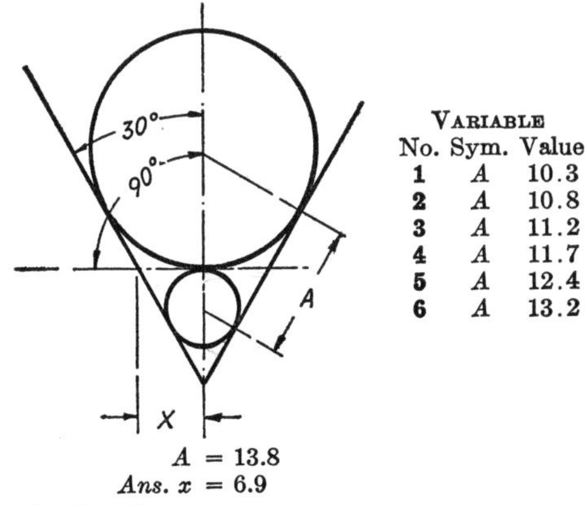

$A = 13.8$
Ans. $x = 6.9$

No.	Sym.	Value
1	A	10.3
2	A	10.8
3	A	11.2
4	A	11.7
5	A	12.4
6	A	13.2

10. Determine the distance x.

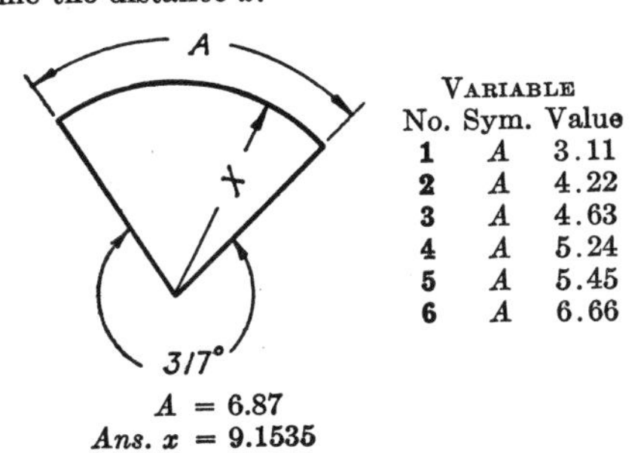

$A = 6.87$
Ans. $x = 9.1535$

No.	Sym.	Value
1	A	3.11
2	A	4.22
3	A	4.63
4	A	5.24
5	A	5.45
6	A	6.66

11. Determine the radius x.

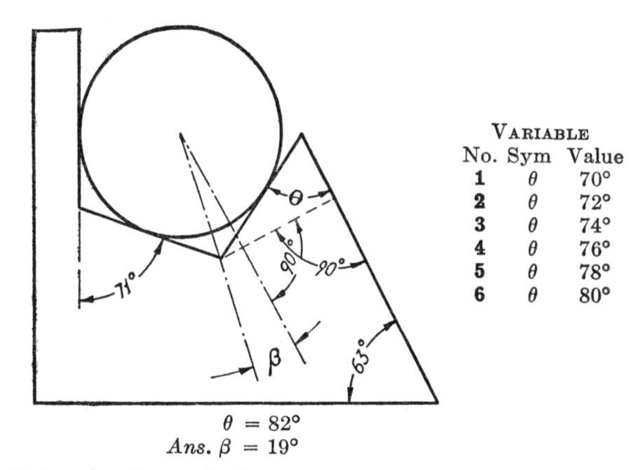

VARIABLE		
No.	Sym	Value
1	θ	70°
2	θ	72°
3	θ	74°
4	θ	76°
5	θ	78°
6	θ	80°

$\theta = 82°$
Ans. $\beta = 19°$

12. Determine the angle β.

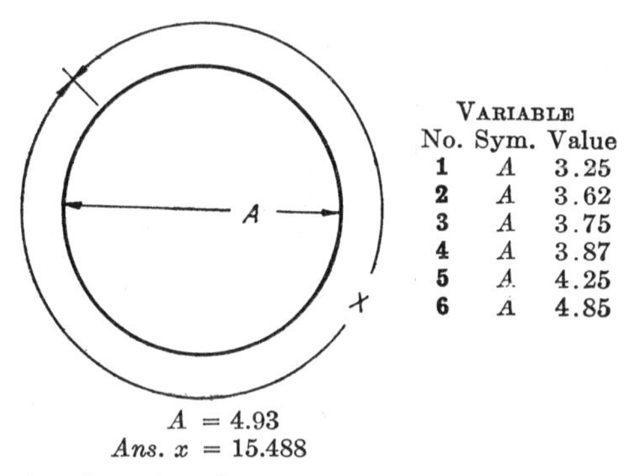

VARIABLE		
No.	Sym.	Value
1	A	3.25
2	A	3.62
3	A	3.75
4	A	3.87
5	A	4.25
6	A	4.85

$A = 4.93$
Ans. $x = 15.488$

13. Determine the value of x.

PROPOSITION 38

The diameter of a circle inscribed in a right triangle is equal to the difference obtained by subtracting the hypotenuse from the sum of the two legs.

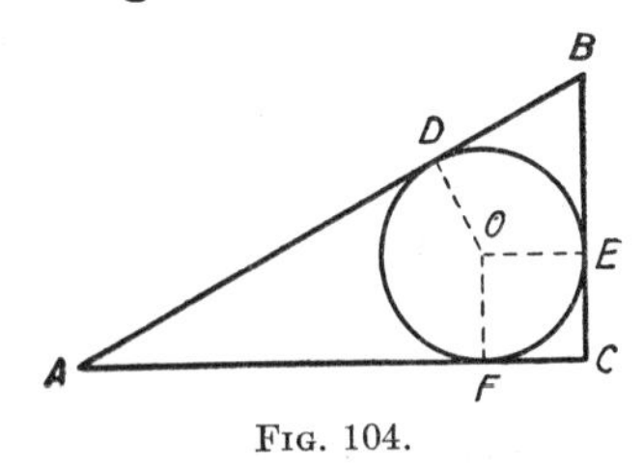

FIG. 104.

Given: Rt. $\triangle ABC$ and inscribed circle DEF.
To prove: $AC + BC - AB =$ diameter.
Draw radii OD, OE, and OF to points of tangency,
Then $AD = AF$, $BD = BE$, and $CF = CE$ (P-37).
EC and OF are parallel and FC and OE are parallel (P-36 and P-5).
$\therefore CE = OF$ and $FC = OE$ (cor. 1 to P-23).
$\therefore FC + CE = OF + OE$ (A-III).
But $OF + OE =$ diameter.
$\therefore FC + CE =$ diameter (A-1).
$AC - AD = FC$
and $BC - BD = EC$.
Adding these two equations:
$AC + BC - (AD + BD) = FC + EC$ (A-III).
$AC + BC - AB = FC + EC$ (A-VIII).
$\therefore AC + BC - AB =$ diameter (A-II).

PROPOSITION 39

Any diameter of a circle, which is perpendicular to a chord, bisects the chord and the arc subtended by it.

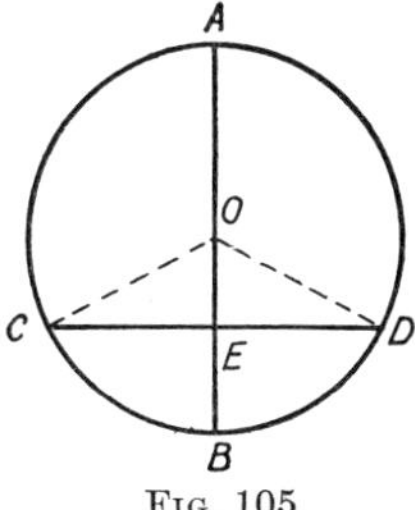

Fig. 105.

Given: Diameter $AB \perp$ chord CD.

To prove: $CE = ED$, $\widehat{CB} = \widehat{BD}$, and $\widehat{AC} = \widehat{AD}$.
Draw radii OC and OD.
Rt. $\triangle CEO \cong$ rt. $\triangle DEO$ (P-21).
$\therefore CE = ED$ (D-25).
$\angle COE = \angle DOE$ (D-25).
$\therefore \widehat{CB} = \widehat{DB}$ (P-33)
and $\widehat{AC} = \widehat{AD}$ (A-IV).

Corollary to Proposition 39.—*The perpendicular bisector of a chord passes through the center of the circle.*

There can be only one perpendicular bisector of a chord, and from (P-39) the diameter is a perpendicular bisector of the chord.

PROPOSITION 40

An inscribed angle is measured by one-half of its intercepted arc.

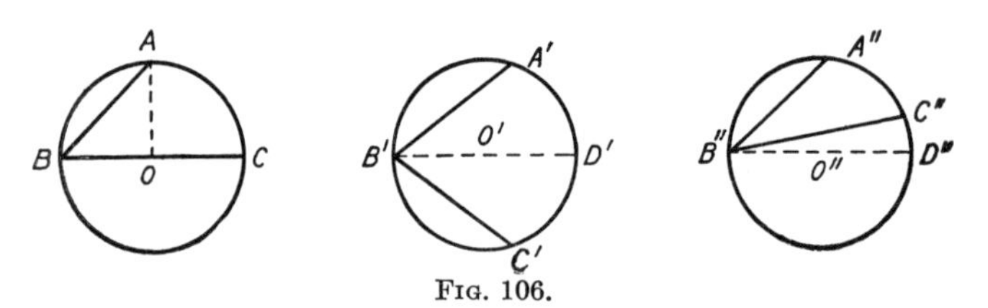

FIG. 106.

Given: Inscribed $\angle ABC$ (or $\angle A'B'C'$ or $\angle A''B''C''$).

To prove: That the inscribed angle is measured by one-half the intercepted arc AC (or $A'C'$ or $A''C''$).

Case I.—One side of the angle is a diameter.

Draw OA.

$\angle AOC$ is measured by $\widehat{AC}$ (cor. to P-34).
$\angle AOC = \angle OAB + \angle CBA$ (P-13).
$\angle OAB = \angle CBA$ (P-17).
$\therefore \angle AOC = 2\angle CBA$ (A-II)
or $\angle CBA = \frac{1}{2}\angle AOC.$

$\therefore \angle CBA$ is measured by $\frac{1}{2}\widehat{AC}$.

Case II.—Sides of the angle are on opposite sides of the center of the circle.

Draw diameter $B'D'$.

Then from Case I, $\angle C'B'D'$ is measured by $\frac{1}{2}\widehat{C'D'}$.

$\angle D'B'A'$ is measured by $\frac{1}{2}\widehat{A'D'}$.

Hence $\angle A'B'C'$ is measured by $\frac{1}{2}\widehat{A'C'}$. (A-III).

Case III.—Sides of the angle are on the same side of the center of the circle.

The proof in this case is left to the student.
(*Hint:* Draw diameter $B''D''$.)

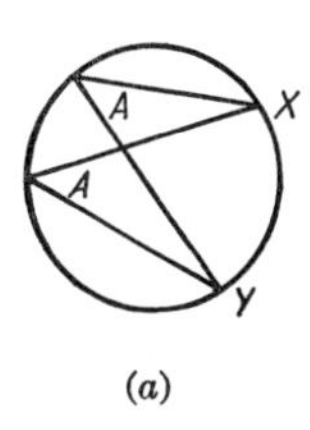

(a)

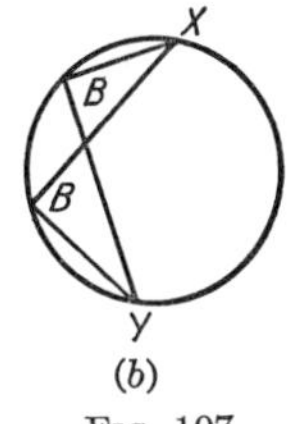

(b)

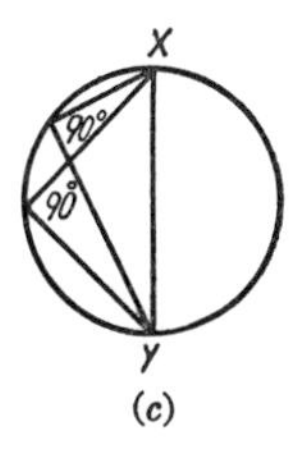

(c)

FIG. 107.

Corollary 1 to Proposition 40.—*All inscribed angles subtending the same arc are equal* (see Figs. 107*a* and *b*).

Corollary 2 to Proposition 40.—*An inscribed angle in a semicircle is a right angle* (see Fig. 107*c*).

PROBLEMS

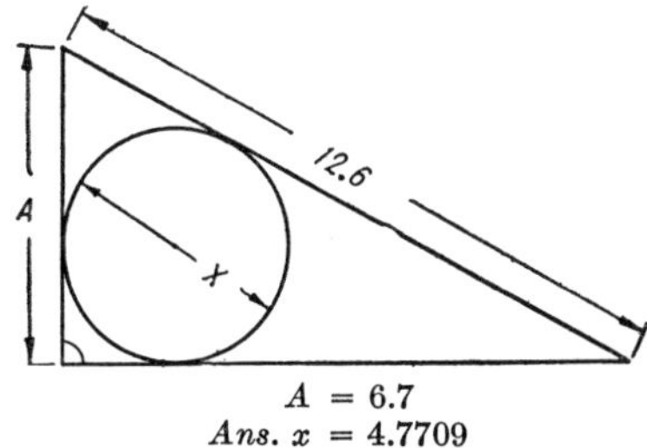

$A = 6.7$
Ans. $x = 4.7709$

VARIABLE		
No.	Sym.	Value
1	A	4.9
2	A	5.1
3	A	5.3
4	A	5.5
5	A	7.1
6	A	6.9

1. Determine the diameter x.

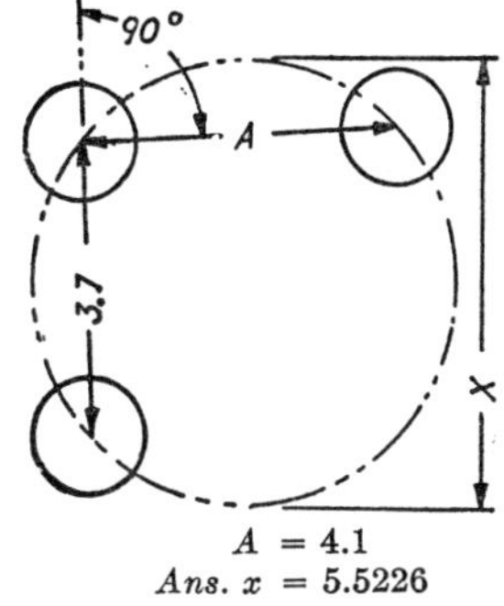

$A = 4.1$
Ans. $x = 5.5226$

VARIABLE		
No.	Sym.	Value
1	A	2.5
2	A	2.8
3	A	3.1
4	A	3.3
5	A	3.5
6	A	3.9

2. Determine the distance x.

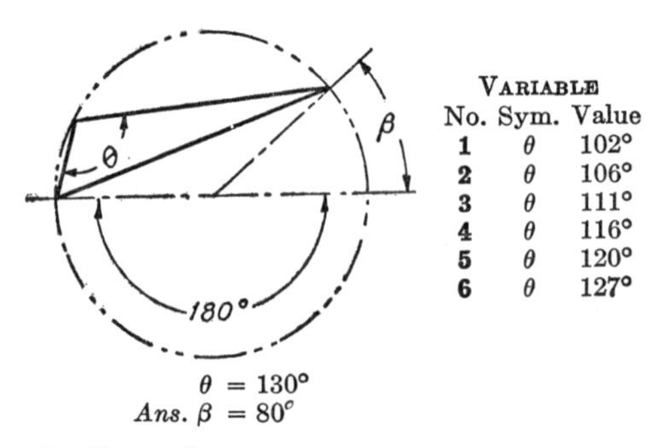

Variable		
No.	Sym.	Value
1	θ	102°
2	θ	106°
3	θ	111°
4	θ	116°
5	θ	120°
6	θ	127°

$\theta = 130°$
Ans. $\beta = 80°$

3. Determine the angle β.

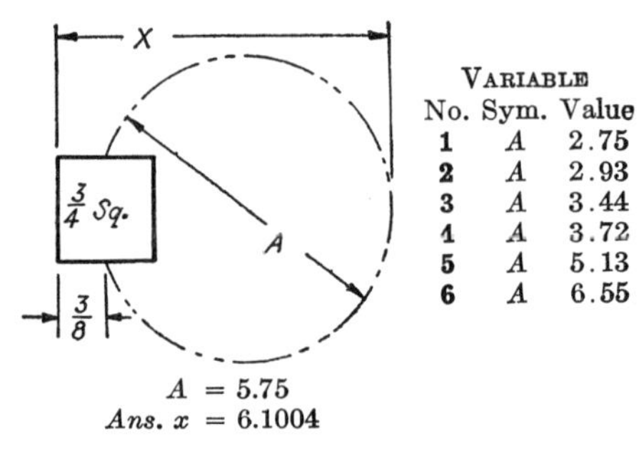

Variable		
No.	Sym.	Value
1	A	2.75
2	A	2.93
3	A	3.44
4	A	3.72
5	A	5.13
6	A	6.55

$A = 5.75$
Ans. $x = 6.1004$

4. Determine the distance x.

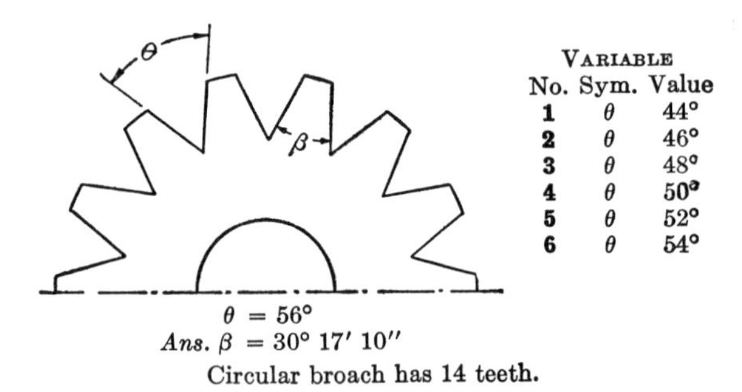

Variable		
No.	Sym.	Value
1	θ	44°
2	θ	46°
3	θ	48°
4	θ	50°
5	θ	52°
6	θ	54°

$\theta = 56°$
Ans. $\beta = 30° \; 17' \; 10''$

Circular broach has 14 teeth.

5. Determine the angle β.

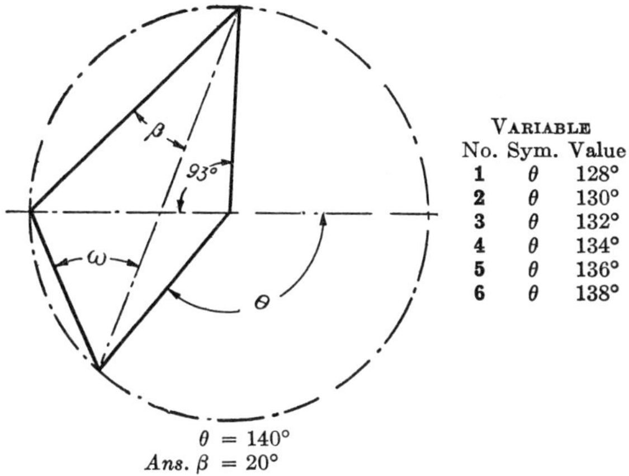

VARIABLE		
No.	Sym.	Value
1	θ	128°
2	θ	130°
3	θ	132°
4	θ	134°
5	θ	136°
6	θ	138°

$\theta = 140°$
Ans. $\beta = 20°$

6. Determine the angle β.

7. Determine the angle ω.

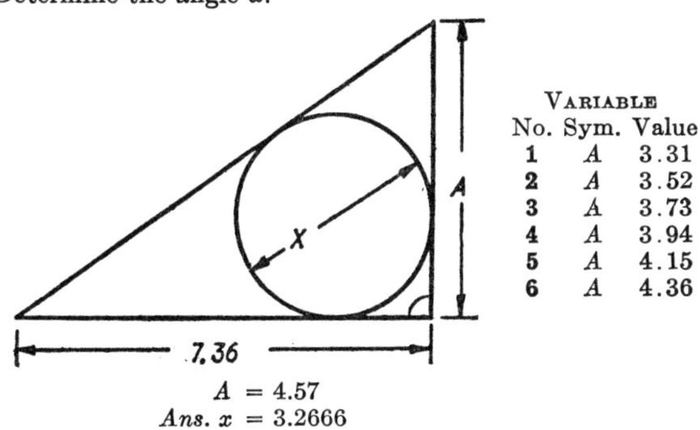

VARIABLE		
No.	Sym.	Value
1	A	3.31
2	A	3.52
3	A	3.73
4	A	3.94
5	A	4.15
6	A	4.36

$A = 4.57$
Ans. $x = 3.2666$

8. Determine the diameter x.

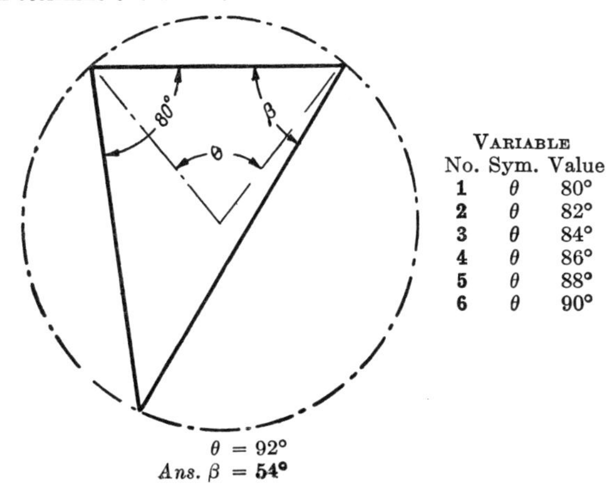

VARIABLE		
No.	Sym.	Value
1	θ	80°
2	θ	82°
3	θ	84°
4	θ	86°
5	θ	88°
6	θ	90°

$\theta = 92°$
Ans. $\beta = \mathbf{54°}$

9. Determine the angle β.

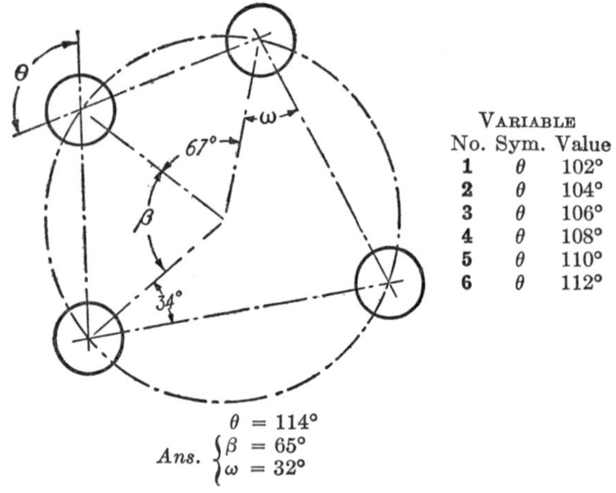

No.	Sym.	Value
1	θ	102°
2	θ	104°
3	θ	106°
4	θ	108°
5	θ	110°
6	θ	112°

$\theta = 114°$

Ans. $\begin{cases} \beta = 65° \\ \omega = 32° \end{cases}$

10. Determine the angle β.
11. Determine the angle ω.

PROPOSITION 41

Construction

To inscribe a square in a given circle.

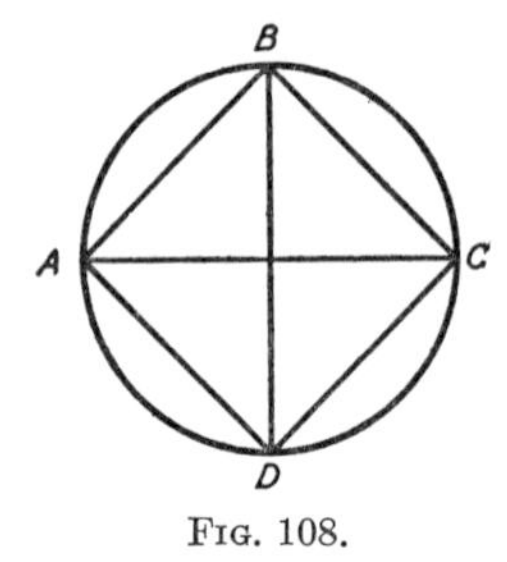

Fig. 108.

Given: Circle $ABCD$.

Required: To inscribe a square, *i.e.*, to draw a square whose vertices lie in the circumference of the circle.

Draw any diameter AC and another diameter BD perpendicular to AC.

Each of ∡ ABC, BCD, CDA, and DAB is inscribed in a semicircle and therefore each is a right angle (cor. 2 to P-40).

Hence, $ABCD$ is a square (D-22).

PROPOSITION 42

Construction

To inscribe a regular hexagon in a given circle.

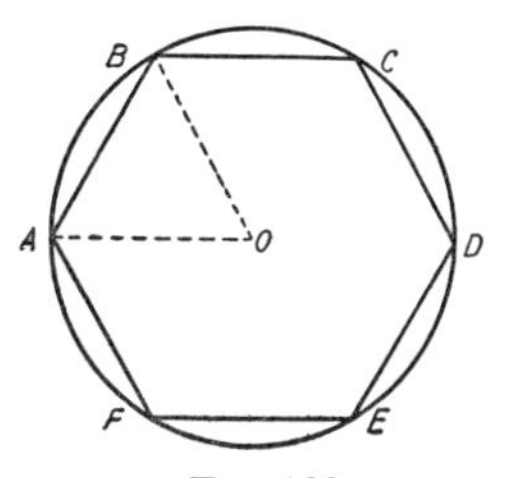

FIG. 109.

Given: Circle $ABCDEFA$.

Required: To inscribe a hexagon in the circle.

From any point A draw an arc using the radius of the circle OA as a radius. This arc will cut the circle at some point B.

Draw OB.

$\triangle AOB$ is equilateral ($AB = OA$ by hyp.).

$\therefore \triangle AOB$ is equiangular (cor. 2 to P-17).

Hence $\angle AOB = 60°$ (P-10).

$60°$ is $\frac{1}{6}$ of $360°$ so arc AB is one-sixth of the circumference (P-34).

Hence AB fits into the circumference just six times.

$\therefore ABCDEFA$ is a regular hexagon (D-20 and D-22).

PROPOSITION 43

If two circles are tangent to each other externally or internally, the line of centers passes through the point of tangency.

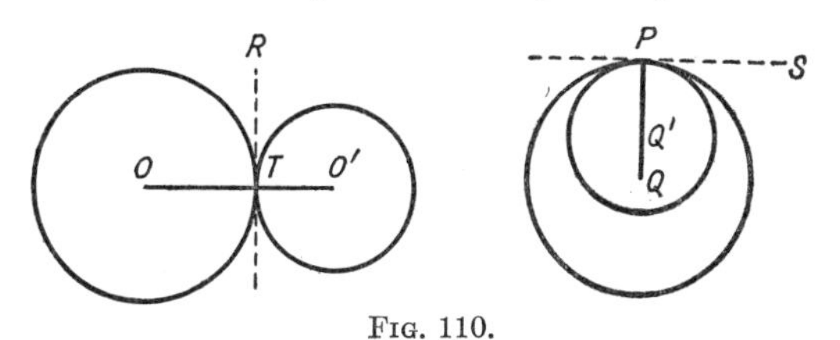

FIG. 110.

Case I. *Given:* Circles O and O' tangent externally at T.

To prove: Line OO' passes through T.

Draw the line of tangency TR and the radii OT and $O'T$.
$\measuredangle OTR$ and $O'TR$ are right angles (P-36).
Hence $\measuredangle OTR$ and $O'TR$ are supplementary (D-11).
$\therefore OTO'$ is a straight line.
That is, OO' passes through T.

Case II. *Given:* Circles Q and Q' tangent internally at P.
To prove: Line QQ' passes through P.
Draw the line of tangency PS and the radii QP and $Q'P$.
$PQ \perp PS$ and $PQ' \perp PS$ (P-36).
Hence PQ and PQ' coincide.
That is, QQ' passes through P.

PROPOSITION 44

The angle between two secants intersecting outside a circumference, the angle between an intersecting tangent and a secant, and the angle between two intersecting tangents are each measured by one-half the difference of the intercepted arcs.

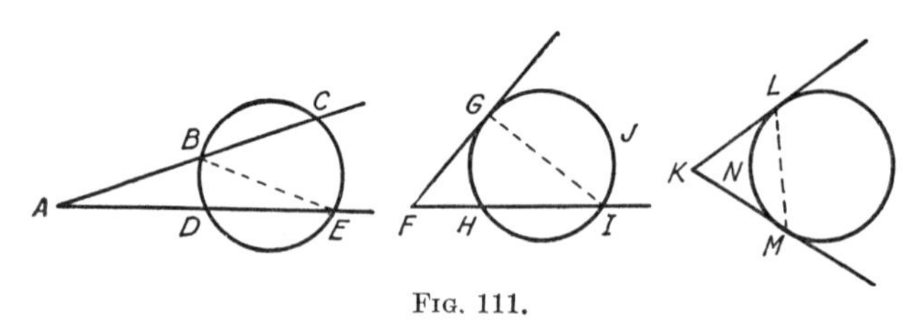

Fig. 111.

Case I. *Given:* Two secants AC and AE intersecting at point A outside the circumference.

To prove: $\angle A$ is measured by $\frac{1}{2}(\overset{\frown}{CE} - \overset{\frown}{BD})$.
Draw BE.

$\angle CBE$ is measured by $\frac{1}{2}\overset{\frown}{CE}$ (P-40).

$\angle DEB$ is measured by $\frac{1}{2}\overset{\frown}{BD}$ (P-40).
$\angle A = \angle CBE - \angle DEB$ (P-13).

$\therefore \angle A$ is measured by $\frac{1}{2}\overset{\frown}{CE} - \frac{1}{2}\overset{\frown}{BD}$ (A-II).

That is, $\angle A$ is measured by $\frac{1}{2}(\overset{\frown}{CE} - \overset{\frown}{BD})$.

Case II. *Given:* Tangent FG and secant FI intersecting at F.

To prove: $\angle F$ is measured by $\frac{1}{2}(\widehat{GJI} - \widehat{GH})$.

The proof is exactly like that of Case I, so it is left to the student.

Case III. Let the student tell what is given, what is to be proved, and supply the proof.

PROPOSITION 45

If two chords of a circle intersect within the circle, the product of the two segments of the one is equal to the product of the two segments of the other.

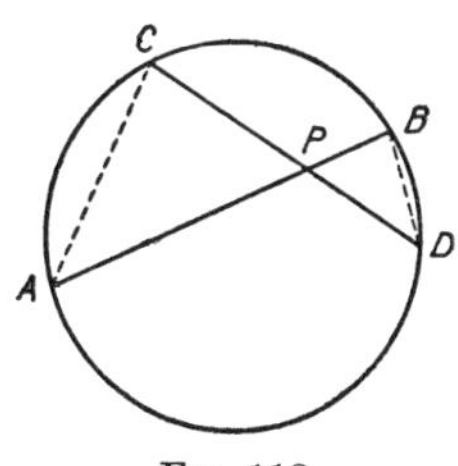

Fig. 112.

Given: Two chords AB and CD intersecting each other within the circle at point P.

To prove: $PA \times PB = PC \times PD$.

Draw AC and BD.

In the $\triangle APC$ and BPD,

$\angle CPA = \angle BPD$ (P-1).

$\angle ACD = \angle ABD$ (cor. 1 to P-40).

$\therefore \triangle APC \sim \triangle BPD$ (cor. 1 to P-26).

Hence $\dfrac{PC}{PB} = \dfrac{PA}{PD}$ (D-34).

$\therefore PA \times PB = PC \times PD$ (Theorem 1 of Chap. V).

PROPOSITION 46

The product of two sides of a triangle is equal to the product of the diameter of a circumscribed circle and the altitude upon the third side.

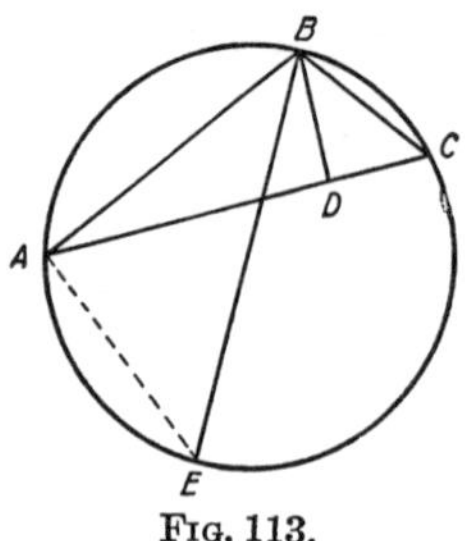

FIG. 113.

Given: $\triangle ABC$ with $BD \perp AC$ and the circumscribed circle $ABCE$ with diameter BE.

To prove: $AB \times BC = BE \times BD$.

Draw AE.

In $\triangle s$ ABE and BCD,

$\angle BAE$ is a right angle (cor. 2 to P-40).

and $\therefore \angle BAE = \angle BDC$ (both rt. $\angle$s).

$\angle BCA = \angle BEA$ (cor. 1 to P-40).

$\therefore \triangle ABE \sim \triangle BCD$ (cor. 2 to P-26).

$\therefore \frac{AB}{BD} = \frac{BE}{BC}$ (D-34)

and $AB \times BC = BE \times BD$ (Theorem I of Chap. V).

PROBLEMS

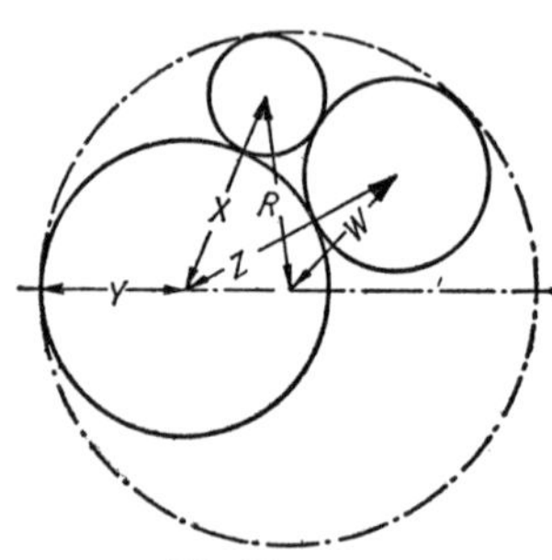

NO VARIABLE

1. State which of the foregoing lines pass through points of contact of the large circle. Extend the lines if necessary.

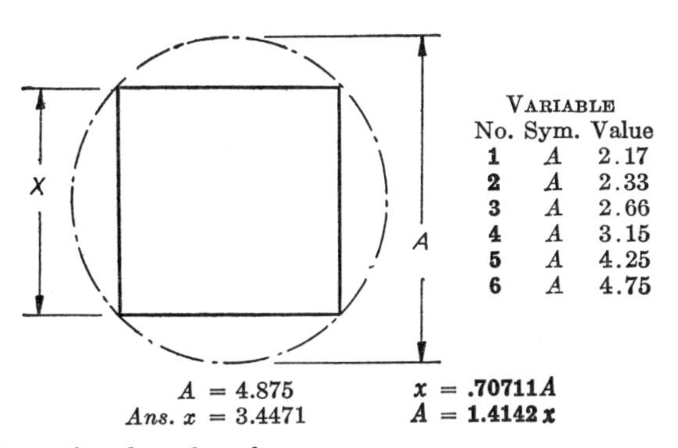

No.	Sym.	Value
Variable		
1	A	2.17
2	A	2.33
3	A	2.66
4	A	3.15
5	A	4.25
6	A	4.75

$A = 4.875$ $\quad$ $x = .70711A$

Ans. $x = 3.4471$ $\quad$ $A = 1.4142x$

2. Determine the value of x.

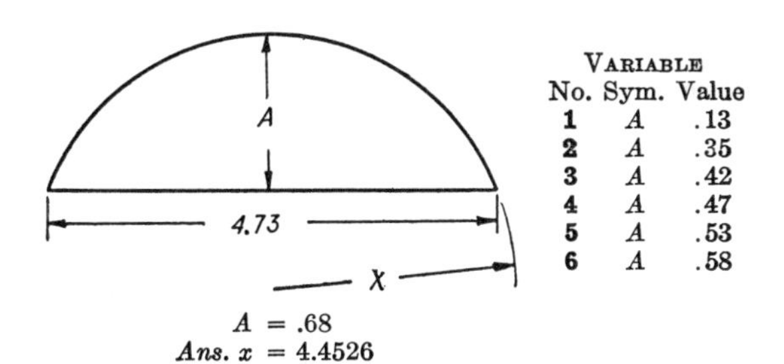

No.	Sym.	Value
Variable		
1	A	.13
2	A	.35
3	A	.42
4	A	.47
5	A	.53
6	A	.58

$A = .68$

Ans. $x = 4.4526$

3. Determine the distance x.

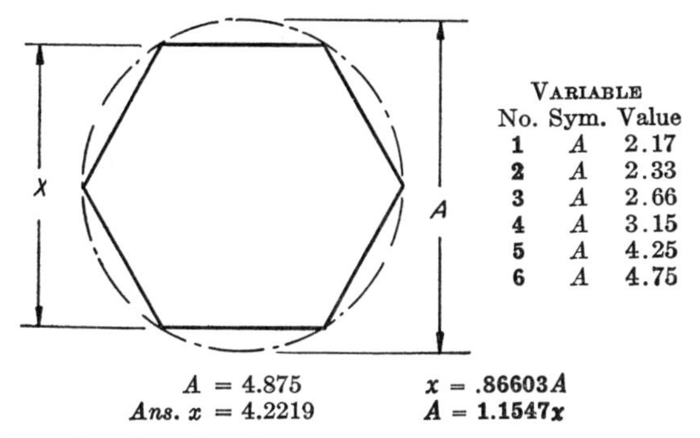

No.	Sym.	Value
Variable		
1	A	2.17
2	A	2.33
3	A	2.66
4	A	3.15
5	A	4.25
6	A	4.75

$A = 4.875$ $\quad$ $x = .86603A$

Ans. $x = 4.2219$ $\quad$ $A = 1.1547x$

4. Determine the value of x.

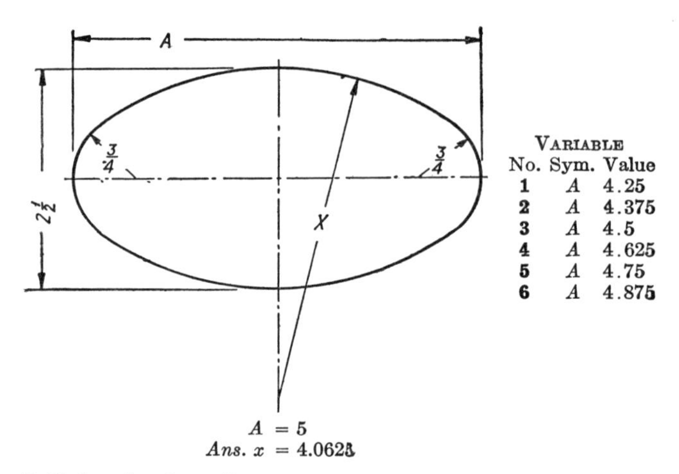

Variable		
No.	Sym.	Value
1	A	4.25
2	A	4.375
3	A	4.5
4	A	4.625
5	A	4.75
6	A	4.875

$A = 5$
Ans. $x = 4.0625$

5. Determine the radius x.

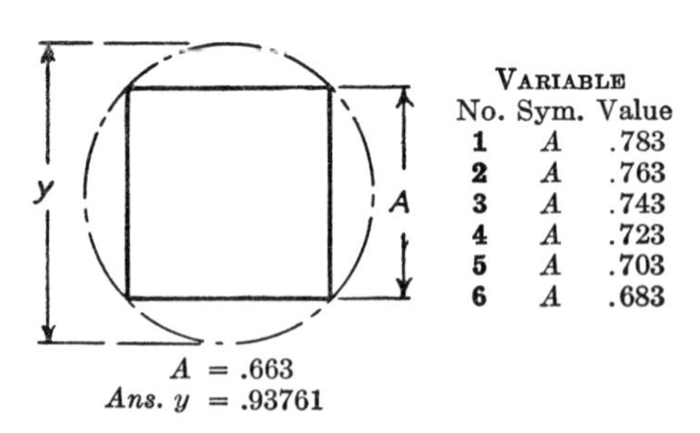

Variable		
No.	Sym.	Value
1	A	.783
2	A	.763
3	A	.743
4	A	.723
5	A	.703
6	A	.683

$A = .663$
Ans. $y = .93761$

6. Determine the value of y.

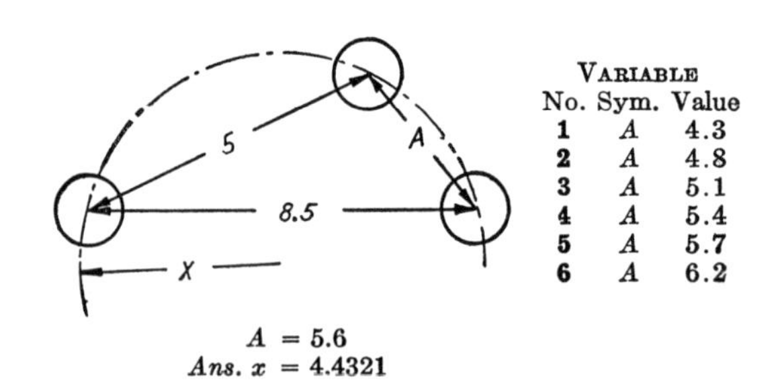

Variable		
No.	Sym.	Value
1	A	4.3
2	A	4.8
3	A	5.1
4	A	5.4
5	A	5.7
6	A	6.2

$A = 5.6$
Ans. $x = 4.4321$

7. Determine the radius x.

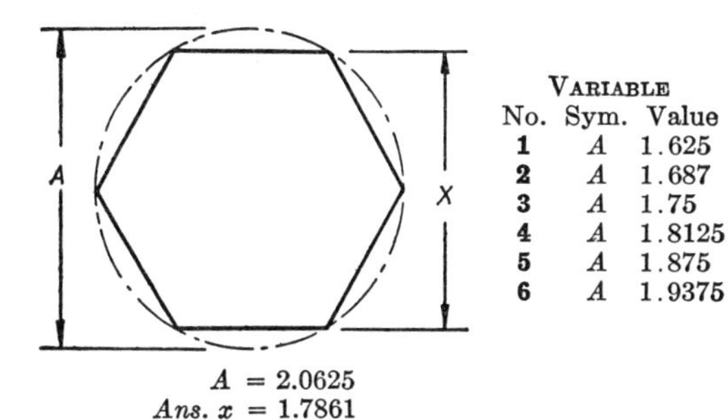

VARIABLE		
No.	Sym.	Value
1	A	1.625
2	A	1.687
3	A	1.75
4	A	1.8125
5	A	1.875
6	A	1.9375

$A = 2.0625$
Ans. $x = 1.7861$

8. Determine the value of x.

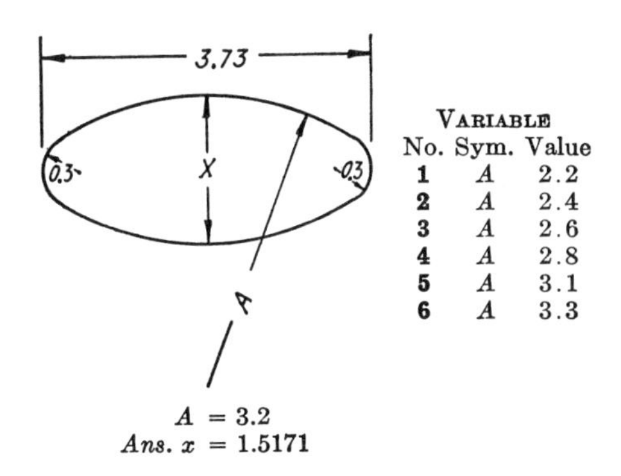

VARIABLE		
No.	Sym.	Value
1	A	2.2
2	A	2.4
3	A	2.6
4	A	2.8
5	A	3.1
6	A	3.3

$A = 3.2$
Ans. $x = 1.5171$

9. Determine the distance x.

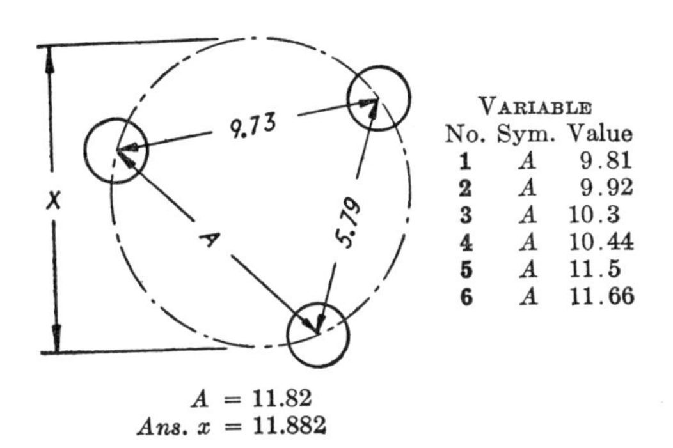

VARIABLE		
No.	Sym.	Value
1	A	9.81
2	A	9.92
3	A	10.3
4	A	10.44
5	A	11.5
6	A	11.66

$A = 11.82$
Ans. $x = 11.882$

10. Determine the distance x.

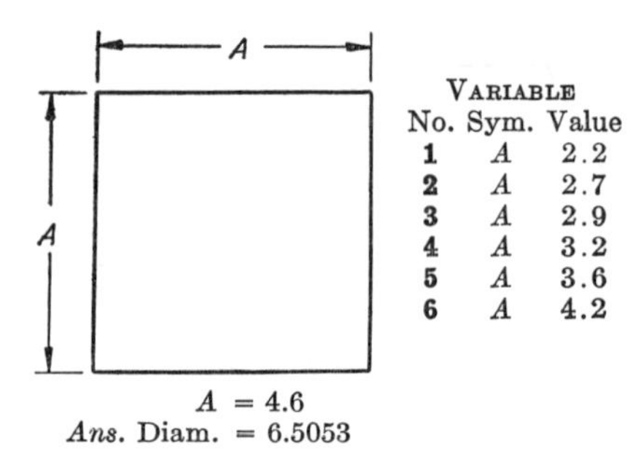

VARIABLE		
No.	Sym.	Value
1	A	2.2
2	A	2.7
3	A	2.9
4	A	3.2
5	A	3.6
6	A	4.2

$A = 4.6$
Ans. Diam. $= 6.5053$

11. Determine the diameter of a circle that will circumscribe the given square.

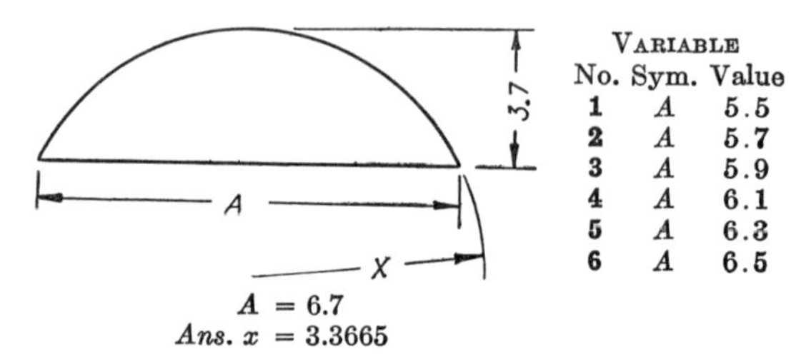

VARIABLE		
No.	Sym.	Value
1	A	5.5
2	A	5.7
3	A	5.9
4	A	6.1
5	A	6.3
6	A	6.5

$A = 6.7$
Ans. $x = 3.3665$

12. Determine the radius x.

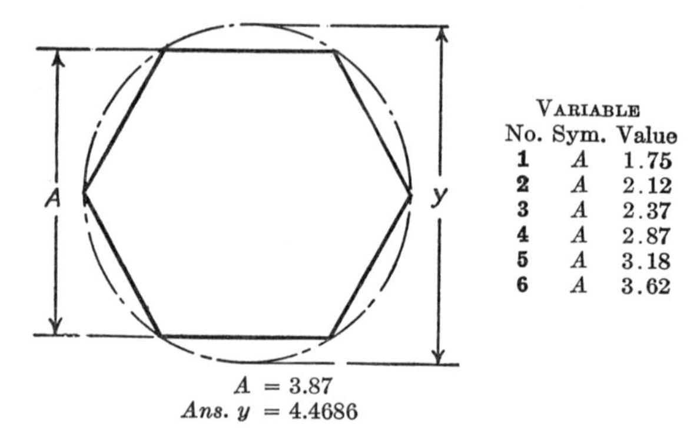

VARIABLE		
No.	Sym.	Value
1	A	1.75
2	A	2.12
3	A	2.37
4	A	2.87
5	A	3.18
6	A	3.62

$A = 3.87$
Ans. $y = 4.4686$

13. Determine the value of y.

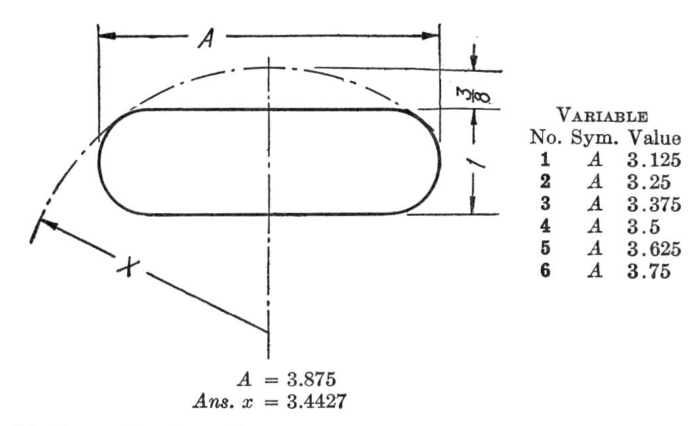

VARIABLE		
No.	Sym.	Value
1	A	3.125
2	A	3.25
3	A	3.375
4	A	3.5
5	A	3.625
6	A	3.75

$A = 3.875$
Ans. $x = 3.4427$

14. Determine the radius x.

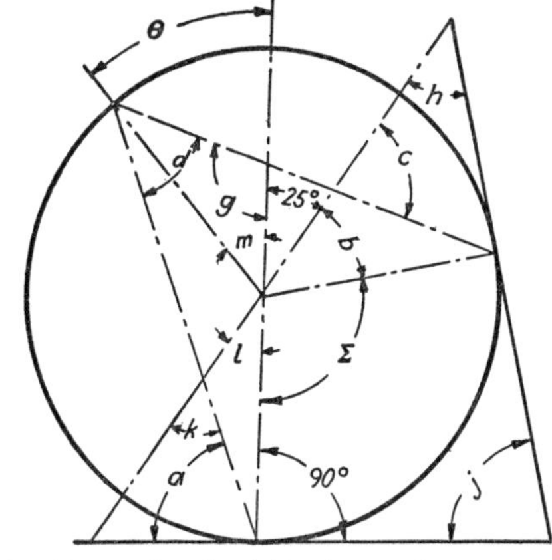

VARIABLE		
No.	Sym.	Value
1	θ	36°
2	θ	38°
3	θ	40°
4	θ	42°
5	θ	44°
6	θ	46°

VARIABLE		
No.	Sym.	Value
1	Σ	95°
2	Σ	97°
3	Σ	99°
4	Σ	101°
5	Σ	103°
6	Σ	105°

15. Determine the following angles.

$a, \quad b, \quad c, \quad d, \quad g, \quad h, \quad j, \quad k, \quad l, \quad m.$

FORMULAS FOR THE AREAS OF VARIOUS PLANE FIGURES

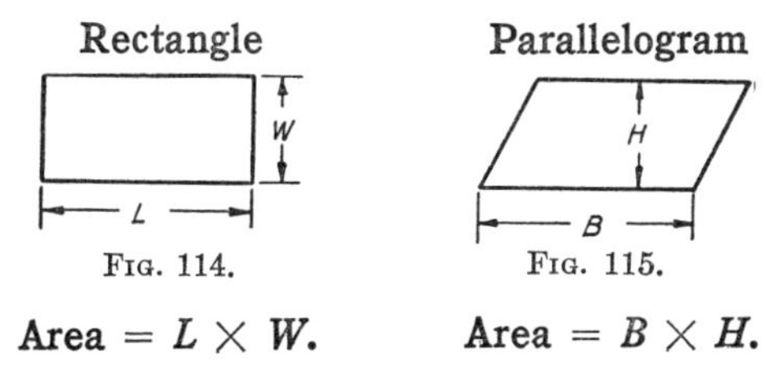

FIG. 114. FIG. 115.

Area $= L \times W$. **Area** $= B \times H$.

Triangle

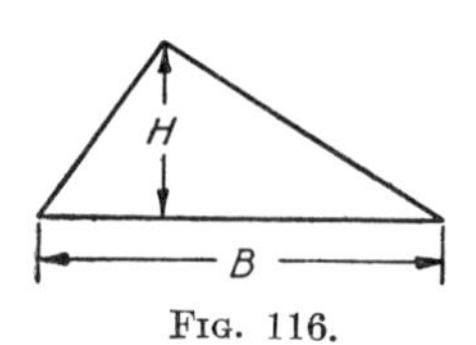

FIG. 116.

Area $= \frac{1}{2}B \times H.$

Trapezoid

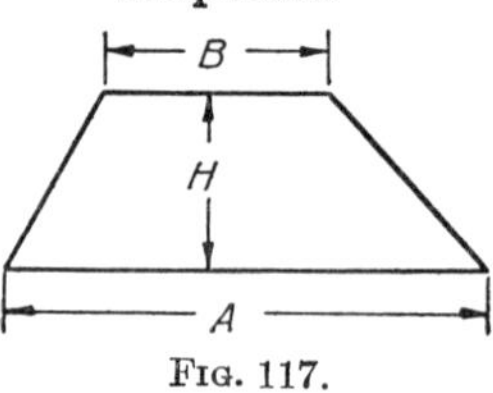

FIG. 117.

Area $= \frac{1}{2}(A + B)H.$

Circle

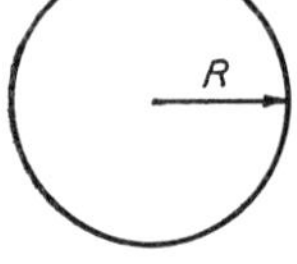

FIG. 118.

$$\text{Area} = \pi R^2 = \pi\left(\frac{D}{2}\right)^2 = \frac{\pi D^2}{4} = .7854D^2.$$

Circular Sector

FIG. 119.

$$\text{Area} = \frac{\beta}{360} \times \pi R^2 = .008727\beta R^2.$$

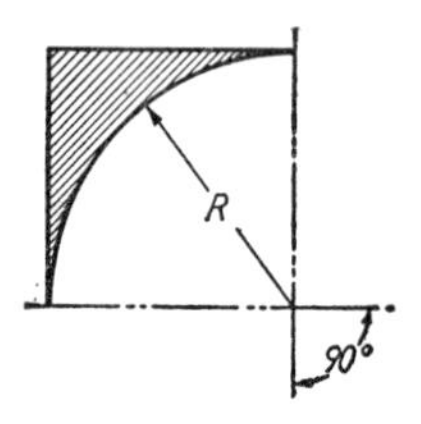

FIG. 120.

$$\text{Area (of shaded portion)} = R^2 - \tfrac{1}{4}\pi R^2 = (1 - \tfrac{1}{4}\pi)R^2 = .2146R^2.$$

FORMULAS FOR THE VOLUMES OF VARIOUS SOLID FIGURES

Rectangular Parallelepiped

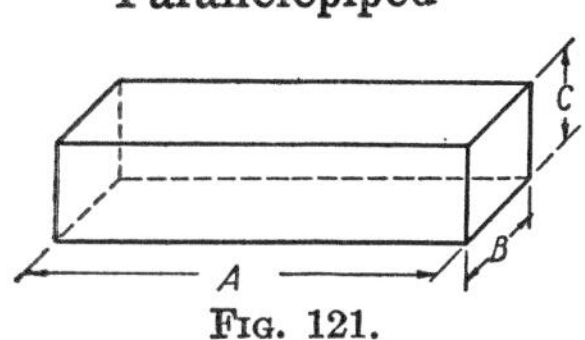

FIG. 121.

Volume = base × height
= ABC.

Oblique Parallelepiped

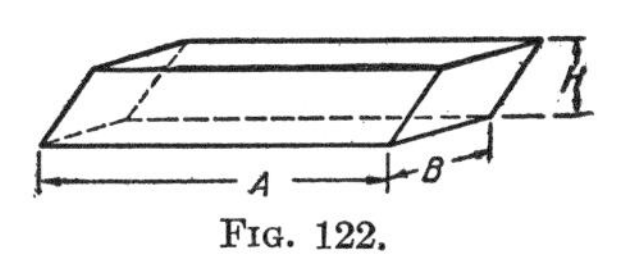

FIG. 122.

Volume = base × height
= ABH.

Right Circular Cylinder

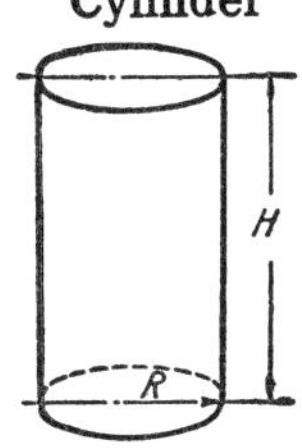

FIG. 123.

Lateral surface area = $2\pi RH$.

Total surface area = $2\pi R^2 + 2\pi RH$
= $2\pi R(R + H)$.

Volume = base × altitude
= πR^2H.

Oblique Circular Cylinder

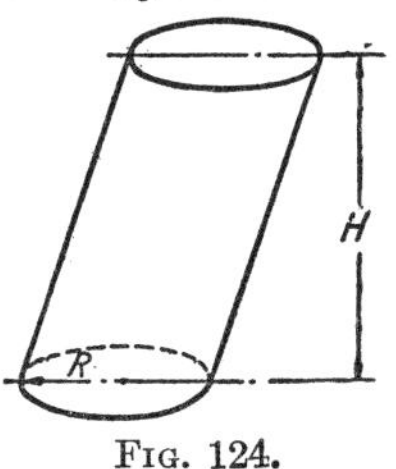

FIG. 124.

Volume = base × altitude
= πR^2H.

Note: The volume of any solid having the lateral elements parallel is equal to the product of the area of the base and the altitude. This applies to all types of rods.

Right Circular Cone

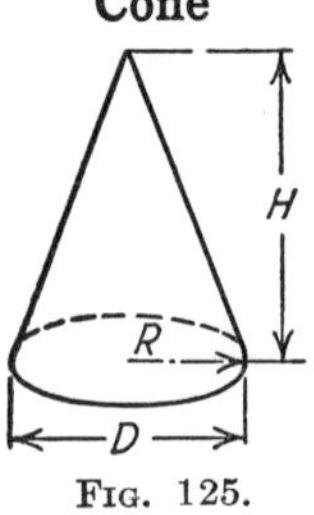

FIG. 125.

$$\text{Vol.} = \tfrac{1}{3}\text{ base} \times \text{altitude}$$
$$= \tfrac{1}{3}\pi R^2 H$$
$$= .2618D^2H.$$

Frustum of a Right Circular Cone

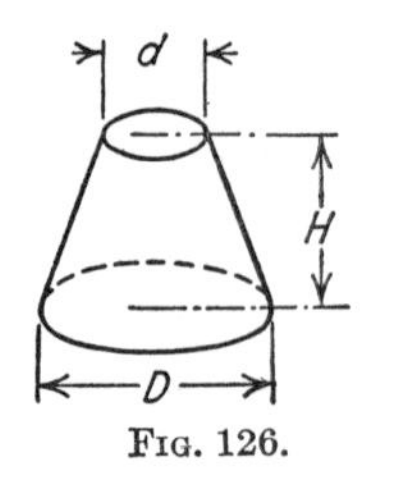

FIG. 126.

$$\text{Vol.} = .2618H(D^2 + d^2 + Dd).$$

Sphere

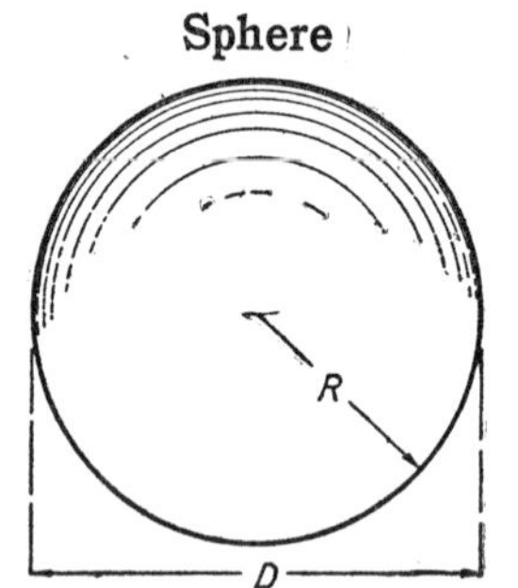

FIG. 127.

$$\text{Surface area} = 4\pi R^2 = 12.566R^2$$
$$\text{Volume} = \tfrac{4}{3}\pi R^3 = 4.1888R^3$$
$$= \tfrac{1}{6}\pi D^3 = .5236D^3.$$

The number of gallons in a given volume is expressed by the following formula.

$$\text{Volume (in gallons)} = \frac{\text{Volume (in cubic inches)}}{231}.$$

PROBLEMS

Some of the following problems are direct applications of the formulas for areas and volumes, but some of the figures are seen to consist of combinations of the single figures, and the total area or volume must be obtained by solving for areas or volumes of the separate simple units and adding.

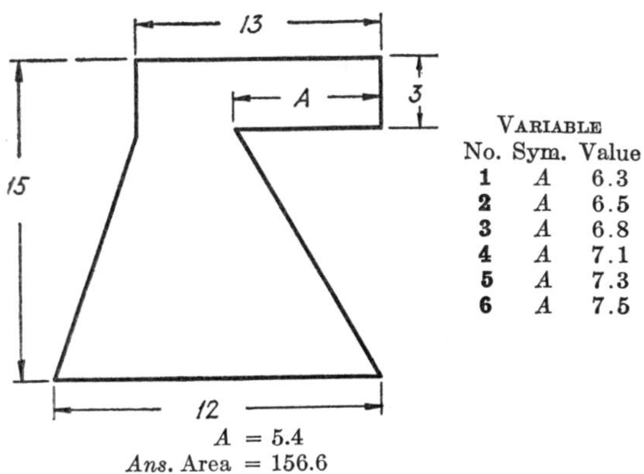

No.	Sym.	Value
1	A	6.3
2	A	6.5
3	A	6.8
4	A	7.1
5	A	7.3
6	A	7.5

$A = 5.4$
Ans. Area $= 156.6$

1. Determine the area.

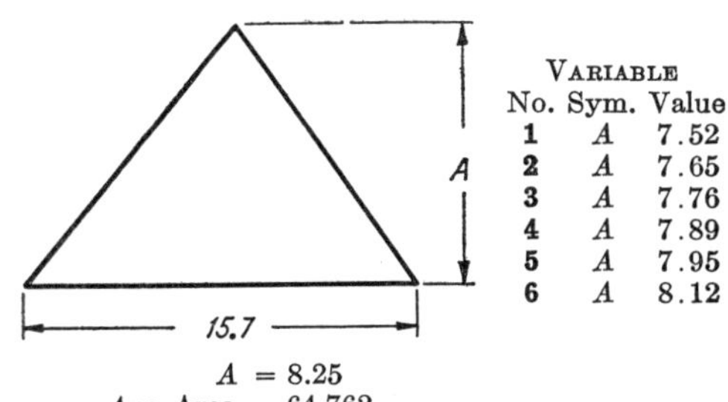

No.	Sym.	Value
1	A	7.52
2	A	7.65
3	A	7.76
4	A	7.89
5	A	7.95
6	A	8.12

$A = 8.25$
Ans. Area $= 64.762$

2. Determine the area.

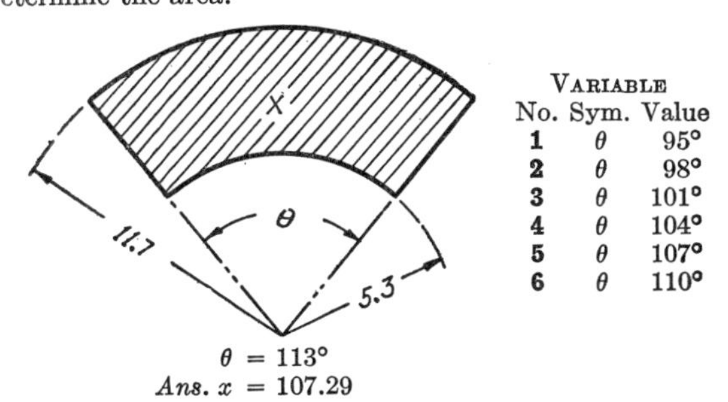

No.	Sym.	Value
1	θ	95°
2	θ	98°
3	θ	101°
4	θ	104°
5	θ	107°
6	θ	110°

$\theta = 113°$
Ans. $x = 107.29$

3. Determine the area x.

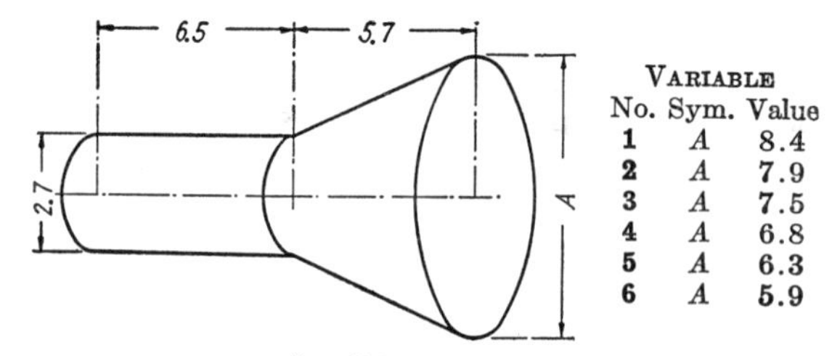

VARIABLE		
No.	Sym.	Value
1	*A*	8.4
2	*A*	7.9
3	*A*	7.5
4	*A*	6.8
5	*A*	6.3
6	*A*	5.9

$A = 5.6$
Ans. Volume = 117.451

4. Determine the volume.

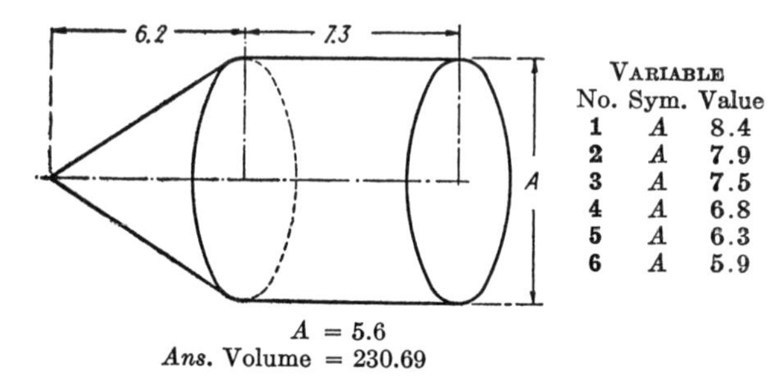

VARIABLE		
No.	Sym.	Value
1	*A*	8.4
2	*A*	7.9
3	*A*	7.5
4	*A*	6.8
5	*A*	6.3
6	*A*	5.9

$A = 5.6$
Ans. Volume = 230.69

5. Determine the volume.

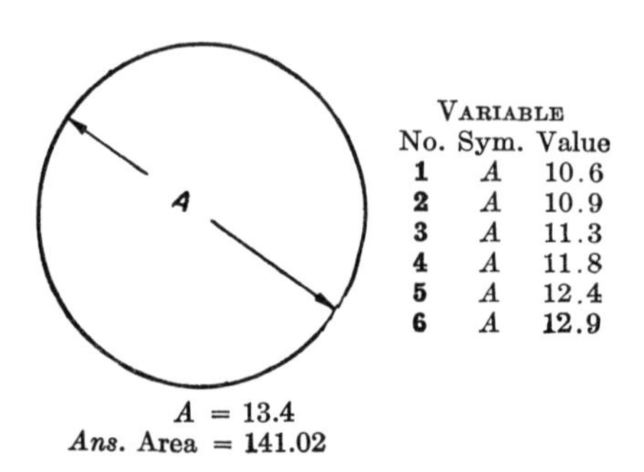

VARIABLE		
No.	Sym.	Value
1	*A*	10.6
2	*A*	10.9
3	*A*	11.3
4	*A*	11.8
5	*A*	12.4
6	*A*	12.9

$A = 13.4$
Ans. Area = 141.02

6. Determine the area.

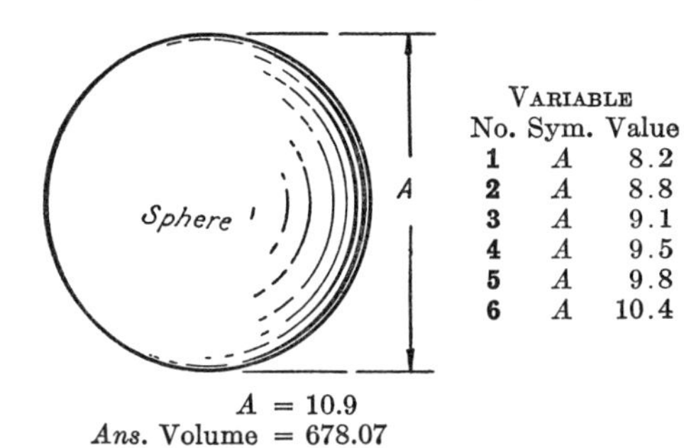

No.	Sym.	Value
1	A	8.2
2	A	8.8
3	A	9.1
4	A	9.5
5	A	9.8
6	A	10.4

$A = 10.9$
Ans. Volume $= 678.07$

7. Determine the volume.

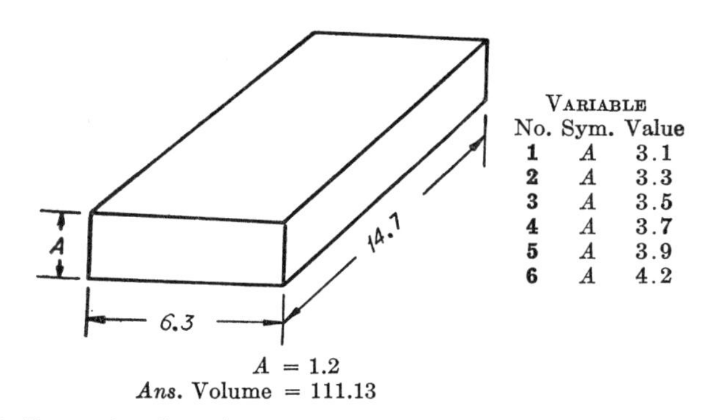

No.	Sym.	Value
1	A	3.1
2	A	3.3
3	A	3.5
4	A	3.7
5	A	3.9
6	A	4.2

$A = 1.2$
Ans. Volume $= 111.13$

8. Determine the volume.

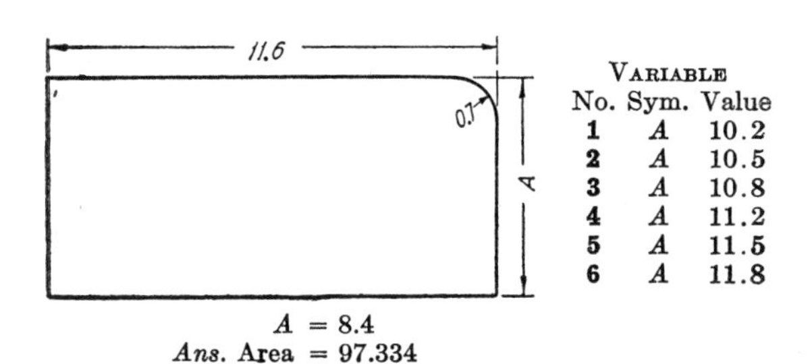

No.	Sym.	Value
1	A	10.2
2	A	10.5
3	A	10.8
4	A	11.2
5	A	11.5
6	A	11.8

$A = 8.4$
Ans. Area $= 97.334$

9. Determine the area.

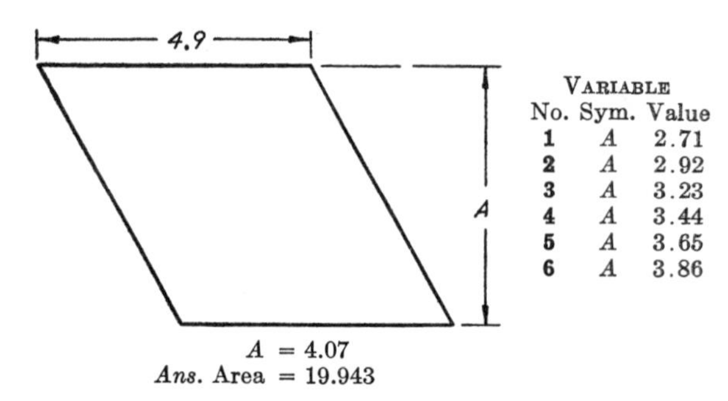

No.	Sym.	Value
1	*A*	2.71
2	*A*	2.92
3	*A*	3.23
4	*A*	3.44
5	*A*	3.65
6	*A*	3.86

(Variable)

$A = 4.07$

Ans. Area $= 19.943$

10. Determine the area.

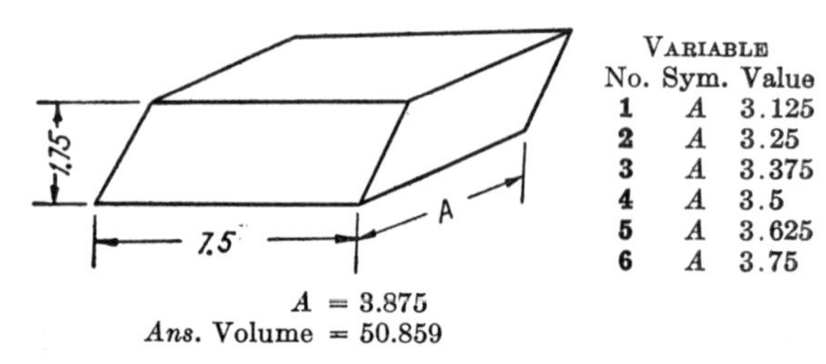

No.	Sym.	Value
1	*A*	3.125
2	*A*	3.25
3	*A*	3.375
4	*A*	3.5
5	*A*	3.625
6	*A*	3.75

(Variable)

$A = 3.875$

Ans. Volume $= 50.859$

11. Determine the volume.

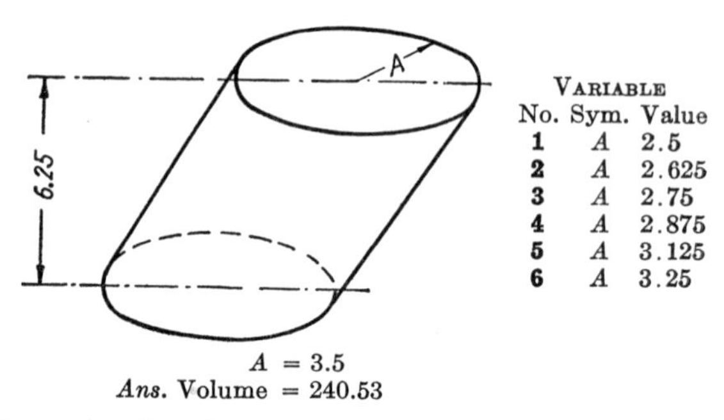

No.	Sym.	Value
1	*A*	2.5
2	*A*	2.625
3	*A*	2.75
4	*A*	2.875
5	*A*	3.125
6	*A*	3.25

(Variable)

$A = 3.5$

Ans. Volume $= 240.53$

12. Determine the volume.

13. A tank having the size and shape of the figure of Problem 8 can contain how many gallons of water?

The geometrical propositions given in this chapter are the basis for the solution of most practical shop problems. In this chapter, problems have been given to enable the student to acquire the ability to use each theorem separately. In the problems of the next chapter, the solution of any one problem will usually involve the use of a combination of several geometrical theorems. The student will gradually, through practice, acquire the ability to recognize and apply the proper combination of geometrical theorems involved in a solution.

CHAPTER X

TRIGONOMETRY

In geometry a triangle is said to be determined when sufficient sides and angles are given so that the triangle may be constructed. For example, a triangle is determined if two sides and the included angle are known, or if two angles and the included side are given. Frequently problems arise which require that the other parts (sides and angles) of a determined triangle shall be computed. This computation often cannot be carried out by geometry.

Trigonometry is a branch of mathematics which enables one to compute the remaining sides and angles of any triangle which has sufficient parts given. In order to do this, use must be made of what are called the trigonometric functions, *viz.*, sine, cosine, tangent, cotangent, secant, and cosecant. These functions, applied to any angle α, are usually written $\sin \alpha$, $\cos \alpha$, $\tan \alpha$, $\cot \alpha$, $\sec \alpha$, and $\csc \alpha$.

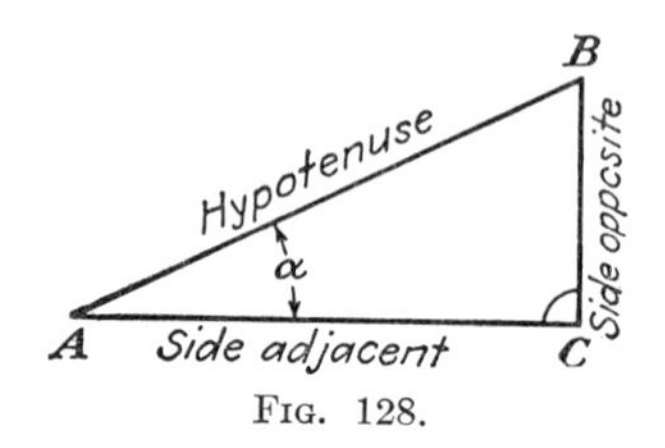

Fig. 128.

If one considers the angle α of the right triangle ABC, having the hypotenuse AB, AC is said to be the side adjacent, and BC the side opposite. The small quarter arc at C is used to denote the right angle and will be used to designate right angles throughout the book. It is very important that the student learn to recognize at a glance which side is the side opposite an angle and which is the side adjacent regardless of the position of the angle. For example, in the triangle DEF, having the hypotenuse DF, DE is the side opposite angle θ, and EF is the side adjacent to angle θ.

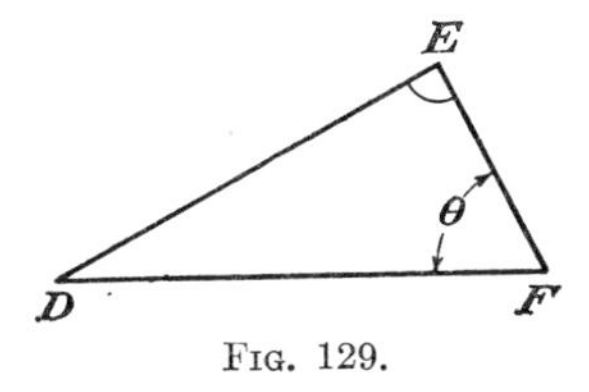

FIG. 129.

DEFINITIONS OF TRIGONOMETRIC FUNCTIONS

There are two methods of defining trigonometric functions: the "ratio method" and the "unity method." The two methods are, of course, equivalent.

RATIO METHOD

Referring to the angle α of the right triangle ABC, the six trigonometric functions are defined as follows:

$$\sin \alpha = \frac{\text{side opposite}}{\text{hypotenuse}} = \frac{BC}{AB}. \qquad \csc \alpha = \frac{\text{hypotenuse}}{\text{side opposite}} = \frac{AB}{BC}.$$

$$\cos \alpha = \frac{\text{side adjacent}}{\text{hypotenuse}} = \frac{AC}{AB}. \qquad \sec \alpha = \frac{\text{hypotenuse}}{\text{side adjacent}} = \frac{AB}{AC}.$$

$$\tan \alpha = \frac{\text{side opposite}}{\text{side adjacent}} = \frac{BC}{AC}. \qquad \cot \alpha = \frac{\text{side adjacent}}{\text{side opposite}} = \frac{AC}{BC}.$$

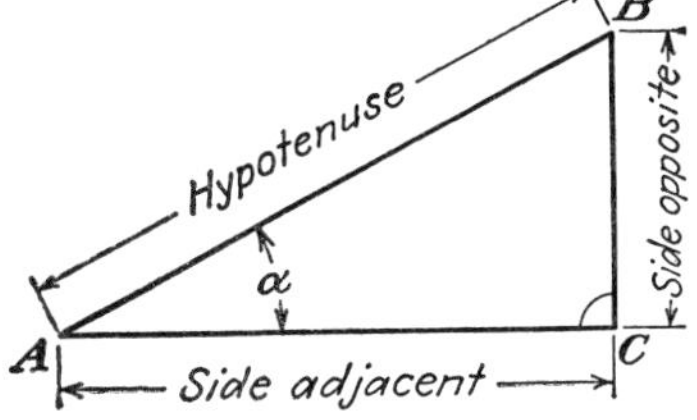

FIG. 130.

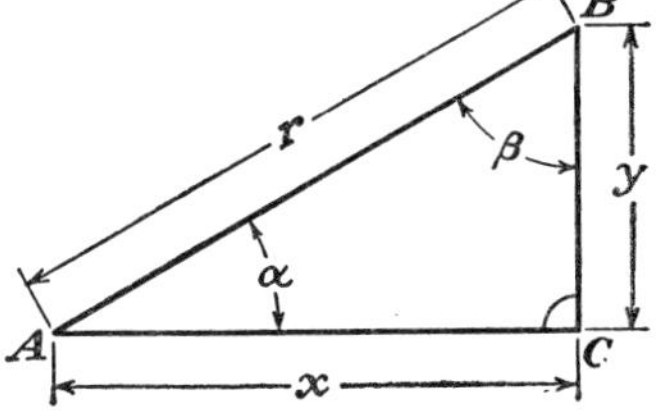

FIG. 131.

RECIPROCAL RELATIONS OF TRIGONOMETRIC FUNCTIONS

By the "ratio" definitions, in Fig. 131, $\sin \alpha = \frac{y}{r}$ and $\csc \alpha = \frac{r}{y}$. Since $\frac{y}{r} = \frac{1}{\frac{r}{y}}$, it follows that

$$\sin \alpha = \frac{1}{\csc \alpha}.$$

Similarly,
$$\cos \alpha = \frac{1}{\sec \alpha}$$

and
$$\tan \alpha = \frac{1}{\cot \alpha}.$$

FUNCTIONS OF COMPLEMENTARY ANGLES

By definition, in right triangle ABC of Fig. 131, $\sin \alpha = \frac{y}{r}$ and $\cos \beta = \frac{y}{r}$. Hence $\sin \alpha = \cos \beta$. However, α and β are complementary angles (*i.e.*, $\alpha + \beta = 90°$). Hence $\beta = 90° - \alpha$.

$$\therefore \sin \alpha = \cos (90° - \alpha).$$

Similarly,
$$\cos \alpha = \frac{x}{r} = \sin \beta$$

or
$$\cos \alpha = \sin (90° - \alpha).$$

Also
$$\tan \alpha = \frac{y}{x} = \cot \beta$$

or
$$\tan \alpha = \cot (90° - \alpha).$$

The above relations mean that

$$\sin 30° = \cos (90° - 30°) = \cos 60°$$
$$\tan 40° = \cot (90° - 40°) = \cot 50°$$
$$\csc 20° = \sec (90° - 20°) = \sec 70°, \text{ etc.}$$

FUNDAMENTAL RELATIONS BETWEEN THE TRIGONOMETRIC FUNCTIONS

In Fig. 132 let α be any acute angle in the right triangle ABC.

$a^2 + b^2 = c^2$ by the Pythagorean theorem.

$$\sin \alpha = \frac{a}{c}, \cos \alpha = \frac{b}{c}.$$

Squaring each equation and adding;

$$\sin^2 \alpha + \cos^2 \alpha = \frac{a^2}{c^2} + \frac{b^2}{c^2} = \frac{a^2 + b^2}{c^2} = \frac{c^2}{c^2} = 1$$

or $$\sin^2 \alpha + \cos^2 \alpha = 1.$$

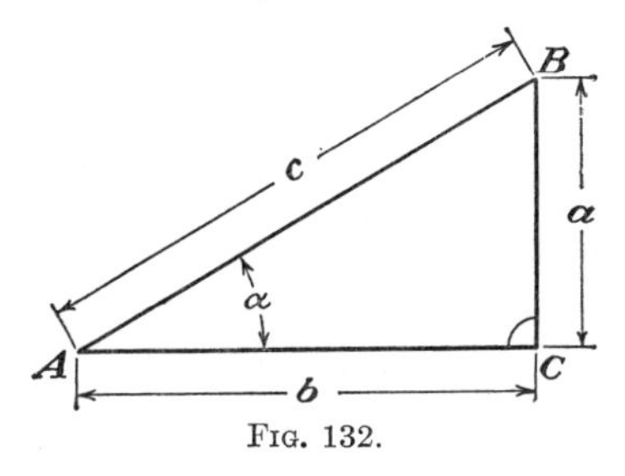

FIG. 132.

Also in Fig. 132,

$$\sec \alpha = \frac{c}{b}, \tan \alpha = \frac{a}{b}.$$

Squaring each equation and subtracting,

$$\sec^2 \alpha - \tan^2 \alpha = \frac{c^2}{b^2} - \frac{a^2}{b^2} = \frac{c^2 - a^2}{b^2} = \frac{b^2}{b^2} = 1$$

or $$\sec^2 \alpha = 1 + \tan^2 \alpha.$$

Similarly, $$\csc \alpha = \frac{c}{a}, \cot \alpha = \frac{b}{a}.$$

Squaring each equation and subtracting;

$$\csc^2 \alpha - \cot^2 \alpha = \frac{c^2}{a^2} - \frac{b^2}{a^2} = \frac{c^2 - b^2}{a^2} = \frac{a^2}{a^2} = 1$$

$$\csc^2 \alpha = 1 + \cot^2 \alpha.$$

$$\frac{\sin \alpha}{\cos \alpha} = \frac{\frac{a}{c}}{\frac{b}{c}} = \frac{a}{\not c} \cdot \frac{\not c}{b} = \frac{a}{b} = \tan \alpha$$

or $$\tan \alpha = \frac{\sin \alpha}{\cos \alpha}.$$

Similarly, $$\frac{\cos \alpha}{\sin \alpha} = \frac{\frac{b}{c}}{\frac{a}{c}} = \frac{b}{\not c} \cdot \frac{\not c}{a} = \frac{b}{a} = \cot \alpha.$$

or
$$\cot \alpha = \frac{\cos \alpha}{\sin \alpha}.$$

UNITY METHOD OF THE TRIGONOMETRIC FUNCTIONS

From the three trigonometric relations, $\sin^2 \alpha + \cos^2 \alpha = 1$, $\sec^2 \alpha - \tan^2 \alpha = 1$, and $\csc^2 \alpha - \cot^2 \alpha = 1$, derived in the preceding section, the following diagrammatic relations of the trigonometric functions may be shown in their respective places in each of the triangles in Figs. 133, 134, and 135. The trigonometric functions now shown diagrammatically may be stated in terms of the unity method as follows:

In Fig. 133, **sin** α is numerically equal to the length of the side opposite (*HI*) when the hypotenuse (*GI*) is unity. Likewise, **cos** α is numerically equal to the length of the side adjacent (*GH*) when the hypotenuse (GI) is unity.

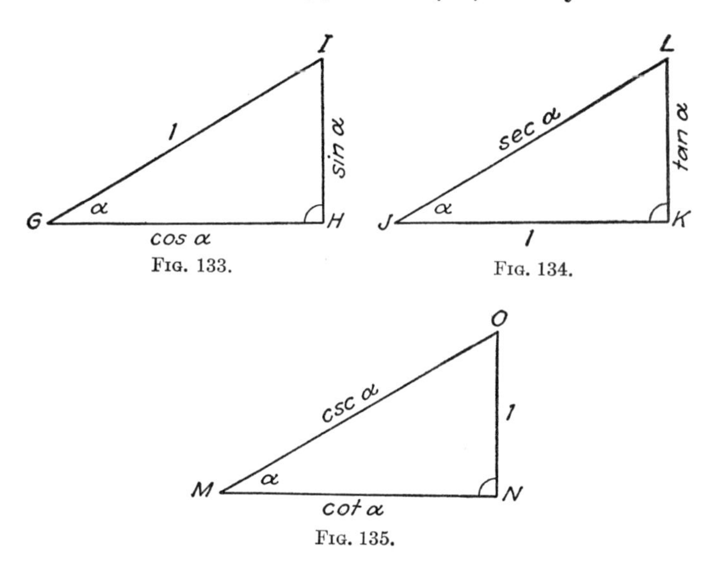

Fig. 133.
Fig. 134.
Fig. 135.

In Fig. 134, **tan** α is numerically equal to the length of the side opposite (*KL*) when the side adjacent (*JK*) is unity. Likewise, **sec** α is numerically equal to the length of the hypotenuse (*JL*) when the side adjacent (*JK*) is unity.

In Fig. 135, **cot** α is numerically equal to the length of the side adjacent (MN) when the side opposite (NO) is unity. Likewise, **csc** α is numerically equal to the length of the hypotenuse (MO) when the side opposite (NO) is unity.

Drills on Trigonometric Functions

A thorough understanding of the meaning of the trigonometric functions as given above is very essential and to attain this the following drill is beneficial:

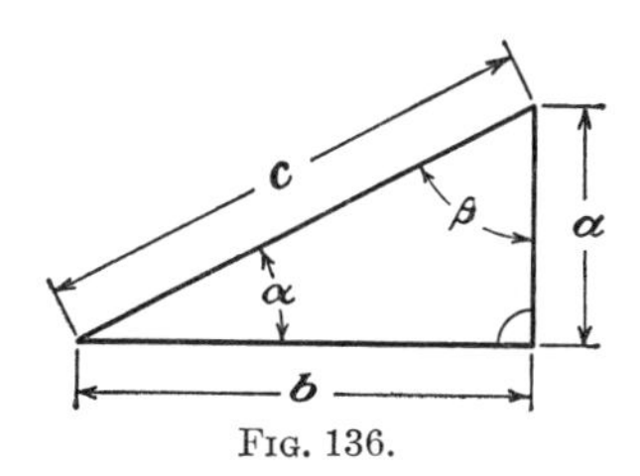

FIG. 136.

When a is 1, b is what function of the angle α?
  Ans.: b is the cotangent of angle α.
When c is 1, a is what function of the angle β?
  Ans.: a is the cosine of angle β.
When b is 1, c is what function of the angle β?
  Ans.: csc β.
When c is 1, a is what function of the angle α?
  Ans.: sin α.
When a is 1, c is what function of the angle α?
  Ans.: csc α.
When b is 1, c is what function of the angle β?
  Ans.: csc β.
When a is 1, c is what function of the angle β?
  Ans.: sec β.
When c is 1, b is what function of the angle α?
  Ans.: cos α.

Drills similar to the foregoing should be practiced frequently until the student is so familiar with this work that, instead of thinking in terms of a rule, he will think in terms of a triangle and will, considering one side to be unity, immediately recognize the other sides as the proper functions of a given angle.

EACH TRIGONOMETRIC FUNCTION EXPRESSED IN TERMS OF THE OTHER FIVE FUNCTIONS

From the reciprocal relations and the fundamental relations derived in the preceding section (all of which should be memorized by the student), each of the six functions may be expressed in terms of each of the other five.

To illustrate this, sin α can be expressed in terms of the others as follows:

From Fig. 133,

$$\sin \alpha = \sqrt{1 - \cos^2 \alpha}$$

Using Fig. 134 and the relation $\sec^2 \alpha = 1 + \tan^2 \alpha$,

$$\sin \alpha = \frac{\text{opp. side}}{\text{hyp.}} = \frac{KL}{JL} = \frac{\tan \alpha}{\sec \alpha} = \frac{\tan \alpha}{\sqrt{1 + \tan^2 \alpha}}.$$

From Fig. 135 and the relation $\csc^2 \alpha = 1 + \cot^2 \alpha$,

$$\sin \alpha = \frac{\text{opp. side}}{\text{hyp.}} = \frac{NO}{MO} = \frac{1}{\csc \alpha} = \frac{1}{\sqrt{1 + \cot^2 \alpha}}.$$

From Fig. 134,

$$\sin \alpha = \frac{KL}{JL} = \frac{\tan \alpha}{\sec \alpha} = \frac{\sqrt{\sec^2 \alpha - 1}}{\sec \alpha}.$$

By the reciprocal relation,

$$\sin \alpha = \frac{1}{\csc \alpha}.$$

The following chart gives the value of each of the trigonometric functions in terms of the other five. The expressions given in the first horizontal row are the values just developed for sin α. The other expressions should be verified by the student and worked out in a manner similar to that given above, if a thorough understanding of the subject is desired.

Trigonometric Functions and Their Relations Shown in Chart Form

Function	In terms of					
	$\sin\alpha$	$\cos\alpha$	$\tan\alpha$	$\cot\alpha$	$\sec\alpha$	$\csc\alpha$
$\sin\alpha$		$\sqrt{1-\cos^2\alpha}$	$\frac{\tan\alpha}{\sqrt{1+\tan^2\alpha}}$	$\frac{1}{\sqrt{\cot^2\alpha+1}}$	$\frac{\sqrt{\sec^2\alpha-1}}{\sec\alpha}$	$\frac{1}{\csc\alpha}$
$\cos\alpha$	$\sqrt{1-\sin^2\alpha}$		$\frac{1}{\sqrt{1+\tan^2\alpha}}$	$\frac{\cot\alpha}{\sqrt{1+\cot^2\alpha}}$	$\frac{1}{\sec\alpha}$	$\frac{\sqrt{\csc^2\alpha-1}}{\csc\alpha}$
$\tan\alpha$	$\frac{\sin\alpha}{\sqrt{1-\sin^2\alpha}}$	$\frac{\sqrt{1-\cos^2\alpha}}{\cos\alpha}$		$\frac{1}{\cot\alpha}$	$\sqrt{\sec^2\alpha-1}$	$\frac{1}{\sqrt{\csc^2\alpha-1}}$
$\cot\alpha$	$\frac{\sqrt{1-\sin^2\alpha}}{\sin\alpha}$	$\frac{\cos\alpha}{\sqrt{1-\cos^2\alpha}}$	$\frac{1}{\tan\alpha}$		$\frac{1}{\sqrt{\sec^2\alpha-1}}$	$\sqrt{\csc^2\alpha-1}$
$\sec\alpha$	$\frac{1}{\sqrt{1-\sin^2\alpha}}$	$\frac{1}{\cos\alpha}$	$\sqrt{1+\tan^2\alpha}$	$\frac{\sqrt{\cot^2\alpha+1}}{\cot\alpha}$		$\frac{\csc\alpha}{\sqrt{\csc^2\alpha-1}}$
$\csc\alpha$	$\frac{1}{\sin\alpha}$	$\frac{1}{\sqrt{1-\cos^2\alpha}}$	$\frac{\sqrt{1+\tan^2\alpha}}{\tan\alpha}$	$\sqrt{1+\cot^2\alpha}$	$\frac{\sec\alpha}{\sqrt{\sec^2\alpha-1}}$	

The relations given in the preceding table, together with the simple relations, enable one to determine the value of any of the remaining functions when one of them is given.

Example: Determine each of the other functions of the angle α if $\sin\alpha = .8$

$$\cos\alpha = \sqrt{1-\sin^2\alpha} = \sqrt{1-.64} = \sqrt{.36} = .6.$$

$$\tan\alpha = \frac{\sin\alpha}{\cos\alpha} = \frac{.8}{.6} = 1.3333$$

$$\cot\alpha = \frac{\cos\alpha}{\sin\alpha} = \frac{.6}{.8} = .75$$

$$\sec\alpha = \frac{1}{\cos\alpha} = \frac{1}{.6} = 1.6667$$

$$\csc\alpha = \frac{1}{\sin\alpha} = \frac{1}{.8} = 1.25$$

VARIATION OF THE TRIGONOMETRIC FUNCTIONS FROM 0° TO 90°

In Fig. 137, let DE be a quadrant of a circle of unit radius.

$$\sin\alpha = \frac{BC}{AB} = \frac{BC}{1} = BC$$

As the angle α decreases and approaches 0 the line BC which represents the sin α approaches 0.

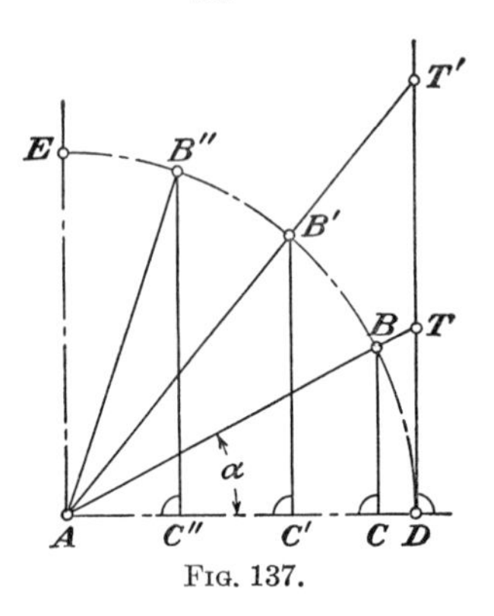

FIG. 137.

As the angle α increases, the sin α increases to $B'C'$, to $B''C''$, and finally to AE, which is 1. Thus the sine of an angle increases from 0 to 1 as the angle increases from 0° to 90°.

The $\cos\alpha = \dfrac{AC}{AB}$ is represented by the line AC which, for $\alpha = 0°$, has the value $AD = 1$ and successively takes the smaller values AC, AC', and AC'' as α gets larger. As α approaches 90°, the line AC approaches 0.

Thus the cosine varies from 1 to 0 as the angle varies from 0° to 90°.

$\tan\alpha = \dfrac{TD}{AD}$ is represented by the line DT, which for $\alpha = 0°$ is 0, and successively takes larger values DT, DT', etc., as the angle α increases. As α approaches 90°, DT gets longer and finally approaches infinity (∞).

Thus the tangent varies from 0 to ∞ as the angle varies from 0° to 90°.

From the reciprocal relation, $\cot\alpha = \dfrac{1}{\tan\alpha}$, the cotangent varies from $\dfrac{1}{0} \to \infty$ to $\dfrac{1}{\infty} \to 0$ as the angle varies from 0° to 90°.

From the reciprocal relation, $\sec\alpha = \dfrac{1}{\cos\alpha}$, the secant varies from $\frac{1}{1} = 1$ to $\frac{1}{0} \rightarrow \infty$ as the angle varies from 0° to 90°.

From the reciprocal relation, $\csc \infty = \frac{1}{\sin \alpha}$, the cosecant varies from $\frac{1}{0} \rightarrow \infty$ to $\frac{1}{1} = 1$ as the angle varies from 0° to 90°.

The variations of the functions may be summarized as follows:

As angle	increases	from	0°	to	90°
sine	increases	from	0	to	1
cosine	decreases	from	1	to	0
tangent	increases	from	0	to	∞
cotangent	decreases	from	∞	to	0
secant	increases	from	1	to	∞
cosecant	decreases	from	∞	to	1

TO FIND THE TRIGONOMETRIC FUNCTIONS OF A GIVEN ANGLE

The numerical values of the six simple trigonometric functions have been accurately worked out for all angles. These values are given to five figures in the table* for all angles in degrees and minutes from 0° to 90°. This table is used as follows: For any angle up to 45°, the degree of the angle is at the top of the page and the minutes of the degree are in the vertical column at the left. The functions for any given angle from 0° to 45° are given in the horizontal rows to the right of the given minute, the names of the functions for each column being read at the top.

Example a: The tangent of 37° 21′ is found on the 37° page (page 358) in the column labeled tangent, in the horizontal row opposite 21′, the value being .76318.

For angles from 45° to 90°, the degree of the angle is at the bottom of the page, the minutes are in the vertical column at the right, reading from bottom to top, and the names of the functions for each column are read at the bottom.

Example b: The cosine of 64° 51′ is found on the 64° page (page 367) in the column labeled cosine (at the bottom) in the horizontal row opposite 51′ (in the *right* vertical column), the value being .42499.

* See table of trigonometric functions at end of book (pp. 349–383).

Determination of an Unknown Side

Consider the type of problem of determining the length of one side of a right-angled triangle when another side and an acute angle are given:

Procedure: Assume the given side to be unity. Then the side in question will be some function of the given angle according to the definitions of the trigonometric functions. For any problem, the side in question will be as many times as great as the function as the given side is of unity.

Example a: In the accompanying figure, determine the length of BC.

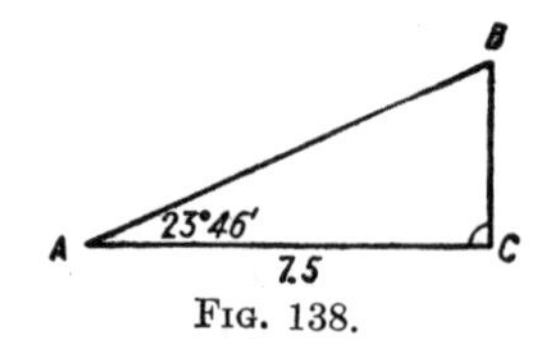

FIG. 138.

Solution: If AC were unity, BC would be, by definition, tan 23° 46′.

Since AC is 7.5 (which is 7.5 times unity), BC will be 7.5 times the tan 23° 46′, or $7.5 \times .44036 = 3.3027$.

Example b: In the figure DEF, determine DF.

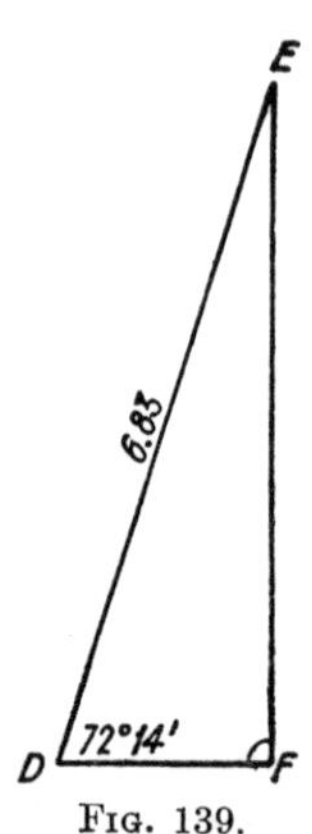

FIG. 139.

Solution: If DE were unity, DF by definition would be cos 72° 14′. But DE is 6.83 and therefore $DF = 6.83 \times \cos$ 72° 14′ or $6.83 \times .30514 = 2.0841$.

Example c: In the figure *GHJ*, determine *X*.

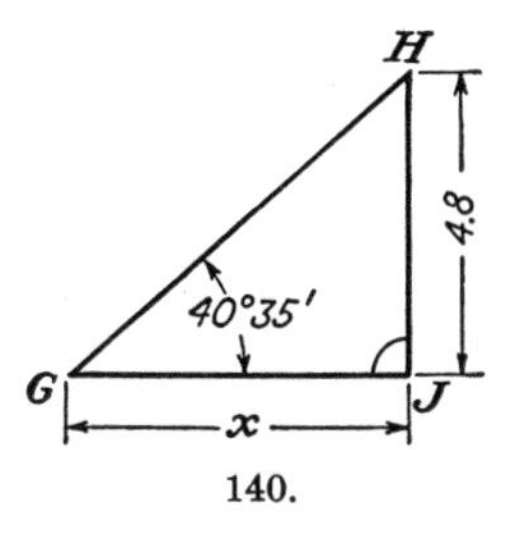

140.

Solution: If *HJ* were unity, *X*, by definition, would be cot 40° 35′. But *HJ* is 4.8, and therefore *X* is 4.8 × cot 40° 35′ = 4.8 × 1.1674 = 5.6035.

PROBLEMS

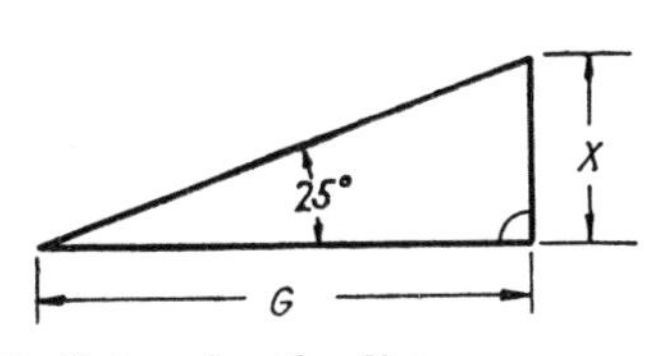

1. Determine the distance *x*.

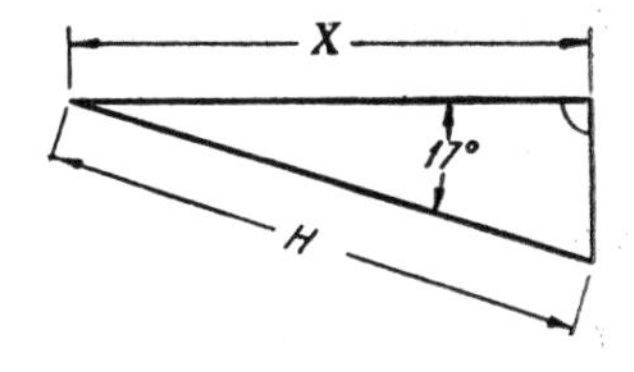

2. Determine the distance *x*.

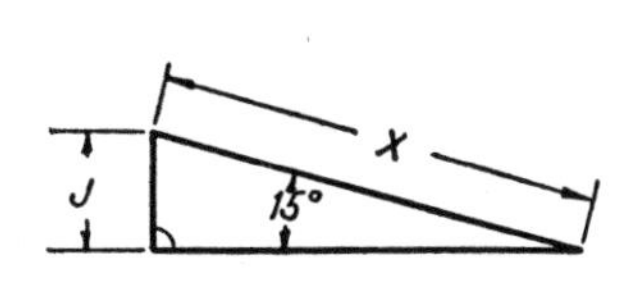

3. Determine the distance *x*.

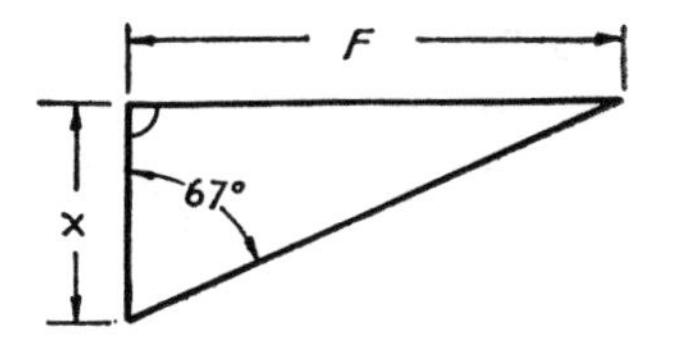

4. Determine the distance *x*.

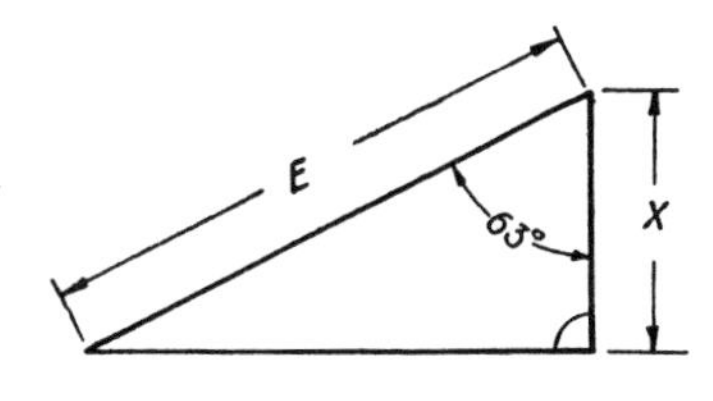

5. Determine the distance *x*.

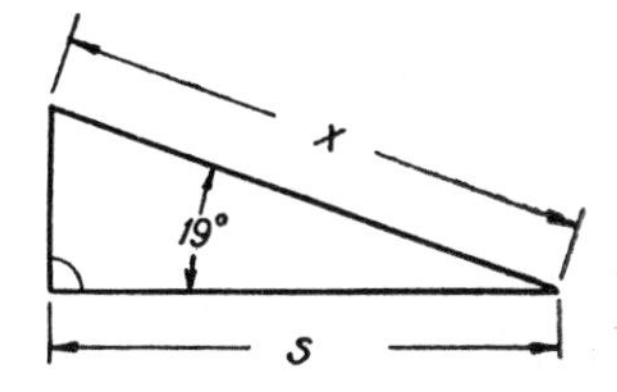

6. Determine the distance *x*.

VARIABLES

Prob.	Sym.	No. 1	No. 2	No. 3	No. 4	No. 5	No. 6
1	*G*	10.106	10.218	10.342	10.412	10.818	10.937
2	*H*	12.135	12.248	12.375	12.492	12.625	12.75
3	*J*	2.225	2.312	2.386	2.468	2.591	2.724
4	*F*	8.875	8.955	9.128	9.322	9.462	9.575
5	*E*	15.25	15.41	15.54	15.63	15.76	15.88
6	*S*	17.32	17.46	17.59	17.72	17.83	17.97

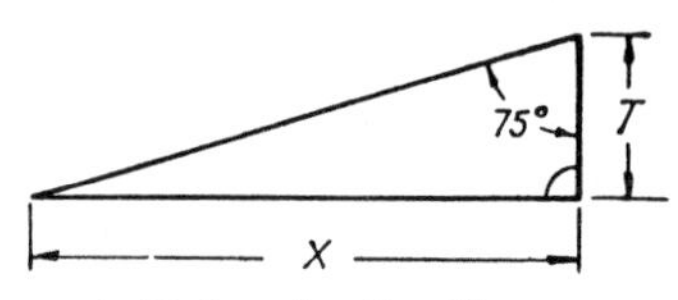

7. Determine the distance x.

8. Determine the distance x.

VARIABLES

Prob.	Sym.	No. 1	No. 2	No. 3	No. 4	No. 5	No. 6
7	*R*	14.46	14.52	14.89	15.08	15.23	15.39
8	*T*	2.751	2.812	2.933	3.104	3.225	3.376

TO FIND THE ANGLE CORRESPONDING TO A GIVEN TRIGONOMETRIC FUNCTION OR COFUNCTION

It should be noticed in the table of the trigonometric functions that the degrees from 0 to 45 are given at the top of the page and the degrees from 45 to 90 are given at the bottom. It should also be noticed that in the same column of the trigonometric tables a column headed at the top of the page by the function of the angle is headed at the bottom of the page by the cofunction of the complement of the angle, and *vice versa*. Hence, to locate the value of a given $\begin{pmatrix}\text{function}\\ \text{cofunction}\end{pmatrix}$ in the trigonometric table, proceed to find the nearest number $\begin{pmatrix}\text{smaller}\\ \text{larger}\end{pmatrix}$ than the given $\begin{pmatrix}\text{function}\\ \text{cofunction}\end{pmatrix}$ in either column headed by the $\begin{pmatrix}\text{function}\\ \text{cofunction}\end{pmatrix}$ or its $\begin{pmatrix}\text{cofunction}\\ \text{function}\end{pmatrix}$. If this nearest $\begin{pmatrix}\text{smaller}\\ \text{larger}\end{pmatrix}$ number is found in the vertical column headed at the top by that $\begin{pmatrix}\text{function}\\ \text{cofunction}\end{pmatrix}$, the degree of the angle is taken from the top of the page and the minutes from the left-hand column horizontally opposite the nearest $\begin{pmatrix}\text{smaller}\\ \text{larger}\end{pmatrix}$ number to the given $\begin{pmatrix}\text{function}\\ \text{cofunction}\end{pmatrix}$; but if this nearest $\begin{pmatrix}\text{smaller}\\ \text{larger}\end{pmatrix}$ number is found in the vertical column headed at the bottom by that $\begin{pmatrix}\text{function}\\ \text{cofunction}\end{pmatrix}$, the degree of the angle is taken from the bottom of the page and the minutes from the right-hand column horizontally opposite the nearest $\begin{pmatrix}\text{smaller}\\ \text{larger}\end{pmatrix}$ number to the given $\begin{pmatrix}\text{function}\\ \text{cofunction}\end{pmatrix}$.

Example a: Find the angle whose sine is .36442.

Solution: First, locate the nearest smaller number in the column headed by either sin or cos. In this case the nearest smaller number to .36442 is .36434, which is found on page 366 in the column headed sin at the top. Hence, the degree of the angle is at the top of the page, and the minute is found in the left-hand column horizontally opposite .36434. Thus, in this case, the angle in degrees and minutes is 21° 22′.

Example b: Find the angle whose tangent is 4.0908.

Solution: Locate the nearest smaller number in the column headed by either tan or cot, which is 4.08666 on page 352 in the column headed cot at the top. Since this column is headed tan at the bottom (and since the angle desired is the angle whose tan is 4.0908), the degree of the angle is read at the bottom of the page and the minute is found in the right-hand column horizontally opposite the value 4.08666. The angle in degrees and minutes is found to be 76° 15′.

Note: In the foregoing problem the nearest degree and minute corresponding to the tangent of 4.0908 is 76° 16′.

However, the student is cautioned to follow the rule of using the nearest smaller number in the case of finding an angle corresponding to a *function,* and the nearest larger number in the case of finding an angle corresponding to a *cofunction,* in order to maintain a definite procedure for the coming work of interpolating for seconds.

Example c: Find the angle whose cosine is .53155.

Solution: Locate the nearest *larger* number (since this is a cofunction) in the column headed either cos or sin, which is .53164 on page 368 in the column headed sin at the top and cos at the bottom. Since the angle desired is the angle whose cos is .53155, the degree of the angle is read at the bottom of the page and the minute is found in the right-hand column horizontally opposite .53164. The angle in degrees and minutes is found to be 57° 53′.

TO DETERMINE AN ANGLE WHEN TWO SIDES OF A RIGHT TRIANGLE ARE GIVEN

In actual practice it is often necessary to obtain an angle of a right triangle when two of its sides are given.

Example: Determine the angle α when the two sides AC and BC are given.

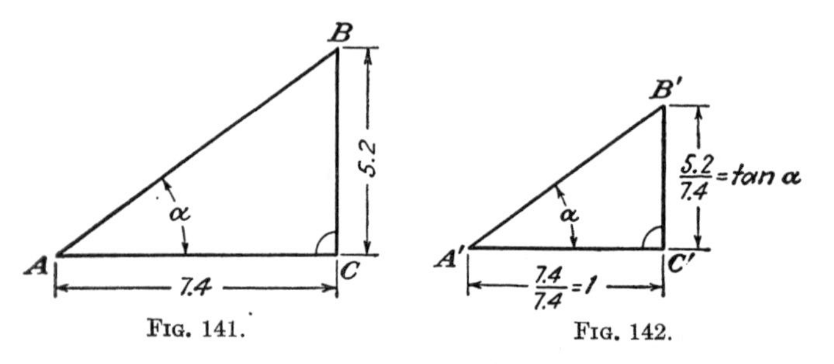

FIG. 141. FIG. 142.

Solution: If AC is made unity, BC becomes the tangent of α. To make AC unity, it must be divided by itself (7.4). However, if AC is divided by 7.4, BC must also be divided by 7.4 in order that the triangle retain a similar shape as in triangle $A'B'C'$. $B'C'$ is now the tangent of α. Hence,

$\tan \alpha = \frac{5.2}{7.4} = .70270$. Find the angle having a tangent of .70270 as in the foregoing example *b*. This angle is 35° 5′.

RULE FOR FINDING FUNCTION OF AN ANGLE

From the foregoing procedure a **general rule** may be formulated for finding a function of an unknown angle: Divide one side of the right-angled triangle by another. The side which is the denominator of the fraction thus formed may be considered as one, and the side which is the numerator of the fraction then represents the function of the angle, and the value of the fraction is equal to the function of the angle.

Note: When one of the sides given is the hypotenuse, always divide a side by the hypotenuse (*not* the hypotenuse by a side). The reason for this is that dividing the hypotenuse by a side gives the secant or the cosecant. These functions, for a large range of angles, have very small differences, which thus makes it difficult to compute the seconds accurately.

Example: Determine the angle α in Fig. 143.

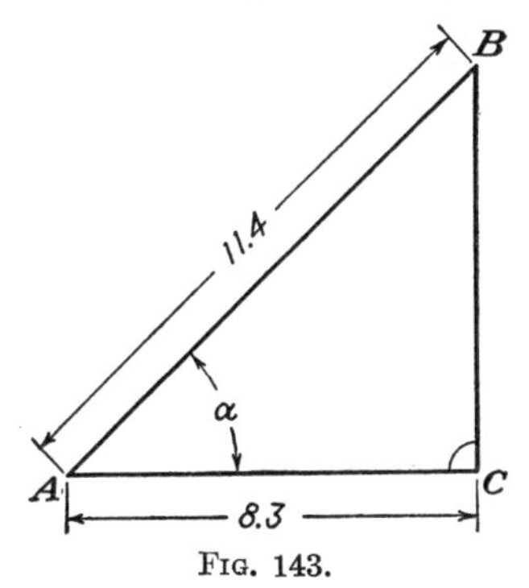

FIG. 143.

Solution: Following the general rule, divide 8.3 by 11.4. Hence 11.4 is the side to be considered unity and 8.3 is the side which represents the function, which in this case is the $\cos \alpha$, and the value of which is $\frac{8.3}{11.4} = .72807$. Find the angle having this cosine as in the foregoing example *c*.

$$\alpha = 43° \ 16'.$$

PROBLEMS

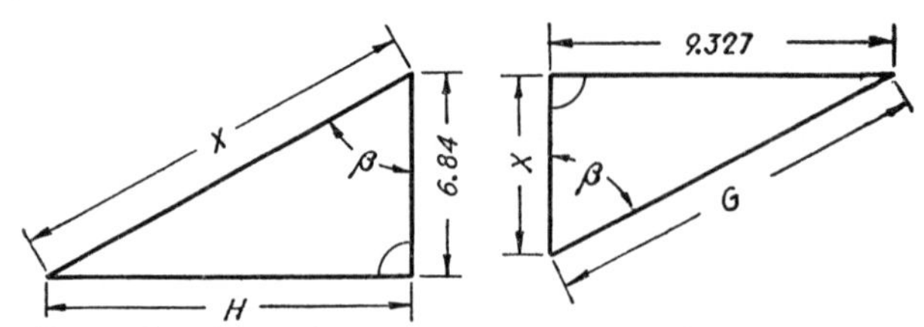

1. Determine the angle β

2. Determine the distance x.

3. Determine the angle β.

4. Determine the distance x.

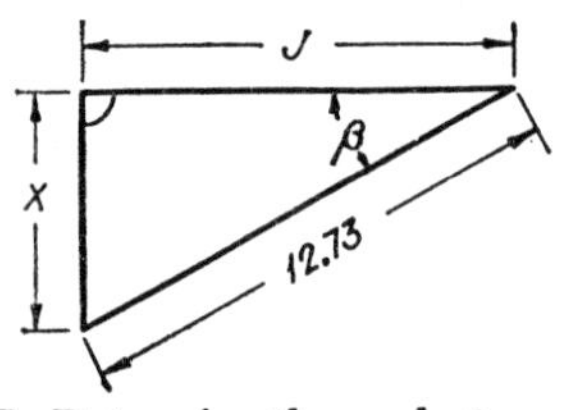

5. Determine the angle β.

6. Determine the distance x.

7. Determine the angle β.

8. Determine the distance x.

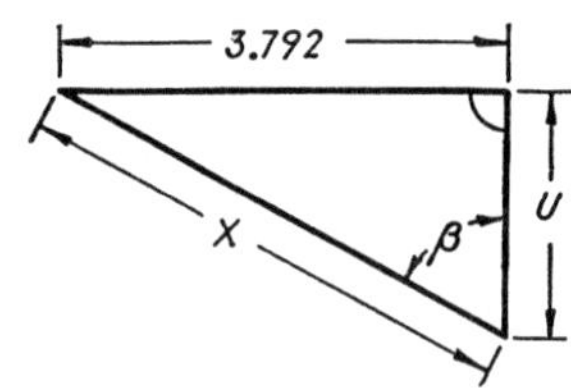

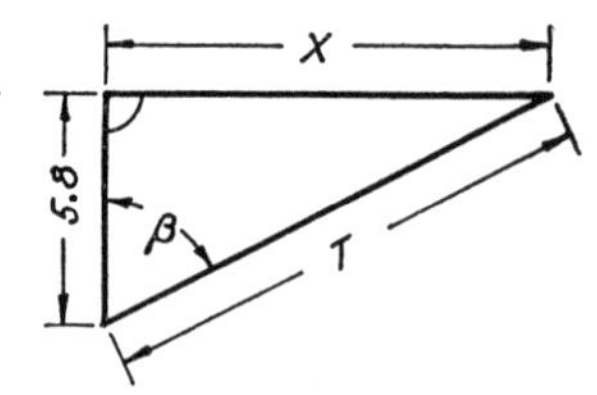

9. Determine the angle β.

10. Determine the distance x.

11. Determine the angle β.

12. Determine the distance x.

VARIABLES

Prob.	Sym.	No. 1	No. 2	No. 3	No. 4	No. 5	No. 6
1	H	9.12	9.25	9.37	9.48	9.56	9.75
2	H	9.12	9.25	9.37	9.48	9.56	9.75
3	G	11.431	11.572	11.613	11.724	11.815	11.976
4	G	11.431	11.572	11.613	11.724	11.815	11.976
5	J	8.55	8.71	8.82	8.96	9.25	9.42
6	J	8.55	8.71	8.82	8.96	9.25	9.42
7	L	4.28	4.37	4.46	4.65	4.84	4.93
8	L	4.28	4.37	4.46	4.65	4.84	4.93
9	U	1.54	1.66	1.78	1.82	1.85	1.95
10	U	1.54	1.66	1.78	1.82	1.85	1.95
11	T	10.3	10.8	11.1	11.5	11.9	12.2
12	T	10.3	10.8	11.1	11.5	11.9	12.2

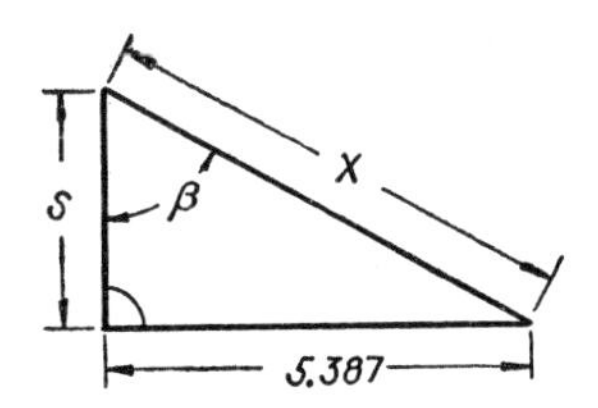

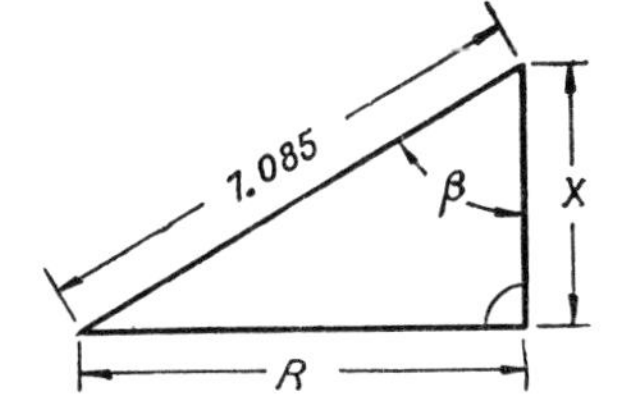

13. Determine the angle β.
14. Determine the distance x.
15. Determine the angle β.
16. Determine the distance x.

VARIABLES

Prob.	Sym.	No. 1	No. 2	No. 3	No. 4	No. 5	No. 6
13	S	3.12	3.25	3.37	3.52	3.65	3.75
14	S	3.12	3.25	3.37	3.52	3.65	3.75
15	R	4.91	5.12	5.23	5.44	5.65	5.86
16	R	4.91	5.12	5.23	5.44	5.65	5.86

INTERPOLATION

General Method

The tables of natural trigonometric functions of angles are usually given for degrees and minutes only. If an angle is required in degrees, minutes, and seconds corresponding to a given function or cofunction, or if the value of a function or cofunction is required for an angle in degrees, minutes, and seconds, a process called **interpolation** must be used.

The process of interpolating the number of seconds, when a function or cofunction of an angle is given, can be best explained by an illustrative problem.

Illustrative Problem: Let it be required to find the angle in degrees, minutes, and seconds corresponding to a tangent of .27038. From the method previously given, the required angle is found to be between 15° 7′ and 15° 8′. The accompanying figure is distorted in order to bring out the procedure more clearly. When $AE = 1$, $DE = \tan\ 15°\ 7' = .27013$ and $BE = \tan\ 15°\ 8' = .27044$ and $BD = BE - DE$, or $BD = .27044 - .27013$ which, disregarding the decimal point, is 31.

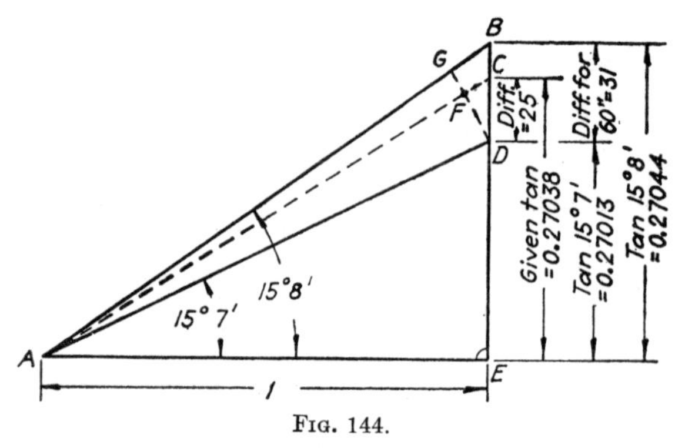

FIG. 144.

The difference between the tangent CE of the required angle and the tangent DE of 15° 7′ is $CD = .27038 - .27013$, which, disregarding the decimal point, is 25. Draw an arc DFG with A as a center and radius AD. $\dfrac{\text{Angle } DAF}{\text{Angle } DAG} = \dfrac{\widehat{DF}}{\widehat{DG}}$ (P-34). Arc DFG is nearly a straight line, so that the figures DFC and DGB approximate two similar triangles and $\dfrac{\widehat{DF}}{\widehat{DG}} = \dfrac{DC}{DB}$ nearly. Hence, $\dfrac{\text{angle } DAF}{\text{angle } DAG} = \dfrac{DC}{DB}$ nearly where $\angle DAF$ is the required angle (θ) and $\angle DAG = 1'$ or $60''$. Hence, $\dfrac{\theta}{60''} = \dfrac{25}{31}$ or $\theta = \dfrac{25}{31} \times 60'' = 48''$ with no appreciable error.

PROCEDURE FOR FINDING AN ANGLE IN DEGREES, MINUTES, AND SECONDS BY INTERPOLATION

1. *Find the number of degrees and minutes by the method previously outlined.*

2. *Find the difference between the next smaller and the next larger values of the function. (Disregard the decimal point.)*

3. *Find the difference between the* $\left(\begin{matrix}\textit{function}\\ \textit{cofunction}\end{matrix}\right)$ *corresponding to the next* $\left(\begin{matrix}\textit{smaller}\\ \textit{larger}\end{matrix}\right)$ *in value and the given value. (Disregard the decimal point.)*

4. *To obtain the number of seconds, multiply* 60 *by a common fraction, the numerator of which is the number obtained in* 3, *and the denominator of which is the number obtained in* 2.

The following examples will illustrate this procedure:

Example a: Find the angle whose sine is .19758.

Solution: By (2), $\sin 11^\circ\, 24' = .19766$

$\sin 11^\circ\, 23' = \underline{.19737}$

The difference is 29

By (3), given value $= .19758$

$\sin 11^\circ\, 23' = \underline{.19737}$

The difference is 21

By (4), $60'' \times \frac{21}{29} = 43''$

Hence the required angle is $11^\circ\, 23'\, 43''$.

Example b: Find the angle whose cotangent is 1.9096.

Solution: By (2), $\cot 27^\circ\, 38' = 1.9101$

$\cot 27^\circ\, 39' = \underline{1.9088}$

The difference is 13

By (3), $\cot 27^\circ\, 38' = 1.9101$

given value $= \underline{1.9096}$

The difference is 5

By (4), $60'' \times \frac{5}{13} = 23''$

Hence, the required angle is $27^\circ\, 38'\, 23''$.

PROBLEMS

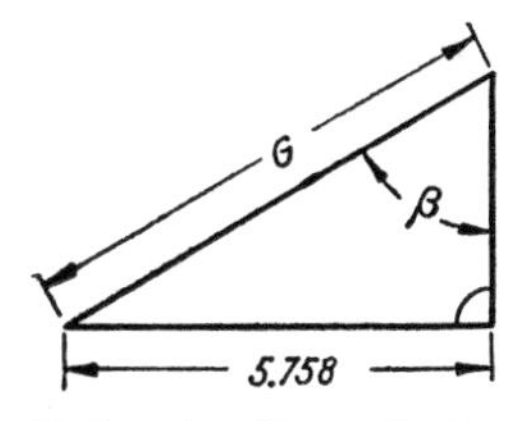

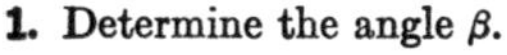

1. Determine the angle β.

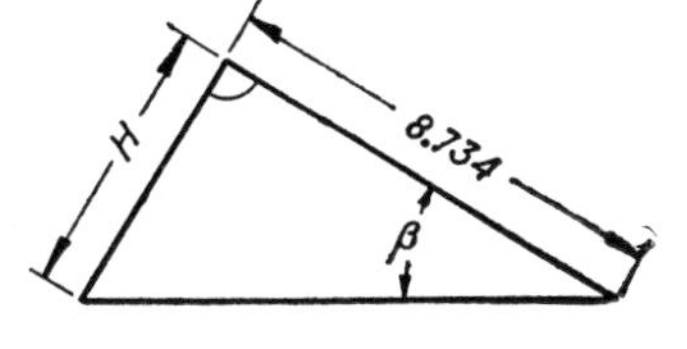

2. Determine the angle β.

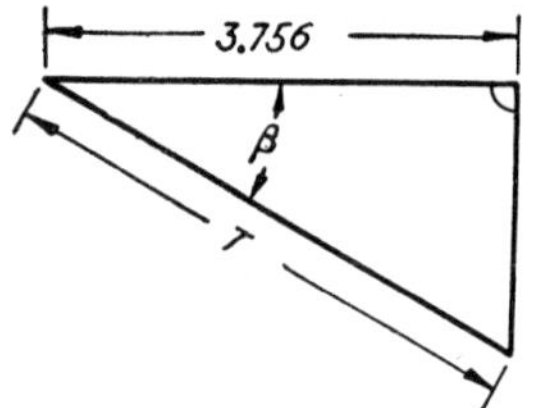

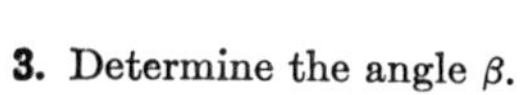
3. Determine the angle β.

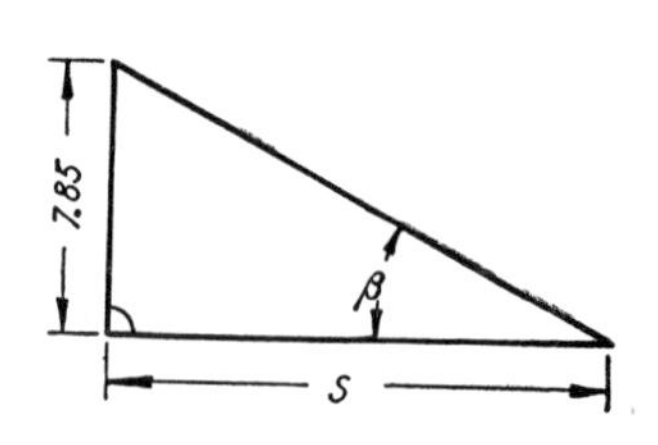

4. Determine the angle β.

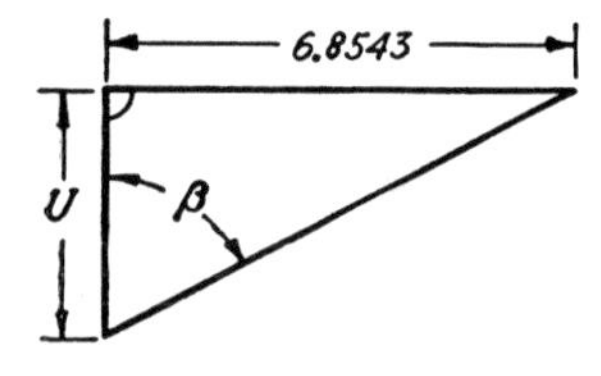

5. Determine the angle β.

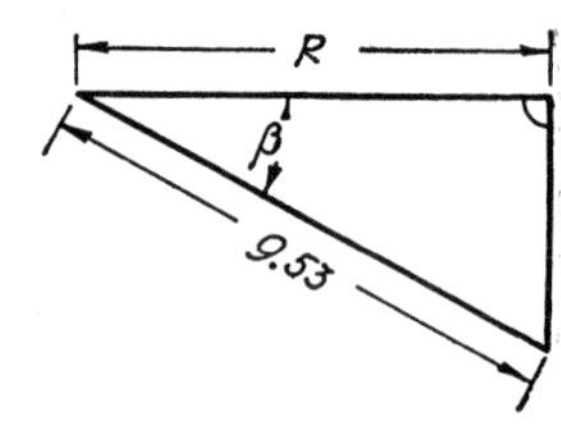

6. Determine the angle β.

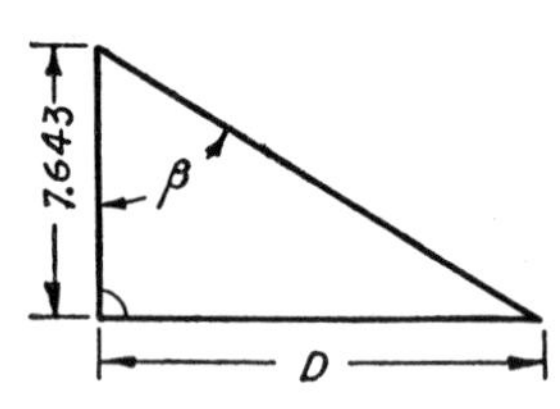

7. Determine the angle β.

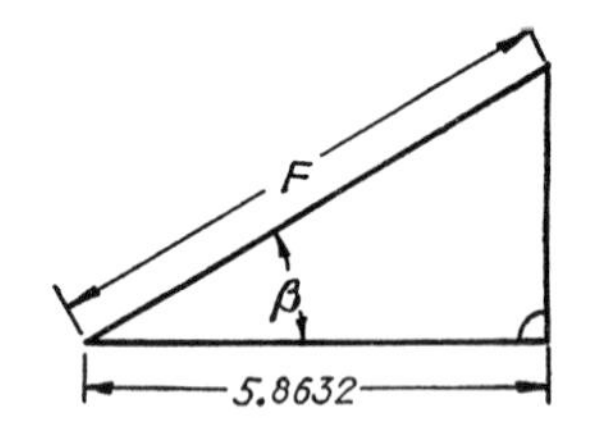

8. Determine the angle β.

VARIABLES

Prob.	Sym.	No. 1	No. 2	No. 3	No. 4	No. 5	No. 6
1	*G*	8.543	8.628	8.764	8.775	8.848	8.965
2	*H*	3.223	3.394	3.430	3.568	3.672	3.828
3	*T*	5.128	5.237	5.382	5.446	5.529	5.644
4	*S*	9.108	9.266	9.391	9.473	9.582	9.712
5	*U*	3.516	3.728	3.935	4.069	4.222	4.475
6	*R*	6.112	6.329	6.546	6.762	6.876	6.989
7	*D*	9.344	9.575	9.773	9.919	10.101	10.362
8	*F*	7.231	7.462	7.663	7.824	7.985	8.226

PROCESS OF FINDING THE FUNCTION OR COFUNCTION OF AN ANGLE GIVEN IN DEGREES, MINUTES AND SECONDS BY INTERPOLATION

1. *Find the value of the* $\begin{pmatrix}\textit{function}\\ \textit{cofunction}\end{pmatrix}$ *of the angle in degrees and minutes by the method previously given.*

2. *Find the value of the* $\begin{pmatrix}\textit{function}\\ \textit{cofunction}\end{pmatrix}$ *for an angle* 1′ *greater.*

3. *Obtain the difference between these two* $\begin{pmatrix}\textit{functions}\\ \textit{cofunctions}\end{pmatrix}$. *Disregard the decimal point.*

4. *Multiply this difference by a fraction, the numerator of which is the number of seconds given and the denominator of which is* 60″.

5. $\begin{pmatrix}\textit{Add}\\ \textit{Subtract}\end{pmatrix}$ *the result obtained in* 4 *to the last digits of the value of the* $\begin{pmatrix}\textit{function}\\ \textit{cofunction}\end{pmatrix}$ *obtained in* 1.

The following problems will illustrate this procedure:

Example a: Find the tangent of 27° 16′ 38″.

Solution: 1. The tangent of 27° 16′ is .51540
2. The tangent of 27° 17′ is .51577
3. The difference is 37
4. $37 \times \frac{38}{60} = 23.4$ or 23.
5. Since the tangent is a *function*, the 23 must be added to the last digits of the figure of step 1.

$$\begin{array}{r} .51540 \\ 23 \\ \hline .51563 \end{array}$$

Thus the tangent of 27° 16′ 38″ is .51563.

Example b: Find the cosecant of 7° 48′ 18″.

Solution: 1. The cosecant of 7° 48′ is 7.3683
2. The cosecant of 7° 49′ is 7.3527
3. The difference is 156
4. $156 \times \frac{18}{60} = 46.8$ or 47.
5. Since the cosecant is a *cofunction*, the 47 must be subtracted from the last digits of the figure of step 1.

$$\begin{array}{r} 7.3683 \\ 47 \\ \hline 7.3636 \end{array}$$

Thus the cosecant of 7° 48′ 18″ is 7.3636.

PROBLEMS

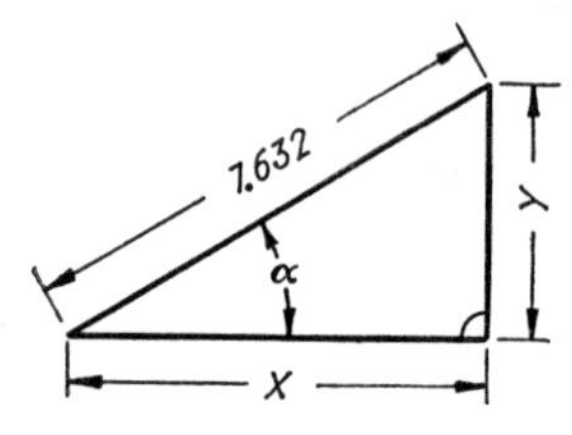

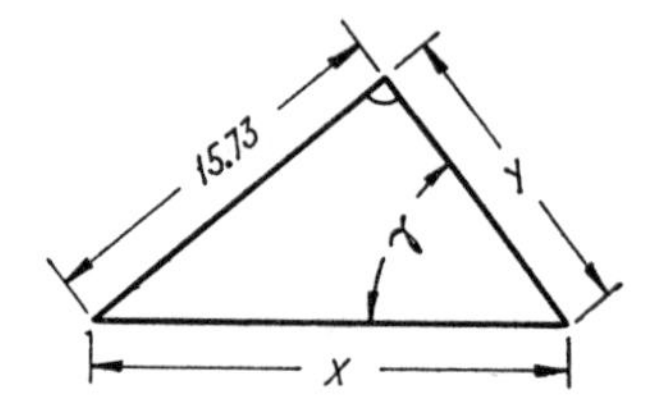

1. Determine the distance x.
2. Determine the distance y.
3. Determine the distance x.
4. Determine the distance y.

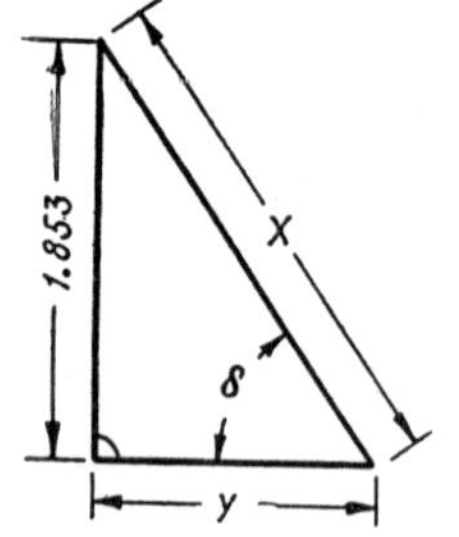

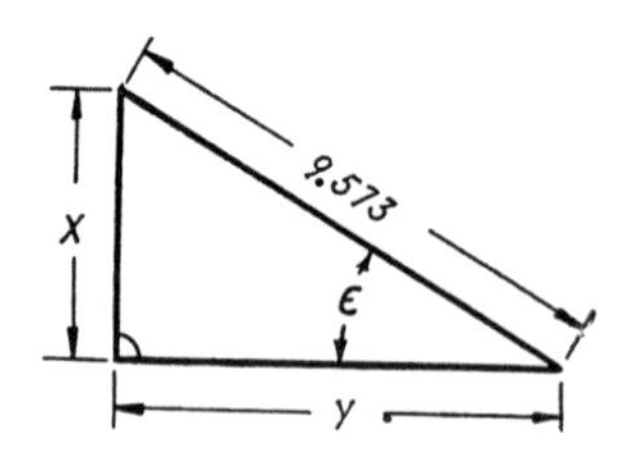

5. Determine the distance x.
6. Determine the distance y.
7. Determine the distance x.
8. Determine the distance y.

VARIABLES

Prob.	Sym.	No. 1	No. 2	No. 3	No. 4	No. 5	No. 6
1	α	32° 12′ 28″	33° 16′ 19″	34° 12′ 33″	35° 42′ 12″	36° 48′ 46″	37° 52′ 48″
2	α	32° 12′ 28″	33° 16′ 19″	34° 12′ 33″	35° 42′ 12″	36° 48′ 46″	37° 52′ 48″
3	γ	46° 42′ 28″	47° 18′ 32″	48° 36′ 10″	49° 15′ 15″	50° 11′ 56″	51° 24′ 14″
4	γ	46° 42′ 28″	47° 18′ 32″	48° 36′ 10″	49° 15′ 15″	50° 11′ 56″	51° 24′ 14″
5	δ	55° 34′ 31″	56° 53′ 46″	57° 18′ 45″	57° 36′ 56″	53° 24′ 18″	48° 13′ 18″
6	δ	55° 34′ 31″	56° 53′ 46″	57° 18′ 45″	57° 36′ 56″	53° 24′ 18″	48° 13′ 18″
7	ε	32° 18′ 36″	56° 25′ 13″	30° 46′ 48″	29° 21′ 27″	28° 56′ 11″	27° 34′ 50″
8	ε	32° 18′ 36″	56° 25′ 13″	30° 46′ 48″	29° 21′ 27″	28° 56′ 11″	27° 34′ 50″

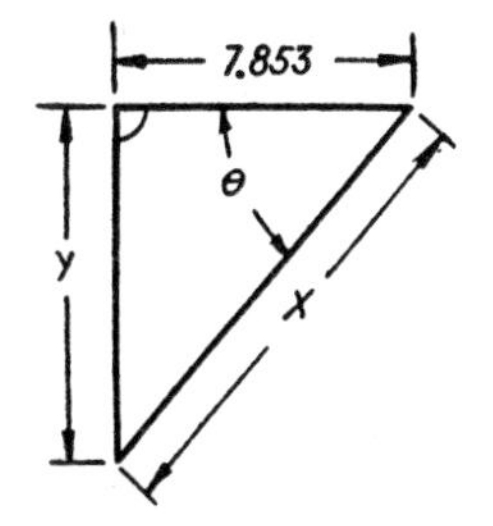

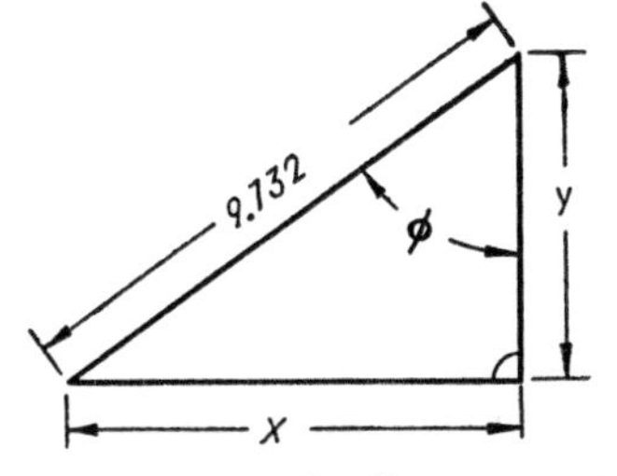

9. Determine the distance x.
10. Determine the distance y.
11. Determine the distance x.
12. Determine the distance y.

VARIABLES

Prob.	Sym.	No. 1	No. 2	No. 3	No. 4	No. 5	No. 6
9	θ	47° 22′ 15″	48° 34′ 40″	49° 16′ 10″	50° 58′ 56″	51° 12′ 14″	52° 45′ 11″
10	θ	47° 22′ 15″	48° 34′ 40″	49° 16′ 10″	50° 58′ 56″	51° 12′ 14″	52° 45′ 11″
11	ϕ	41° 11′ 31″	42° 19′ 42″	43° 32′ 13″	44° 33′ 24″	45° 35′ 15″	46° 52′ 6″
12	ϕ	41° 11′ 31″	42° 19′ 42″	43° 32′ 13″	44° 33′ 24″	45° 35′ 15″	46° 52′ 6″

INTERPOLATION

Special Method

The authors of this text are introducing a new system of interpolating trigonometric functions analogous to that frequently used in connection with logarithms. The advantage of this system is that it eliminates many arithmetical computations, thus resulting in a saving of time and an increase in accuracy.

In especially prepared trigonometric tables,* the differences of the values of the functions for successive minutes have been computed and placed in columns to the right of the functions.

On the page opposite the functions or on the margin below, each of these differences has been divided into 12 equal parts, which may be referred to as proportional parts, corresponding to twelfths of minutes (*i.e.*, 5″, 10″, 15″, etc.). The proportional parts for 1″, 2″, 3″, and 4″ may be obtained by dividing the proportional parts for 10″, 20″, 30″, and 40″ by 10. This

* "Mechanics Vest Pocket Reference Book" by J. H. Wolfe and E. R. Phelps, Prentice-Hall, Inc., Englewood Cliff, New Jersey.

may be done by moving the decimal point one place to the left in the table. Then the proportional part for 10″ becomes the value for 1″, the proportional part for 20″ becomes the value for 2″, etc.

The use of this system may be best explained by two illustrative examples.

Example a: Find the sine of 33° 29′ 34″.

Solution: 1. Find the sine of 33° 29′ as previously explained (.55169).

2. Note the tabulated difference between the sin 33° 29′ and sin 33° 30′, which in this case is 25.

3. On the right-hand page under proportional parts, locate this difference of 25 in a horizontal row opposite 60″.

4. Since 34″ is between 30″ and 35″, note the difference corresponding to 30″, which is 12.5.

5. To this difference of 12.5, add the difference corresponding to 4″. This value may be obtained by moving the decimal point for 40″ and its proportional part one place to the left. This gives 1.7 as the proportional part for 4″. Thus the difference for 34″ is 12.5 + 1.7 = 14.2 or 14.

6. Since the sine is a *function*, the result of (5) is added to the value obtained from (1). Thus sin 33° 29′ 34″ is .55169 + 14 = .55183.

Example b: Find the angle in degrees, minutes, and seconds when its cotangent is 1.6395.

Solution: 1. Find this angle in degrees and minutes as previously explained (31° 22′).

2. Note the tabulated difference between the cot 31° 22′ and cot 31° 23′, which in this case is 11.

3. Obtain the difference between 1.6395 and the next larger value, (since this is a *cofunction*) 1.6404, which is 9.

4. Referring to the table of proportional parts, in the horizontal row opposite 11, the next number smaller than 9 is 8.3 which corresponds to 45″.

5. Subtracting 8.3 from 9 leaves .7. To find the number of seconds corresponding to this .7, move the decimal point one place to the left for both the angles and the differences. The nearest number to .7 which corresponds to a whole number of

seconds is then 7.3, which corresponds to 40″, so the additional angle for .7 is 4″. Thus the total number of seconds is 45″ + 4″ = 49″. Hence, the angle is 31° 22′ 49″.

PROBLEMS

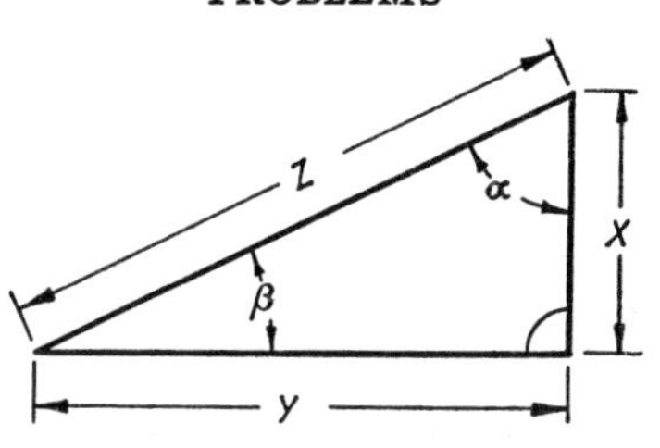

Insert the values given in the following tabular form in their proper places according to the foregoing diagram and solve for the distance, or angle in question. When solving for a distance, the result must be correct to five significant figures. When solving for an angle, the result must be correct to degrees, minutes, and seconds.

Prob.	X	Y	Z	β	α	Determine
1	3.567		E			Angle β
2		F		36° 17′ 18″		Distance Z
3	5.763	G				Angle α
4	5.783	D				Distance Z
5			H		67° 27′ 38″	Distance Y
6	J		7.892			Angle β
7		K	8.291			Angle α

Variables

Prob.	Sym.	No. 1	No. 2	No. 3	No. 4	No. 5	No. 6
1	E	6.258	6.879	7.138	7.386	7.897	8.207
2	F	2.789	3.569	3.896	4.689	4.973	5.289
3	G	12.87	13.24	13.59	13.96	14.42	14.78
4	D	8.875	8.923	9.134	9.356	9.785	9.982
5	H	4.679	4.876	5.136	5.297	5.587	5.956
6	J	1.876	2.196	2.375	2.869	3.158	3.621
7	K	5.347	5.682	5.913	6.147	6.258	6.873

PROBLEMS (*Continued*)

Prob.	X	Y	Z	β	α	Determine
8	L			25° 31′ 42″		Distance Y
9		M			53° 52′ 43″	Distance Z

VARIABLES

Prob.	Sym.	No. 1	No. 2	No. 3	No. 4	No. 5	No. 6
8	L	14.34	14.89	15.13	15.67	15.89	16.14
9	M	3.675	3.875	3.146	4.472	4.735	4.963

PRACTICAL PROBLEMS OCCURRING IN TOOL AND DIE ROOMS, MACHINE CONSTRUCTION, MACHINE REPAIR, AND DRAFTING ROOMS

The general run of the trigonometric problems occurring daily in industry are similar to those shown in this text. The problems in this text fall in several classifications and involve a wide range of the geometrical propositions. The knowledge and thorough understanding of this wide range of geometrical propositions and the ability to apply the correct geometrical proposition to practical shop problems are necessary in order to cope with the trigonometric problems that occur in tool, die, machine construction, machine repair, and drafting rooms.

There are two distinctly different types of trigonometric problems that occur frequently in industry. One of the types is of a standard nature and occurs quite frequently, for example, the dovetail, V block, and many other standard forms all of which are identical in nature and are generally formulated. The other type, equally as important, consists of problems that are a little more difficult and that are not likely to occur again in an identical manner. However, a thorough knowledge of the solutions of the plane trigonometric problems given in this text, since they involve several geometrical classifications, often enables one to find a solution to most other trigonometric problems.

All trigonometric problems occurring in industry fall into certain geometrical classifications, most of which are given in this text. This text contains a wide array of practical shop problems taken directly from tool, die, machine construction, machine repair, and drafting rooms and arranged in a systematical order according to their difficulty.

The previous problems in trigonometry were of the simplest type in which the student was asked to solve for an unknown angle or side of a right triangle. In the following problems, the student must become familiar with a method of locating right triangles by drawing construction lines wherever necessary. The method of drawing the proper construction lines wherever necessary is acquired by diligent practice and from the mastered solutions of the preceding problems.

The forthcoming problems have been taken from blue prints that were drawn by either engineers or draftsmen. As a general rule, the draftsman presents the object to be machined in the form of a blue print that shows all the necessary dimensions of the concrete object. However, in most cases the given dimensions by the engineer or draftsman are not sufficient for the mechanic to set up the job properly on machines or on adapters especially designed for machining certain types of operations. The trigonometric problems given in this text have been drawn according to the general procedures and customs used in most drafting rooms. That is, if a line has the appearance of being parallel or perpendicular to other given lines, it must be understood, unless otherwise stated, that they are either parallel or perpendicular, respectively. The right angles formed by construction lines are indicated with a small arc at the base of the line.

The first 45 problems are relatively simple and may all be reduced to right triangles by drawing only a few construction lines. The partial solution will be given for several of these problems in order to teach the student how to construct properly the necessary auxiliary lines. On others, the construction lines have been drawn to assist the student in solving them, and the rest are left entirely to the student in order to promote self-reliance.

Lathe Center

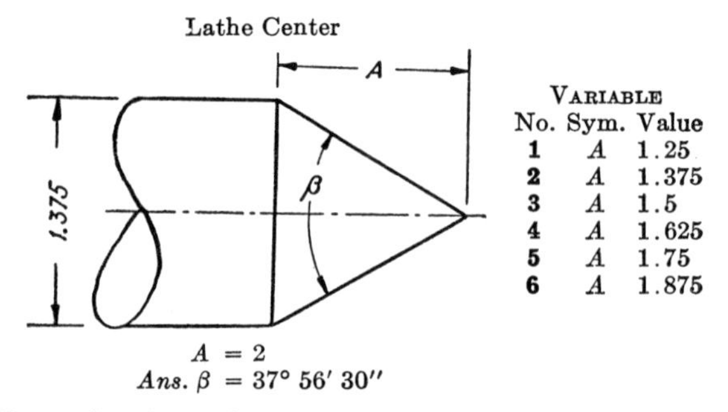

No.	Sym.	Value
1	A	1.25
2	A	1.375
3	A	1.5
4	A	1.625
5	A	1.75
6	A	1.875

(Variable)

$A = 2$

Ans. $\beta = 37° 56' 30''$

1. Determine the angle β.

Flange

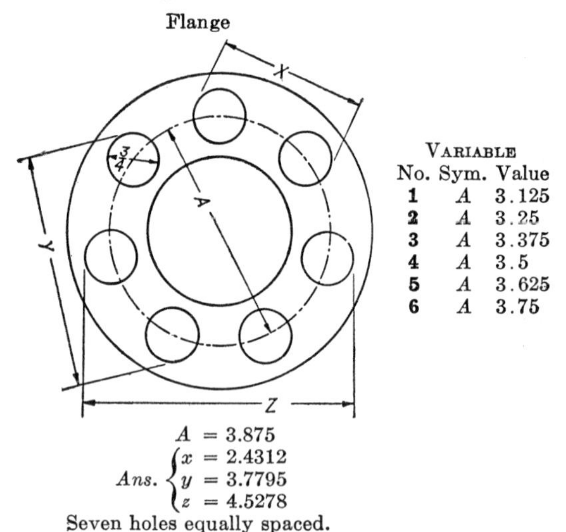

No.	Sym.	Value
1	A	3.125
2	A	3.25
3	A	3.375
4	A	3.5
5	A	3.625
6	A	3.75

(Variable)

$A = 3.875$

Ans. $\begin{cases} x = 2.4312 \\ y = 3.7795 \\ z = 4.5278 \end{cases}$

Seven holes equally spaced.

2. Determine the distance x.

3. Determine the distance y.

4. Determine the distance z.

Note. To express decimal degrees in degrees, minutes, and seconds, multiply the fractional part of the degree by 60 to obtain the minutes and multiply the resulting fractional part of a minute by 60 to obtain the seconds.

Thus: $25.3673° = 25° 22' 2''$

$$\begin{array}{r} 60 \\ \hline 22.0380' \\ 60 \\ \hline 2.2800'' \end{array}$$

Gage

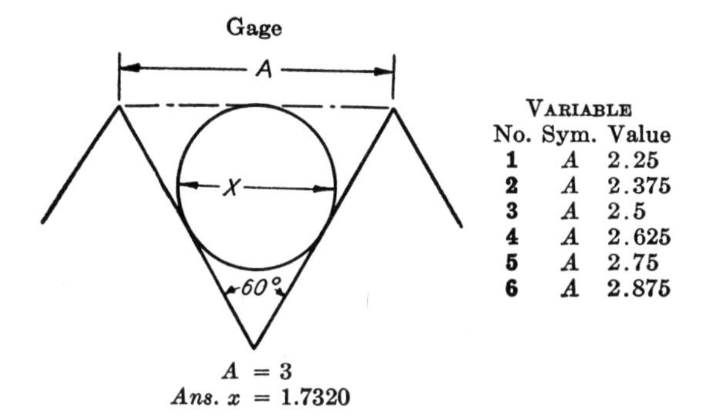

VARIABLE		
No.	Sym.	Value
1	A	2.25
2	A	2.375
3	A	2.5
4	A	2.625
5	A	2.75
6	A	2.875

$A = 3$
Ans. $x = 1.7320$

5. Determine the diameter x.

Gage

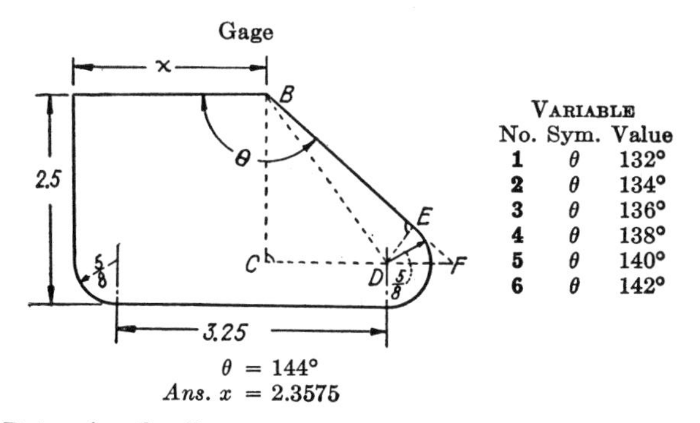

VARIABLE		
No.	Sym.	Value
1	θ	132°
2	θ	134°
3	θ	136°
4	θ	138°
5	θ	140°
6	θ	142°

$\theta = 144°$
Ans. $x = 2.3575$

6. Determine the distance x.

Solution:

$BC = 2.5 - .625.$

$\angle CBF = \theta - 90°.$

In $\triangle CFB$, solve for CF.

$\angle EDF = \theta - 90°.$ Why?

In $\triangle DFE$, solve for DF.

$CD = CF - DF.$

V-Block

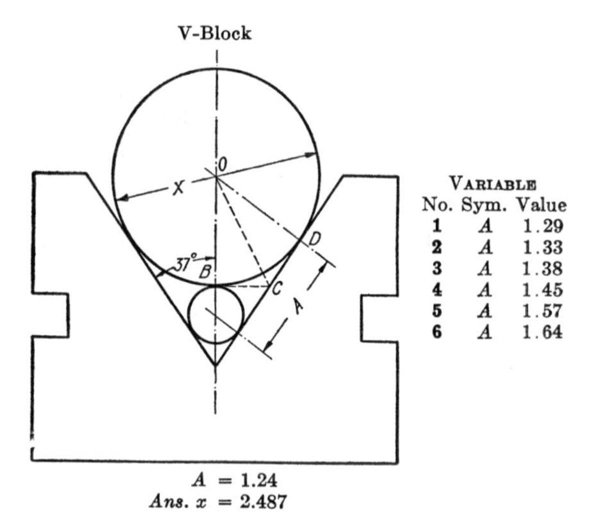

Variable		
No.	Sym.	Value
1	A	1.29
2	A	1.33
3	A	1.38
4	A	1.45
5	A	1.57
6	A	1.64

$A = 1.24$
Ans. $x = 2.487$

7. Determine the diameter x.

Solution:

$\angle DOB = 90° - 37°$.

The tangents BC and CD are equal.

$\therefore BC = A \div 2$.

$\angle BOC = \angle DOB \div 2$. Why?

In $\triangle BOC$, solve for OB.

$OB = x \div 2$. Why?

End View of Gib

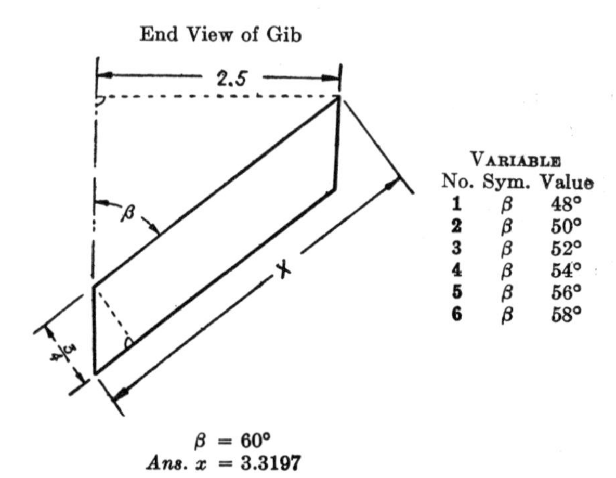

Variable		
No.	Sym.	Value
1	β	48°
2	β	50°
3	β	52°
4	β	54°
5	β	56°
6	β	58°

$\beta = 60°$
Ans. $x = 3.3197$

8. Determine the distance x.

Gage

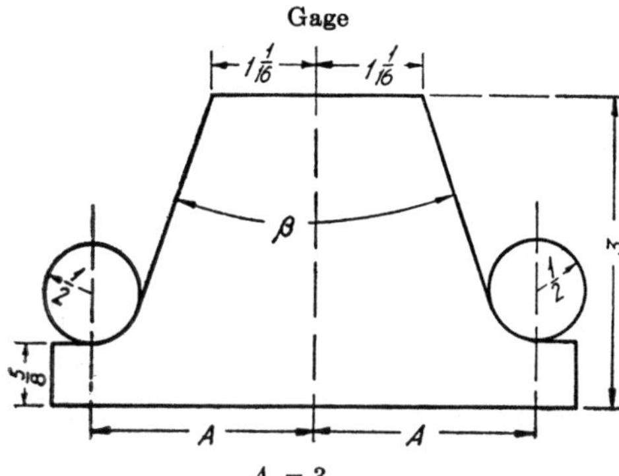

Variable		
No.	Sym.	Value
1	*A*	2.25
2	*A*	2.375
3	*A*	2.5
4	*A*	2.625
5	*A*	2.75
6	*A*	2.875

$A = 3$

Ans. $\beta = 70° \ 30' \ 16''$

9. Determine the angle β.

Worm or Acme Thread

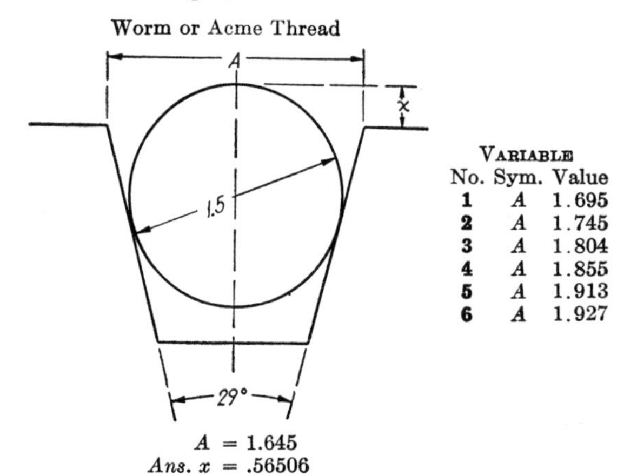

Variable		
No.	Sym.	Value
1	*A*	1.695
2	*A*	1.745
3	*A*	1.804
4	*A*	1.855
5	*A*	1.913
6	*A*	1.927

$A = 1.645$

Ans. $x = .56506$

10. Determine the distance x.

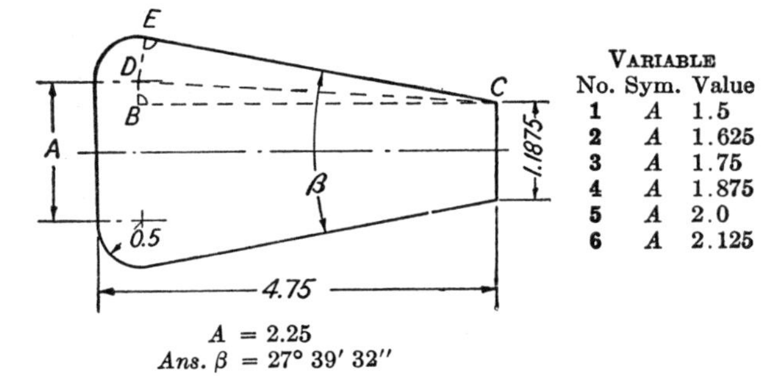

Variable		
No.	Sym.	Value
1	*A*	1.5
2	*A*	1.625
3	*A*	1.75
4	*A*	1.875
5	*A*	2.0
6	*A*	2.125

$A = 2.25$

Ans. $\beta = 27° \ 39' \ 32''$

11. Determine the angle β. (*Solution on next page.*)

Solution for preceding problem:

$BD = (A - 1.1875) \div 2.$

In $\triangle BCD$, solve for $\angle DCB$ and CD.

In $\triangle DCE$, solve for $\angle DCE$.

$\beta = 2(\angle BCD + \angle DCE).$

Dovetail

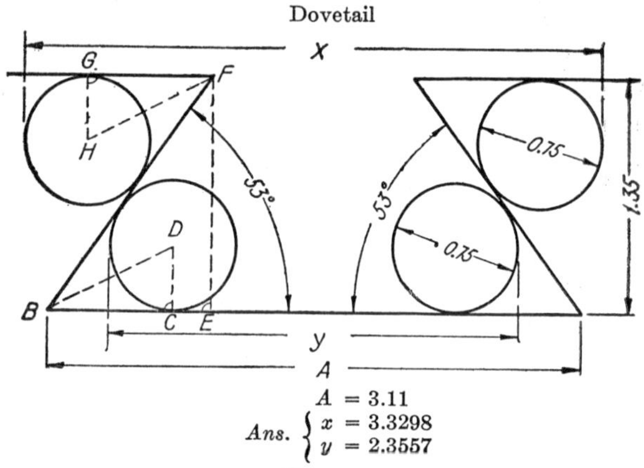

$A = 3.11$

Ans. $\begin{cases} x = 3.3298 \\ y = 2.3557 \end{cases}$

VARIABLE

1. $A = 3.18$	**2.** $A = 3.26$	**3.** $A = 3.31$
4. $A = 3.39$	**5.** $A = 3.46$	**6.** $A = 3.58$

12. Determine the distance x.

13. Determine the distance y.

Solution:

In $\triangle BEF$, solve for BE.

$\angle DBC = 53° \div 2.$

In $\triangle DBC$, solve for BC.

$GF = BC$. Why?

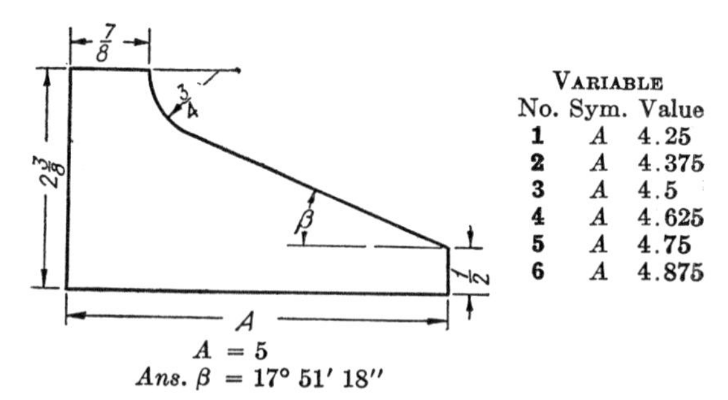

$A = 5$

Ans. $\beta = 17° \; 51' \; 18''$

VARIABLE

No.	Sym.	Value
1	A	4.25
2	A	4.375
3	A	4.5
4	A	4.625
5	A	4.75
6	A	4.875

14. Determine the angle β.

Die Punch

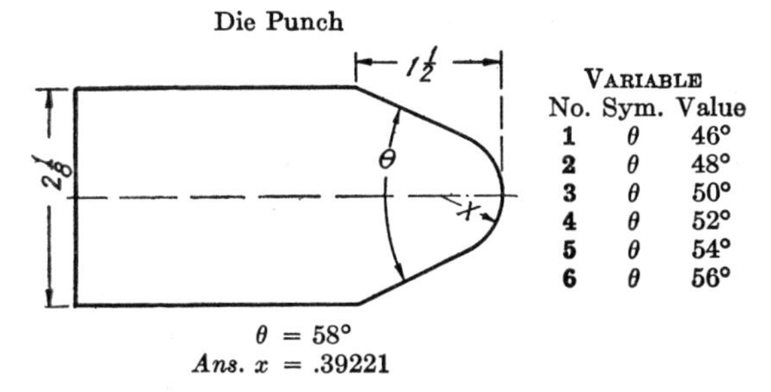

VARIABLE

No.	Sym.	Value
1	θ	46°
2	θ	48°
3	θ	50°
4	θ	52°
5	θ	54°
6	θ	56°

$\theta = 58°$
Ans. $x = .39221$

15. Determine the radius x.

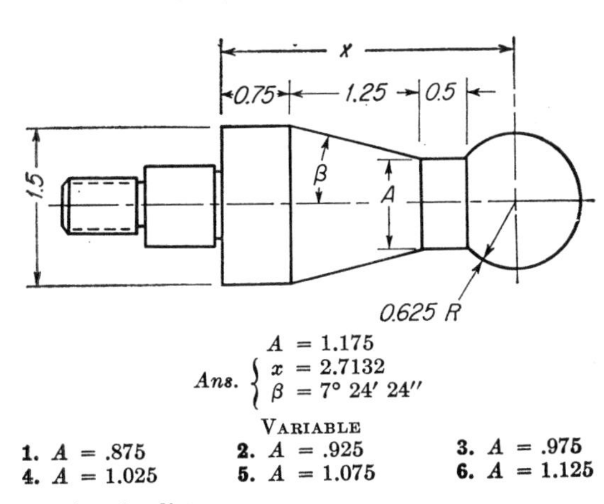

$A = 1.175$

Ans. $\begin{cases} x = 2.7132 \\ \beta = 7° \ 24' \ 24'' \end{cases}$

VARIABLE

1. $A = .875$ **2.** $A = .925$ **3.** $A = .975$
4. $A = 1.025$ **5.** $A = 1.075$ **6.** $A = 1.125$

16. Determine the distance x.

Gage

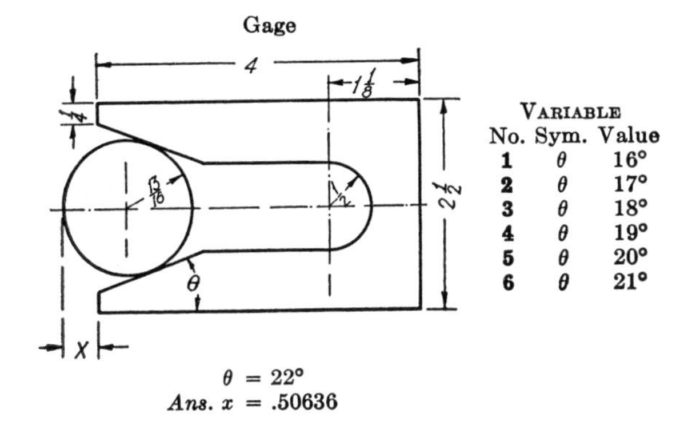

VARIABLE

No.	Sym.	Value
1	θ	16°
2	θ	17°
3	θ	18°
4	θ	19°
5	θ	20°
6	θ	21°

$\theta = 22°$
Ans. $x = .50636$

17. Determine the distance x.

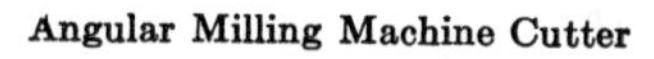

Angular Milling Machine Cutter

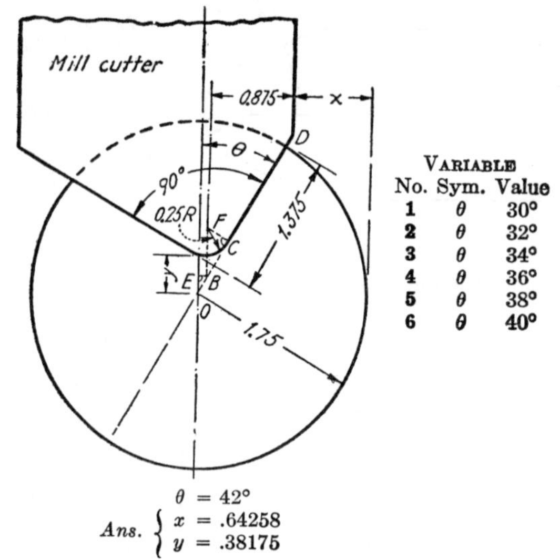

VARIABLE		
No.	Sym.	Value
1	θ	30°
2	θ	32°
3	θ	34°
4	θ	36°
5	θ	38°
6	θ	40°

$\theta = 42°$

Ans. $\begin{cases} x = .64258 \\ y = .38175 \end{cases}$

18. Determine the distance x.

19. Determine the distance y.

Solution:

$CD = 1.375 - CF.$

In $\triangle BCF$, solve for BC and BF.

$OB = 1.75 - BC - CD.$

In $\triangle OBE$, solve for BE and EO

$x = 1.75 - .875 - BE.$

$y = EO + BF - .25.$

Gage

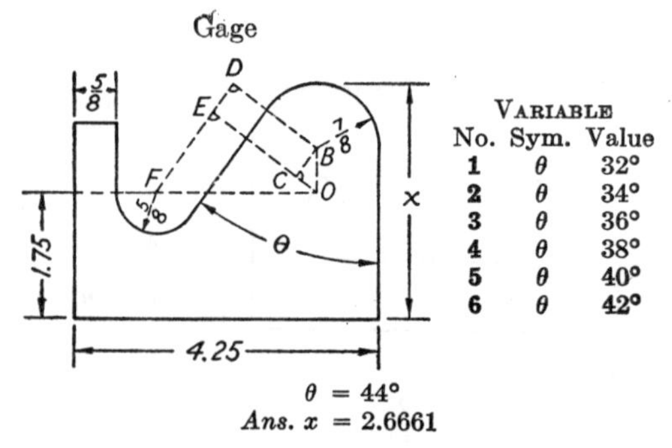

VARIABLE		
No.	Sym.	Value
1	θ	32°
2	θ	34°
3	θ	36°
4	θ	38°
5	θ	40°
6	θ	42°

$\theta = 44°$

Ans. $x = 2.6661$

20. Determine the distance x.

Dotted lines show diagrammatic hint.

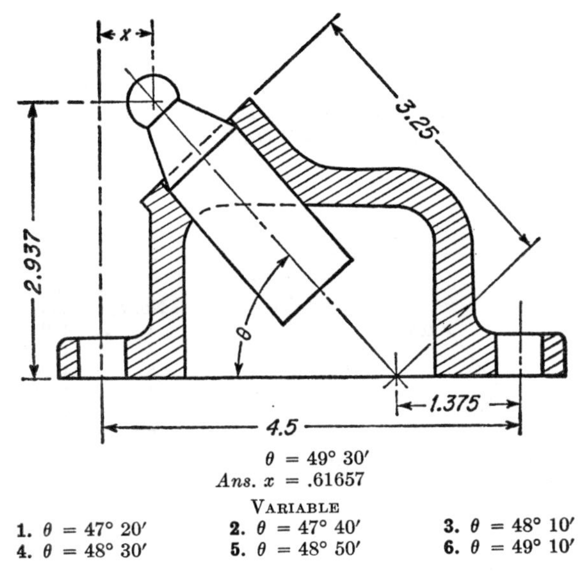

$\theta = 49° 30'$
Ans. $x = .61657$

VARIABLE

1. $\theta = 47° 20'$	**2.** $\theta = 47° 40'$	**3.** $\theta = 48° 10'$
4. $\theta = 48° 30'$	**5.** $\theta = 48° 50'$	**6.** $\theta = 49° 10'$

21. Determine the distance x.
Note: The distance 2.937 is checked with a height gage.

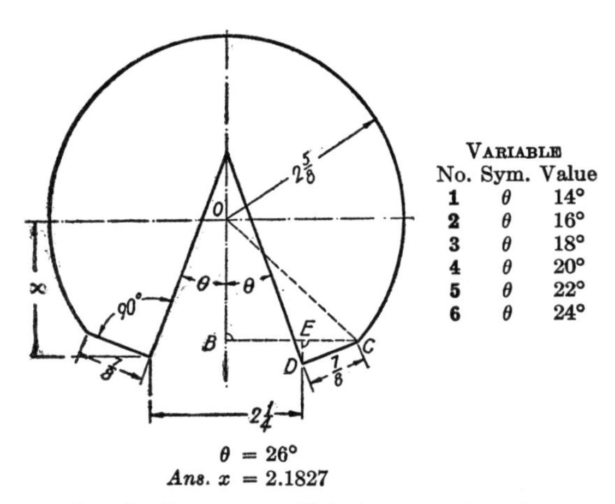

VARIABLE

No.	Sym.	Value
1	θ	14°
2	θ	16°
3	θ	18°
4	θ	20°
5	θ	22°
6	θ	24°

$\theta = 26°$
Ans. $x = 2.1827$

22. Determine the distance x. (*Solution on next page.*)

Solution for preceding problem:

$\angle ECD = \theta$. Why?
In $\triangle DEC$, solve for CE and DE.
$BE = 2.25 \div 2$. $BC = BE + CE$.
In $\triangle BCO$, solve for $\angle BOC$ and OB.
$OB + DE = x$.

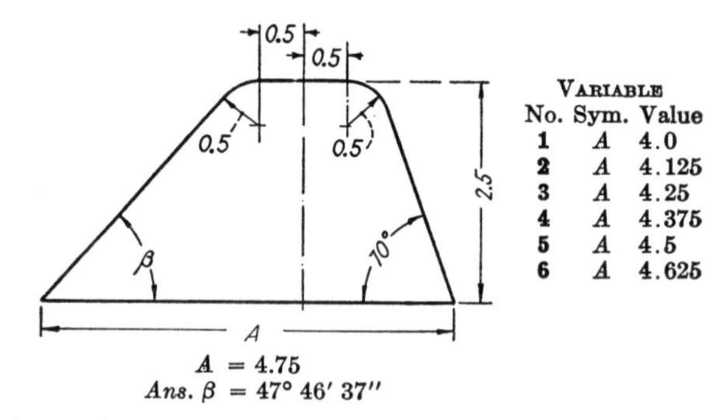

No.	Sym.	Value
1	A	4.0
2	A	4.125
3	A	4.25
4	A	4.375
5	A	4.5
6	A	4.625

$A = 4.75$
Ans. $\beta = 47° 46' 37''$

23. Determine the angle β.

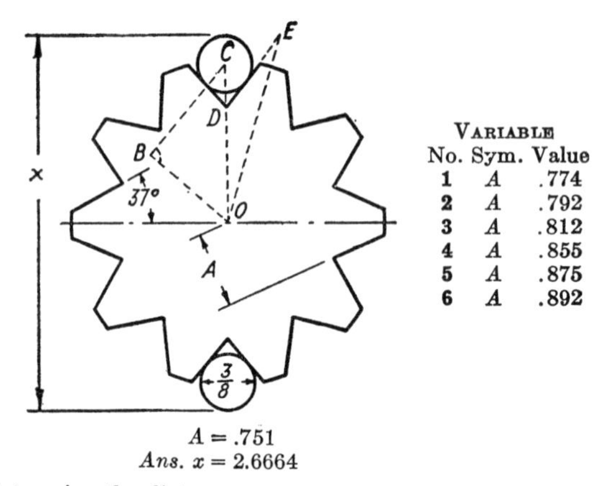

No.	Sym.	Value
1	A	.774
2	A	.792
3	A	.812
4	A	.855
5	A	.875
6	A	.892

$A = .751$
Ans. $x = 2.6664$

24. Determine the distance x.
Solution:

$\angle DEO = 37°$. $\angle DOE = 180° \div 10$. Why?
$\angle CDE = \angle DEO + \angle DOE$. Why?
$\angle BCO = \angle CDE$. Why?
$BO = A + .1875$.
In $\triangle BOC$, solve for CO.

Die Section

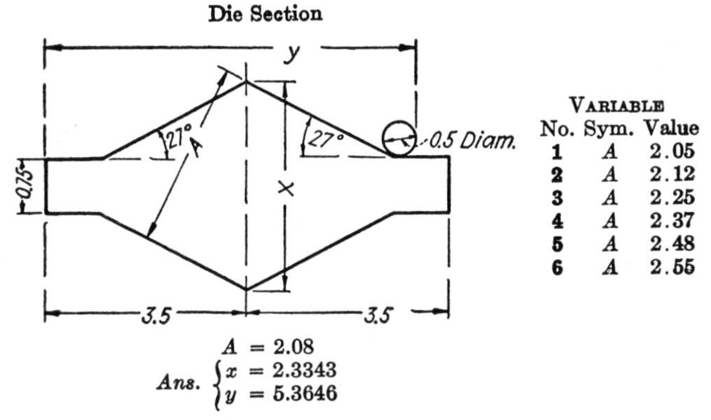

Variable		
No.	Sym.	Value
1	A	2.05
2	A	2.12
3	A	2.25
4	A	2.37
5	A	2.48
6	A	2.55

$A = 2.08$

Ans. $\begin{cases} x = 2.3343 \\ y = 5.3646 \end{cases}$

25. Determine the value of x.

26. Determine the value of y.

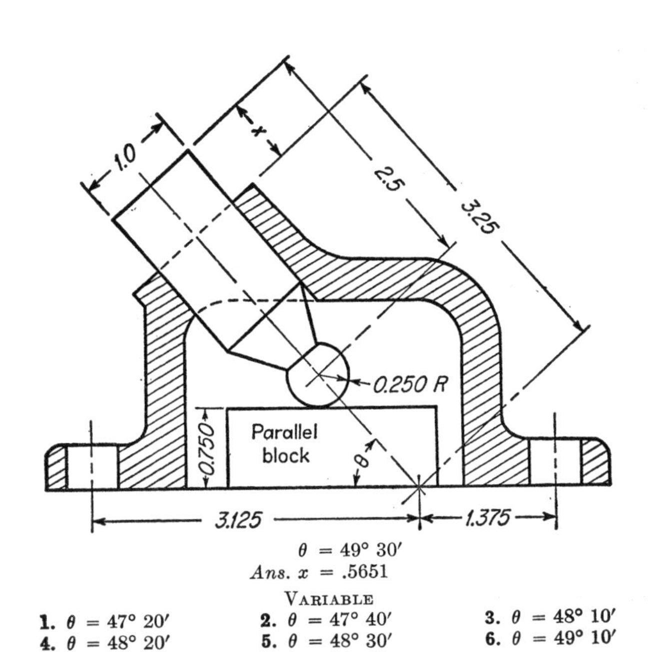

$\theta = 49° 30'$

Ans. $x = .5651$

Variable

1. $\theta = 47° 20'$ **2.** $\theta = 47° 40'$ **3.** $\theta = 48° 10'$

4. $\theta = 48° 20'$ **5.** $\theta = 48° 30'$ **6.** $\theta = 49° 10'$

27. Determine the distance x.

Form Roller

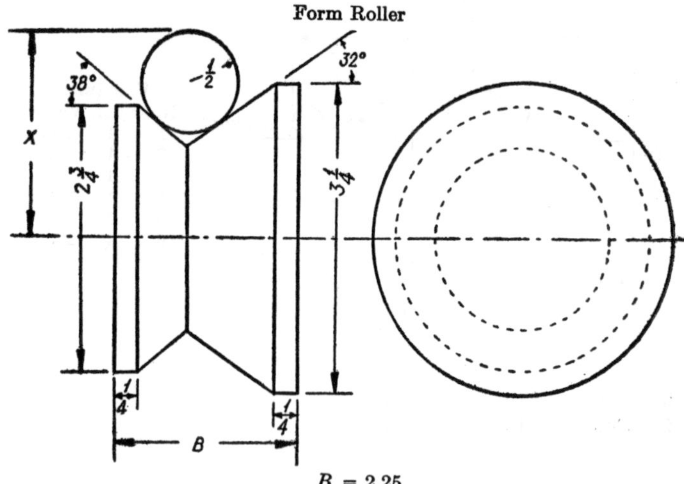

$B = 2.25$
Ans. $x = 2.0159$

VARIABLE

1. $B = 1.5$	**2.** $B = 1.625$	**3.** $B = 1.75$
4. $B = 1.875$	**5.** $B = 2.0$	**6.** $B = 2.125$

28. Determine the distance x.

Gage

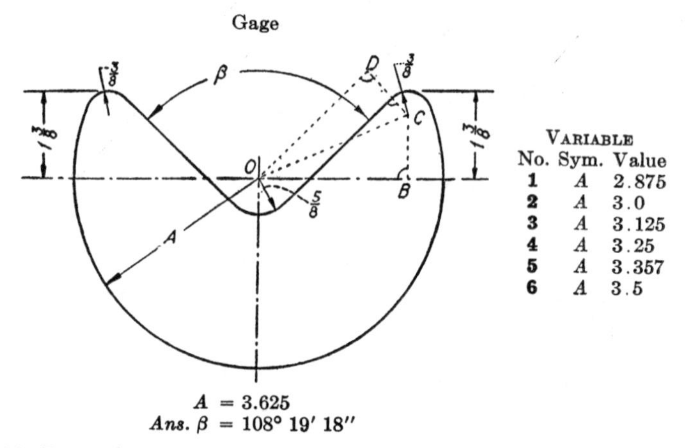

VARIABLE

No.	Sym.	Value
1	A	2.875
2	A	3.0
3	A	3.125
4	A	3.25
5	A	3.357
6	A	3.5

$A = 3.625$
Ans. $\beta = 108° \; 19' \; 18''$

29. Determine the angle β.

Solution:

$OC = A - .375$. $CB = 1.375 - .375$.

In $\triangle OBC$, solve for $\angle COB$.

$CD = .625 + .375$. Why? (*Continued on next page.*)

Solution continued:

In $\triangle ODC$, solve for $\angle DOC$.

$\beta = 2(90° - \angle COB - \angle DOC)$.

Gage

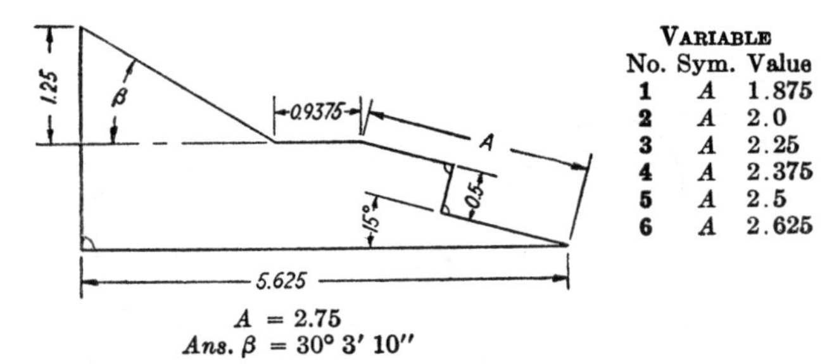

No.	Sym.	Value
1	A	1.875
2	A	2.0
3	A	2.25
4	A	2.375
5	A	2.5
6	A	2.625

$A = 2.75$

Ans. $\beta = 30° 3' 10''$

30. Determine the angle β.

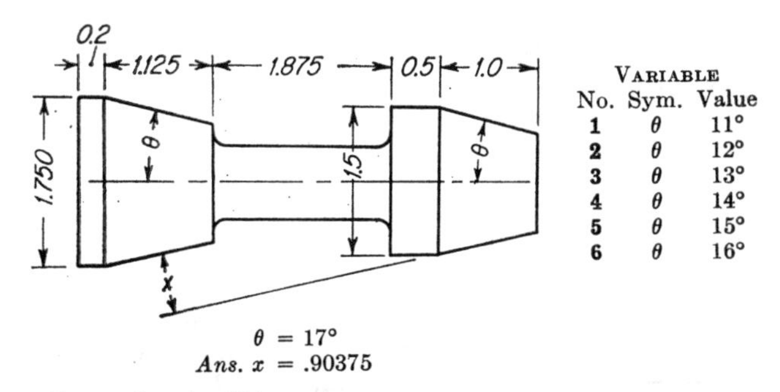

No.	Sym.	Value
1	θ	11°
2	θ	12°
3	θ	13°
4	θ	14°
5	θ	15°
6	θ	16°

$\theta = 17°$

Ans. $x = .90375$

31. Determine the distance x.

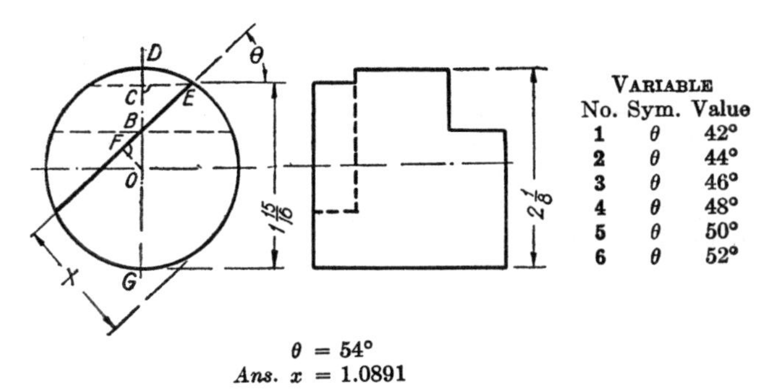

No.	Sym.	Value
1	θ	42°
2	θ	44°
3	θ	46°
4	θ	48°
5	θ	50°
6	θ	52°

$\theta = 54°$

Ans. $x = 1.0891$

32. Determine the distance x. (*Solution on next page.*)

Solution for preceding problem:

$CG = 1.9375$. $CD = 2.125 - 1.9375$.

In circle GED, solve for CE by geometry by P-39 and P-45.

In $\triangle BCE$, solve for BC. $OC = 1.9375 - (2.125 \div 2)$.

$OB = OC - BC$.

In $\triangle OFB$, solve for OF.

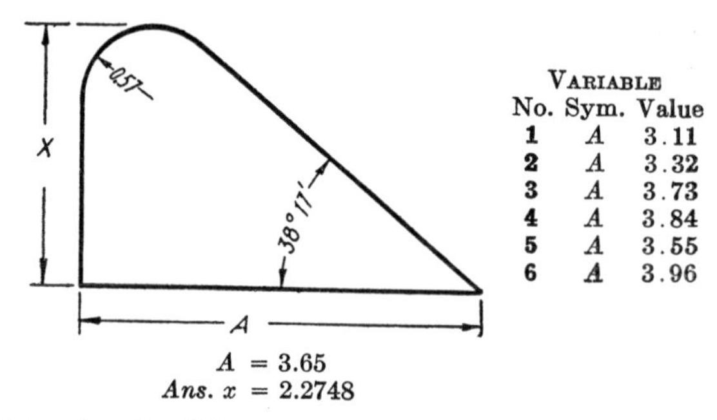

VARIABLE		
No.	Sym.	Value
1	A	3.11
2	A	3.32
3	A	3.73
4	A	3.84
5	A	3.55
6	A	3.96

$A = 3.65$

Ans. $x = 2.2748$

33. Determine the distance x.

V-Block

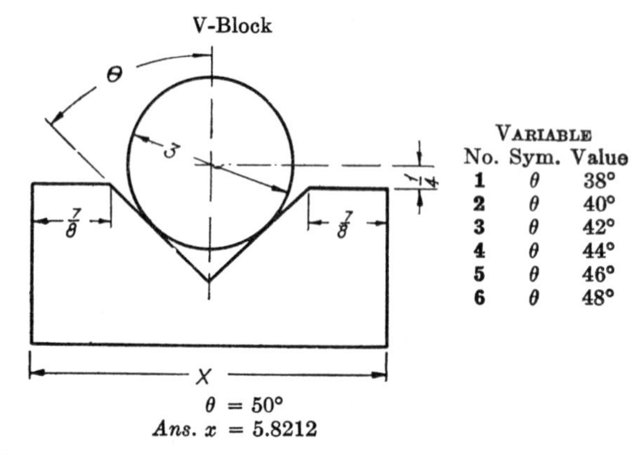

VARIABLE		
No.	Sym.	Value
1	θ	38°
2	θ	40°
3	θ	42°
4	θ	44°
5	θ	46°
6	θ	48°

$\theta = 50°$

Ans. $x = 5.8212$

34. Determine the distance x.

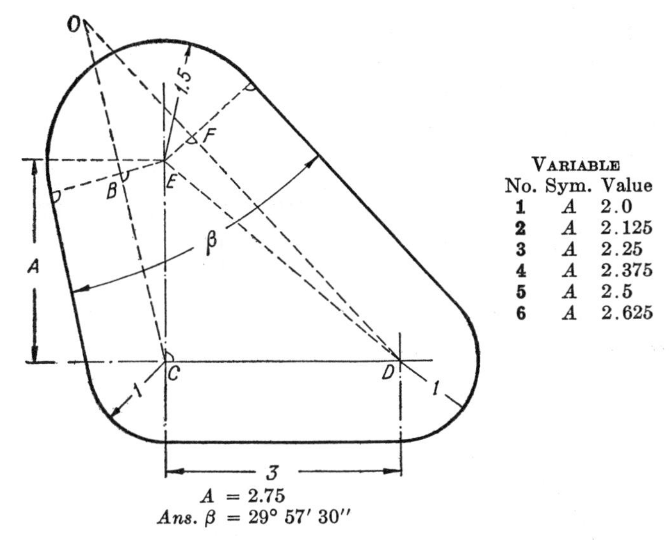

No.	Sym.	Value
1	A	2.0
2	A	2.125
3	A	2.25
4	A	2.375
5	A	2.5
6	A	2.625

$A = 2.75$
Ans. $\beta = 29° \ 57' \ 30''$

35. Determine the angle β.
Solution:

$BE = 1.5 - 1$. $EF = 1.5 - 1$. $\angle COD = \beta$. Why?
In $\triangle BCE$, solve for $\angle BCE$. In $\triangle ECD$, solve for $\angle CDE$ and DE.
$\angle BCD = \angle BCE + 90°$. In $\triangle EDF$, solve for $\angle EDF$.
$\angle CDF = \angle CDE + \angle EDF$.
$\beta = \angle COD = 180° - \angle BCD - \angle CDF$.

Dovetail Section

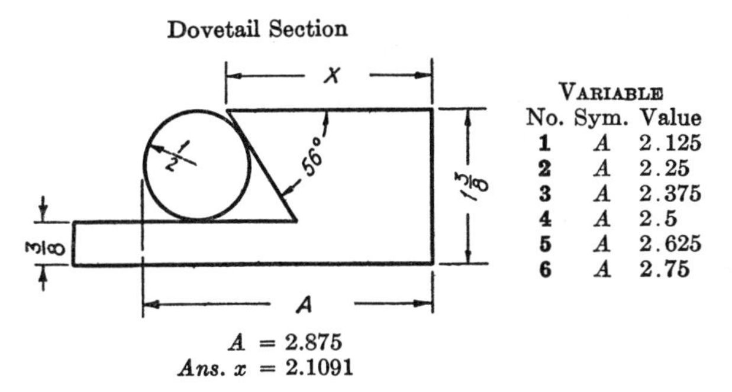

No.	Sym.	Value
1	A	2.125
2	A	2.25
3	A	2.375
4	A	2.5
5	A	2.625
6	A	2.75

$A = 2.875$
Ans. $x = 2.1091$

36. Determine the distance x.

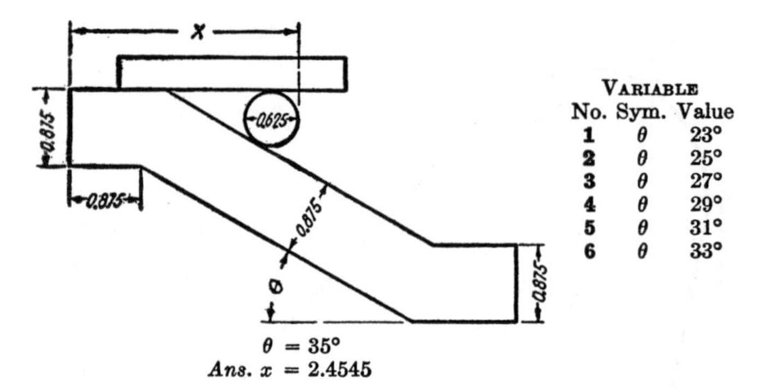

$\theta = 35°$
Ans. $x = 2.4545$

No.	Sym.	Value
1	θ	23°
2	θ	25°
3	θ	27°
4	θ	29°
5	θ	31°
6	θ	33°

37. Determine the distance x.

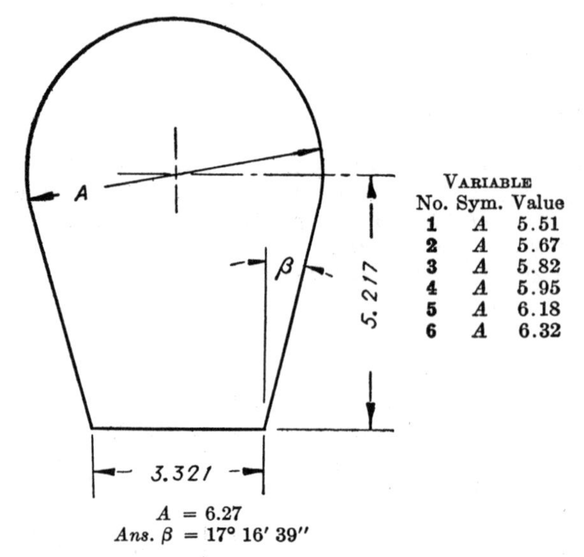

$A = 6.27$
Ans. $\beta = 17° 16' 39''$

No.	Sym.	Value
1	A	5.51
2	A	5.67
3	A	5.82
4	A	5.95
5	A	6.18
6	A	6.32

38. Determine the angle β.

Fixture Section

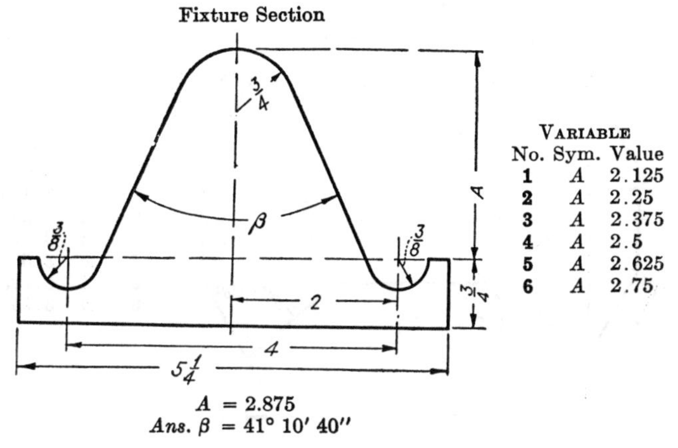

VARIABLE		
No.	Sym.	Value
1	A	2.125
2	A	2.25
3	A	2.375
4	A	2.5
5	A	2.625
6	A	2.75

$A = 2.875$
Ans. $\beta = 41° 10' 40''$

39. Determine the angle β.

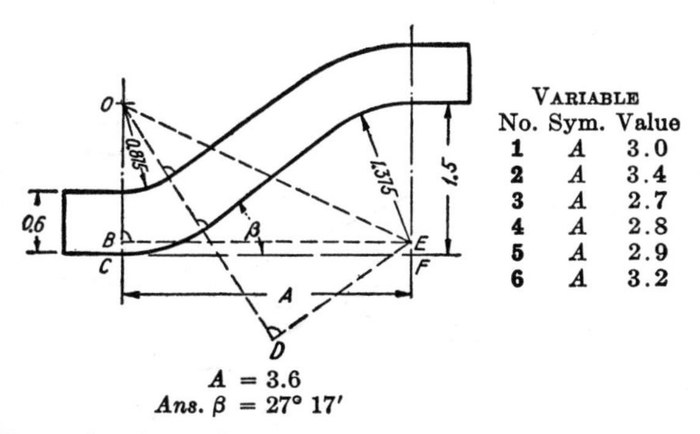

VARIABLE		
No.	Sym.	Value
1	A	3.0
2	A	3.4
3	A	2.7
4	A	2.8
5	A	2.9
6	A	3.2

$A = 3.6$
Ans. $\beta = 27° 17'$

40. Determine the angle β.

Solution:

$EF = BC = 1.5 - 1.375$. $OB = OC - BC$.
In $\triangle OBE$, solve for $\angle BOE$ and OE.
$OD = .875 + .6 + 1.375$. Why?
In $\triangle ODE$, solve for $\angle DOE$.
$\beta = \angle BOE - \angle DOE$. Why?

Gage

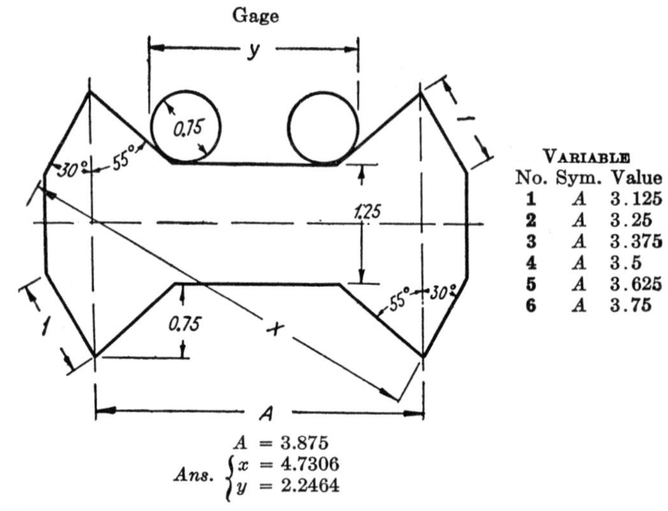

No.	Sym.	Value
1	A	3.125
2	A	3.25
3	A	3.375
4	A	3.5
5	A	3.625
6	A	3.75

Variable

$A = 3.875$

Ans. $\begin{cases} x = 4.7306 \\ y = 2.2464 \end{cases}$

41. Determine the distance x.

42. Determine the distance y.

Gage

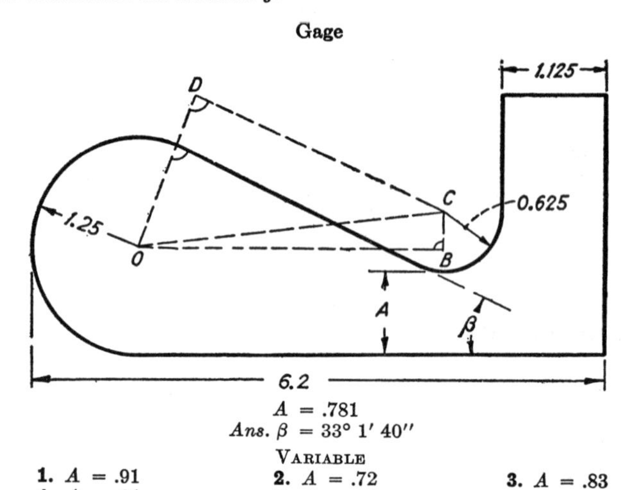

$A = .781$

Ans. $\beta = 33° 1' 40''$

Variable

1. $A = .91$ **2.** $A = .72$ **3.** $A = .83$

4. $A = .94$ **5.** $A = .65$ **6.** $A = .76$

43. Determine the angle β.

Solution:

$BC = A + .625 - 1.25.$

In $\triangle OBC$, solve for $\angle BOC$ and CO.

In $\triangle ODC$, solve for $\angle DOC$.

$\beta = 90° - \angle BOC - \angle DOC.$ Why?

Spline Shaft

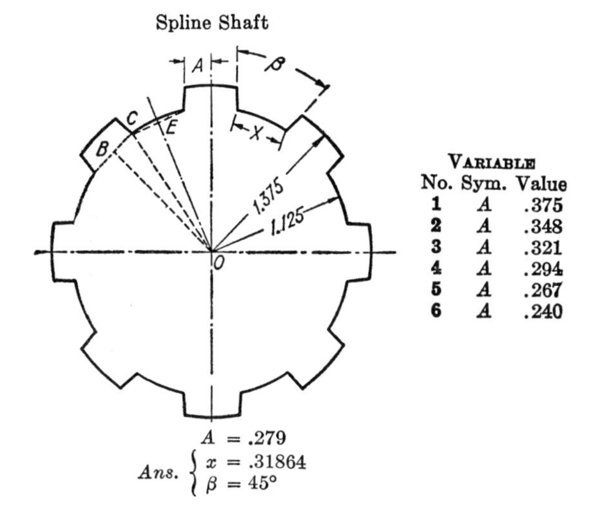

No.	Sym.	Value
1	A	.375
2	A	.348
3	A	.321
4	A	.294
5	A	.267
6	A	.240

Variable

$A = .279$

Ans. $\begin{cases} x = .31864 \\ \beta = 45° \end{cases}$

44. Determine the distance x.

45. Determine the angle β.

Solution:

$$\angle BOE = \frac{180°}{\text{number of splines}}.$$

In $\triangle BCO$, solve for $\angle BOC$. $\angle COE = \angle BOE - \angle BOC$.

In $\triangle COE$, solve for CE. $x = 2(CE)$

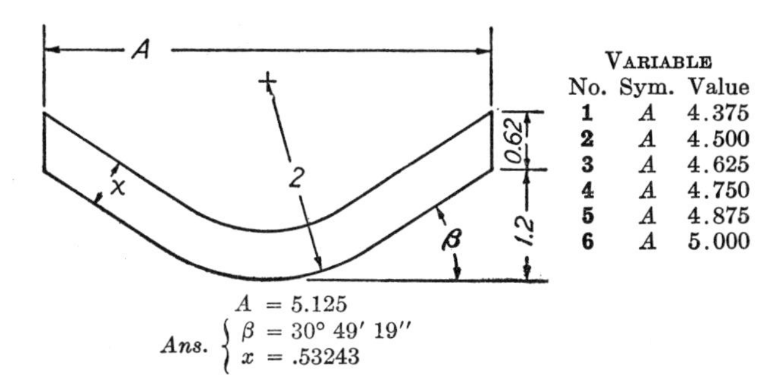

No.	Sym.	Value
1	A	4.375
2	A	4.500
3	A	4.625
4	A	4.750
5	A	4.875
6	A	5.000

Variable

$A = 5.125$

Ans. $\begin{cases} \beta = 30° \, 49' \, 19'' \\ x = .53243 \end{cases}$

46. Determine the angle β.

47. Determine the distance x.

Pivoting Block Device

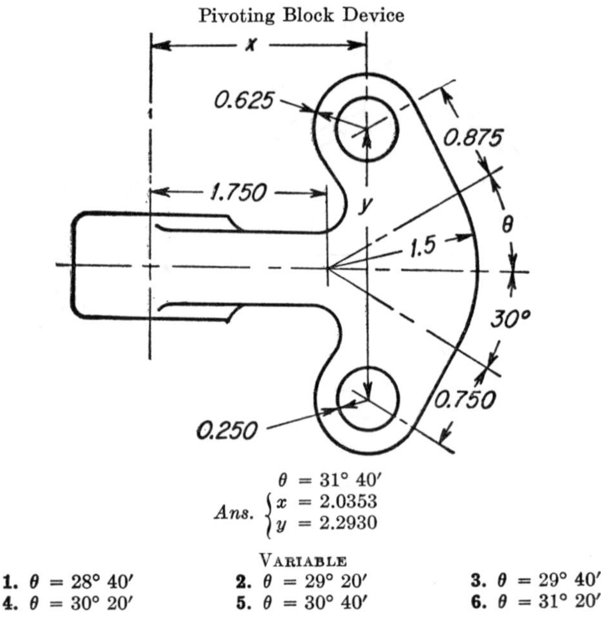

$$Ans. \begin{cases} \theta = 31^\circ 40' \\ x = 2.0353 \\ y = 2.2930 \end{cases}$$

VARIABLE

1. $\theta = 28^\circ 40'$	**2.** $\theta = 29^\circ 20'$	**3.** $\theta = 29^\circ 40'$
4. $\theta = 30^\circ 20'$	**5.** $\theta = 30^\circ 40'$	**6.** $\theta = 31^\circ 20'$

48. Determine the distance x.
49. Determine the distance y.

OBLIQUE TRIANGLES

Many problems may be reduced directly to right triangles (as those of the previous group), but others must first be reduced to oblique triangles (D-29).

These oblique triangles may, in turn, be reduced to right triangles. If the three altitudes (D-30) of an oblique triangle are drawn, six right triangles are formed, as shown in Figs. 145 and 146.

The six right triangles formed in both the acute and the obtuse triangles are ADC, BDC, AFB, AFC, BEC, and BEA.

The **law of sines** and the **law of cosines** may also be applied to the solution of oblique triangles (see pages 251 and 258).

Whenever it is desired to solve for any part of an oblique triangle by reducing it to right triangles, only two of the six right triangles are necessary. The two to be used depend upon what is given and required. The procedure for selecting the proper two right triangles is as follows:

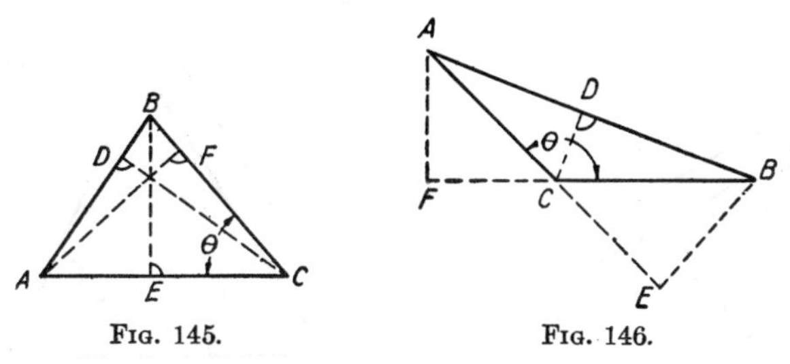

FIG. 145. FIG. 146.

Given: AB, AC, and $\angle\theta$.

To solve for $\angle ABC$ and side BC.

Since the $\angle\theta$ is given, the altitude from C cannot be used since it would divide the known angle into two unknown parts, thus giving two rt. ⊿s BDC and CDA, the first having only the right angle known and the second having the right angle and only one other part known. Hence, these right triangles cannot be used to find any unknown parts (see statement on page 202).

Drawing the altitudes from B would give two rt. ⊿s BEC and BEA each of which contains a right angle and only one other part known. The altitude AF forms two rt. ⊿s AFB and AFC, the first of which cannot be used (insufficient parts). The $\triangle AFC$, however, is determined because it has three known parts, AC, rt. $\angle AFC$ and $\angle ACF$ (which is θ or $180° - \theta$), and any unknown part of it may be obtained. Thus the side AF may be computed. Then the desired angle ABC may be obtained since in the rt. $\triangle AFB$, the parts AB, AF and the rt. $\angle AFB$ are given. In obtaining BC, note that in the acute $\triangle ABC$, $BC = BF + FC$ and in the obtuse $\triangle ABC$, $BC = BF - FC$. BF may be obtained from rt. $\triangle BFA$ which now contains the three known parts, AB, rt. $\angle BFA$ and $\angle ABF$. FC may be obtained from the rt. $\triangle AFC$ which now contains the three known parts, AF, rt. $\angle AFC$, and $\angle\theta$ (or $180° - \theta$). Thus BC can be obtained.

PROBLEMS

Simple Oblique Triangles

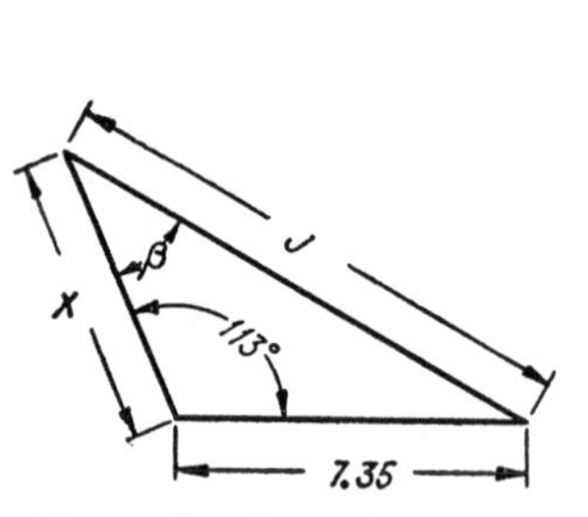

1. Determine the angle β.
2. Determine the distance x.
3. Determine the distance x.
4. Determine the distance y.

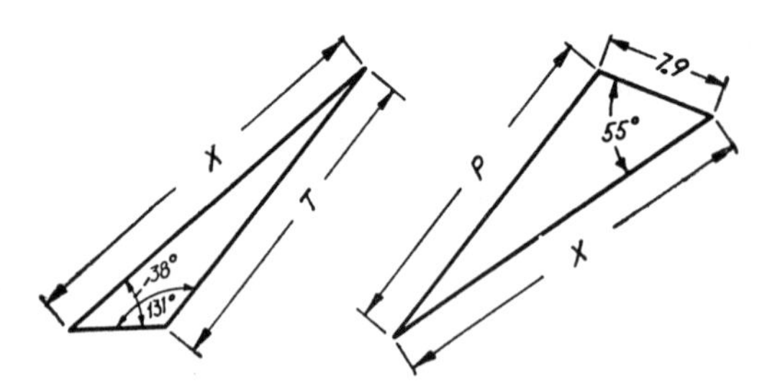

5. Determine the distance x.
6. Determine the distance x.

VARIABLES

Prob.	Sym.	No. 1	No. 2	No. 3	No. 4	No. 5	No. 6
1	J	10.2	10.4	10.6	10.8	11.1	11.3
2	J	10.2	10.4	10.6	10.8	11.1	11.3
3	U	4.65	4.93	5.42	5.68	5.89	6.24
4	U	4.65	4.93	5.42	5.68	5.89	6.24
5	T	8.21	8.43	8.65	8.78	9.12	9.37
6	P	9.12	9.25	9.37	9.48	9.62	9.75

LAW OF SINES

An important relation between the sides and angles of a triangle called the **law of sines** may be stated as follows: *In any triangle the sines of the angles are proportional to the lengths of the opposite sides.* This relation is arrived at in the following manner. Let the angles of either triangle be indicated by capitals A, B, C, and let the sides opposite the angles be called a, b, c, respectively.

First, consider an oblique triangle, having only acute angles as in Fig. 147a.

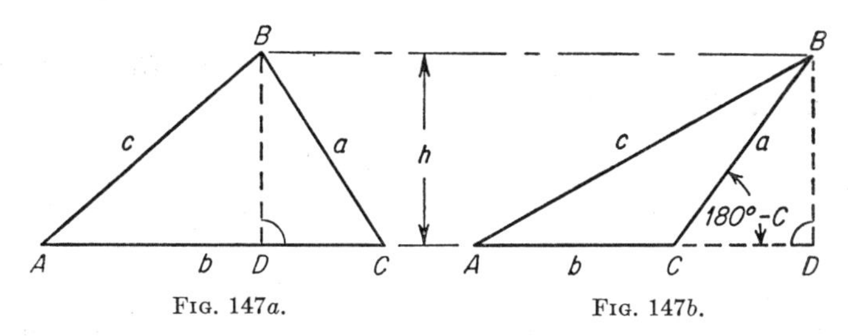

FIG. 147a. FIG. 147b.

Drop a perpendicular from vertex B to the opposite side b and call its length h. Then,

$$\sin A = \frac{h}{c} \quad \text{and} \quad \sin C = \frac{h}{a}$$

Divide the left-hand equation by the right-hand equation,

$$\frac{\sin A}{\sin C} = \frac{a}{c} \quad \text{or} \quad \frac{\sin C}{c} = \frac{\sin A}{a}$$

Similarly by dropping a perpendicular from vertex A to the opposite side a,

$$\frac{\sin B}{\sin C} = \frac{b}{c} \quad \text{or} \quad \frac{\sin C}{c} = \frac{\sin B}{b}$$

These two relations may be combined to give the law of sines in the usual form:

$$\frac{\sin A}{a} = \frac{\sin B}{b} = \frac{\sin C}{c}$$

Before considering an oblique triangle having an obtuse angle, as in Fig. 147*b*, the method of expressing the sine of an angle as a function of the supplementary angle will be developed.

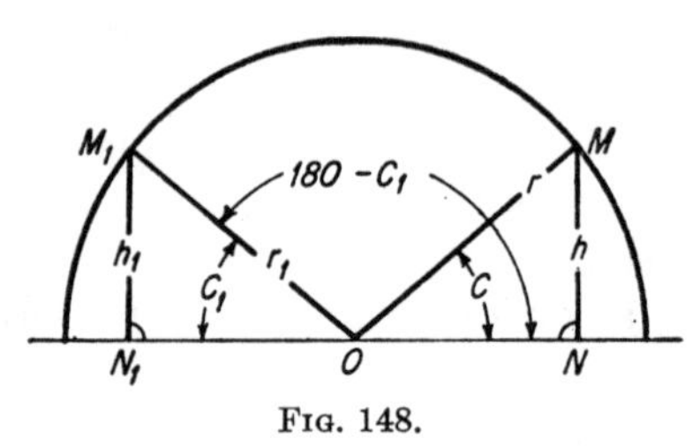

Fig. 148.

In the semicircle of Fig. 148, construct angle C_1 equal to angle C. Then in triangle OMN and OM_1N_1

$$r_1 = r \quad \text{and} \quad h_1 = h$$

$$\sin(180° - C_1) = \frac{h_1}{r_1} = \frac{h}{r} = \sin C$$

Since angle C_1 = angle C then $\sin(180° - C) = \sin C$, which means that the sine of any obtuse angle is equal to the sine of its supplement.

Now in Fig. 147*b* where the perpendicular is dropped from the vertex B to the extension of b:

$$\sin A = \frac{h}{c} \quad \text{and} \quad \sin(180° - C) = \sin C = \frac{h}{a}$$

and the proof proceeds as for the case of all angles acute.

Example:

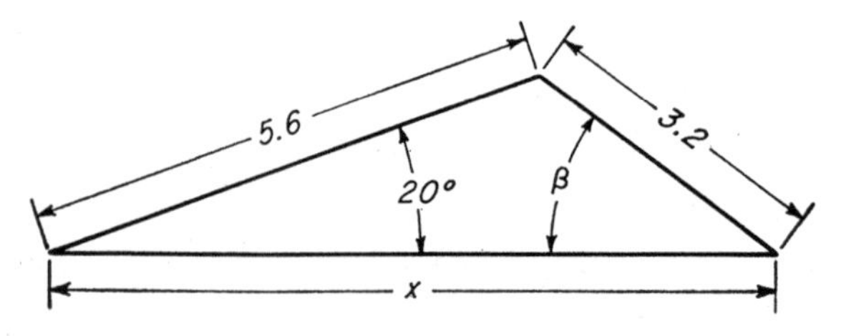

(*a*) Determine the angle β.
(*b*) Determine the distance x.

Solution (*a*): Label the triangle as in Fig. 147*a* where $a = 3.2$, $c = 5.6$, $A = 20°$, and $C = \beta$.

$$\frac{\sin A}{a} = \frac{\sin C}{c} \quad \text{or} \quad \sin C = \frac{c \sin A}{a}$$

$$\sin C = \frac{c \sin A}{a} = \frac{5.6 \sin 20°}{3.2} = .59853,\ C = 36°\ 45'\ 55''.$$

Solution (*b*): Where $x = b$

$$\frac{\sin A}{a} = \frac{\sin B}{b} \quad \text{or} \quad b = \frac{a \sin B}{\sin A} = a \sin B \csc A$$

Using the relation just developed that the sine of an angle is equal to the sine of its supplement and proposition 13,

$$\sin B = \sin (180° - B) = \sin (A + C) = \sin 56°\ 45'\ 55''$$
$$x = b = 3.2 \sin 56°\ 45'\ 55'' \csc 20° = 7.8265$$

PROBLEMS

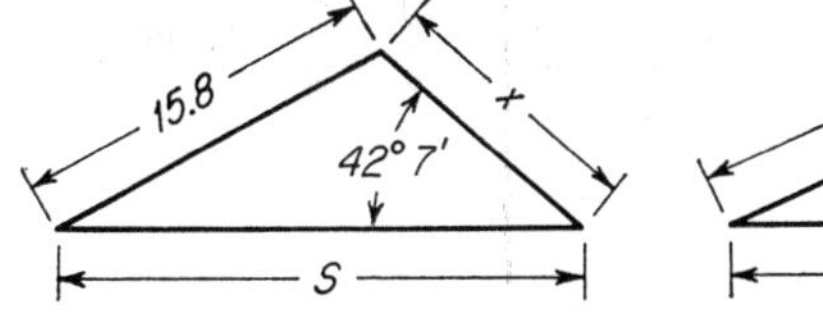

7. Determine the distance x.

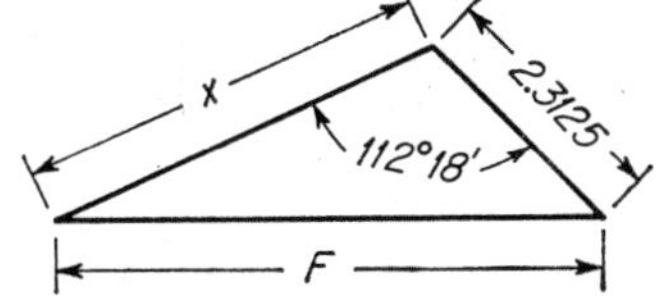

8. Determine the distance x.

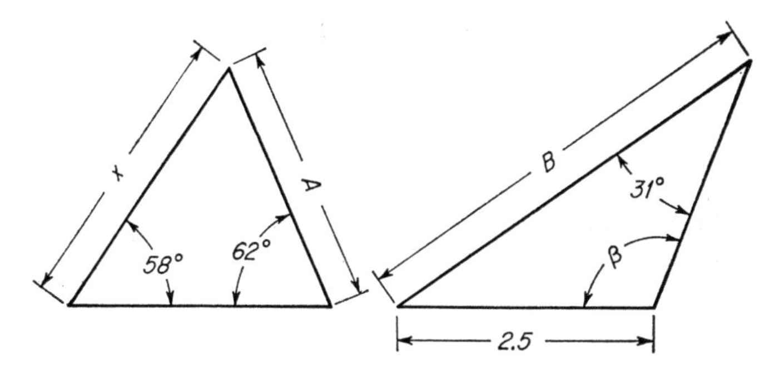

9. Determine the distance x.

10. Determine the angle β.

VARIABLE

Prob.	Sym.	No. 1	No. 2	No. 3	No. 4	No. 5	No. 6
7	S	21.4	21.5	21.6	21.7	21.8	21.9
8	F	4.8	4.9	5.1	5.2	5.3	5.4
9	A	2.4	2.5	2.6	2.7	2.8	2.9
10	B	4.2	4.3	4.4	4.5	4.6	4.7

PRACTICAL PROBLEMS

Involving Oblique Triangles

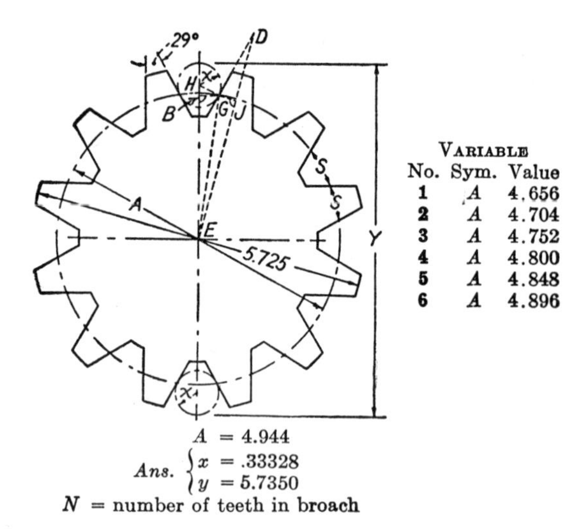

VARIABLE

No.	Sym.	Value
1	A	4.656
2	A	4.704
3	A	4.752
4	A	4.800
5	A	4.848
6	A	4.896

$A = 4.944$

Ans. $\begin{cases} x = .33328 \\ y = 5.7350 \end{cases}$

N = number of teeth in broach

1. Determine the distance y.

Solution:

$\angle BED = 180° \div N$.

$\angle BEG = \angle GEJ$. Why?

In $\triangle BEG$, solve for BG and BE.

In $\triangle BGH$, solve for GH and BH.

$y = HG$. $y = 2(BE + BH + x)$.

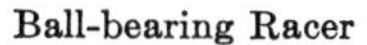

Ball-bearing Racer

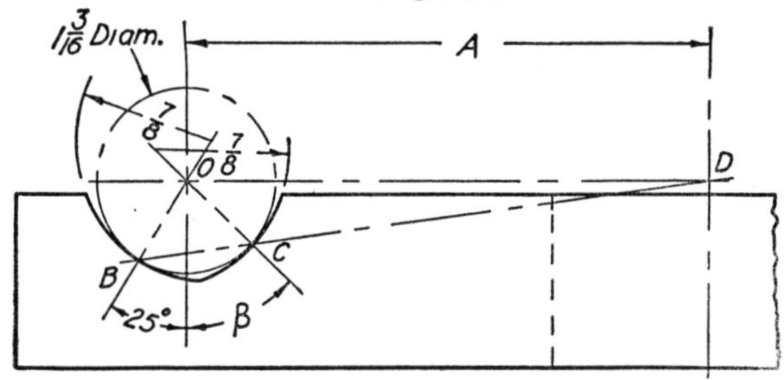

$A = 3.8125$

Ans. $\beta = 40°\ 5'\ 16''$

Variable

1. $A = 3.0625$	**2.** $A = 3.1875$	**3.** $A = 3.3125$
4. $A = 3.4375$	**5.** $A = 3.5625$	**6.** $A = 3.6875$

2. Determine the angle β.

Solution:

B and C are the points of tangency of the ball.

$OB = OC = 1.1875 \div 2$. $\angle BOD = 90° + 25°$

In $\triangle BOD$, solve for $\angle OBD$.

$\angle BOC = 180° - 2\angle OBD$. Why?

$\beta = \angle BOC - 25°$.

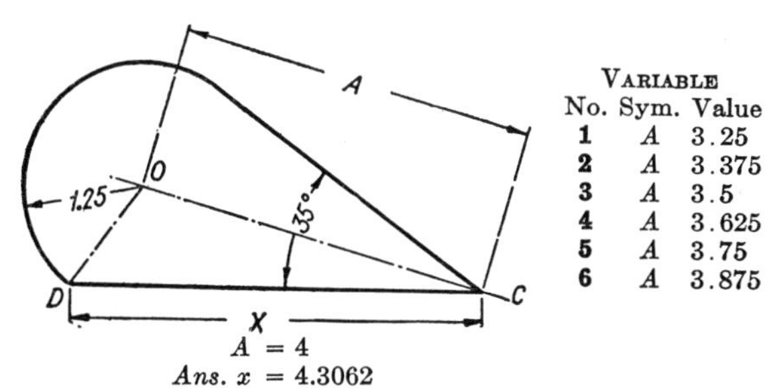

$A = 4$

Ans. $x = 4.3062$

Variable

No.	Sym.	Value
1	A	3.25
2	A	3.375
3	A	3.5
4	A	3.625
5	A	3.75
6	A	3.875

3. Determine the distance x.

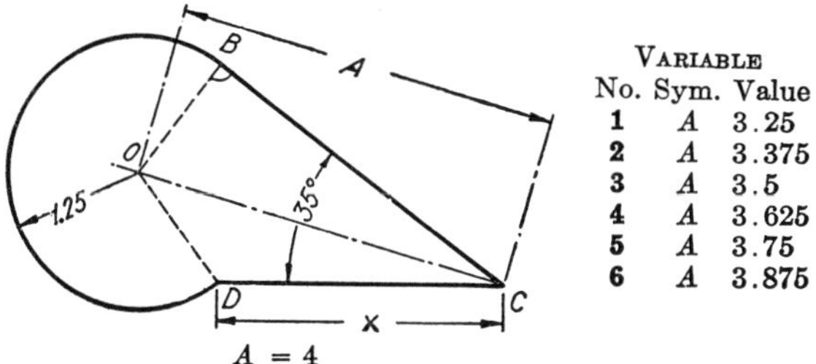

$A = 4$

Ans. $x = 3.3521$

Variable

No.	Sym.	Value
1	A	3.25
2	A	3.375
3	A	3.5
4	A	3.625
5	A	3.75
6	A	3.875

4. Determine the distance x. (*Solution on next page.*)

Solution for preceding problem:

$OB = 1.25$. In $\triangle OBC$, solve for $\angle BCO$.

$\angle OCD = 35° - \angle BCO$.

In $\triangle OCD$, solve for DC.

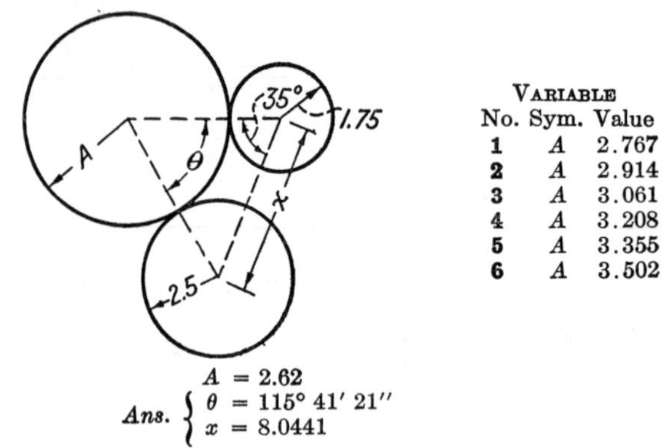

Variable		
No.	Sym.	Value
1	A	2.767
2	A	2.914
3	A	3.061
4	A	3.208
5	A	3.355
6	A	3.502

Ans. $\begin{cases} A = 2.62 \\ \theta = 115° \ 41' \ 21'' \\ x = 8.0441 \end{cases}$

5. Determine the angle θ.

6. Determine the distance x.

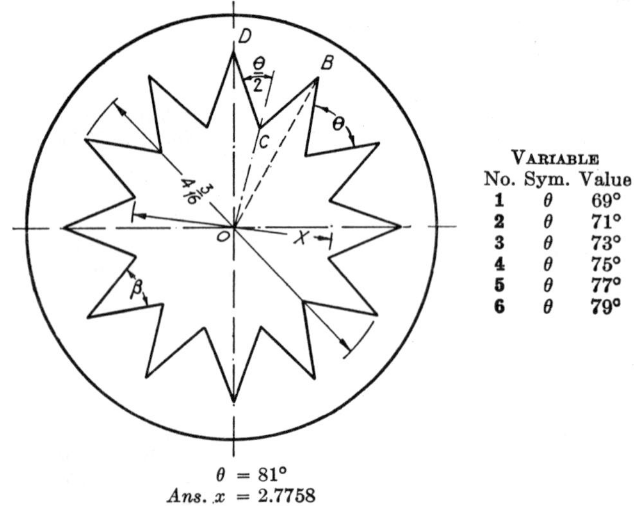

Variable		
No.	Sym.	Value
1	θ	69°
2	θ	71°
3	θ	73°
4	θ	75°
5	θ	77°
6	θ	79°

$\theta = 81°$

Ans. $x = 2.7758$

7. Determine the angle β.

8. Determine the distance x.

Solution:

$\angle DOB = 360°$ divided by the number of points. Why?

$\angle DOC = \angle DOB \div 2$. Why?

The balance of the problem is left to the student.

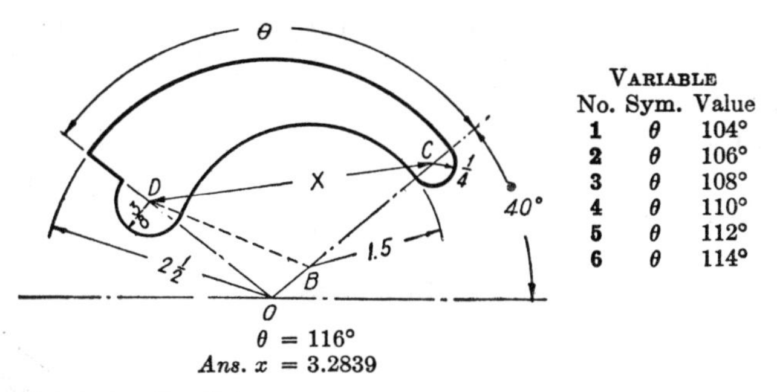

Variable		
No.	Sym.	Value
1	θ	104°
2	θ	106°
3	θ	108°
4	θ	110°
5	θ	112°
6	θ	114°

$\theta = 116°$
Ans. $x = 3.2839$

9. Determine the distance x.

Solution:

$OB = 2.5 - .5 - 1.5.$ $BD = 1.5 + .375.$

In $\triangle DOB$, solve for DO.

In $\triangle DOC$, solve for CD.

PROJECTION FORMULAS

One of the special types of oblique triangle problems is that in which the lengths of the three sides are given, and one or more of the angles are required.

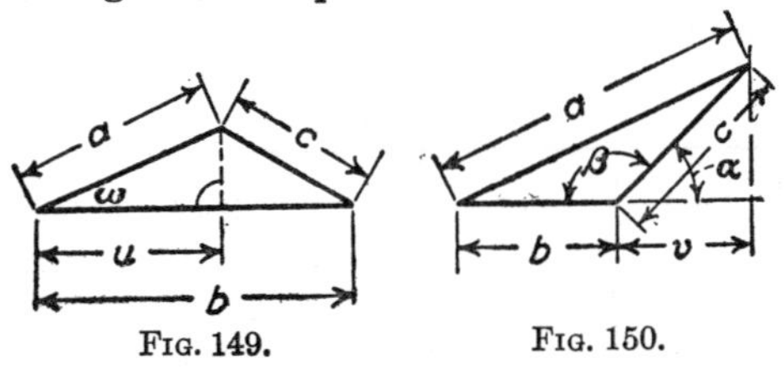

Fig. 149. Fig. 150.

The projection formulas of Proposition 32 give the values for u and v as follows:

$$u = \frac{a^2 + b^2 - c^2}{2b} \quad \text{and} \quad v = \frac{a^2 - b^2 - c^2}{2b}$$

Hence

$$\cos \omega = \frac{u}{a} = \frac{a^2 + b^2 - c^2}{2ba}$$

and

$$\cos \alpha = \frac{v}{c} = \frac{a^2 - b^2 - c^2}{2bc}$$

and $\beta = 180° - \alpha$ *Note:* The projection formulas can be used only when three sides are given.

The above relation involving the cosine of an angle in terms of three sides is commonly referred to as the **law of cosines.** The projection formula or law of cosines also may be applied when two sides and the included angle are given in order to determine the third side. The **law of sines** may then be applied to determine a second angle.

PROBLEMS

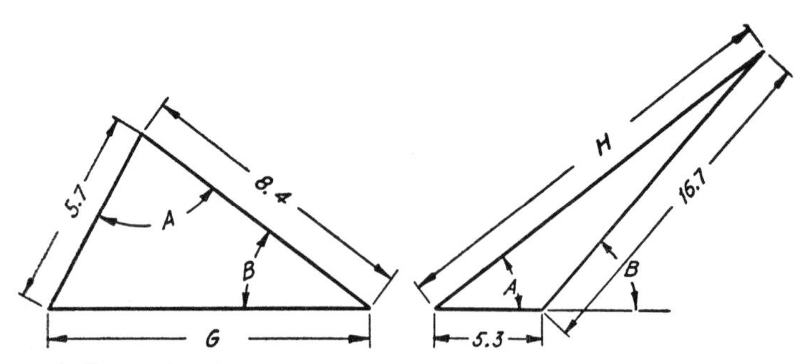

1. Determine the angle A.
2. Determine the angle B.

3. Determine the angle A.
4. Determine the angle B.

VARIABLES

Prob.	Sym.	No. 1	No. 2	No. 3	No. 4	No. 5	No. 6
1	G	9.5	9.7	9.9	10.1	10.3	10.5
2	G	9.5	9.7	9.9	10.1	10.3	10.5
3	H	19.6	19.9	20.1	20.4	20.6	20.9
4	H	19.6	19.9	20.1	20.4	20.6	20.9

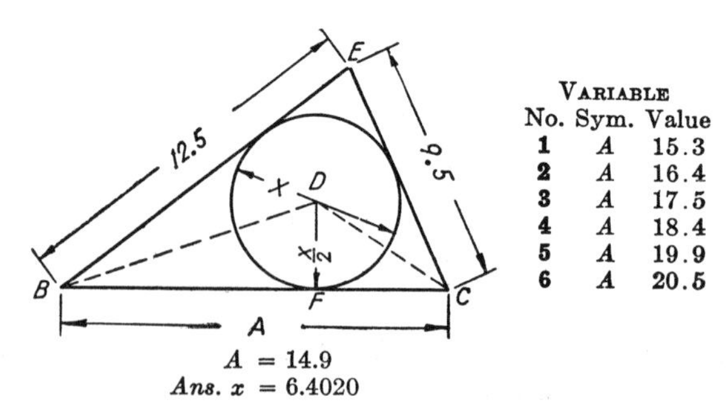

$A = 14.9$
Ans. $x = 6.4020$

VARIABLE

No.	Sym.	Value
1	A	15.3
2	A	16.4
3	A	17.5
4	A	18.4
5	A	19.9
6	A	20.5

5. Determine the diameter x. *(Solution on next page.)*

Solution for preceding problem:

In $\triangle BCE$, solve for $\angle EBC$ and $\angle ECB$.

$\angle DBC = \angle EBC \div 2$. $\angle DCB = \angle ECB \div 2$.

In $\triangle BCD$, solve for DF.

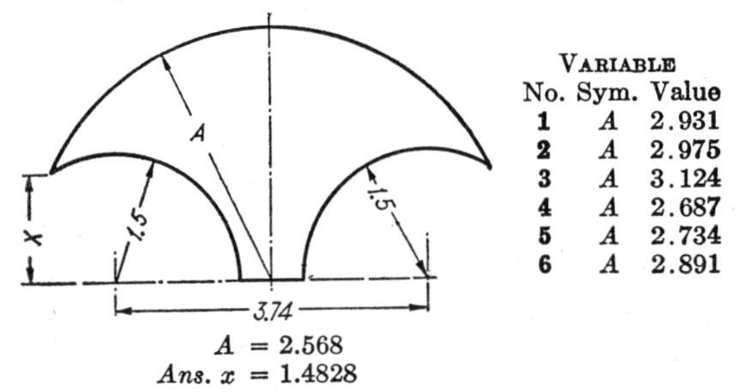

No.	Sym.	Value
1	A	2.931
2	A	2.975
3	A	3.124
4	A	2.687
5	A	2.734
6	A	2.891

VARIABLE (table above)

$A = 2.568$

Ans. $x = 1.4828$

6. Determine the distance x.

Locating the Center of a Gear Which Is in Mesh with Three Definitely Placed Gears Having Equal Numbers of Teeth.

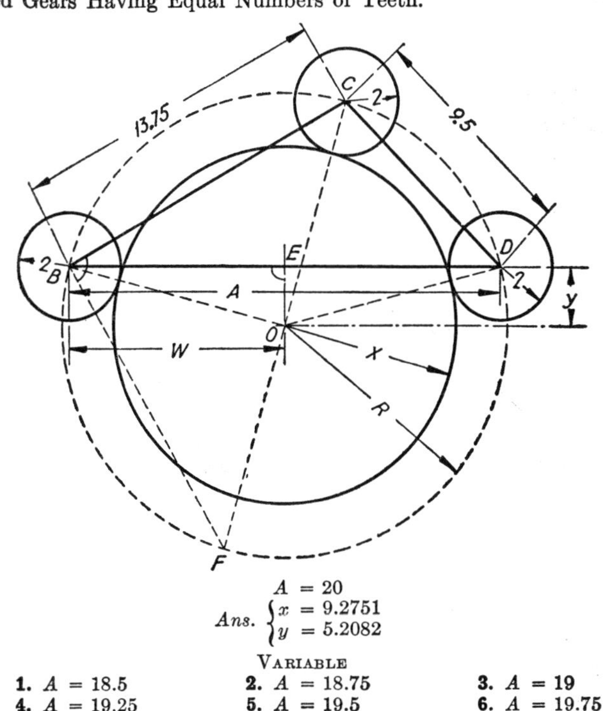

$A = 20$

Ans. $\begin{cases} x = 9.2751 \\ y = 5.2082 \end{cases}$

VARIABLE

1. $A = 18.5$ **2.** $A = 18.75$ **3.** $A = 19$
4. $A = 19.25$ **5.** $A = 19.5$ **6.** $A = 19.75$

7. Determine the radius x.

8. Determine the distance y. *(Solution on next page.)*

Solution for preceding problem:

In $\triangle BCD$, solve for $\angle CDB$.
$\angle BFC = \angle CDB$. Why?
In $\triangle BFC$, solve for CF. $R = CF \div 2$.
$ED = A \div 2$. In $\triangle OED$, solve for EO.

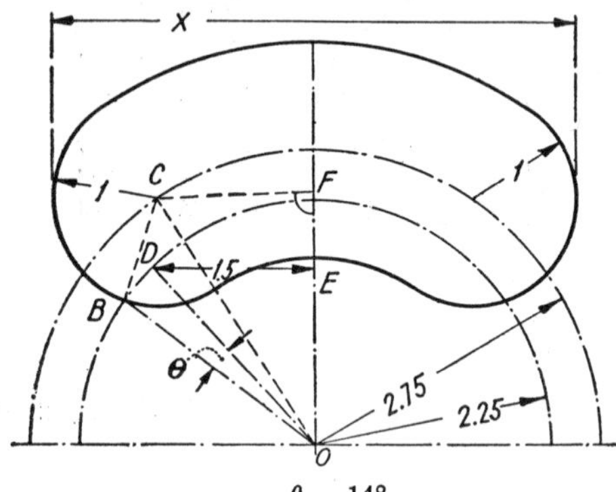

VARIABLE		
No.	Sym.	Value
1	θ	2°
2	θ	4°
3	θ	6°
4	θ	8°
5	θ	10°
6	θ	12°

$\theta = 14°$
Ans. $x = 5.2140$

9. Determine the distance x.
Solution:

In $\triangle BOC$, solve for $\angle BOC$.
$\angle DOC = \angle BOC - \theta$.
In $\triangle DOE$, solve for $\angle DOE$.
$\angle COF = \angle DOE - \angle DOC$.
In $\triangle COF$, solve for CF.

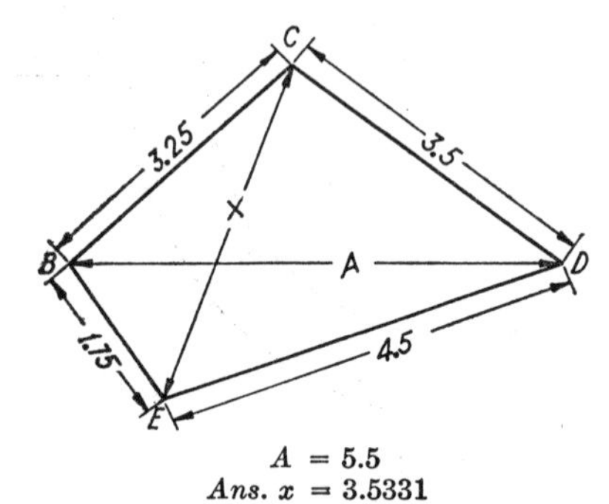

VARIABLE		
No.	Sym.	Value
1	A	4.625
2	A	4.75
3	A	4.875
4	A	5.0
5	A	5.125
6	A	5.25

$A = 5.5$
Ans. $x = 3.5331$

10. Determine the distance x.
Solution:

In $\triangle BDC$, solve for $\angle DBC$.
In $\triangle BDE$, solve for $\angle DBE$.

(*Solution continued on next page.*)

Solution continued:

$\angle CBE = \angle DBC + \angle DBE.$

In $\triangle CBE$, solve for x.

COTANGENT FORMULAS

Another of the special types of oblique triangle problems is that in which a side and the two adjacent angles are given.

A special formula which will be referred to as the "cotangent formula" will be used in solving problems of this type and will now be developed.

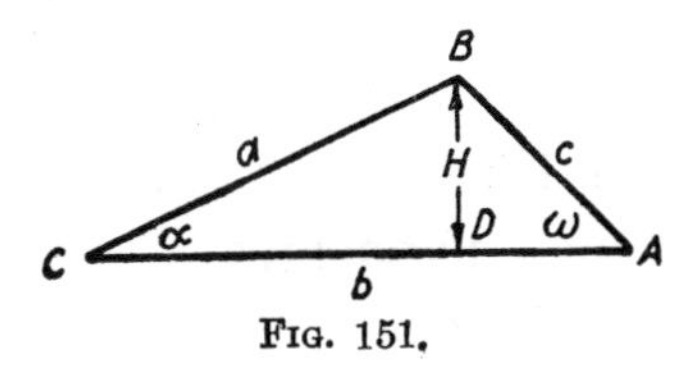

FIG. 151.

Given: Side b and the adjacent $\measuredangle$ α and ω of the $\triangle ABC$.

Draw a similar $\triangle A'B'C'$ whose altitude is unity.

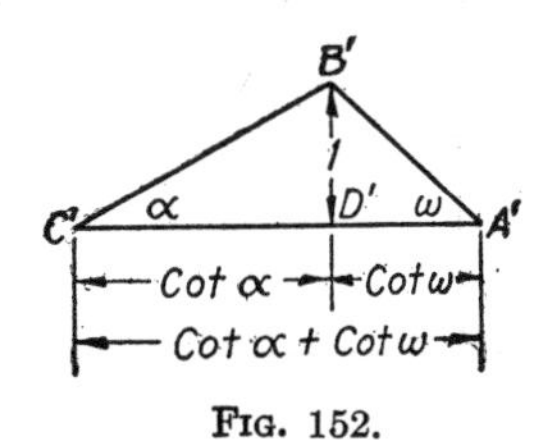

FIG. 152.

Since $B'D' = 1$, $A'D' = \cot\ \omega$ and $C'D' = \cot\ \alpha$ and $A'C' = \cot \omega + \cot \alpha$. Since the triangles in Figs. 151 and 152 are similar,

$$\frac{H}{1} = \frac{b}{\cot \alpha + \cot \omega} \qquad \text{(D-34)}.$$

or

$$H = \frac{b}{\cot \alpha + \cot \omega}.$$

This relation may be stated in the form of a rule as follows:

When a side and two adjacent angles are given, the altitude to that side is equal to the length of the side divided by the sum of the cotangents of the two adjacent angles.

If AB (Fig. 151) is required, it can be obtained by multiplying the altitude by csc ω. Similarly BC can be obtained.

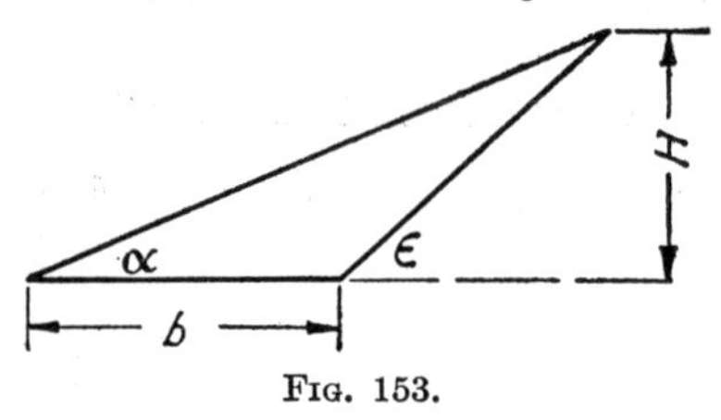

FIG. 153.

In the case of an obtuse triangle, the cotangent formula becomes

$$H = \frac{b}{\cot \alpha - \cot \varepsilon}.$$

The proof of this is left to the student.

PROBLEMS

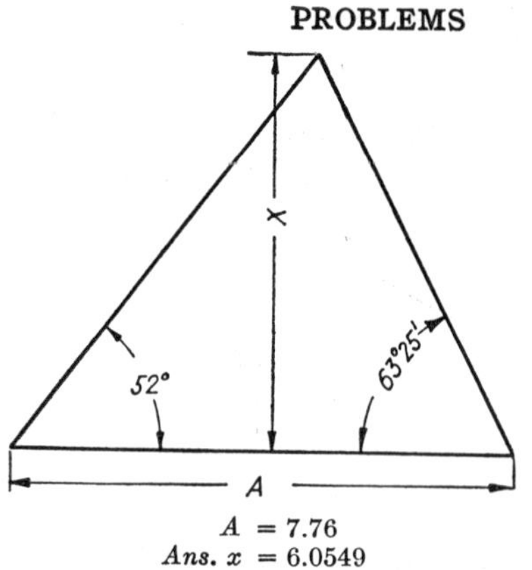

VARIABLE		
No.	Sym.	Value
1	A	4.68
2	A	5.09
3	A	5.85
4	A	6.43
5	A	6.94
6	A	7.65

$A = 7.76$
Ans. $x = 6.0549$

1. Determine the distance x.

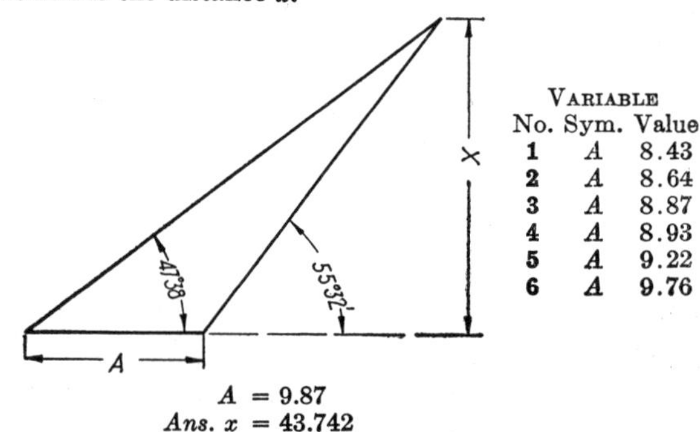

VARIABLE		
No.	Sym.	Value
1	A	8.43
2	A	8.64
3	A	8.87
4	A	8.93
5	A	9.22
6	A	9.76

$A = 9.87$
Ans. $x = 43.742$

2. Determine the distance x.

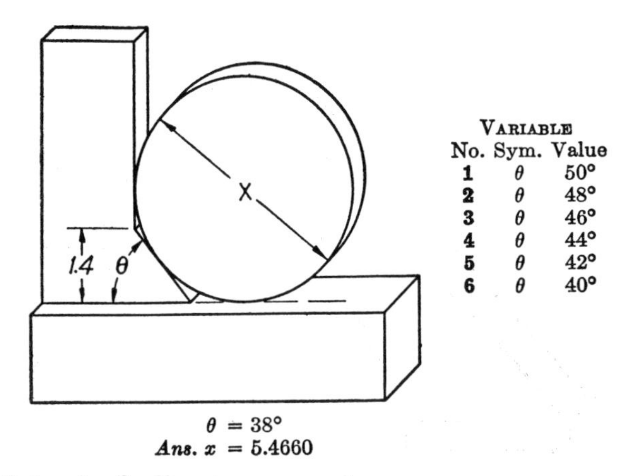

VARIABLE		
No.	Sym.	Value
1	θ	50°
2	θ	48°
3	θ	46°
4	θ	44°
5	θ	42°
6	θ	40°

$\theta = 38°$
Ans. $x = 5.4660$

3. Determine the diameter x.

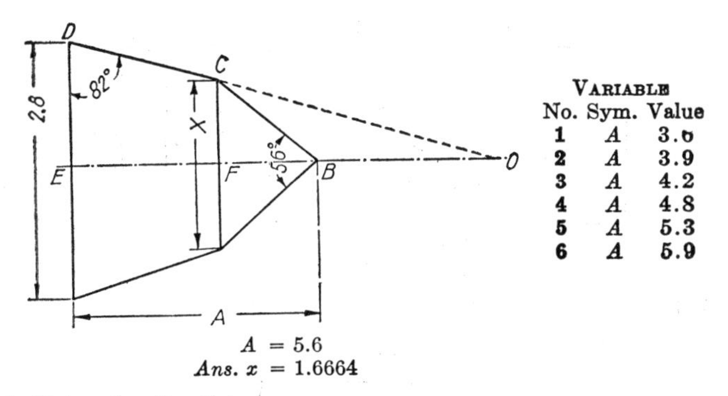

VARIABLE		
No.	Sym.	Value
1	A	**3.6**
2	A	**3.9**
3	A	**4.2**
4	A	**4.8**
5	A	**5.3**
6	A	**5.9**

$A = 5.6$
Ans. $x = 1.6664$

4. Determine the distance x.

Solution:

$DE = 2.8 \div 2$. In $\triangle EOD$, solve for EO.
$BO = EO - A$. $\angle EOD = 90° - 82°$. $\angle CBE = 56° \div 2$.
In $\triangle BOC$, solve for CF.

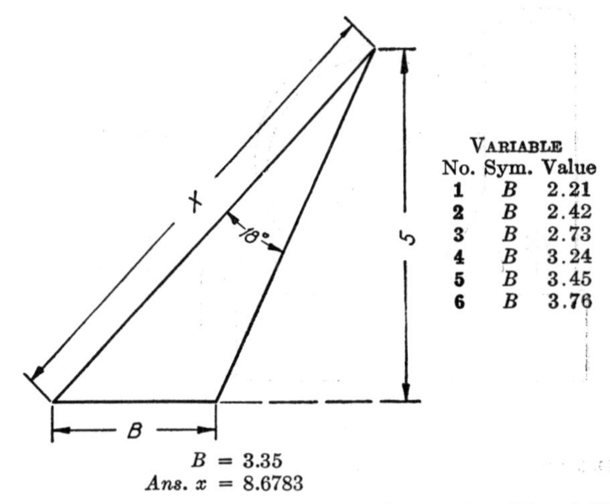

VARIABLE		
No.	Sym.	Value
1	B	2.21
2	B	2.42
3	B	2.73
4	B	3.24
5	B	3.45
6	B	3.76

$B = 3.35$
Ans. $x = 8.6783$

5. Determine the distance x. *Hint: See Propositions* 35 *and* 40.

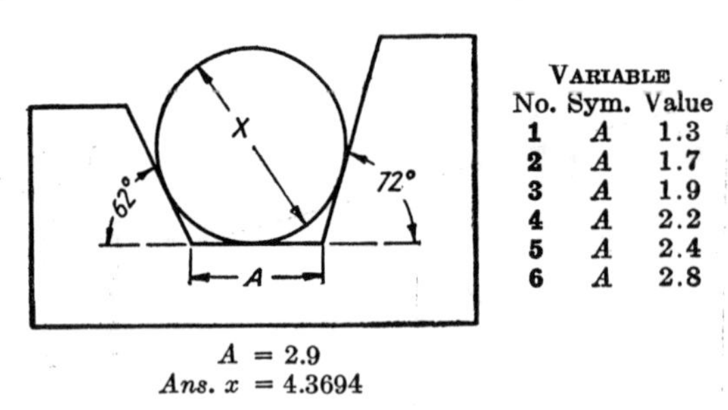

VARIABLE		
No.	Sym.	Value
1	A	1.3
2	A	1.7
3	A	1.9
4	A	2.2
5	A	2.4
6	A	2.8

$A = 2.9$
Ans. $x = 4.3694$

6. Determine the diameter x.

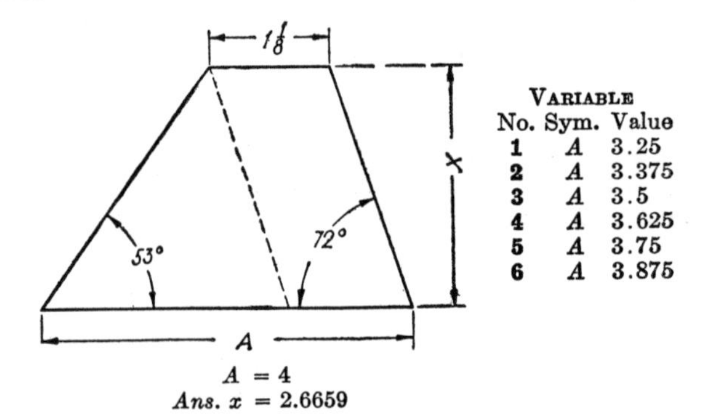

VARIABLE		
No.	Sym.	Value
1	A	3.25
2	A	3.375
3	A	3.5
4	A	3.625
5	A	3.75
6	A	3.875

$A = 4$
Ans. $x = 2.6659$

7. Determine the distance x.

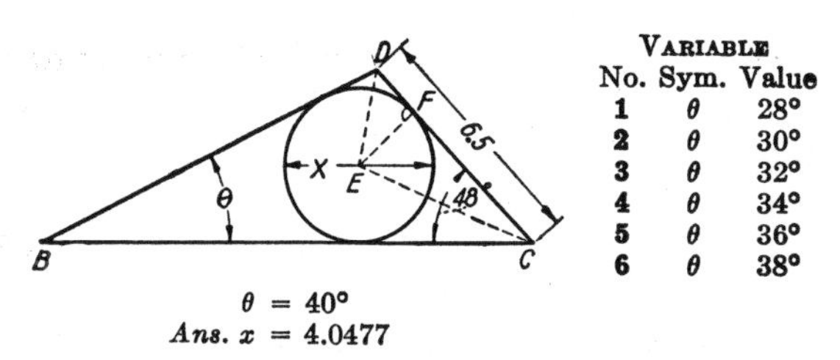

VARIABLE		
No.	Sym.	Value
1	θ	28°
2	θ	30°
3	θ	32°
4	θ	34°
5	θ	36°
6	θ	38°

$\theta = 40°$

Ans. $x = 4.0477$

8. Determine the diameter x.

Solution:

$\angle EDC = (180° - \theta - 48°) \div 2.$ Why?

$\angle ECD = 48° \div 2.$ Why?

Solve for EF by the cotangent formula.

GENERAL METHOD OF PROCEDURE IN SOLVING TRIGONOMETRIC PROBLEMS

Problems 1 to 199 are all practical problems taken from tool rooms, die rooms, or drawing rooms. They are arranged in approximate order of complexity. In solving these problems there are certain methods of procedure with which the student should become thoroughly familiar. First, a drawing should be made which shows all the given dimensions and the required distance or angle. To determine an unknown distance or angle, a triangle (right or oblique) should be searched for, which contains the desired part and has sufficient other parts known to enable the student to determine the required side or angle. If no such triangle exists, auxiliary lines should be drawn to form one. In general, these auxiliary lines should consist of given lines produced or new lines drawn parallel or perpendicular to given dimensions, and usually these lines will be drawn through vertices or through centers of circles already drawn or tangent to given circles. Frequently, the auxiliary lines are simply lines connecting given vertices and centers of given circles, etc.

Often no triangle can be drawn which will have enough given parts to lead directly to a solution of the required side or angle. In that case it will be necessary to draw a second triangle which will include one of the sides or angles of the first triangle (or a line or an angle equal to a side or angle of the first triangle), and which will contain enough given parts to allow a solution. A third, and even a fourth, triangle may be necessary before a triangle is finally reached which contains sufficient known parts. The method is thus to start with the side or angle in question and to continue forming related triangles until one is found which can be solved. Then work in the reverse order through these same triangles to obtain finally the required side or angle.

Illustrative Problem:

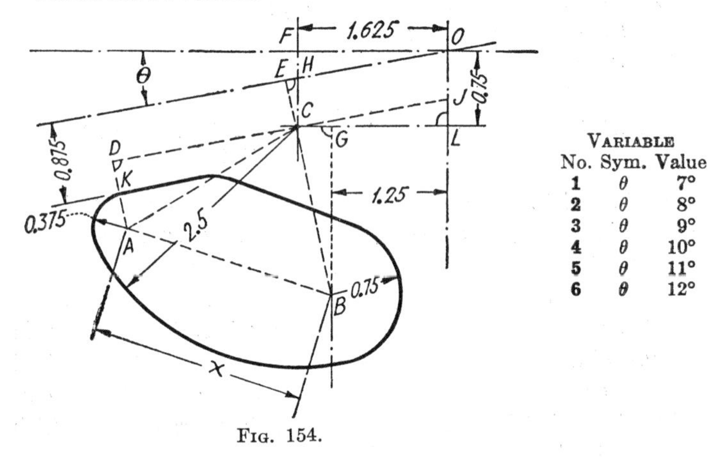

VARIABLE		
No.	Sym.	Value
1	θ	7°
2	θ	8°
3	θ	9°
4	θ	10°
5	θ	11°
6	θ	12°

FIG. 154.

Determine the distance x.

Draw the known auxiliary lines CA and CB to form a triangle containing the required distance x. In this $\triangle ABC$, if $\angle ACB$ were known, x could be determined. Thus $\angle ACB$ must be determined. Draw line $DJ \parallel EO$ through C. If $\measuredangle JCB$ and DCA were determined, $\angle ACB$ would equal 180° minus their sum. $\angle JCG = \angle \theta$. In the rt. $\triangle GCB$, $GC = 1.625 - 1.25$ and CB is known, so $\angle GCB$ may be determined. Thus $\angle JCB$ is determined. To obtain $\angle DCA$, draw $AD \perp DC$. If AD were known, $\angle DCA$ would be known. If KD were known, AD would be known. Draw $CE \parallel KD$. If CE were known, KD would be known. If CH were known, CE would be known. If FH were known, CH would be known. But FH of the right $\triangle FHO$ can be computed since $\angle \theta$ is given. Starting from this triangle, work in the reverse order to obtain the $\angle ACB$ and the distance x.

There are certain types of problems that can be best solved by special methods. Thus, in many problems involving two lines tangent to a given circle, it is often necessary to determine the location of the point of intersection of the two tangents with respect to some set of perpendicular lines as axes. (Sometimes only the distance from one axis is necessary.) The distances of this point of intersection from the two axes can then be used in determining the unknown distance or angle (see Problems 29, 33, 89, 99, 119, etc.).

Solutions will be given for many of the following problems. This will help the student to acquire the ability to analyze a given problem and to draw the proper auxiliary lines necessary for the solution.

Too much space would be required to describe each of the actual problems involved in the following figures, but tool makers, die makers, and draftsmen will recognize them as problems similar to those that they have been confronted with in their work, and it is hoped that the student will solve all of these problems in order to obtain the practice and experience necessary to enable him to solve other problems which he will meet with in his own work.

PRACTICAL PROBLEMS (*continued*)

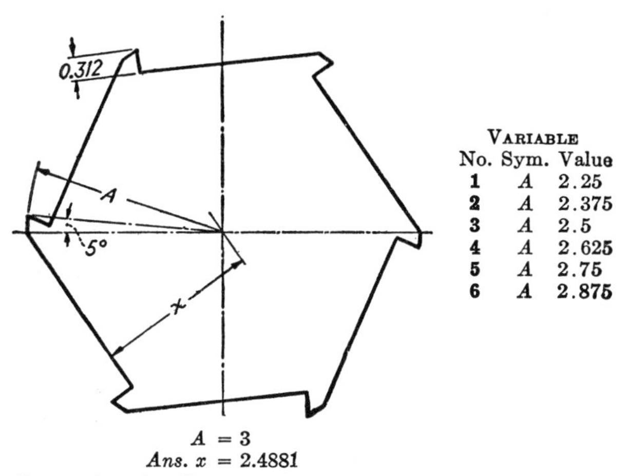

VARIABLE		
No.	Sym.	Value
1	A	2.25
2	A	2.375
3	A	2.5
4	A	2.625
5	A	2.75
6	A	2.875

$A = 3$

Ans. $x = 2.4881$

1. Determine the distance x.

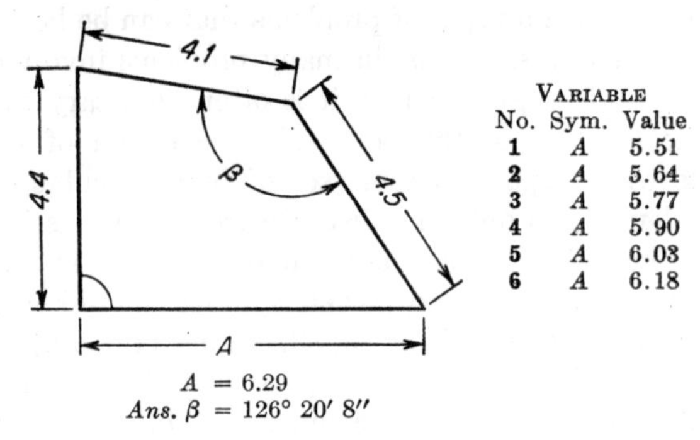

VARIABLE		
No.	Sym.	Value
1	A	5.51
2	A	5.64
3	A	5.77
4	A	5.90
5	A	6.03
6	A	6.18

$A = 6.29$
Ans. $\beta = 126° \; 20' \; 8''$

2. Determine the angle β.

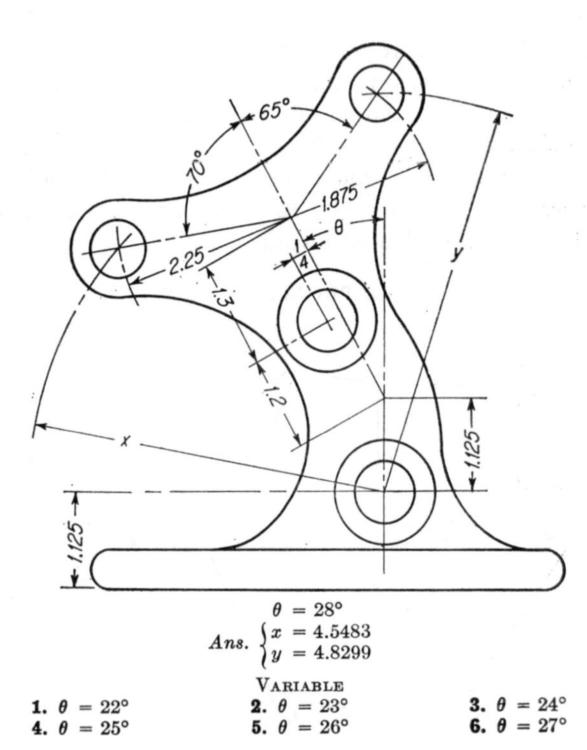

$\theta = 28°$

Ans. $\begin{cases} x = 4.5483 \\ y = 4.8299 \end{cases}$

VARIABLE

1. $\theta = 22°$	**2.** $\theta = 23°$	**3.** $\theta = 24°$
4. $\theta = 25°$	**5.** $\theta = 26°$	**6.** $\theta = 27°$

3. Determine the distance x.
4. Determine the distance y.

Dovetail

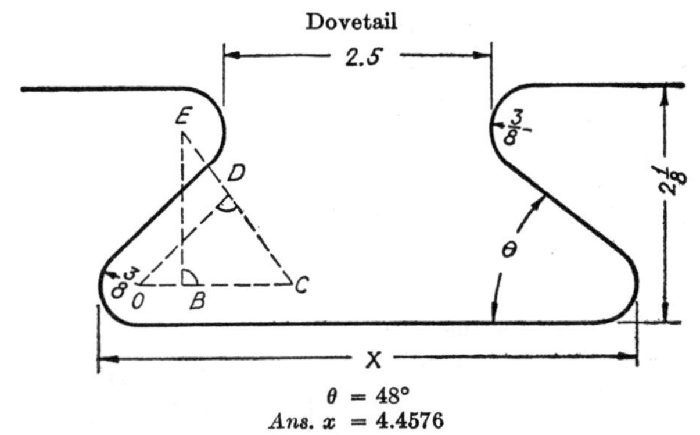

$\theta = 48°$
Ans. $x = 4.4576$

VARIABLE

1. $\theta = 35°$	**2.** $\theta = 38°$	**3.** $\theta = 40°$
4. $\theta = 42°$	**5.** $\theta = 44°$	**6.** $\theta = 46°$

5. Determine the distance x.

Solution:

$BE = 2.125 - .375 - .375.$
$\angle BEC = \theta.$ Why?
In $\triangle EBC$, solve for BC and EC.
$CD = EC - .375 - .375.$
$\angle DOC = \theta.$ In $\triangle ODC$, solve for OC.
$OB = OC - BC.$

Gage

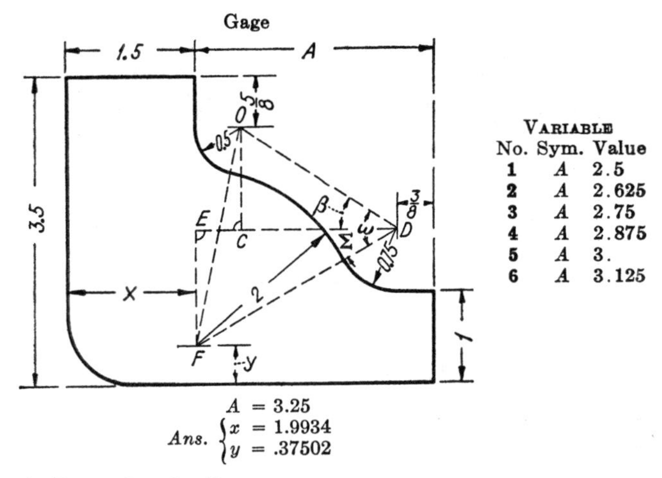

VARIABLE

No.	Sym.	Value
1	A	2.5
2	A	2.625
3	A	2.75
4	A	2.875
5	A	3.
6	A	3.125

$A = 3.25$
Ans. $\begin{cases} x = 1.9934 \\ y = .37502 \end{cases}$

6. Determine the distance x.
7. Determine the distance y.

(*Solution on next page.*)

Solution for preceding problem:

$OC = 3.5 - 1 - .75 - .625.$

$CD = A - .5 - .375.$

In $\triangle OCD$, solve for β and OD.

In $\triangle OFD$, solve for ω.

$\Sigma = \omega - \beta.$

In $\triangle EDF$, solve for DE and EF.

V-Block

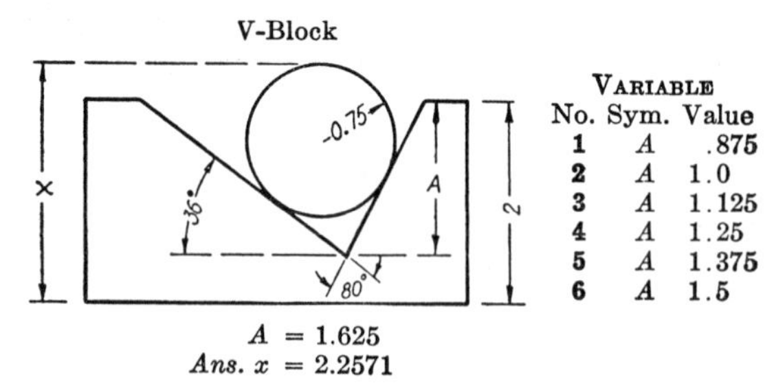

Variable		
No.	Sym.	Value
1	A	.875
2	A	1.0
3	A	1.125
4	A	1.25
5	A	1.375
6	A	1.5

$A = 1.625$

Ans. $x = 2.2571$

8. Determine the distance x.

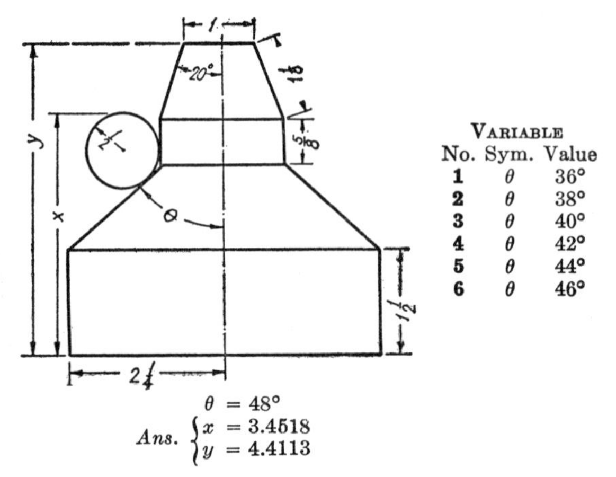

Variable		
No.	Sym.	Value
1	θ	36°
2	θ	38°
3	θ	40°
4	θ	42°
5	θ	44°
6	θ	46°

$\theta = 48°$

Ans. $\begin{cases} x = 3.4518 \\ y = 4.4113 \end{cases}$

9. Determine the distance x.

10. Determine the distance y.

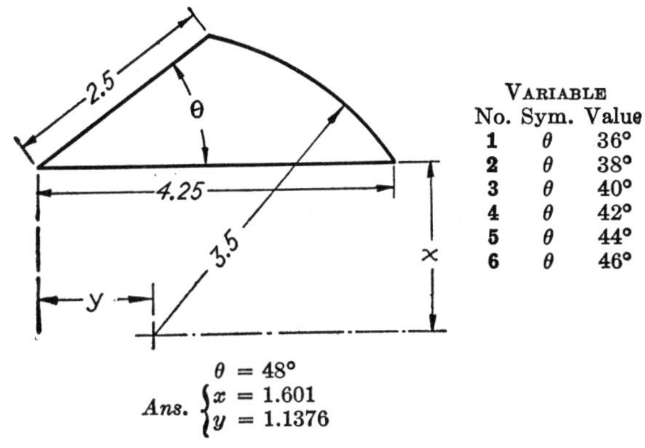

No.	Sym.	Value
1	θ	36°
2	θ	38°
3	θ	40°
4	θ	42°
5	θ	44°
6	θ	46°

$\theta = 48°$

Ans. $\begin{cases} x = 1.601 \\ y = 1.1376 \end{cases}$

11. Determine the distance x.
12. Determine the distance y.

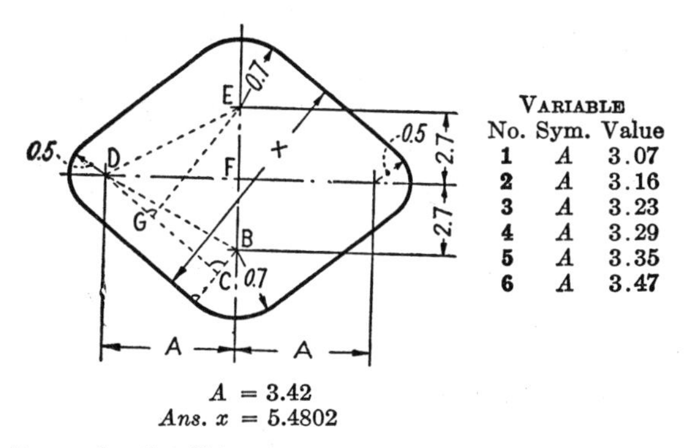

No.	Sym.	Value
1	A	3.07
2	A	3.16
3	A	3.23
4	A	3.29
5	A	3.35
6	A	3.47

$A = 3.42$

Ans. $x = 5.4802$

13. Determine the distance x.

Solution:

In $\triangle EDF$, solve for $\angle EDF$ and DE.

$\angle EDF = \angle FDB$. $DE = BD$. $BC = .7 - .5$.

In $\triangle BDC$, solve for $\angle BDC$.

$\angle EDC = \angle EDF + \angle FDB + \angle BDC$.

In $\triangle EDG$, solve for EG.

$x = EG + .7 + .5$.

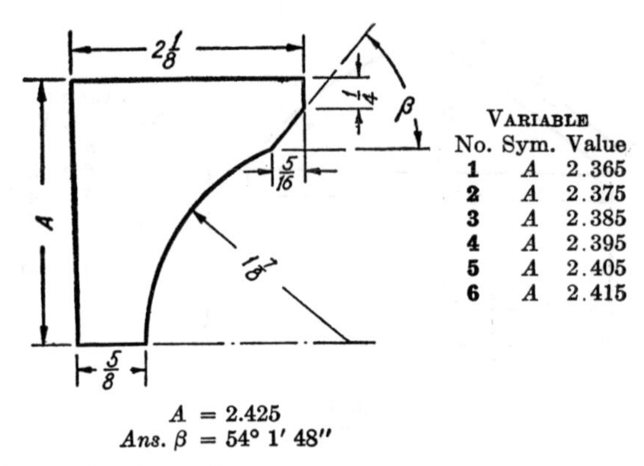

VARIABLE		
No.	Sym.	Value
1	A	2.365
2	A	2.375
3	A	2.385
4	A	2.395
5	A	2.405
6	A	2.415

$A = 2.425$

Ans. $\beta = 54° 1' 48''$

14. Determine the angle β.

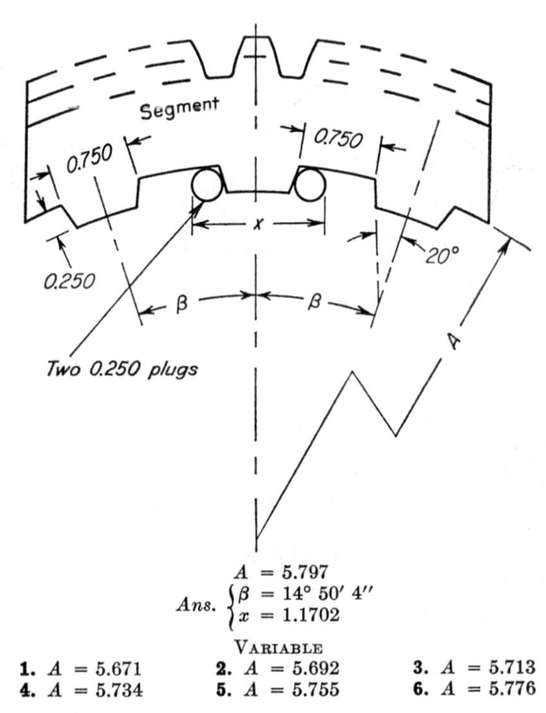

$A = 5.797$

Ans. $\begin{cases} \beta = 14° 50' 4'' \\ x = 1.1702 \end{cases}$

VARIABLE

1. $A = 5.671$ **2.** $A = 5.692$ **3.** $A = 5.713$

4. $A = 5.734$ **5.** $A = 5.755$ **6.** $A = 5.776$

15. Determine the angle β.

16. Determine the distance x.

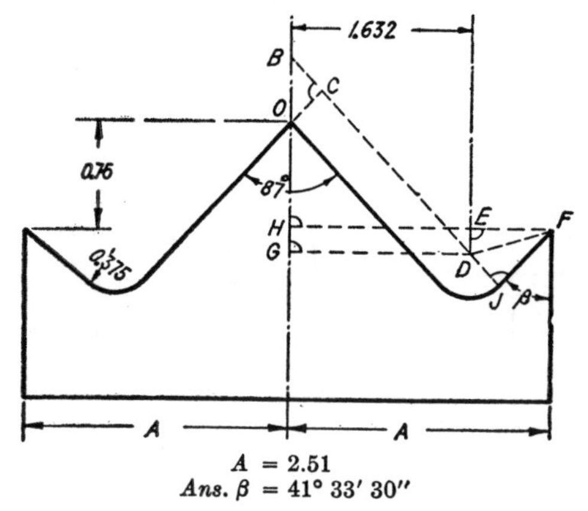

No.	Sym.	Value
1	A	2.72
2	A	2.93
3	A	3.24
4	A	3.35
5	A	3.56
6	A	3.67

$A = 2.51$
Ans. $\beta = 41° 33' 30''$

17. Determine the angle β.

Solution:

$OC = .375.$
In $\triangle OBC$, solve for OB.
In $\triangle BGD$, solve for GB.
$OH = .75.\quad GH = GB - OB - OH = ED.$
$EF = A - HE.$
In $\triangle DEF$, solve for $\angle EFD$ and DF.
$DJ = .375.$
In $\triangle DJF$, solve for $\angle DFJ$.
$\beta = 90° - \angle EFD - \angle DFJ.$

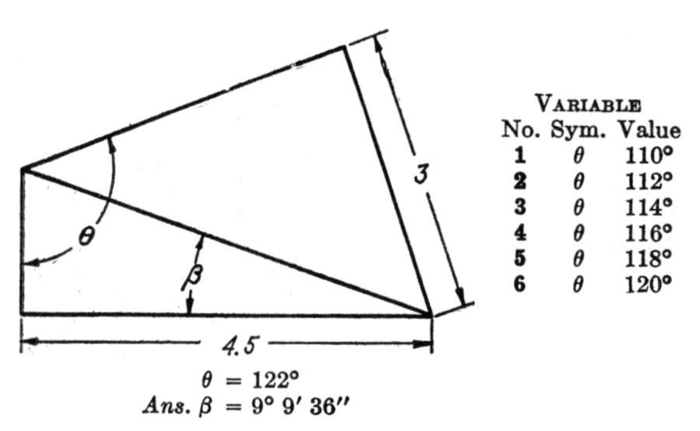

No.	Sym.	Value
1	θ	110°
2	θ	112°
3	θ	114°
4	θ	116°
5	θ	118°
6	θ	120°

$\theta = 122°$
Ans. $\beta = 9° 9' 36''$

18. Determine the angle β.

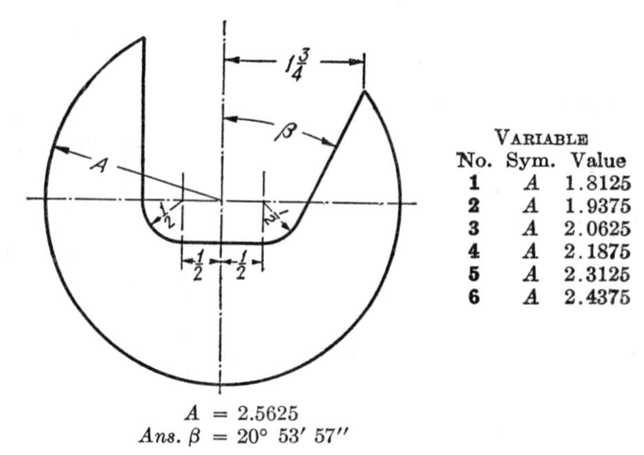

VARIABLE		
No.	Sym.	Value
1	A	1.8125
2	A	1.9375
3	A	2.0625
4	A	2.1875
5	A	2.3125
6	A	2.4375

$$A = 2.5625$$
$$Ans.\ \beta = 20°\ 53'\ 57''$$

19. Determine the angle β.

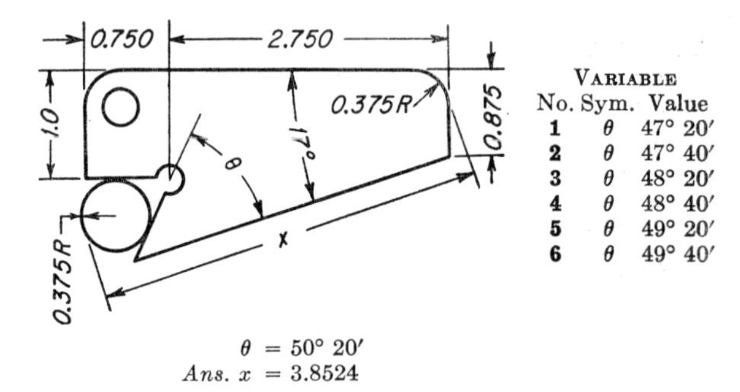

VARIABLE		
No.	Sym.	Value
1	θ	47° 20′
2	θ	47° 40′
3	θ	48° 20′
4	θ	48° 40′
5	θ	49° 20′
6	θ	49° 40′

$$\theta = 50°\ 20'$$
$$Ans.\ x = 3.8524$$

20. Determine the distance x.

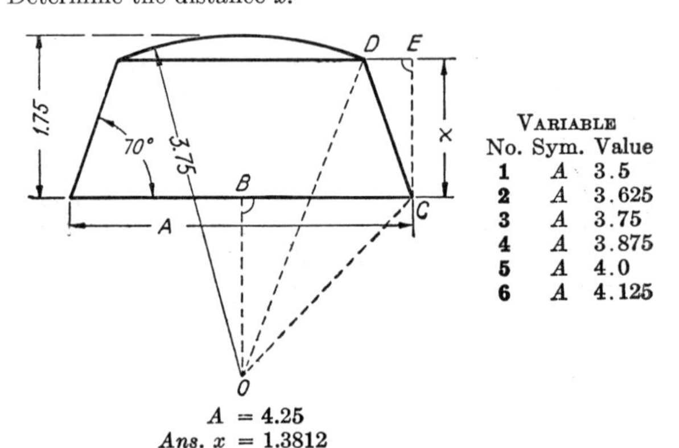

VARIABLE		
No.	Sym.	Value
1	A	3.5
2	A	3.625
3	A	3.75
4	A	3.875
5	A	4.0
6	A	4.125

$$A = 4.25$$
$$Ans.\ x = 1.3812$$

21. Determine the distance x.

(*Solution on next page.*)

Solution for preceding problem:

$BC = A \div 2.$
$BO = 3.75 - 1.75.$
In $\triangle OBC$, solve for $\angle BCO$ and OC.
$\angle DCO = 70° + \angle BCO. \quad OD = 3.75.$
In $\triangle ODC$, solve for CD.
$\angle DCE = 90° - 70°.$
In $\triangle DEC$, solve for CE. $CE = x$.

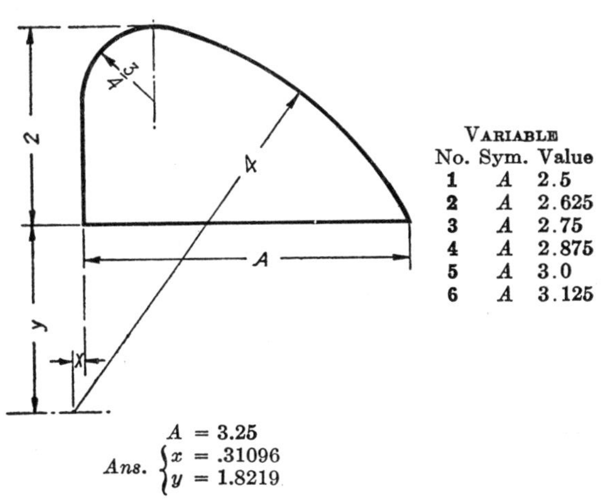

VARIABLE		
No.	Sym.	Value
1	A	2.5
2	A	2.625
3	A	2.75
4	A	2.875
5	A	3.0
6	A	3.125

$A = 3.25$

Ans. $\begin{cases} x = .31096 \\ y = 1.8219 \end{cases}$

22. Determine the distance x.
23. Determine the distance y.

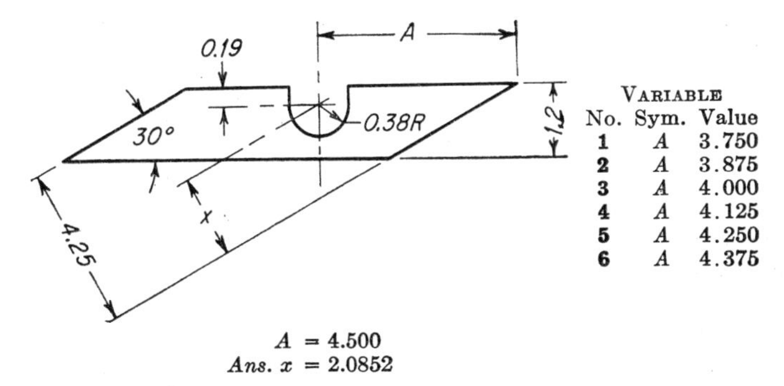

VARIABLE		
No.	Sym.	Value
1	A	3.750
2	A	3.875
3	A	4.000
4	A	4.125
5	A	4.250
6	A	4.375

$A = 4.500$
Ans. $x = 2.0852$

24. Determine the distance x.

Nozzle Spray

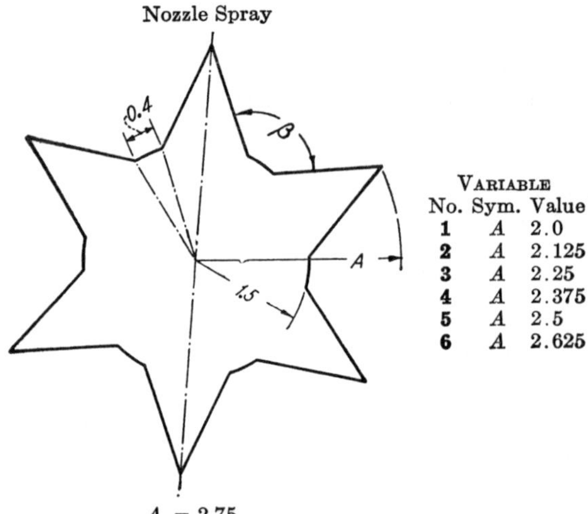

No.	Sym.	Value
1	A	2.0
2	A	2.125
3	A	2.25
4	A	2.375
5	A	2.5
6	A	2.625

$A = 2.75$

Ans. $\beta = 105° 26' 18''$ where 0.4 equals length of arc.

25. Determine the angle β.

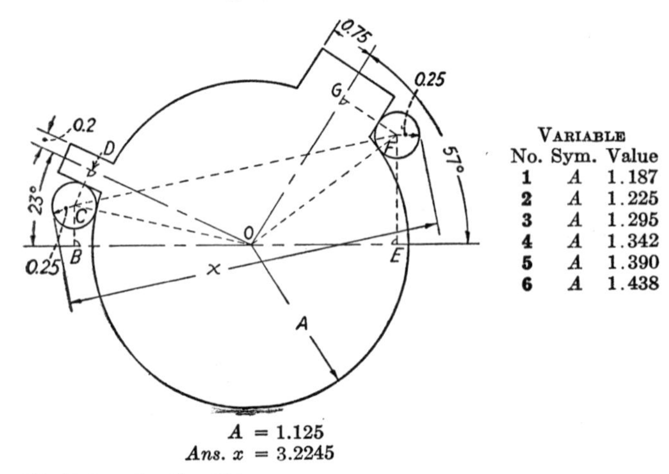

No.	Sym.	Value
1	A	1.187
2	A	1.225
3	A	1.295
4	A	1.342
5	A	1.390
6	A	1.438

$A = 1.125$

Ans. $x = 3.2245$

26. Determine the distance x.

Solution:

$CD = .45$. $FG = 1$.

$OC = OF = A + .25$.

In $\triangle OCD$, solve for $\angle COD$.

$\angle COB = 23° - \angle COD$.

In $\triangle OFG$, solve for $\angle GOF$.

Solution continued:

$\angle FOE = 57° - \angle GOF.$
$\angle COF = 180° - \angle COB - \angle FOE.$
In $\triangle COF$, solve for CF.

Checking the Position of Holes

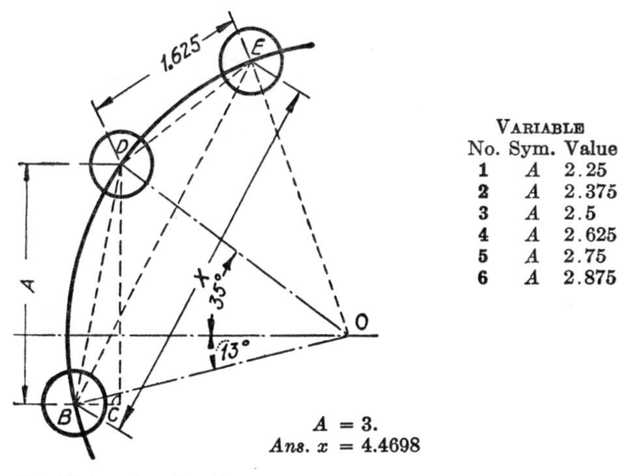

Variable		
No.	Sym.	Value
1	A	2.25
2	A	2.375
3	A	2.5
4	A	2.625
5	A	2.75
6	A	2.875

$A = 3.$
Ans. $x = 4.4698$

27. Determine the distance x.

Solution:

$\angle DBO = (180° - 13° - 35°) \div 2.$
In $\triangle BCD$, solve for BD.
In $\triangle DOE$, solve for $\angle DOE$.
In $\triangle BOE$, solve for BE.

$\angle DBC = \angle DBO + 13°.$
In $\triangle DBO$, solve for OD.
$\angle BOE = \angle DOE + 13° + 35°.$

Gage

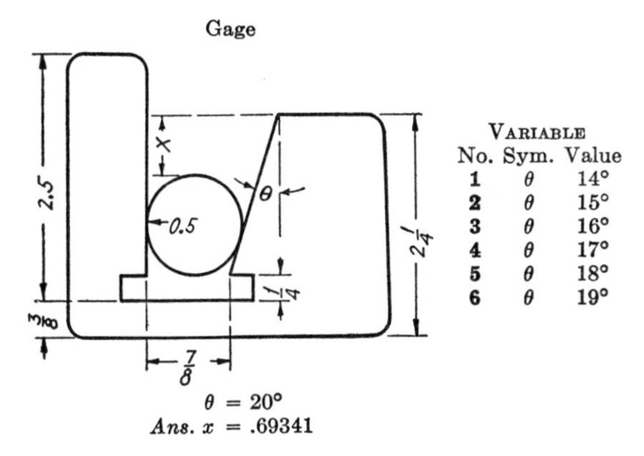

Variable		
No.	Sym.	Value
1	θ	14°
2	θ	15°
3	θ	16°
4	θ	17°
5	θ	18°
6	θ	19°

$\theta = 20°$
Ans. $x = .69341$

28. Determine the distance x.

Checking Angular Rings

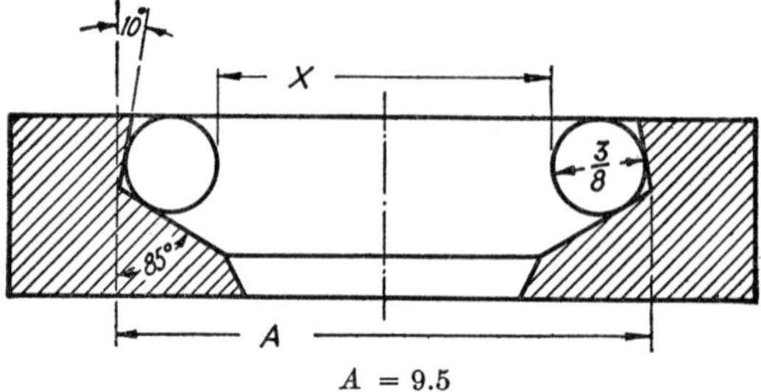

$A = 9.5$
Ans. $x = 8.6846$

VARIABLE

1. $A = 9.8$ **2.** $A = 10.2$ **3.** $A = 10.7$
4. $A = 11.2$ **5.** $A = 11.8$ **6.** $A = 12.4$

29. Determine the distance x.

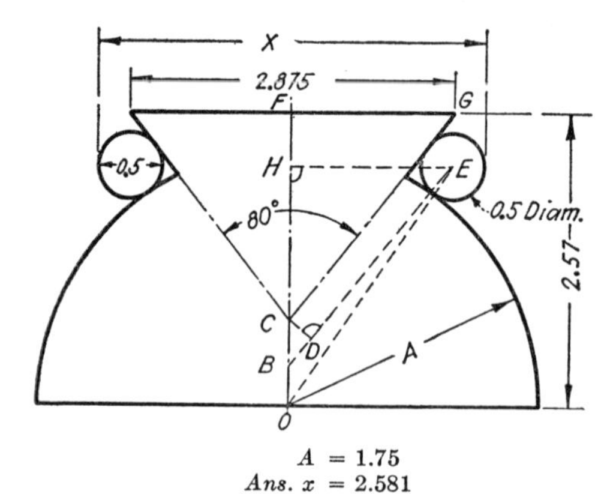

VARIABLE		
No.	Sym.	Value
1	A	1.81
2	A	1.87
3	A	1.95
4	A	2.12
5	A	2.41
6	A	2.62

$A = 1.75$
Ans. $x = 2.581$

30. Determine the distance x.

Solution:

$FG = 2.875 \div 2$. $\angle FCG = 80° \div 2$.
In $\triangle CFG$, solve for CF. $CD = .25$.
$\angle CBD = \angle FCG$. In $\triangle CBD$, solve for CB.
$OB = 2.57 - CF - CB$. $OE = A + .25$.
In $\triangle OBE$, solve for $\angle BOE$.
In $\triangle OEH$, solve for EH.

Die Section

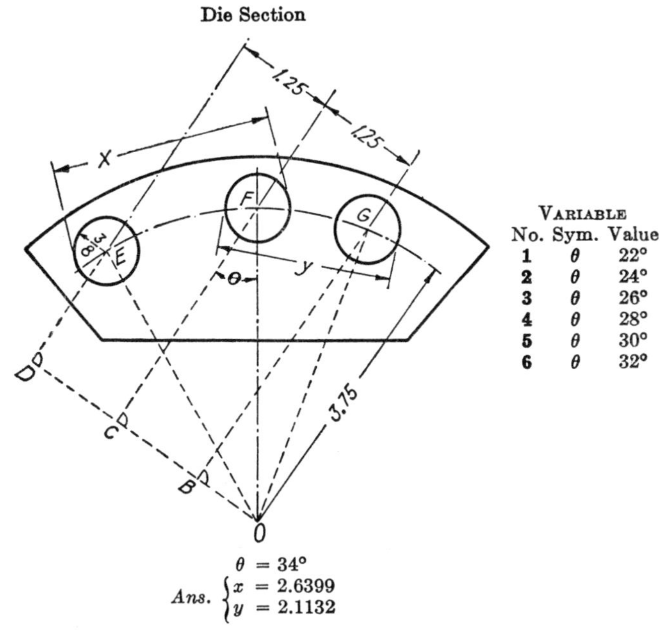

VARIABLE

No.	Sym.	Value
1	θ	22°
2	θ	24°
3	θ	26°
4	θ	28°
5	θ	30°
6	θ	32°

$\theta = 34°$

Ans. $\begin{cases} x = 2.6399 \\ y = 2.1132 \end{cases}$

31. Determine the distance x.

32. Determine the distance y.

Solution for x:

$OE = OF = OG = 3.75.$

$\angle COF = 90° - \theta.$

In $\triangle COF$, solve for CO.

$OD = CO + 1.25.$

In $\triangle DOE$, solve for $\angle DOE$.

$\angle EOF = \angle COF - \angle DOE.$

In sector EOF, solve for chord EF.

Solution for y:

$OB = OC - 1.25.$

In $\triangle BOG$, solve for $\angle BOG$.

$\angle FOG = \angle BOG - \angle COF.$

In sector FOG, solve for chord FG.

V-Block

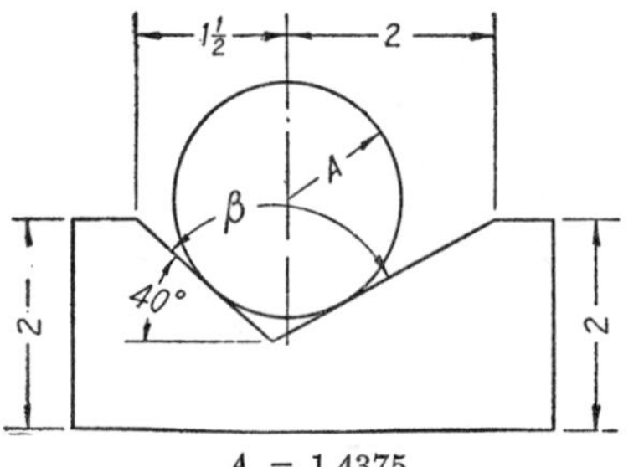

VARIABLE		
No.	Sym.	Value
1	A	1.0625
2	A	1.125
3	A	1.1875
4	A	1.25
5	A	1.3125
6	A	1.375

$A = 1.4375$

Ans. $\beta = 113° 47' 50''$

33. Determine the angle β.

Chain Gear Teeth

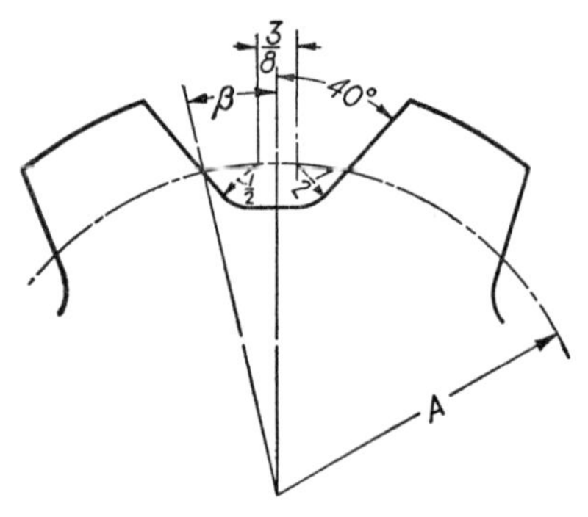

VARIABLE		
No.	Sym.	Value
1	A	2.5
2	A	2.625
3	A	2.75
4	A	2.875
5	A	3.
6	A	3.125

$A = 3.25$

Ans. $\beta = 13° 39' 30''$

34. Determine the angle β.

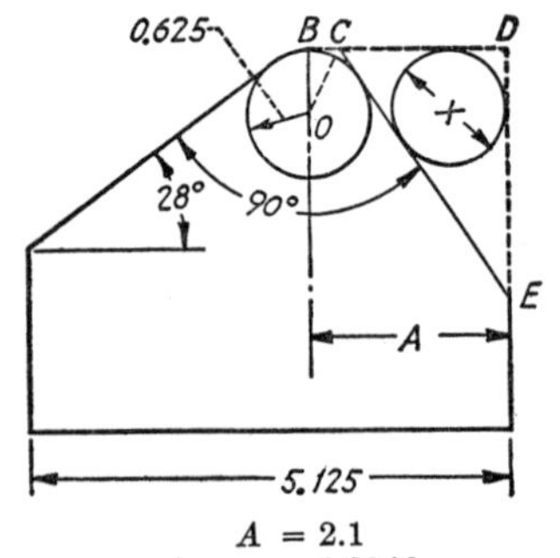

VARIABLE		
No.	Sym.	Value
1	A	2.2
2	A	2.3
3	A	2.4
4	A	2.5
5	A	2.6
6	A	2.7

$A = 2.1$

Ans. $x = 1.2946$

35. Determine the distance x.

(*Solution on next page.*)

Solution for preceding problem:

$OB = .625$. $\angle BCO = (90° + 28°) \div 2$. Why?
In $\triangle OBC$, solve for BC. $CD = A - BC$.
$\angle CED = 28°$. Why?
In $\triangle CDE$, solve for CE and DE.
By geometry solve for x.

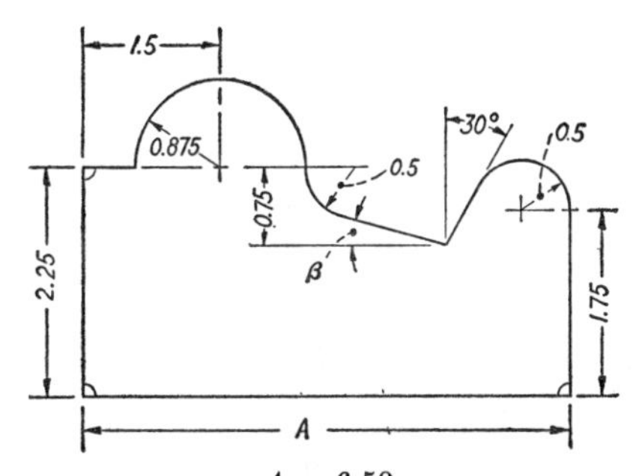

$A = 6.50$
Ans. $\beta = 5° 52' 32''$

VARIABLE

1. $A = 5.00$	**2.** $A = 5.25$	**3.** $A = 5.50$
4. $A = 5.75$	**5.** $A = 6.00$	**6.** $A = 6.25$

36. Determine the angle β.

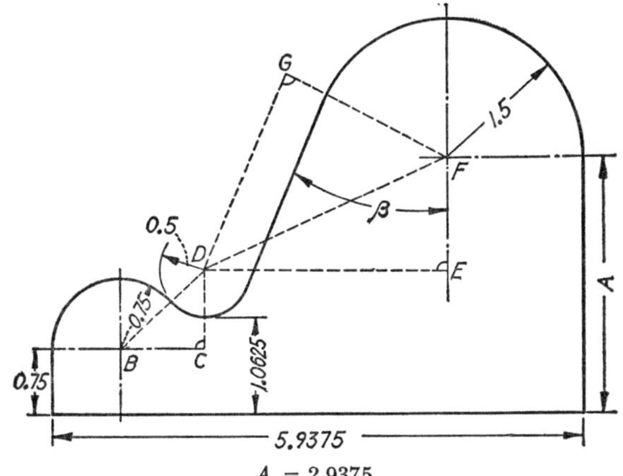

$A = 2.9375$
Ans. $\beta = 22° 34' 17''$

VARIABLE

1. $A = 2.25$	**2.** $A = 2.375$	**3.** $A = 2.5$
4. $A = 2.625$	**5.** $A = 2.75$	**6.** $A = 2.875$

37. Determine the angle β. (*Solution on next page.*)

Solution for preceding problem:

$CD = 1.0625 - .75 + .5$. Why? $BD = .75 + .5$.
In $\triangle BCD$, solve for BC. $DE = 5.9375 - .75 - BC - 1.5$.
$EF = A - 1.0625 - .5$. In $\triangle DEF$, solve for $\angle EDF$ and DF.
$FG = 1.5 + .5$. Why?
In $\triangle DFG$, solve for $\angle GDF$.
$\beta = 90° - \angle EDF - \angle GDF$.

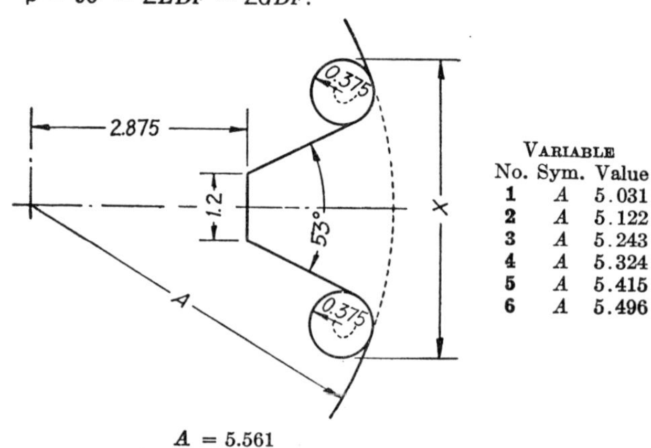

Variable		
No.	Sym.	Value
1	A	5.031
2	A	5.122
3	A	5.243
4	A	5.324
5	A	5.415
6	A	5.496

$A = 5.561$
Ans. $x = 4.7022$

38. Determine the distance x.

Gage

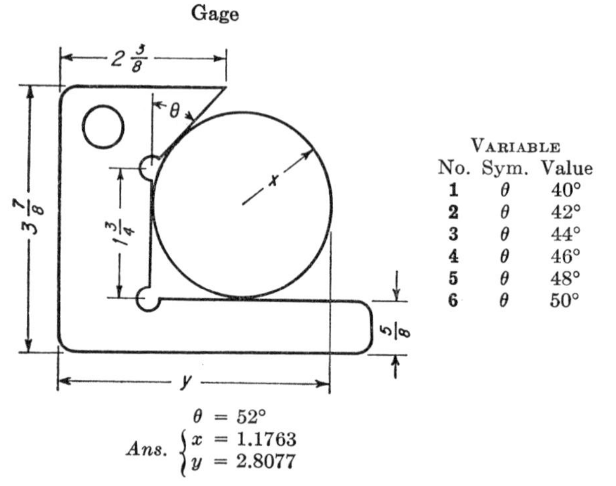

Variable		
No.	Sym.	Value
1	θ	40°
2	θ	42°
3	θ	44°
4	θ	46°
5	θ	48°
6	θ	50°

$\theta = 52°$
Ans. $\begin{cases} x = 1.1763 \\ y = 2.8077 \end{cases}$

39. Determine the distance x.
40. Determine the distance y.

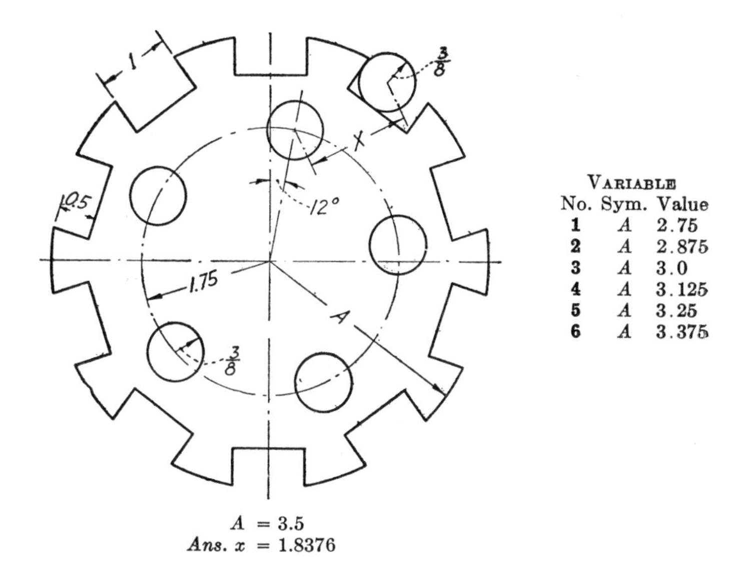

No.	Sym.	Value
1	A	2.75
2	A	2.875
3	A	3.0
4	A	3.125
5	A	3.25
6	A	3.375

$A = 3.5$
Ans. $x = 1.8376$

41. Determine the distance x.

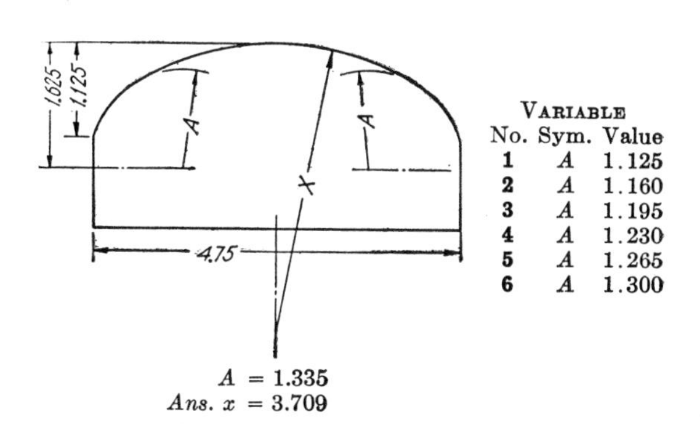

No.	Sym.	Value
1	A	1.125
2	A	1.160
3	A	1.195
4	A	1.230
5	A	1.265
6	A	1.300

$A = 1.335$
Ans. $x = 3.709$

42. Determine the distance x.

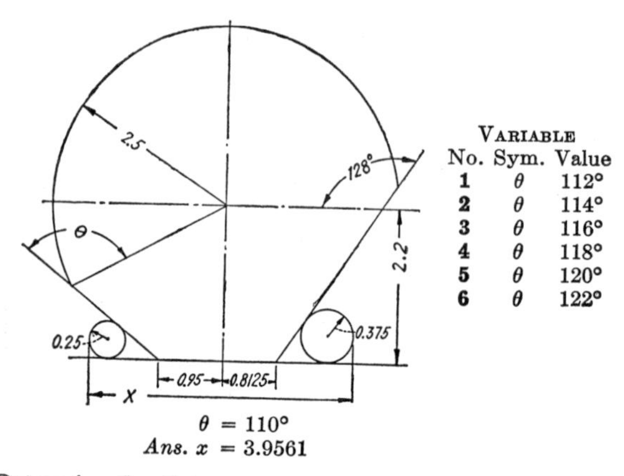

No.	Sym.	Value
1	θ	112°
2	θ	114°
3	θ	116°
4	θ	118°
5	θ	120°
6	θ	122°

$\theta = 110°$
Ans. $x = 3.9561$

43. Determine the distance x.

Combination Gear and Clutch

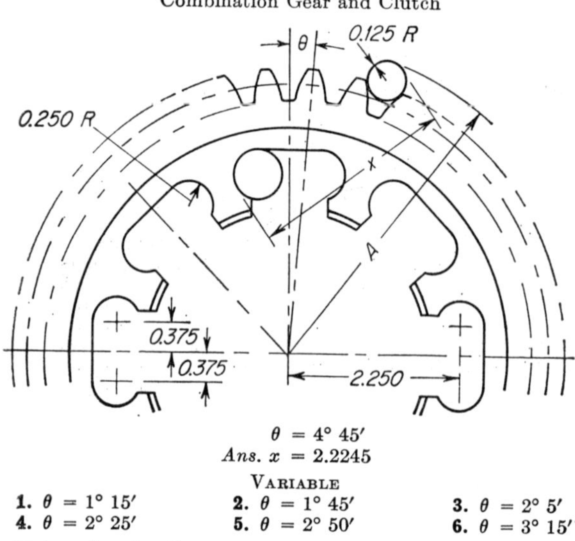

$\theta = 4° \ 45'$
Ans. $x = 2.2245$

VARIABLE

1. $\theta = 1° \ 15'$ **2.** $\theta = 1° \ 45'$ **3.** $\theta = 2° \ 5'$
4. $\theta = 2° \ 25'$ **5.** $\theta = 2° \ 50'$ **6.** $\theta = 3° \ 15'$

44. Determine the distance x.

Note: The teeth in the gear shown above are located with plug gages before the teeth are machined to a finish size. It may be necessary to check the distances A and x two or three times, with a micrometer, before the teeth are properly located. The value A for this checking is 3.687.

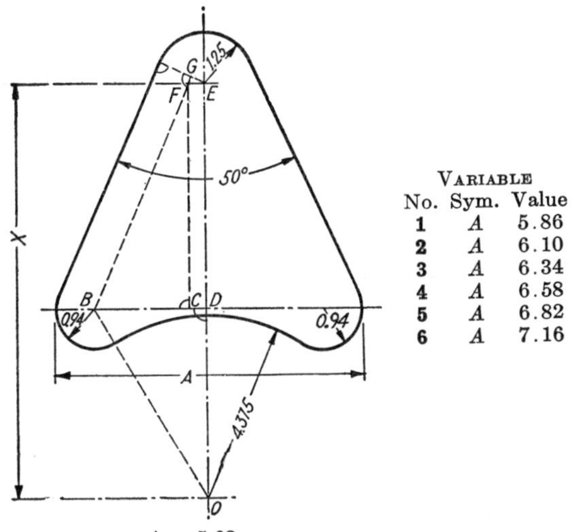

No.	Sym.	Value
1	*A*	5.86
2	*A*	6.10
3	*A*	6.34
4	*A*	6.58
5	*A*	6.82
6	*A*	7.16

Variable

$A = 5.62$
Ans. $x = 8.2518$

45. Determine the distance x.

Solution:

$GE = 1.25 - .94$. Why? $\angle GEF = 50° \div 2$. Why?
In $\triangle FEG$, solve for FE. $BD = (A - .94 - .94) \div 2$.
$BC = BD - CD$. $CD = FE$. In $\triangle BCF$, solve for FC.
In $\triangle OBD$, solve for DO.
$x = DO + FC$.

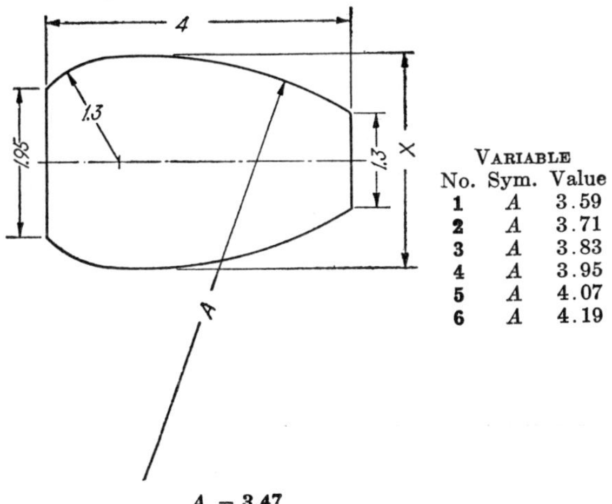

Variable

No.	Sym.	Value
1	*A*	3.59
2	*A*	3.71
3	*A*	3.83
4	*A*	3.95
5	*A*	4.07
6	*A*	4.19

$A = 3.47$
Ans. $x = 2.9733$

46. Determine the distance x.

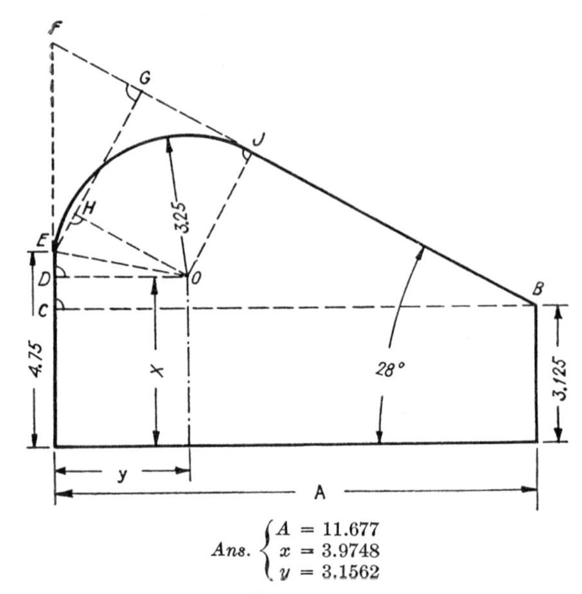

$$Ans. \begin{cases} A = 11.677 \\ x = 3.9748 \\ y = 3.1562 \end{cases}$$

VARIABLE

1. $A = 10.351$	2. $A = 10.572$	3. $A = 10.793$
4. $A = 11.014$	5. $A = 11.235$	6. $A = 11.456$

47. Determine the distance x.

48. Determine the distance y.

Solution:

$\angle CBF = \angle FEG = 28°$. $OJ = 3.25$.
In $\triangle BCF$, solve for CF. $EF = CF + 3.125 - 4.75$.
In $\triangle EFG$, solve for EG. $EH = EG - OJ$.
$EO = 3.25$. In $\triangle EOH$, solve for $\angle EOH$.
$\angle DOH = \angle FEG = 28°$. Why? $\angle EOD = 28° - \angle EOH$.
In $\triangle EOD$, solve for ED and DO.

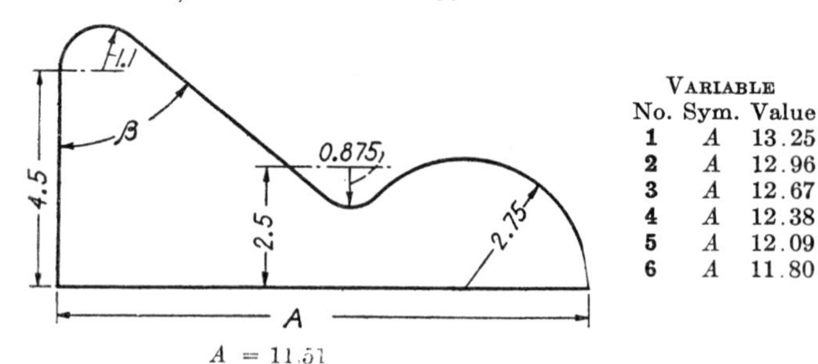

VARIABLE

No.	Sym.	Value
1	A	13.25
2	A	12.96
3	A	12.67
4	A	12.38
5	A	12.09
6	A	11.80

$A = 11.51$
Ans. $\beta = 46° 57' 24''$

49. Determine the angle β.

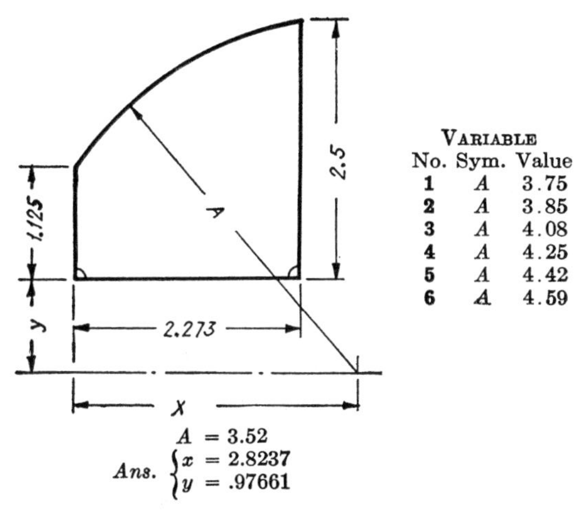

No.	Sym.	Value
1	A	3.75
2	A	3.85
3	A	4.08
4	A	4.25
5	A	4.42
6	A	4.59

Variable

$A = 3.52$

Ans. $\begin{cases} x = 2.8237 \\ y = .97661 \end{cases}$

50. Determine the distance x.
51. Determine the distance y.

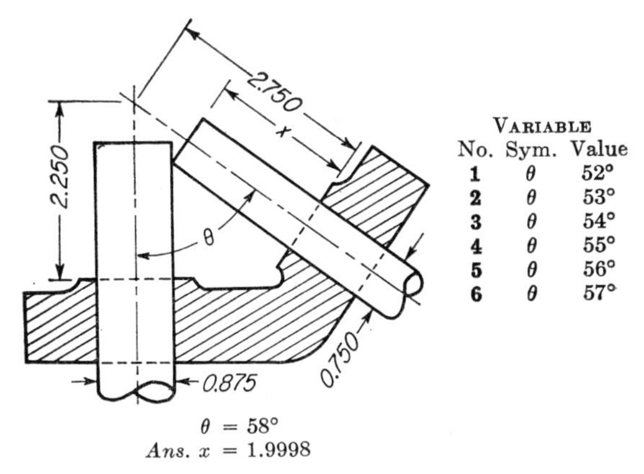

No.	Sym.	Value
1	θ	52°
2	θ	53°
3	θ	54°
4	θ	55°
5	θ	56°
6	θ	57°

Variable

$\theta = 58°$
Ans. $x = 1.9998$

52. Determine the distance x.

Note: The above problem shows how to check properly and accurately the hub distance in a gear box designed for a pair of bevel gears. This is accomplished by the aid of a pair of plug gages whose top edges are sharp. The hub distance for both the bevel gear and bevel gear box is an important factor when considering the efficiency for a pair of bevel gears.

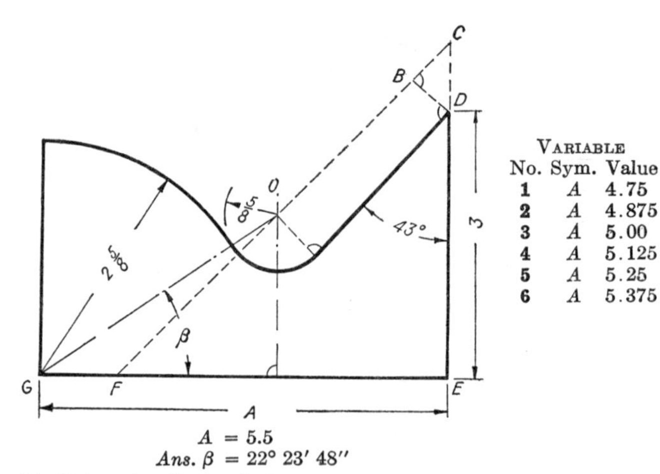

VARIABLE		
No.	Sym.	Value
1	A	4.75
2	A	4.875
3	A	5.00
4	A	5.125
5	A	5.25
6	A	5.375

$A = 5.5$
Ans. $\beta = 22° \ 23' \ 48''$

53. Determine the angle β.

Solution:

$\angle BCD = 43°$. In $\triangle BCD$, solve for CD.
In $\triangle CEF$, solve for EF.
$GF = A - EF$. $GO = 2.625 + .625$.
$\angle CFE = 90° - 43°$.
In $\triangle GOF$, solve for β.

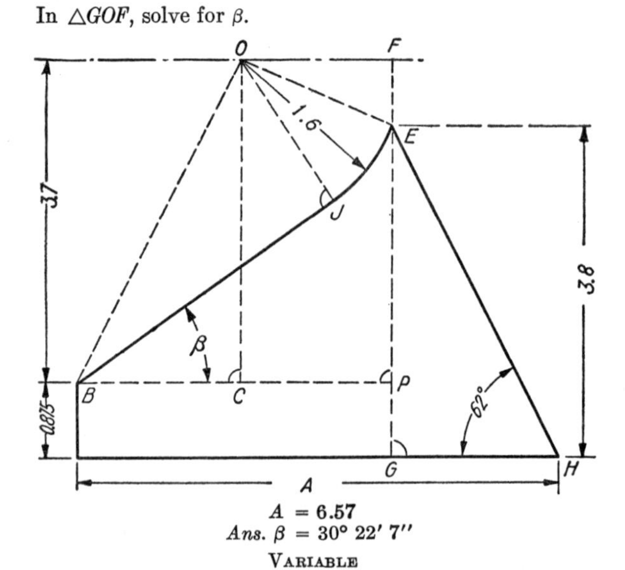

$A = 6.57$
Ans. $\beta = 30° \ 22' \ 7''$

VARIABLE

1. $A = 5.31$ **2.** $A = 5.52$ **3.** $A = 5.73$
4. $A = 5.94$ **5.** $A = 6.15$ **6.** $A = 6.36$

54. Determine the angle β. (*Solution on next page.*)

Solution for preceding problem:

In $\triangle GHE$, solve for GH. $BP = A - GH$. $OE = OJ = 6.1$
$PE = 3.8 - .875$. $EF = .875 + 3.7 - 3.8$
In $\triangle EOF$, solve for OF. $BC = BP - OF$.
$OC = 3.7$. In $\triangle OBC$, solve for $\angle OBC$ and OB.
In $\triangle OBJ$, solve for $\angle OBJ$
$\beta = \angle OBC - \angle OBJ$.

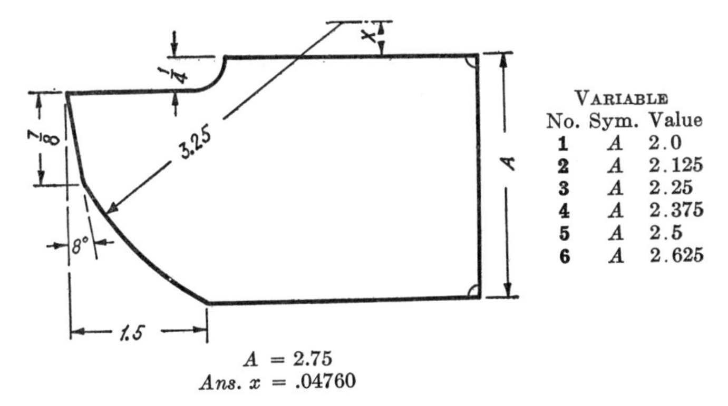

No.	Sym.	Value
1	A	2.0
2	A	2.125
3	A	2.25
4	A	2.375
5	A	2.5
6	A	2.625

VARIABLE

$A = 2.75$
Ans. $x = .04760$

55. Determine the distance x.

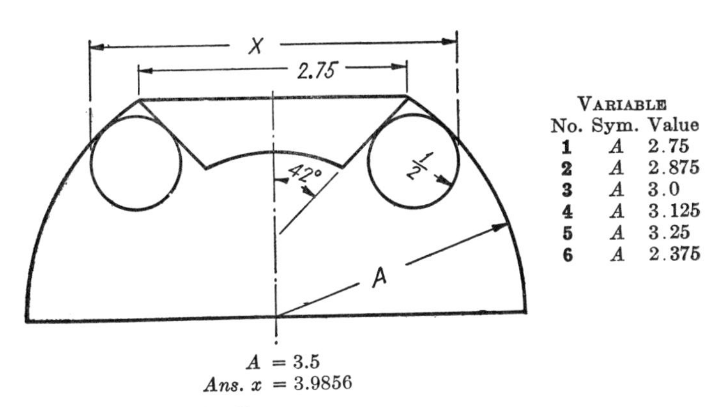

VARIABLE

No.	Sym.	Value
1	A	2.75
2	A	2.875
3	A	3.0
4	A	3.125
5	A	3.25
6	A	2.375

$A = 3.5$
Ans. $x = 3.9856$

56. Determine the distance x.

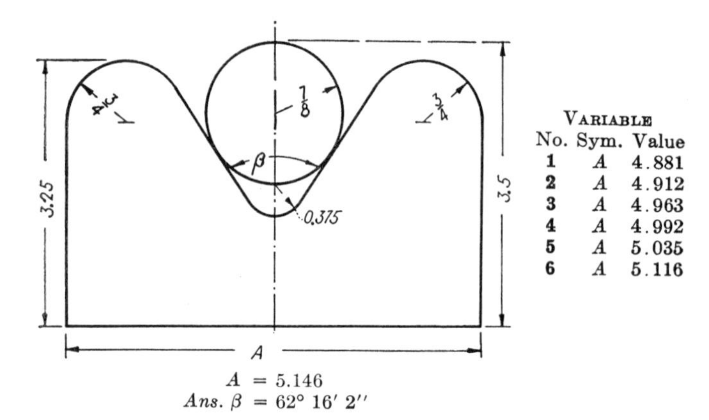

VARIABLE		
No.	Sym.	Value
1	A	4.881
2	A	4.912
3	A	4.963
4	A	4.992
5	A	5.035
6	A	5.116

$A = 5.146$

Ans. $\beta = 62° 16' 2''$

57. Determine the angle β.

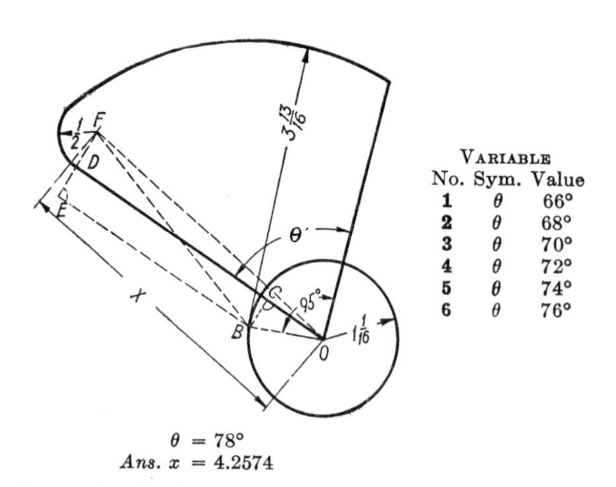

VARIABLE		
No.	Sym.	Value
1	θ	66°
2	θ	68°
3	θ	70°
4	θ	72°
5	θ	74°
6	θ	76°

$\theta = 78°$

Ans. $x = 4.2574$

58. Determine the distance x.

Solution:

$\angle BOC = 95° - \theta$. $BC = ED$. In $\triangle BOC$, solve for BC and OC.

$EF = ED + .5$. $BF = 3.8125 - .5$.

In $\triangle EBF$, solve for $\angle EBF$ and EB. $EB = DC$.

$OD = DC + OC$. $DF = .5$.

In $\triangle FDO$, solve for $\angle DOF$ and OF.

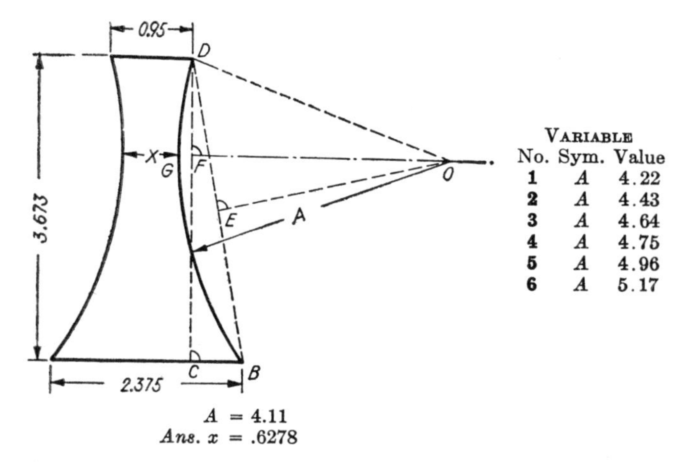

VARIABLE		
No.	Sym.	Value
1	A	4.22
2	A	4.43
3	A	4.64
4	A	4.75
5	A	4.96
6	A	5.17

$A = 4.11$
Ans. $x = .6278$

59. Determine the distance x.

Solution:

$BC = 2.375 \div 2 - .95 \div 2.$
In $\triangle BCD$, solve for $\angle BDC$ and BD.
$DE = BD \div 2.$ $DO = A.$
In $\triangle DEO$, solve for $\angle DOE$.
$\angle EOF = \angle BDC.$ Why?
$\angle FOD = \angle DOE - \angle EOF.$
In $\triangle DOF$, solve for OF.
$x = .95 - 2(A - OF).$ Why?

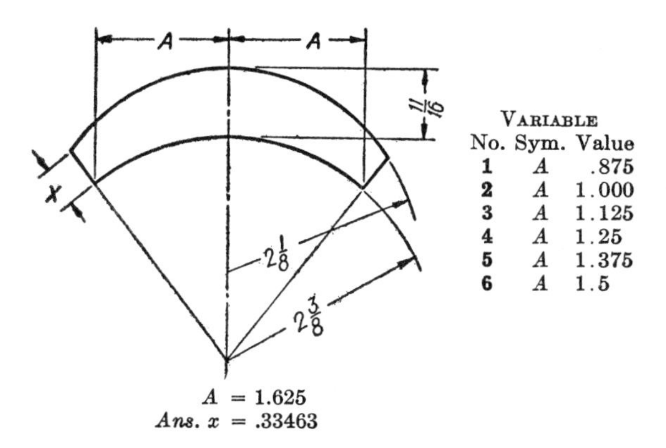

VARIABLE		
No.	Sym.	Value
1	A	.875
2	A	1.000
3	A	1.125
4	A	1.25
5	A	1.375
6	A	1.5

$A = 1.625$
Ans. $x = .33463$

60. Determine the distance x.

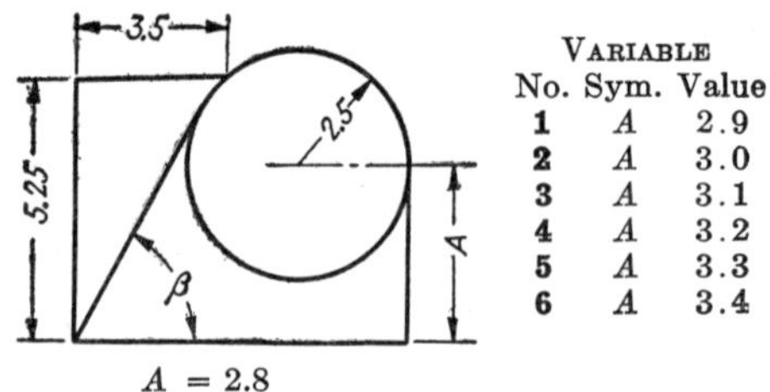

VARIABLE		
No.	Sym.	Value
1	A	2.9
2	A	3.0
3	A	3.1
4	A	3.2
5	A	3.3
6	A	3.4

$A = 2.8$

Ans. $\beta = 65° 49' 33''$

61. Determine the angle β.

Die Punch

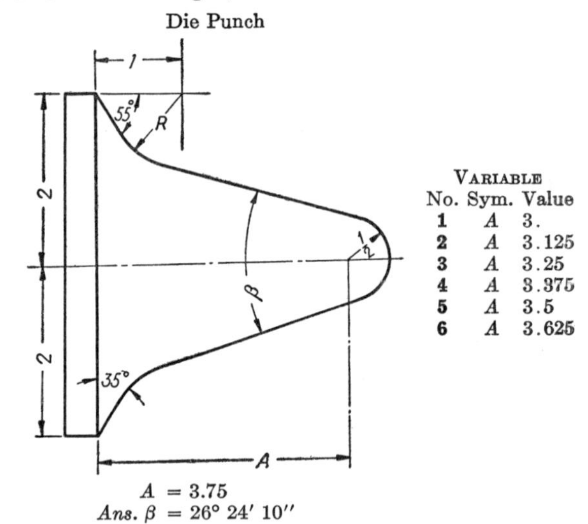

VARIABLE		
No.	Sym.	Value
1	A	3.
2	A	3.125
3	A	3.25
4	A	3.375
5	A	3.5
6	A	3.625

$A = 3.75$

Ans. $\beta = 26° 24' 10''$

62. Determine the angle β.

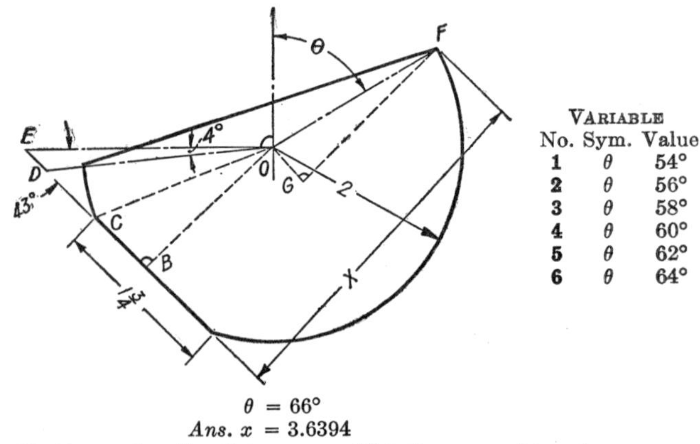

VARIABLE		
No.	Sym.	Value
1	θ	54°
2	θ	56°
3	θ	58°
4	θ	60°
5	θ	62°
6	θ	64°

$\theta = 66°$

Ans. $x = 3.6394$

63. Determine the distance x. (*Solution on next page.*)

Solution for preceding problem:

$BC = 1.75 \div 2$. Why? $OC = OF = 2$.
In $\triangle BCO$, solve for OB.
$\angle DOB = 90° - 43° - 4°$. Why?
$\angle FOG = 270° - \angle DOB - 4° - 90° - \theta$. Why?
In $\triangle OFG$, solve for FG.
$x = OB + FG$.

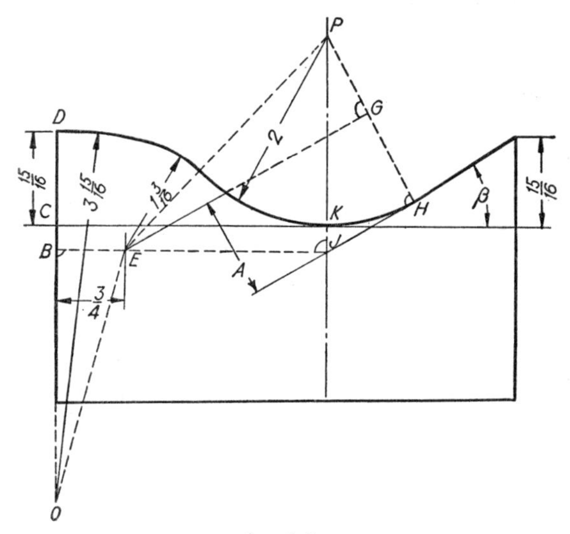

$A = 1.5$
Ans. $\beta = 38° \ 35' \ 14''$

VARIABLE

1. $A = .75$	**2.** $A = .875$	**3.** $A = 1.00$
4. $A = 1.125$	**5.** $A = 1.25$	**6.** $A = 1.375$

64. Determine the angle β.

Solution:

$OE = 3.9375 - 1.1875$. In $\triangle BOE$, solve for OB.
$BC = JK = 3.9375 - .9375 - OB$.
$EP = 1.1875 + 2$. $PG = PH - A$.
In $\triangle EPG$, solve for $\angle EPG$.
In $\triangle EPJ$, solve for $\angle EPJ$.
$\angle KPH = \angle EPG - \angle EPJ$.
$\beta = \angle KPH$. Why?

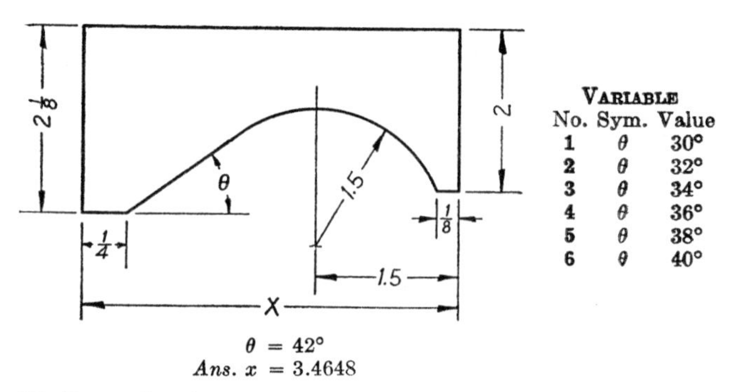

VARIABLE		
No.	Sym.	Value
1	θ	30°
2	θ	32°
3	θ	34°
4	θ	36°
5	θ	38°
6	θ	40°

$\theta = 42°$
Ans. $x = 3.4648$

65. Determine the distance x.

Spring Retainer Frame

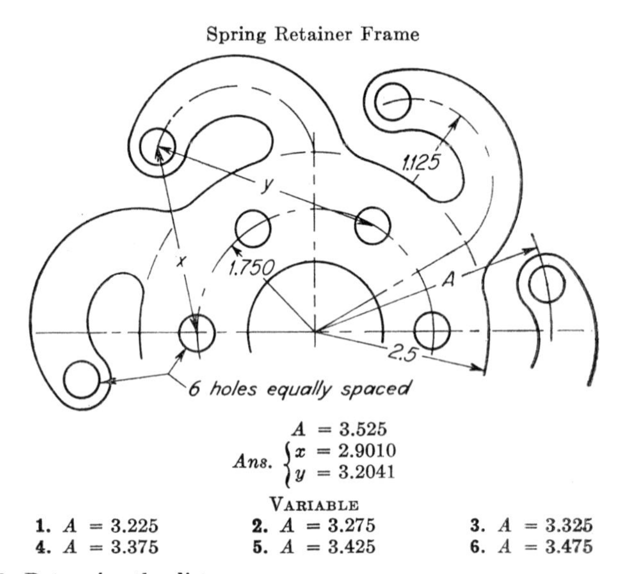

$A = 3.525$

Ans. $\begin{cases} x = 2.9010 \\ y = 3.2041 \end{cases}$

VARIABLE

1. $A = 3.225$	**2.** $A = 3.275$	**3.** $A = 3.325$
4. $A = 3.375$	**5.** $A = 3.425$	**6.** $A = 3.475$

66. Determine the distance x.
67. Determine the distance y.

Note: This was a punch press job. It was necessary to make a punch and die as shown above. Later the circular lugs were turned up into the shape of a cup. In order to machine the punch and die properly, the x and y distances, with others, were very important dimensions.

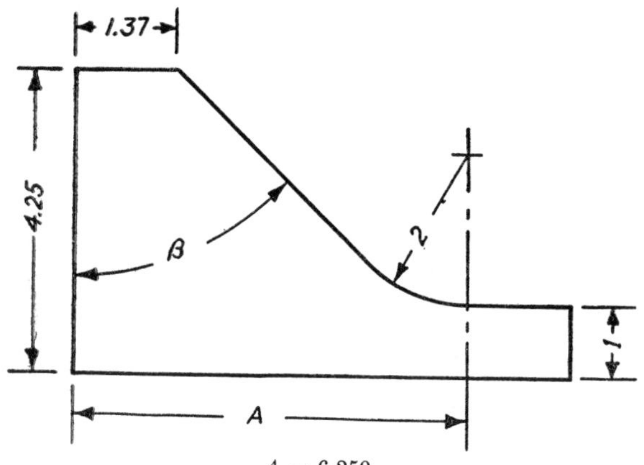

$A = 6.250$
Ans. $\beta = 52° \ 14' \ 28''$

VARIABLE

1. $A = 5.500$	**2.** $A = 5.625$	**3.** $A = 5.750$
4. $A = 5.875$	**5.** $A = 6.000$	**6.** $A = 6.125$

68. Determine the angle β.

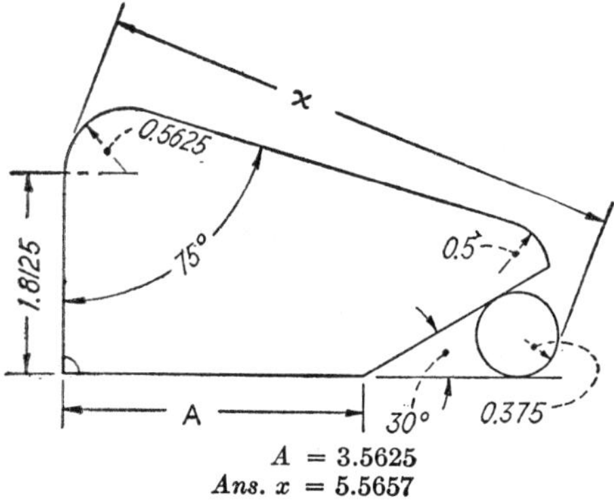

$A = 3.5625$
Ans. $x = 5.5657$

VARIABLE

1. $A = 2.8125$	**2.** $A = 2.9375$	**3.** $A = 3.0625$
4. $A = 3.1875$	**5.** $A = 3.3125$	**6.** $A = 3.4375$

69. Determine the distance x.

Gage

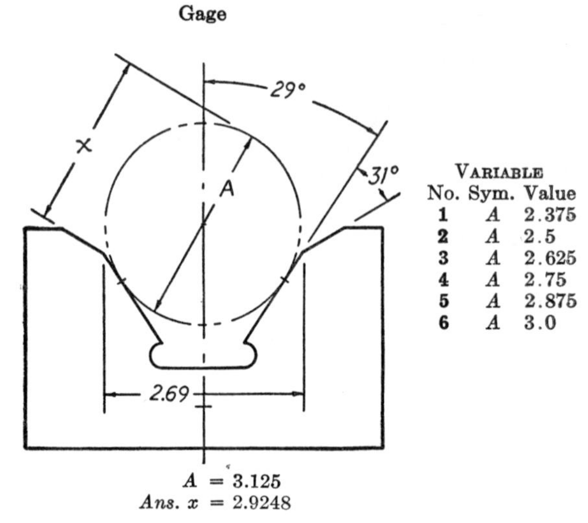

VARIABLE		
No.	Sym.	Value
1	*A*	2.375
2	*A*	2.5
3	*A*	2.625
4	*A*	2.75
5	*A*	2.875
6	*A*	3.0

$A = 3.125$
Ans. $x = 2.9248$

70. Determine the distance x.

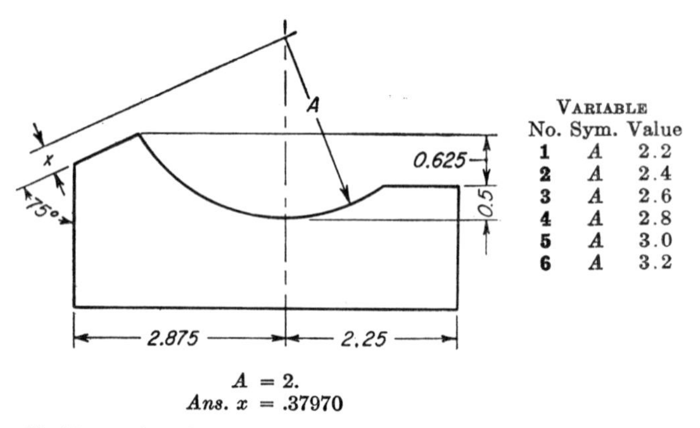

VARIABLE		
No.	Sym.	Value
1	*A*	2.2
2	*A*	2.4
3	*A*	2.6
4	*A*	2.8
5	*A*	3.0
6	*A*	3.2

$A = 2.$
Ans. $x = .37970$

71. Determine the distance x.

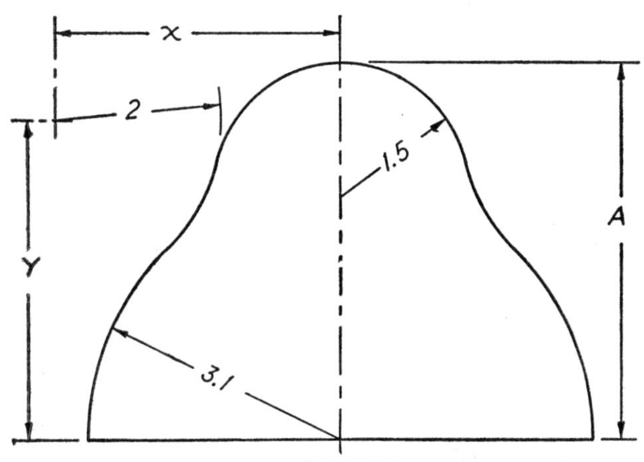

$A = 4.250$

Ans. $\begin{cases} x = 3.3136 \\ y = 3.8768 \end{cases}$

VARIABLE

1. $A = 3.500$ **2.** $A = 3.625$ **3.** $A = 3.750$
4. $A = 3.875$ **5.** $A = 4.000$ **6.** $A = 4.125$

72. Determine the distance x.
73. Determine the distance y.

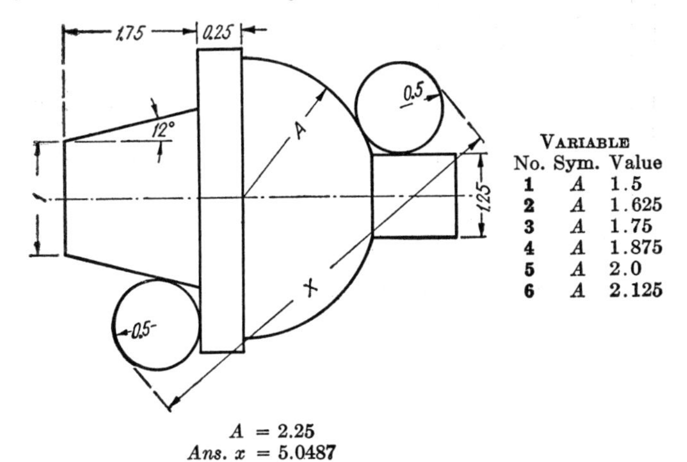

VARIABLE

No.	Sym.	Value
1	A	1.5
2	A	1.625
3	A	1.75
4	A	1.875
5	A	2.0
6	A	2.125

$A = 2.25$

Ans. $x = 5.0487$

74. Determine the distance x.

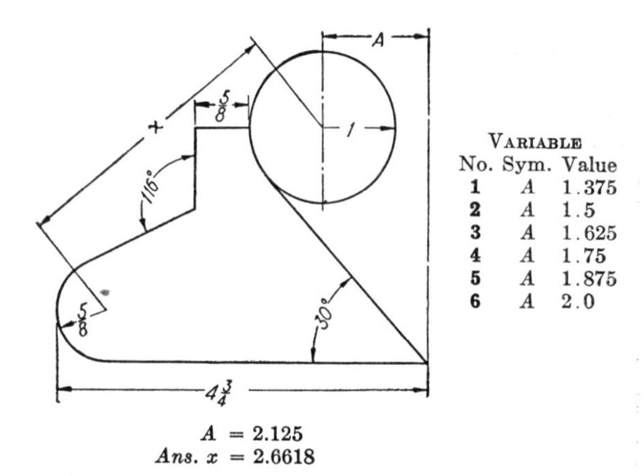

VARIABLE		
No.	Sym.	Value
1	*A*	1.375
2	*A*	1.5
3	*A*	1.625
4	*A*	1.75
5	*A*	1.875
6	*A*	2.0

$A = 2.125$
Ans. $x = 2.6618$

75. Determine the distance x.

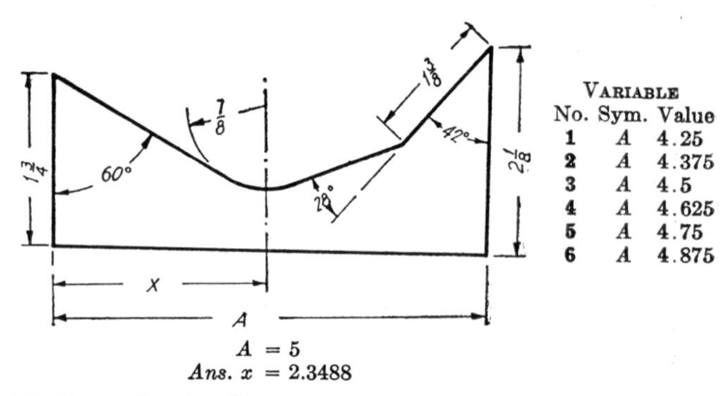

VARIABLE		
No.	Sym.	Value
1	*A*	4.25
2	*A*	4.375
3	*A*	4.5
4	*A*	4.625
5	*A*	4.75
6	*A*	4.875

$A = 5$
Ans. $x = 2.3488$

76. Determine the distance x.

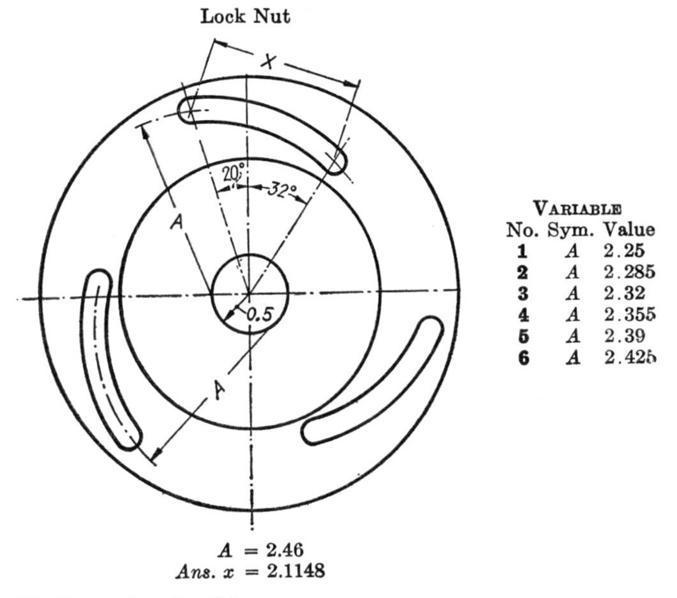

VARIABLE		
No.	Sym.	Value
1	A	2.25
2	A	2.285
3	A	2.32
4	A	2.355
5	A	2.39
6	A	2.425

$A = 2.46$
Ans. $x = 2.1148$

77. Determine the distance x.

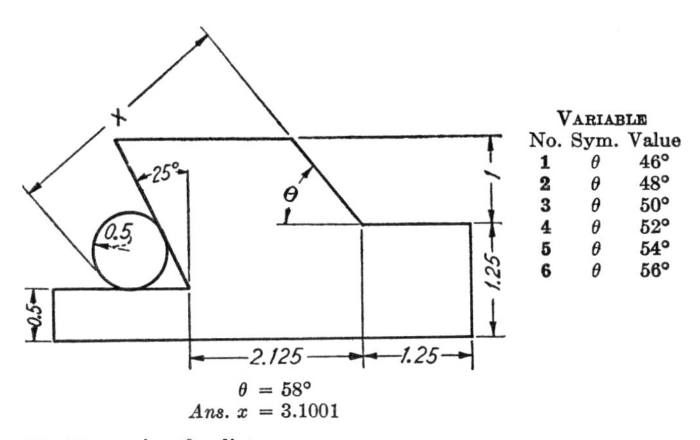

VARIABLE		
No.	Sym.	Value
1	θ	46°
2	θ	48°
3	θ	50°
4	θ	52°
5	θ	54°
6	θ	56°

$\theta = 58°$
Ans. $x = 3.1001$

78. Determine the distance x.

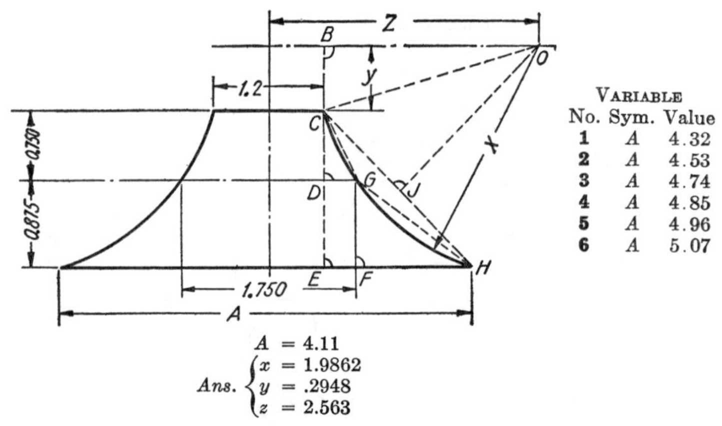

VARIABLE		
No.	Sym.	Value
1	A	4.32
2	A	4.53
3	A	4.74
4	A	4.85
5	A	4.96
6	A	5.07

$$A = 4.11$$

$$\textit{Ans.} \begin{cases} x = 1.9862 \\ y = .2948 \\ z = 2.563 \end{cases}$$

79. Determine the distance x.

80. Determine the distance y.

81. Determine the distance z.

Solution:

$DG = 1.75 \div 2 - 1.2 \div 2$. In $\triangle DCG$, solve for $\angle DCG$ and CG.
In $\triangle FGH$, solve for GH. In $\triangle CEH$, solve for CH.
In $\triangle CGH$, solve for $\angle GCH$ and x.
$CJ = CH \div 2$. $CO = x$.
In $\triangle CJO$, solve for $\angle JCO$.
$\angle BCO = 180° - \angle DCG - \angle GCH - \angle JCO$.
In $\triangle BCO$, solve for BC and BO.

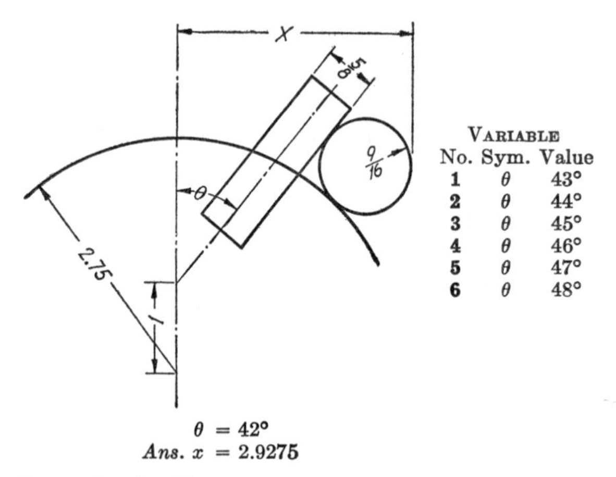

VARIABLE		
No.	Sym.	Value
1	θ	43°
2	θ	44°
3	θ	45°
4	θ	46°
5	θ	47°
6	θ	48°

$$\theta = 42°$$

$$\textit{Ans. } x = 2.9275$$

82. Determine the distance x.

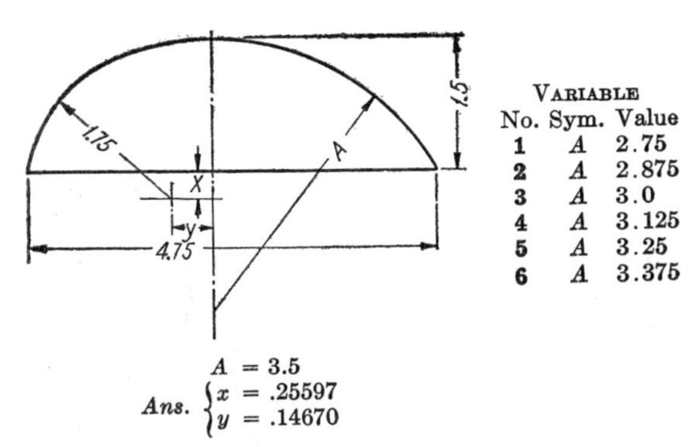

No.	Sym.	Value
1	A	2.75
2	A	2.875
3	A	3.0
4	A	3.125
5	A	3.25
6	A	3.375

(VARIABLE)

$A = 3.5$

Ans. $\begin{cases} x = .25597 \\ y = .14670 \end{cases}$

83. Determine the distance x.

84. Determine the distance y.

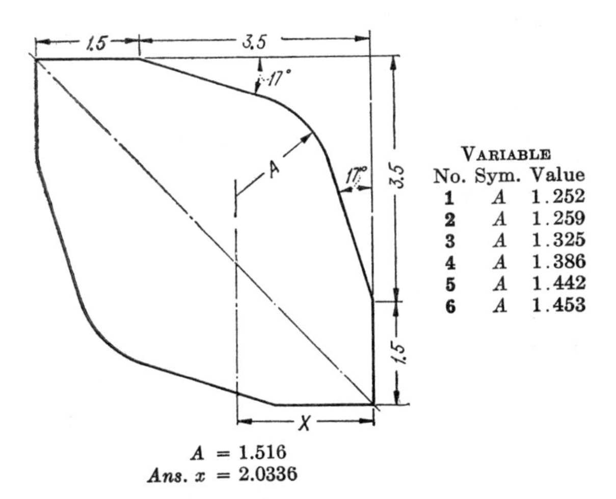

No.	Sym.	Value
1	A	1.252
2	A	1.259
3	A	1.325
4	A	1.386
5	A	1.442
6	A	1.453

(VARIABLE)

$A = 1.516$

Ans. $x = 2.0336$

85. Determine the distance x.

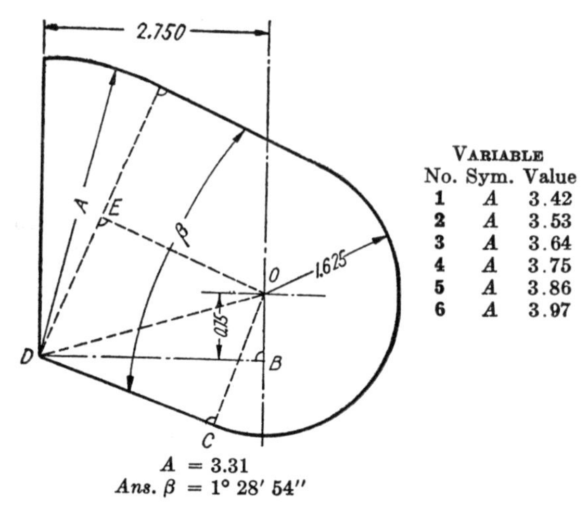

VARIABLE		
No.	Sym.	Value
1	A	3.42
2	A	3.53
3	A	3.64
4	A	3.75
5	A	3.86
6	A	3.97

$A = 3.31$
Ans. $\beta = 1° 28' 54''$

86. Determine the angle β.

Solution:

In $\triangle DBO$, solve for DO.
$DE = A - 1.625$. Why?
In $\triangle DOE$, solve for $\angle DOE$. $CO = 1.625$.
In $\triangle DOC$, solve for $\angle DOC$.
$\beta = \angle DOE + \angle DOC - 90°$. Why?

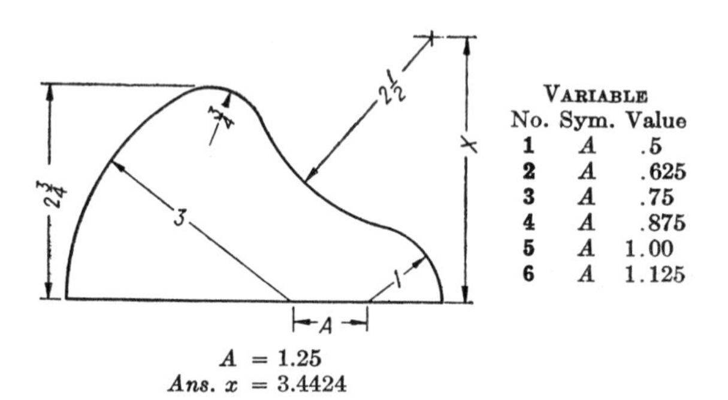

VARIABLE		
No.	Sym.	Value
1	A	.5
2	A	.625
3	A	.75
4	A	.875
5	A	1.00
6	A	1.125

$A = 1.25$
Ans. $x = 3.4424$

87. Determine the distance x.

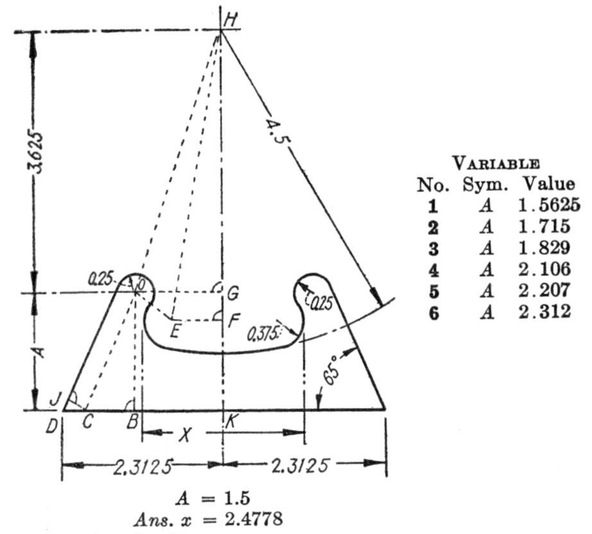

No.	Sym.	Value
1	A	1.5625
2	A	1.715
3	A	1.829
4	A	2.106
5	A	2.207
6	A	2.312

$A = 1.5$
Ans. $x = 2.4778$

88. Determine the distance x.

Solution:

$\angle JDC = \angle OCB = 65°$. $CJ = .25$. $BO = A - .25$.
In $\triangle JDC$, solve for DC. In $\triangle OCB$, solve for CB.
$DK - DC - CB = BK = OG$. $HG = 3.625$.
In $\triangle HOG$, solve for $\angle OHG$ and OH. $OE = .25 + .375$.
$EH = 4.5 - .375$. In $\triangle OHE$, solve for $\angle OHE$.
$\angle EHF = \angle OHG - \angle OHE$. In $\triangle HEF$, solve for EF.
$x = 2(EF + .375)$.

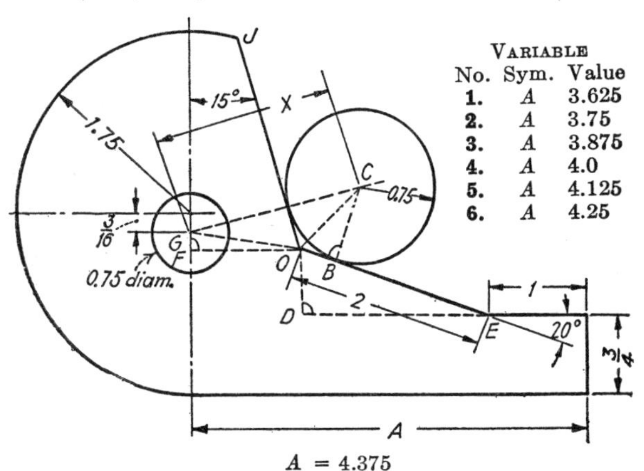

No.	Sym.	Value
1.	A	3.625
2.	A	3.75
3.	A	3.875
4.	A	4.0
5.	A	4.125
6.	A	4.25

$A = 4.375$
Ans. $x = 2.1687$

89. Determine the distance x. (*Solution on next page.*)

Solution for preceding problem:

$\angle OED = 20°$. In $\triangle OED$, solve for DE and DO.
$FO = A - DE - 1$. $FG = 1.75 - .1875 - DO - .75$.
In $\triangle GFO$, solve for $\angle GOF$ and GO.
$\angle GOJ = 90° - 15° - \angle GOF$.
$\angle JOE = 90° + 15° + 20°$. Why?
$\angle JOC = \angle COB = \angle JOE \div 2$. In $\triangle COB$, solve for CO.
$\angle GOC = \angle GOJ + \angle JOC$.
In $\triangle GOC$, solve for CG.

Checking Position of Holes

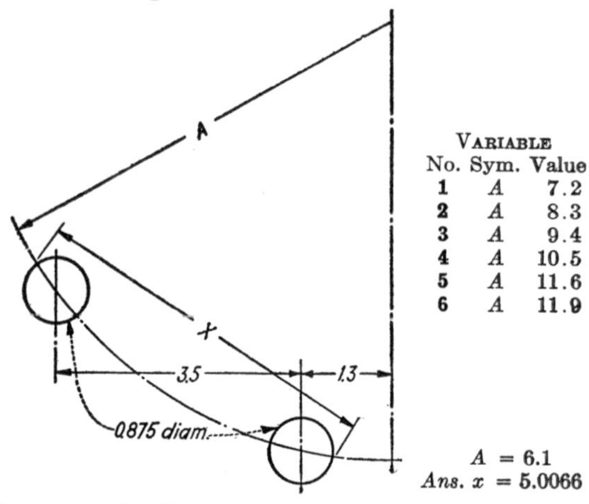

VARIABLE		
No.	Sym.	Value
1	A	7.2
2	A	8.3
3	A	9.4
4	A	10.5
5	A	11.6
6	A	11.9

$A = 6.1$
Ans. $x = 5.0066$

90. Determine the distance x.

Checking Angular Holes

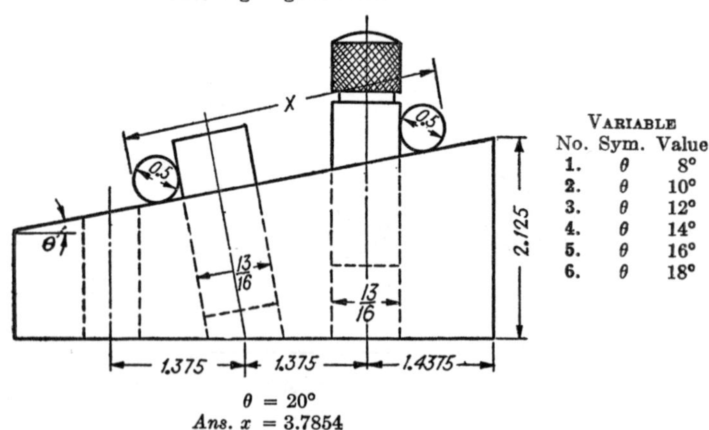

VARIABLE		
No.	Sym.	Value
1.	θ	8°
2.	θ	10°
3.	θ	12°
4.	θ	14°
5.	θ	16°
6.	θ	18°

$\theta = 20°$
Ans. $x = 3.7854$

91. Determine the distance x.

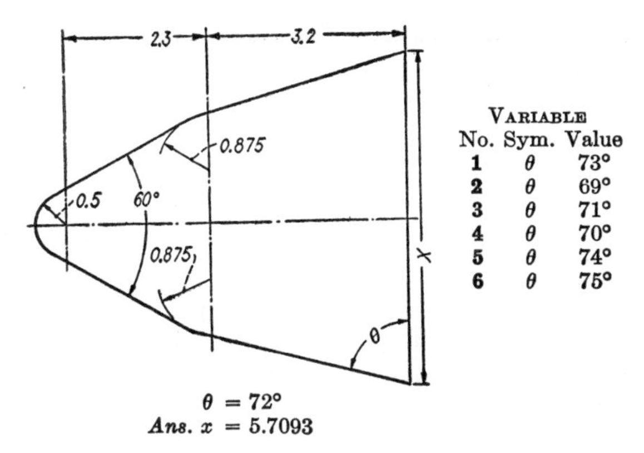

Variable		
No.	Sym.	Value
1	θ	73°
2	θ	69°
3	θ	71°
4	θ	70°
5	θ	74°
6	θ	75°

$\theta = 72°$
Ans. $x = 5.7093$

92. Determine the distance x.

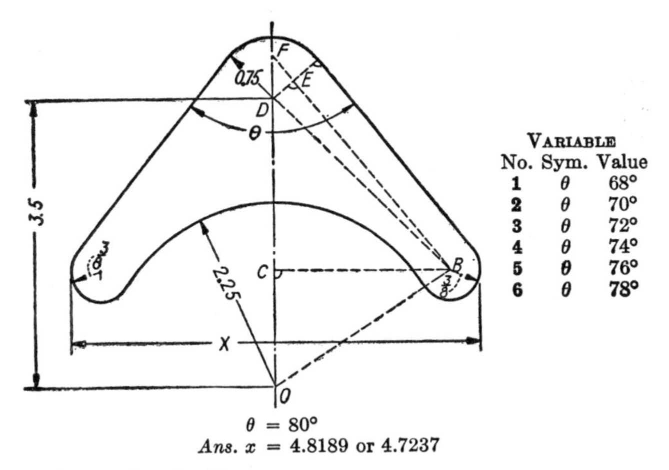

Variable		
No.	Sym.	Value
1	θ	68°
2	θ	70°
3	θ	72°
4	θ	74°
5	θ	76°
6	θ	78°

$\theta = 80°$
Ans. $x = 4.8189$ or 4.7237

93. Determine the distance x.

Solution:

$\angle DFE = \theta \div 2$. $DE = .75 - .375$.
In $\triangle DFE$, solve for DF. $FO = DO + DF$.
$BO = 2.25 + .375$. In $\triangle FOB$, solve for $\angle FOB$.
In $\triangle OCB$, solve for BC.

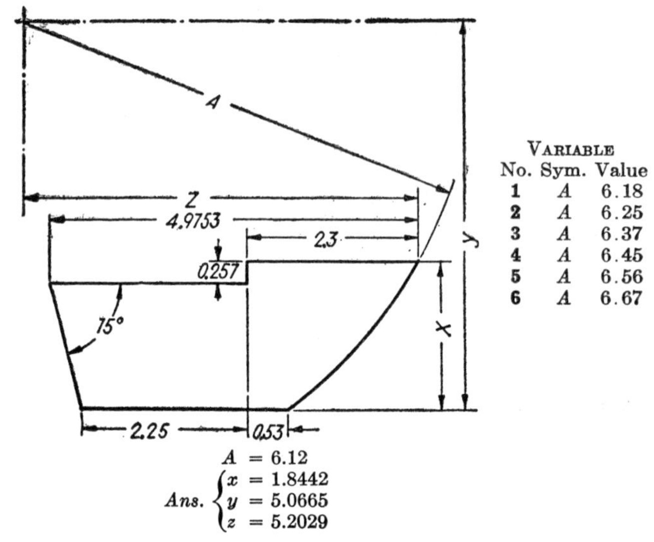

No.	Sym.	Value
1	A	6.18
2	A	6.25
3	A	6.37
4	A	6.45
5	A	6.56
6	A	6.67

VARIABLE

$A = 6.12$

Ans. $\begin{cases} x = 1.8442 \\ y = 5.0665 \\ z = 5.2029 \end{cases}$

94. Determine the distance x.
95. Determine the distance y.
96. Determine the distance z.

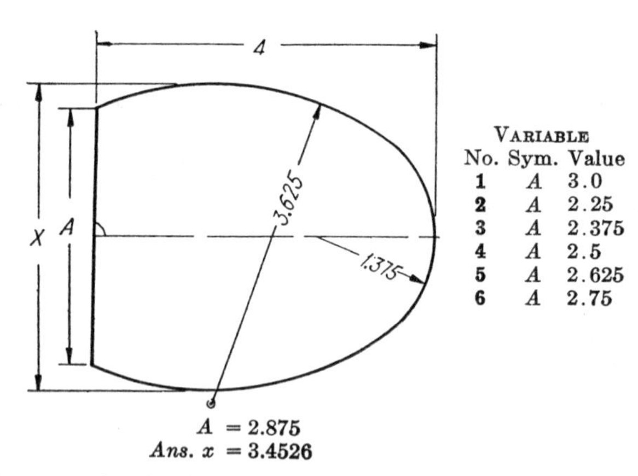

VARIABLE

No.	Sym.	Value
1	A	3.0
2	A	2.25
3	A	2.375
4	A	2.5
5	A	2.625
6	A	2.75

$A = 2.875$
Ans. $x = 3.4526$

97. Determine the distance x.

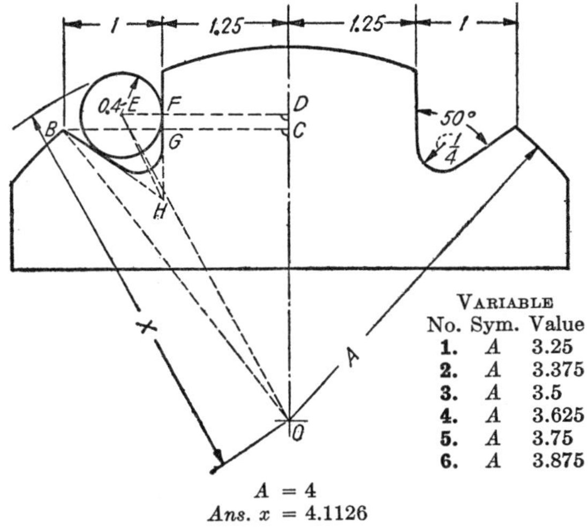

VARIABLE

No.	Sym.	Value
1.	A	3.25
2.	A	3.375
3.	A	3.5
4.	A	3.625
5.	A	3.75
6.	A	3.875

$A = 4$
Ans. $x = 4.1126$

98. Determine the distance x.

Solution:

$BC = 1 + 1.25$, $BO = A$. In $\triangle BOC$, solve for CO.
$\angle BHG = 50°$. In $\triangle BHG$, solve for GH.
$\angle EHF = 50° \div 2$. In $\triangle EHF$, solve for FH.
$FG = FH - GH = CD$. $DO = CO + CD$.
$DE = 1.25 + .4$.
In $\triangle EDO$, solve for EO.

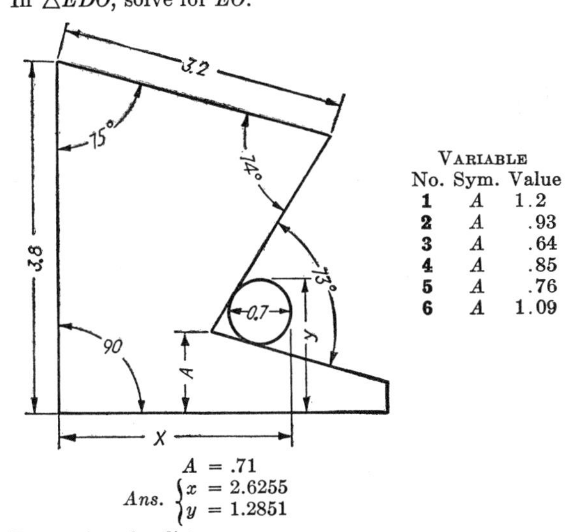

VARIABLE

No.	Sym.	Value
1	A	1.2
2	A	.93
3	A	.64
4	A	.85
5	A	.76
6	A	1.09

$A = .71$
Ans. $\begin{cases} x = 2.6255 \\ y = 1.2851 \end{cases}$

99. Determine the distance x.
100. Determine the distance y.

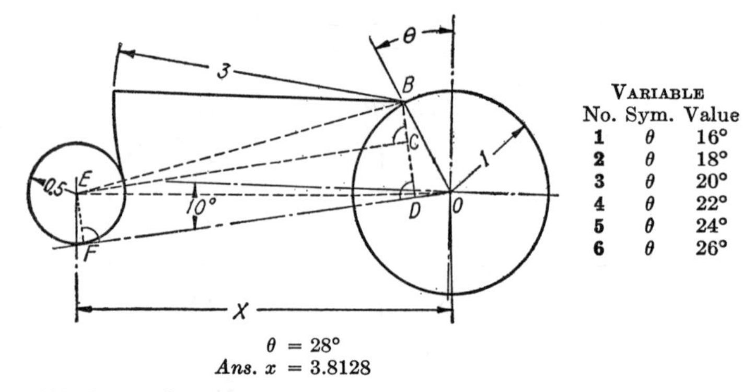

No.	Sym.	Value
1	θ	16°
2	θ	18°
3	θ	20°
4	θ	22°
5	θ	24°
6	θ	26°

Variable

$\theta = 28°$
Ans. $x = 3.8128$

101. Determine the distance x.
Solution:

$\angle DBO = \theta - 10°$. Why?
In $\triangle DOB$, solve for BD. $BC = BD - CD$.
$EF = .5 = CD$. $BE = 3 + .5$.
In $\triangle BEC$, solve for $\angle EBC$.
$\angle EBO = \angle EBC + \angle DBO$.
In $\triangle EBO$, solve for EO.

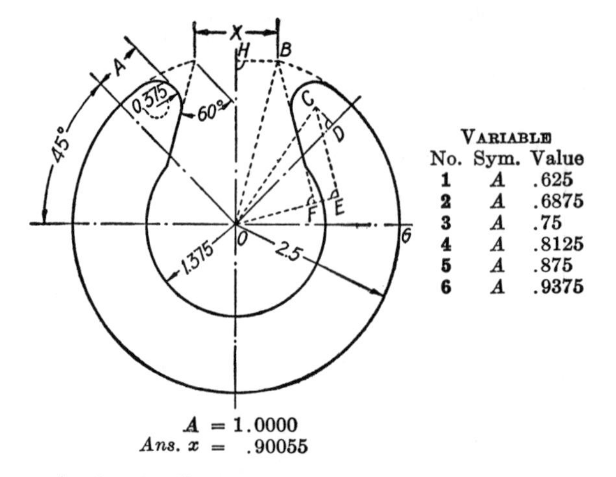

No.	Sym.	Value
1	A	.625
2	A	.6875
3	A	.75
4	A	.8125
5	A	.875
6	A	.9375

Variable

$A = 1.0000$
Ans. $x = .90055$

102. Determine the distance x.
Solution:

In $\triangle COD$, solve for $\angle COD$. $\angle DOE = 90° - 60°$. Why?
In $\triangle COE$, solve for OE. $OF = OE - .375$. Why?
In $\triangle BOF$, solve for $\angle BOF$. $\angle EOG = 15°$. Why?
$\angle HOB = 90° - \angle BOF - \angle EOG$.
In $\triangle HOB$, solve for BH.

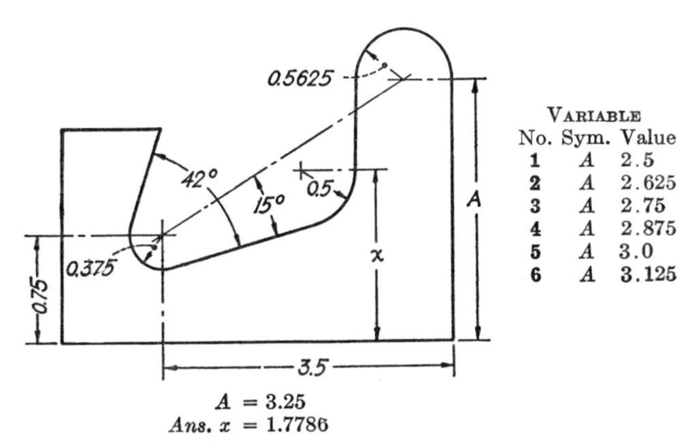

VARIABLE		
No.	Sym.	Value
1	A	2.5
2	A	2.625
3	A	2.75
4	A	2.875
5	A	3.0
6	A	3.125

$A = 3.25$

Ans. $x = 1.7786$

103. Determine the distance x.

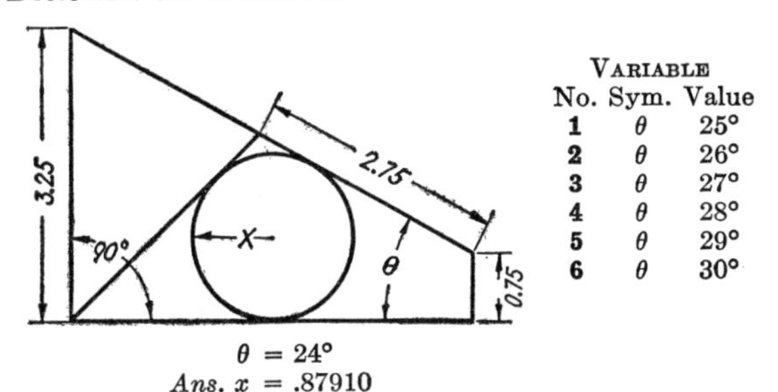

VARIABLE		
No.	Sym.	Value
1	θ	25°
2	θ	26°
3	θ	27°
4	θ	28°
5	θ	29°
6	θ	30°

$\theta = 24°$

Ans. $x = .87910$

104. Determine the radius x.

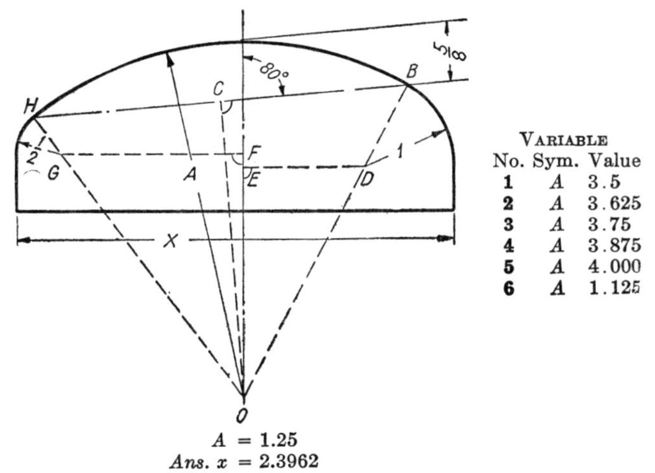

VARIABLE		
No.	Sym.	Value
1	A	3.5
2	A	3.625
3	A	3.75
4	A	3.875
5	A	4.000
6	A	1.125

$A = 1.25$

Ans. $x = 2.3962$

105. Determine the distance x. *(Solution on next page.)*

Solution for preceding problem:

$CO = A - .625$. In $\triangle COB$, solve for $\angle COB$.
$\angle COF = 10°$. Why? $\angle FOB = \angle COB - 10°$.
$DO = A - 1$. In $\triangle EOD$, solve for DE.
$\angle HOC = \angle COB$. $\angle GOF = \angle HOC + 10°$.
In $\triangle GOF$, solve for FG.

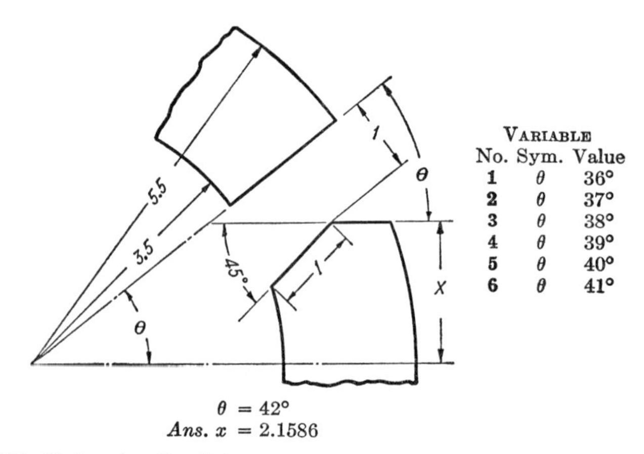

VARIABLE		
No.	Sym.	Value
1	θ	36°
2	θ	37°
3	θ	38°
4	θ	39°
5	θ	40°
6	θ	41°

$\theta = 42°$
Ans. $x = 2.1586$

106. Determine the distance x.

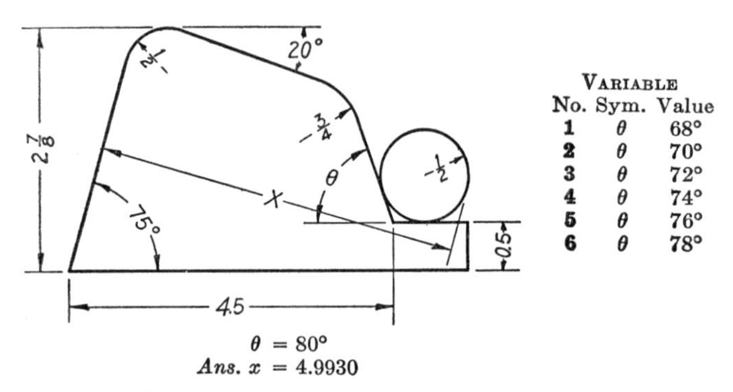

VARIABLE		
No.	Sym.	Value
1	θ	68°
2	θ	70°
3	θ	72°
4	θ	74°
5	θ	76°
6	θ	78°

$\theta = 80°$
Ans. $x = 4.9930$

107. Determine the distance x.

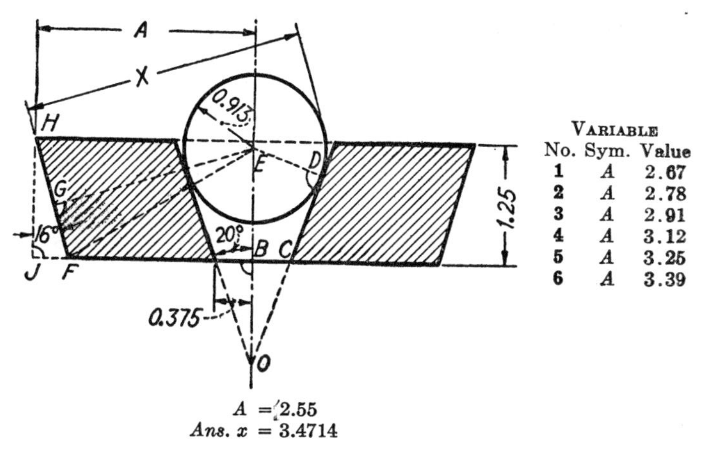

VARIABLE

No.	Sym.	Value
1	A	2.67
2	A	2.78
3	A	2.91
4	A	3.12
5	A	3.25
6	A	3.39

$A = 2.55$
Ans. $x = 3.4714$

108. Determine the distance x.

Solution:

$BC = .375$. In $\triangle BCO$, solve for BO.
$DE = .913$. In $\triangle EDO$, solve for EO.
$BE = EO - BO$. In $\triangle HJF$, solve for FJ.
$BF = A - FJ$. In $\triangle FBE$, solve for $\angle EFB$ and EF.
$\angle GFE = 180° - \angle HFJ - \angle EFB$.
In $\triangle GFE$, solve for EG.

Die-sinking Templet

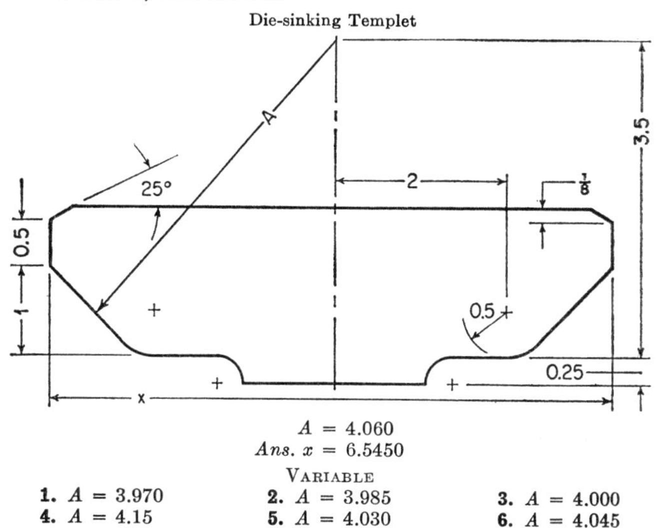

$A = 4.060$
Ans. $x = 6.5450$

VARIABLE

1. $A = 3.970$ **2.** $A = 3.985$ **3.** $A = 4.000$
4. $A = 4.15$ **5.** $A = 4.030$ **6.** $A = 4.045$

109. Determine the distance x.

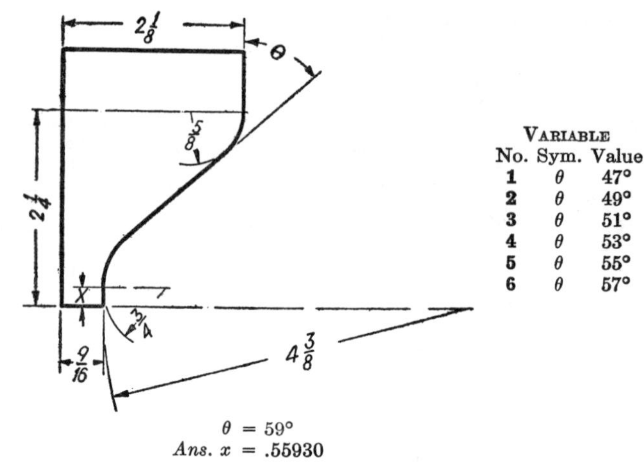

VARIABLE		
No.	Sym.	Value
1	θ	47°
2	θ	49°
3	θ	51°
4	θ	53°
5	θ	55°
6	θ	57°

$$\theta = 59°$$
Ans. $x = .55930$

110. Determine the distance x.

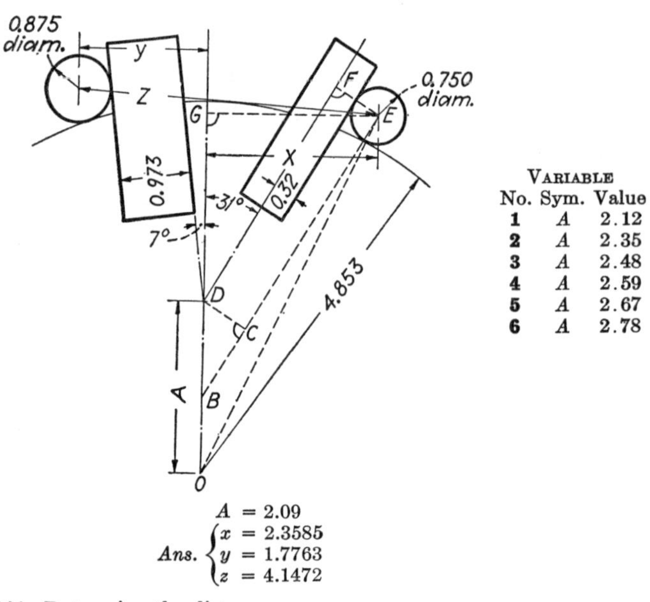

VARIABLE		
No.	Sym.	Value
1	A	2.12
2	A	2.35
3	A	2.48
4	A	2.59
5	A	2.67
6	A	2.78

$$A = 2.09$$
Ans. $\begin{cases} x = 2.3585 \\ y = 1.7763 \\ z = 4.1472 \end{cases}$

111. Determine the distance x.
112. Determine the distance y.
113. Determine the distance z.

Solution:

$$EF = CD = .32 + .375.$$
$$\angle CBD = 31°.$$

(*Solution continued on next page.*)

Solution continued:

In $\triangle CBD$, solve for BD.

$BO = A - BD$. $EO = 4.853 + .375$. In $\triangle BOE$, solve for $\angle BOE$.

In $\triangle OEG$, solve for EG.

The solution for y is similar to that of x.

The solution for z is left to the student.

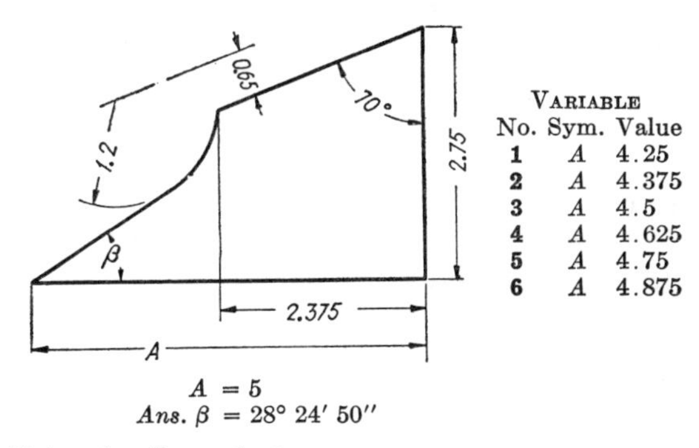

VARIABLE		
No.	Sym.	Value
1	A	4.25
2	A	4.375
3	A	4.5
4	A	4.625
5	A	4.75
6	A	4.875

$A = 5$
Ans. $\beta = 28° 24' 50''$

114. Determine the angle β.

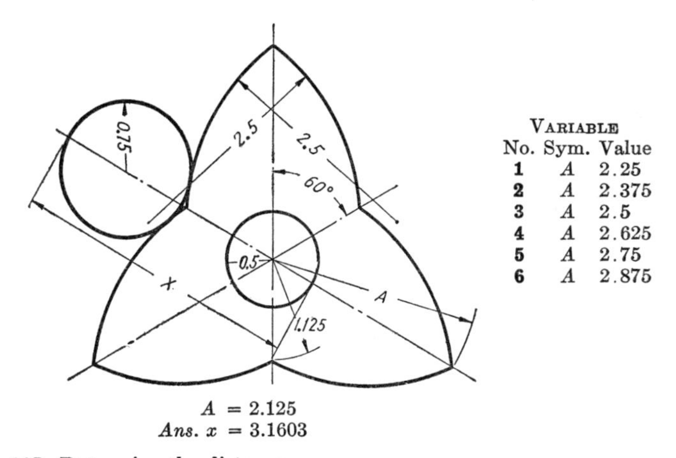

VARIABLE		
No.	Sym.	Value
1	A	2.25
2	A	2.375
3	A	2.5
4	A	2.625
5	A	2.75
6	A	2.875

$A = 2.125$
Ans. $x = 3.1603$

115. Determine the distance x.

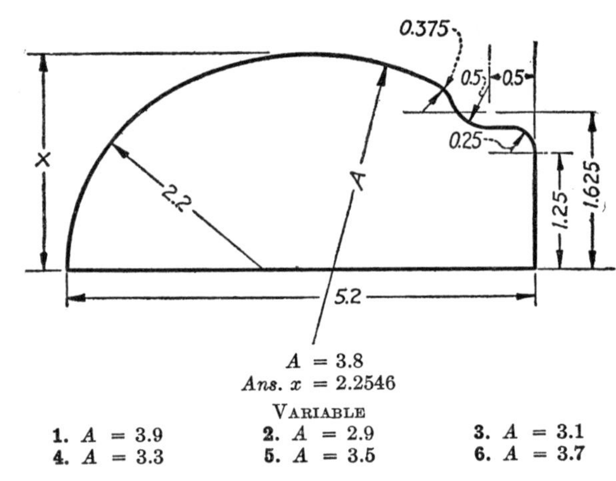

$A = 3.8$
Ans. $x = 2.2546$

VARIABLE

1. $A = 3.9$	**2.** $A = 2.9$	**3.** $A = 3.1$
4. $A = 3.3$	**5.** $A = 3.5$	**6.** $A = 3.7$

116. Determine the distance x

Checking by Means of a Sine Bar

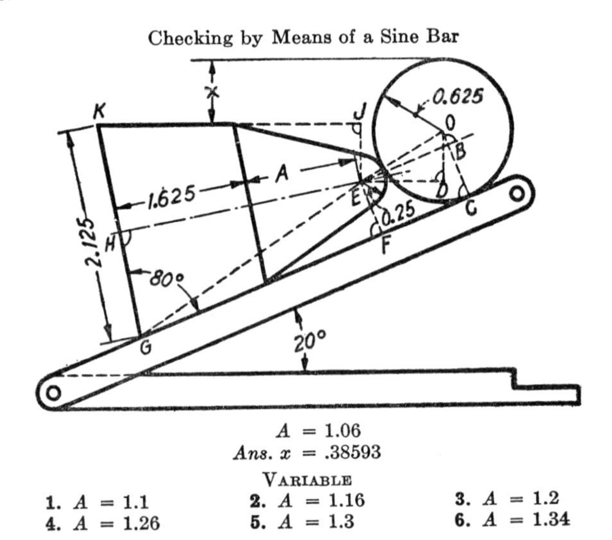

$A = 1.06$
Ans. $x = .38593$

VARIABLE

1. $A = 1.1$	**2.** $A = 1.16$	**3.** $A = 1.2$
4. $A = 1.26$	**5.** $A = 1.3$	**6.** $A = 1.34$

117. Determine the distance x.

Solution:

BE is parallel to GF.

DE is parallel to JK. Hence $\angle BED = 20°$. Why?

In $\triangle HEG$, solve for $\angle HGE$ and EG. $\angle EGF = 80° - \angle HGE$.

In $\triangle GEF$, solve for EF. $EF = BC$. $CO = .625$.

$BO = CO - BC$. In $\triangle OEB$, solve for $\angle OEB$.

$\angle OED = \angle OEB + \angle BED$. (*Continued on next page.*)

Solution continued:

In $\triangle OED$, solve for DO. $EJ = EF$. Why?
$x = DO + .625 - EJ$.

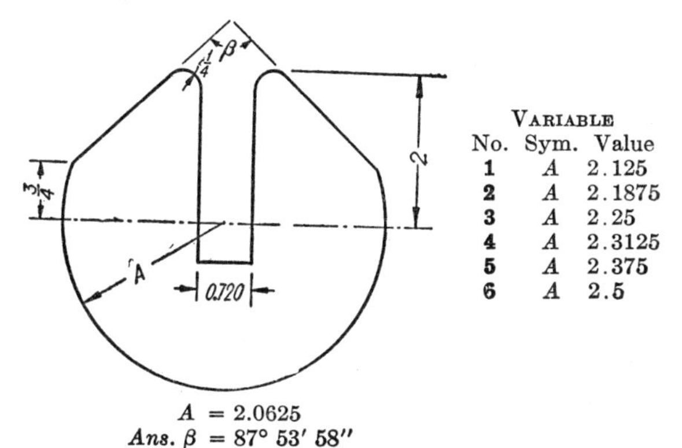

No.	Sym.	Value
1	A	2.125
2	A	2.1875
3	A	2.25
4	A	2.3125
5	A	2.375
6	A	2.5

$A = 2.0625$
Ans. $\beta = 87° 53' 58''$

118. Determine the angle β.

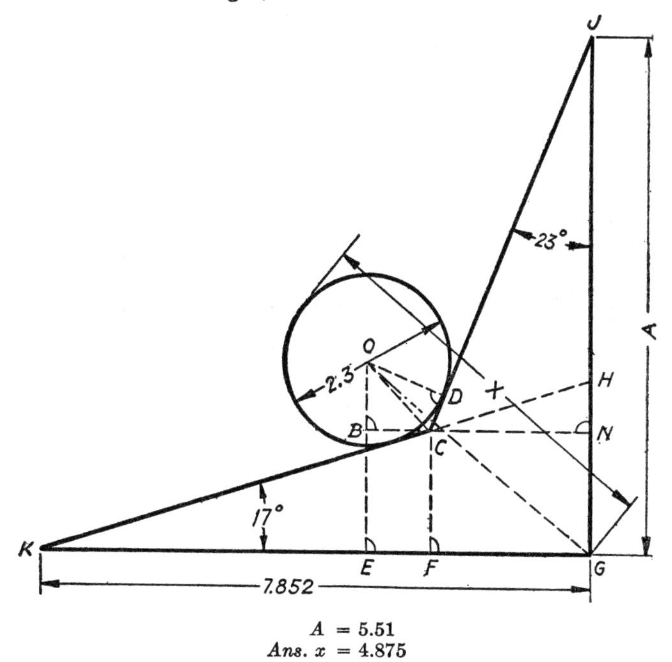

$A = 5.51$
Ans. $x = 4.875$

VARIABLE

1. $A = 5.72$	**2.** $A = 5.93$	**3.** $A = 6.14$
4. $A = 6.25$	**5.** $A = 6.35$	**6.** $A = 6.45$

119. Determine the distance x. (*Solution on next page.*)

Solution for preceding problem:

In $\triangle HKG$, solve for GH. $\angle KHG = 90° - 17°$. $JH = GJ - GH$.

In $\triangle JCH$, solve for CN. $CN = FG$. $FK = GK - FG$.

In $\triangle CKF$, solve for CF. $CF = BE$.

$\angle KCJ = 90° + 17° + 23°$. $\angle OCD = \angle OCK = \angle KCJ \div 2$.

$DO = 2.3 \div 2$. In $\triangle OCD$, solve for CO.

$\angle OCB = \angle OCK - 17°$. In $\triangle OCB$, solve for BO and BC.

$EG = BC + FG$. $EO = BE + BO$.

In $\triangle OEG$, solve for GO.

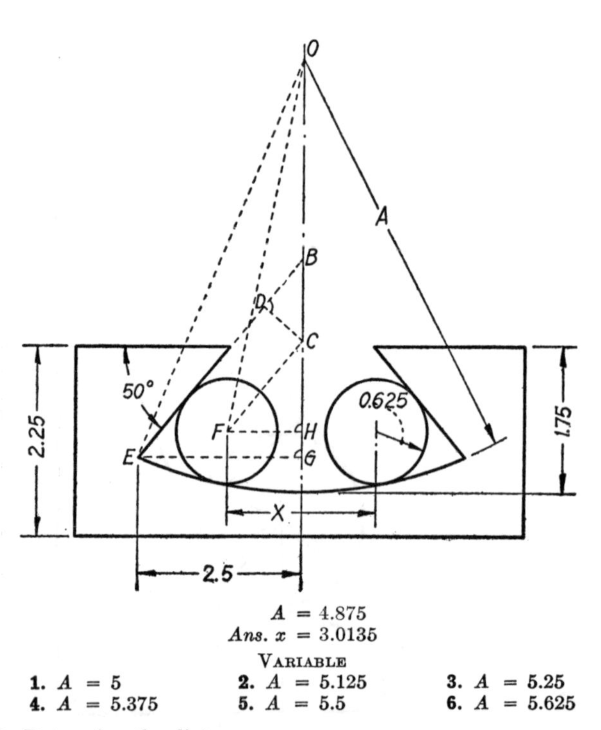

$A = 4.875$

Ans. $x = 3.0135$

VARIABLE

1. $A = 5$	**2.** $A = 5.125$	**3.** $A = 5.25$
4. $A = 5.375$	**5.** $A = 5.5$	**6.** $A = 5.625$

120. Determine the distance x.

Solution:

In $\triangle OEG$, solve for GO. $\angle BEG = \angle BCD = 50°$. Why?

In $\triangle BEG$, solve for GB. $CD = .625$. In $\triangle DCB$, solve for BC.

$CO = GO - GB + BC$. $FO = A - .625$.

In $\triangle OFC$, solve for FH.

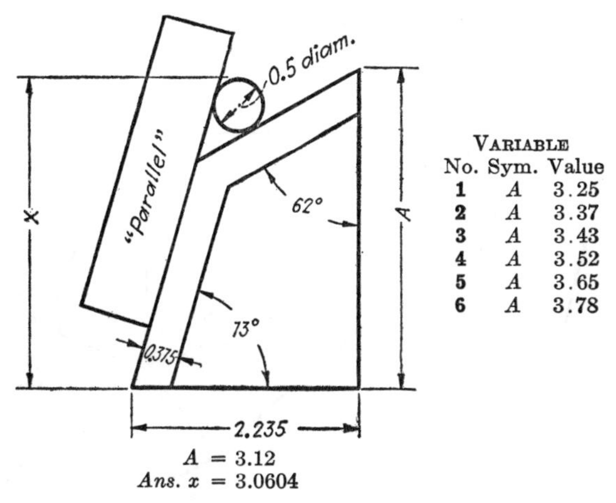

No.	Sym.	Value
1	A	3.25
2	A	3.37
3	A	3.43
4	A	3.52
5	A	3.65
6	A	3.78

$A = 3.12$

Ans. $x = 3.0604$

121. Determine the distance x.

Broach

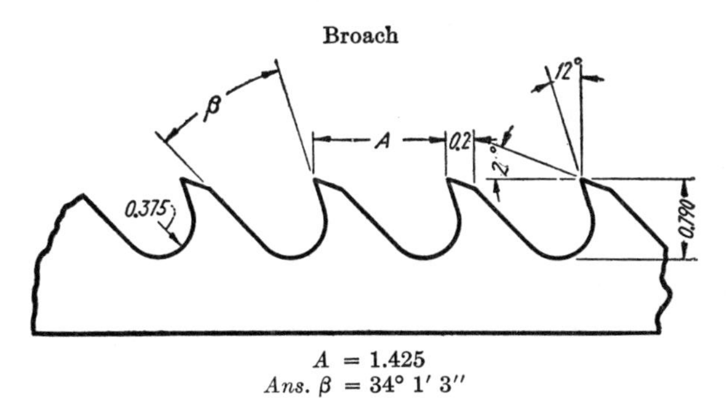

$A = 1.425$

Ans. $\beta = 34° 1' 3''$

VARIABLE

1. $A = 1.125$ **2.** $A = 1.175$ **3.** $A = 1.225$

4. $A = 1.275$ **5.** $A = 1.325$ **6.** $A = 1.375$

122. Determine the angle β. (*Solution on next page.*)

Solution for preceding problem:

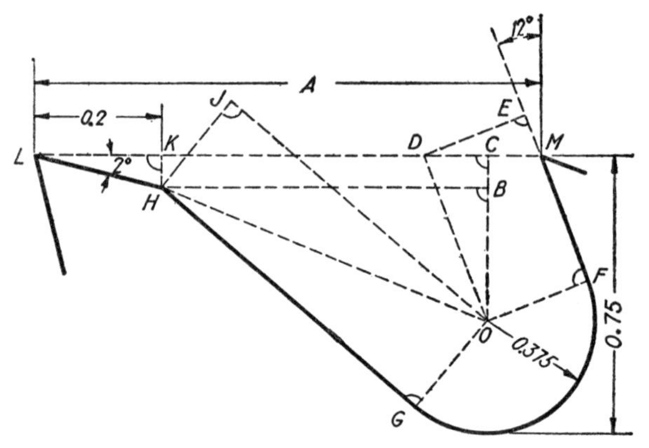

$\angle DOC = \angle EDM = 12°$. $CO = .75 - .375$.
$DE = FO = .375$. In $\triangle DOC$, solve for CD.
In $\triangle DEM$, solve for DM. $CM = DM - CD$.
In $\triangle LKH$, solve for LK and HK.
$CK = BH = A - CM - LK$. $HK = CB$.
$BO = CO - CB$. In $\triangle HOB$, solve for $\angle HOB$ and HO.
$OG = HJ = .375$. In $\triangle HOJ$, solve for $\angle HOJ$.
$\beta = \angle JOD = \angle HOB - 12° - \angle HOJ$.

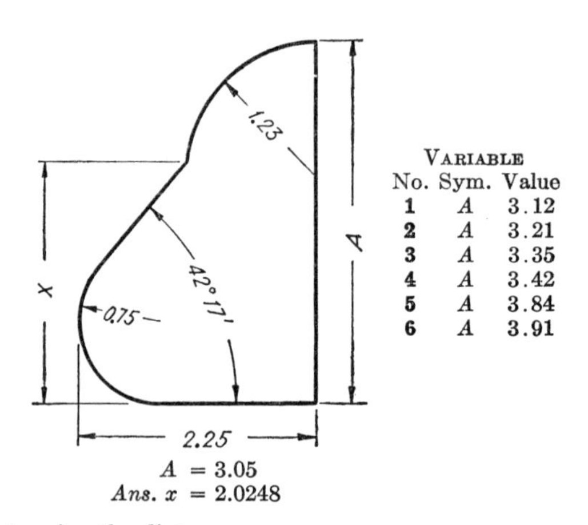

$A = 3.05$
Ans. $x = 2.0248$

Variable		
No.	Sym.	Value
1	A	3.12
2	A	3.21
3	A	3.35
4	A	3.42
5	A	3.84
6	A	3.91

123. Determine the distance x.

Die Section

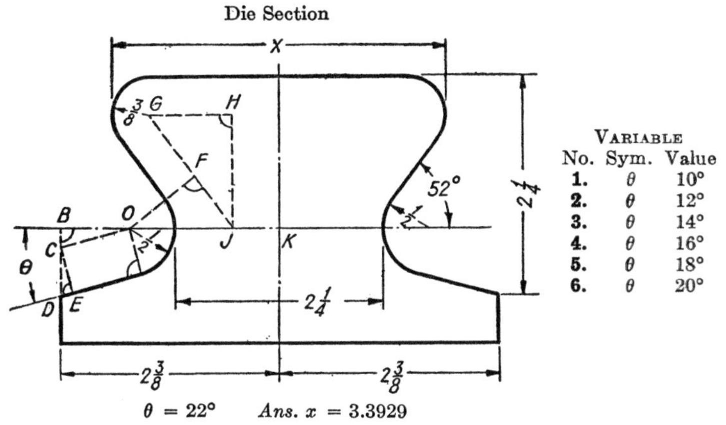

No.	Sym.	Value
1.	θ	10°
2.	θ	12°
3.	θ	14°
4.	θ	16°
5.	θ	18°
6.	θ	20°

$\theta = 22°$ *Ans.* $x = 3.3929$

124. Determine the distance x.

Solution:

$\angle DCE = \theta$. $CE = .5$. $BO = 2.375 - 1.125 - .5$.
In $\triangle ECD$, solve for CD. In $\triangle OBC$, solve for BC.
$FO = .5 + .375$. $\angle OJF = 52°$.
In $\triangle OJF$, solve for JO. $JK = KO = JO$.
$JH = 2.25 - .375 - BC - CD$. $\angle HGJ = 52°$. Why?
In $\triangle JHG$, solve for HG.

Valve Head

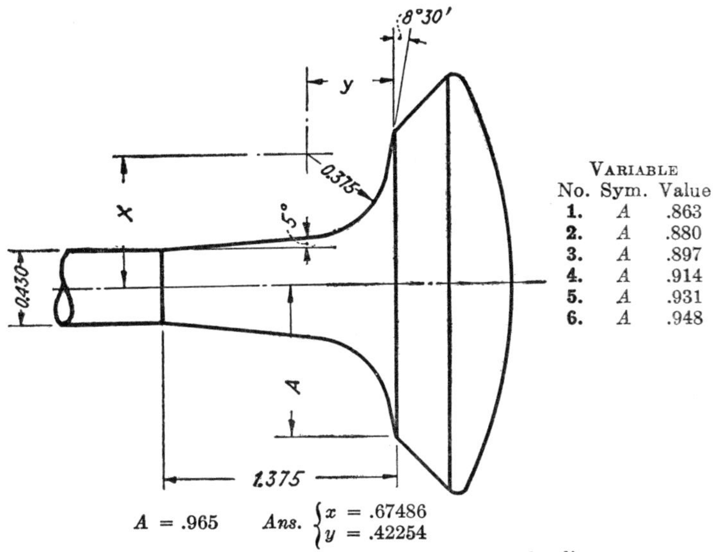

No.	Sym.	Value
1.	A	.863
2.	A	.880
3.	A	.897
4.	A	.914
5.	A	.931
6.	A	.948

$A = .965$ *Ans.* $\begin{cases} x = .67486 \\ y = .42254 \end{cases}$

125. Determine the distance x. **126.** Determine the distance y.

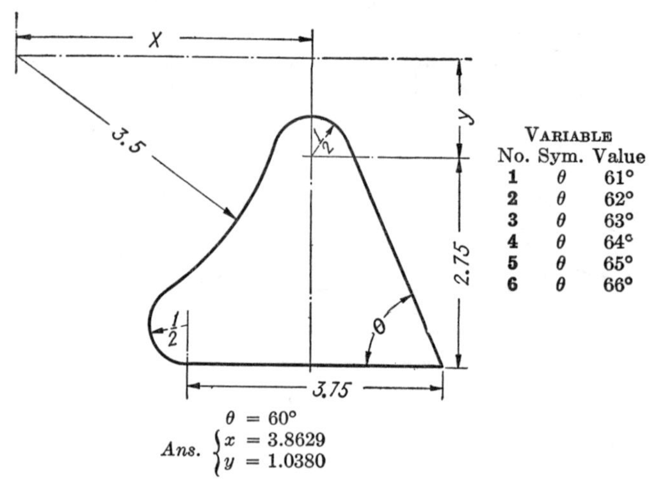

No.	Sym.	Value
1	θ	61°
2	θ	62°
3	θ	63°
4	θ	64°
5	θ	65°
6	θ	66°

$\theta = 60°$

Ans. $\begin{cases} x = 3.8629 \\ y = 1.0380 \end{cases}$

127. Determine the distance x.
128. Determine the distance y.

Gage

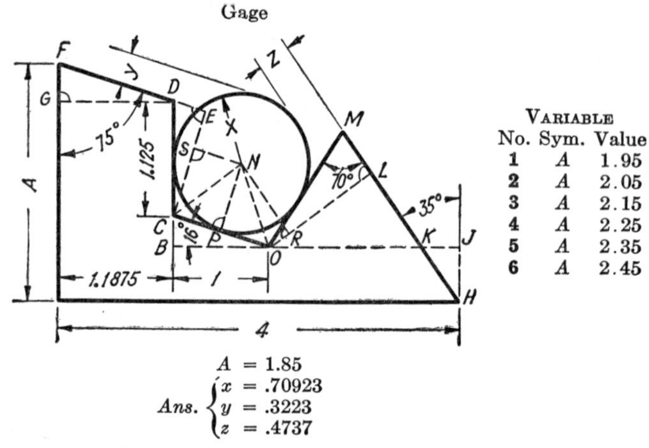

No.	Sym.	Value
1	A	1.95
2	A	2.05
3	A	2.15
4	A	2.25
5	A	2.35
6	A	2.45

$A = 1.85$

Ans. $\begin{cases} x = .70923 \\ y = .3223 \\ z = .4737 \end{cases}$

129. Determine the distance x.
130. Determine the distance y.
131. Determine the distance z.

Solution:

$\angle NCO = \angle DCO \div 2.$

$\angle MOC = 180° - \angle MOK - 16°.$

$\angle NOC = \angle MOC \div 2 - \angle NOM.$

(*Continued on next page.*)

Solution continued:

In $\triangle CBO$, solve for CO and BC.
In $\triangle NCO$, solve for NP, NC, and NO.
$x = NP$.
$\angle DCE = 90° - 75°$.
$\angle SCN = \angle NCO - \angle DCE$.
In $\triangle SCN$, solve for SC.
In $\triangle DCE$, solve for CE.
$y = x + SC - CE$.
In $\triangle FDG$, solve for DG and FG.
$HJ = A - FG - 1.125 - BC$.
In $\triangle KHJ$, solve for KJ.
$OK = 4 - DG - BO - KJ$.
In $\triangle OKL$, solve for OL.
$\angle NOR = \angle NOM + 20°$. Why?
In $\triangle NOR$, solve for OR.
$z = OL - OR - x$.

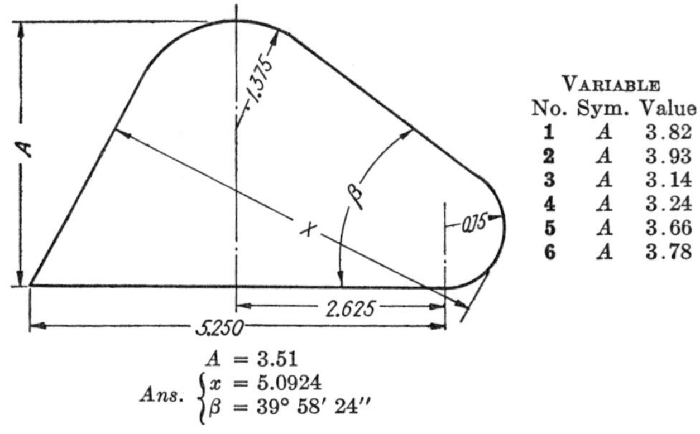

Variable No.	Sym.	Value
1	A	3.82
2	A	3.93
3	A	3.14
4	A	3.24
5	A	3.66
6	A	3.78

$A = 3.51$

Ans. $\begin{cases} x = 5.0924 \\ \beta = 39° \, 58' \, 24'' \end{cases}$

132. Determine the distance x.
133. Determine the angle β.

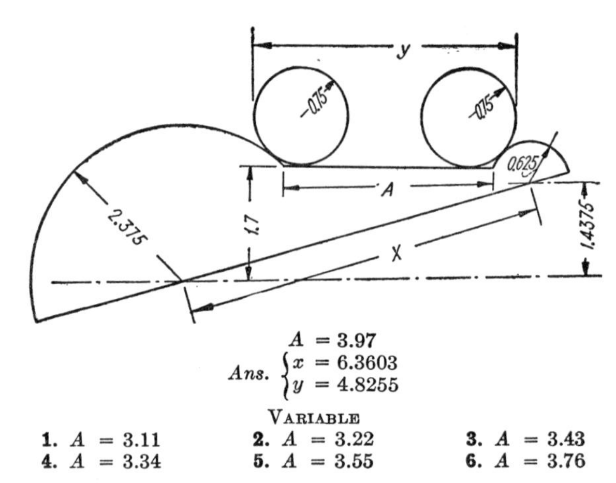

$A = 3.97$

Ans. $\begin{cases} x = 6.3603 \\ y = 4.8255 \end{cases}$

Variable

1. $A = 3.11$	**2.** $A = 3.22$	**3.** $A = 3.43$
4. $A = 3.34$	**5.** $A = 3.55$	**6.** $A = 3.76$

134. Determine the distance x.

135. Determine the distance y.

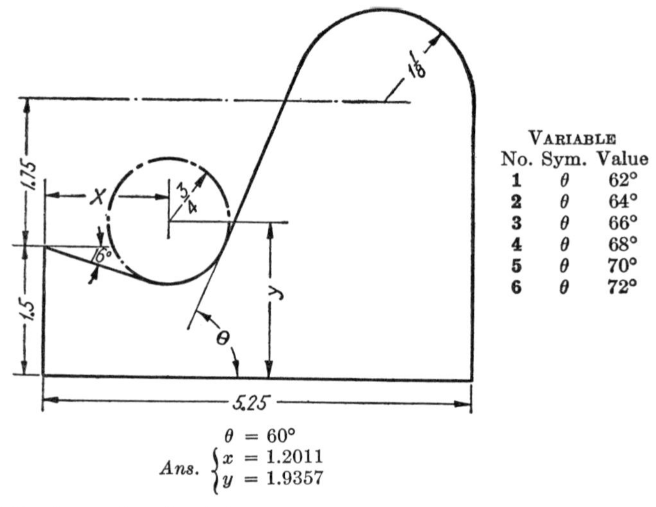

Variable

No.	Sym.	Value
1	θ	62°
2	θ	64°
3	θ	66°
4	θ	68°
5	θ	70°
6	θ	72°

$\theta = 60°$

Ans. $\begin{cases} x = 1.2011 \\ y = 1.9357 \end{cases}$

136. Determine the distance x.

137. Determine the distance y.

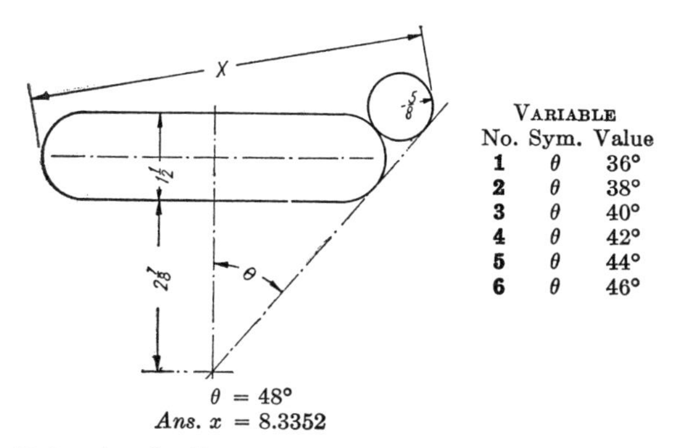

No.	Sym.	Value
1	θ	36°
2	θ	38°
3	θ	40°
4	θ	42°
5	θ	44°
6	θ	46°

VARIABLE

$\theta = 48°$
Ans. $x = 8.3352$

138. Determine the distance x.

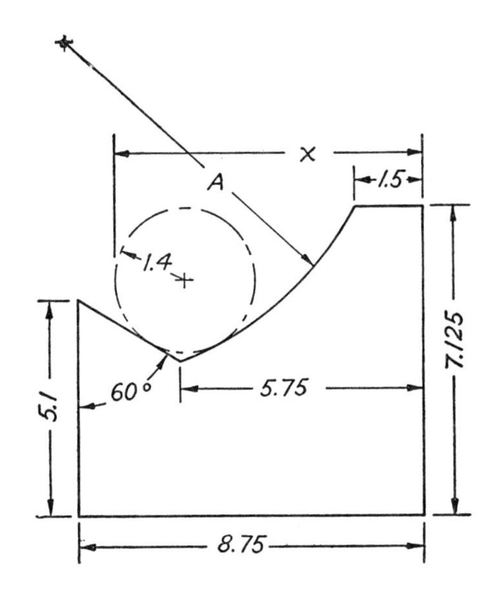

$A = 9.287$
Ans. $x = 7.0724$

VARIABLE

1. $A = 8.501$ **2.** $A = 8.632$ **3.** $A = 8.763$
4. $A = 8.894$ **5.** $A = 9.025$ **6.** $A = 9.156$

139. Determine the distance x.

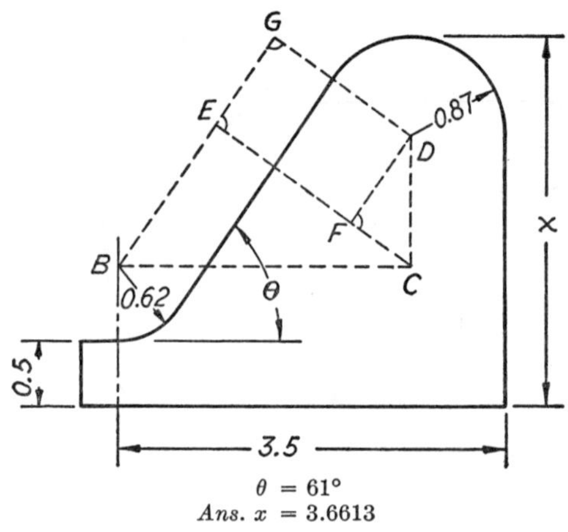

$\theta = 61°$
Ans. $x = 3.6613$

VARIABLE

1. $\theta = 55°$	**2.** $\theta = 56°$	**3.** $\theta = 57°$
4. $\theta = 58°$	**5.** $\theta = 59°$	**6.** $\theta = 60°$

140. Determine the distance x.
Solution:

$\angle EBC = \theta$. Why?
In $\triangle BCE$, solve for EC.
$EF = .62 + .87$. Why?
In $\triangle CFD$, solve for CD.

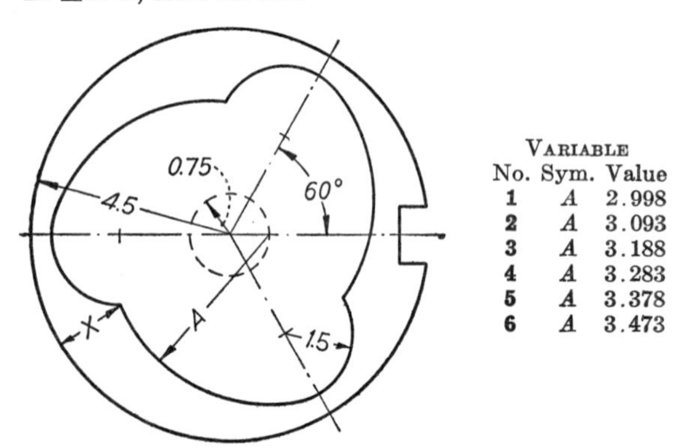

VARIABLE

No.	Sym.	Value
1	A	2.998
2	A	3.093
3	A	3.188
4	A	3.283
5	A	3.378
6	A	3.473

$A = 3.948$
Ans. $x = 1.2342$

141. Determine the distance x.

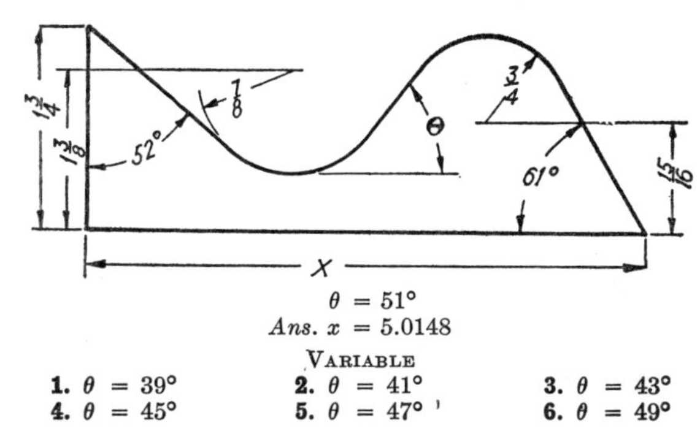

$\theta = 51°$
Ans. $x = 5.0148$

VARIABLE

1. $\theta = 39°$	**2.** $\theta = 41°$	**3.** $\theta = 43°$
4. $\theta = 45°$	**5.** $\theta = 47°$	**6.** $\theta = 49°$

142. Determine the distance x.

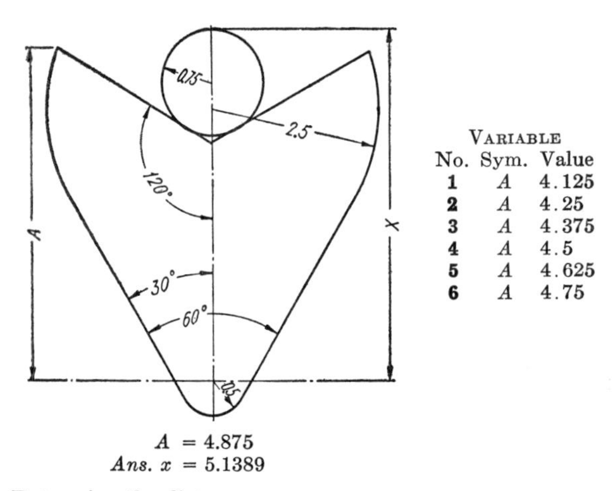

VARIABLE

No.	Sym.	Value
1	A	4.125
2	A	4.25
3	A	4.375
4	A	4.5
5	A	4.625
6	A	4.75

$A = 4.875$
Ans. $x = 5.1389$

143. Determine the distance x.

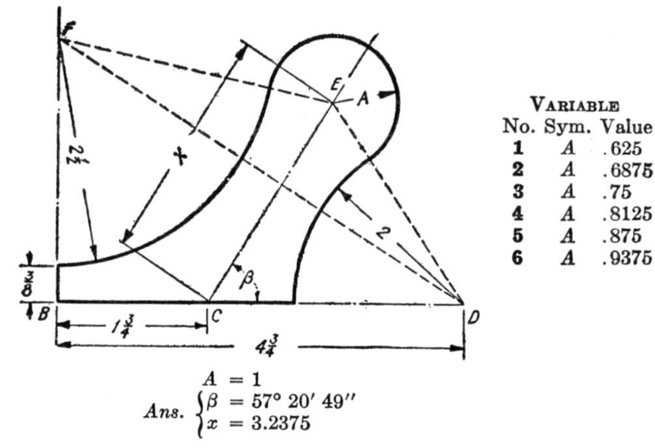

No.	Sym.	Value
	Variable	
1	A	.625
2	A	.6875
3	A	.75
4	A	.8125
5	A	.875
6	A	.9375

$A = 1$

Ans. $\begin{cases} \beta = 57° \, 20' \, 49'' \\ x = 3.2375 \end{cases}$

144. Determine the angle β.

145. Determine the distance x.

Solution:

In $\triangle FBD$, solve for $\angle BDF$ and DF.

$EF = 2.5 + A$. $DE = 2 + A$

In $\triangle FDE$, solve for $\angle FDE$.

$\angle CDE = \angle BDF + \angle FDE$.

In $\triangle CDE$, solve for $\angle ECD$ and CE.

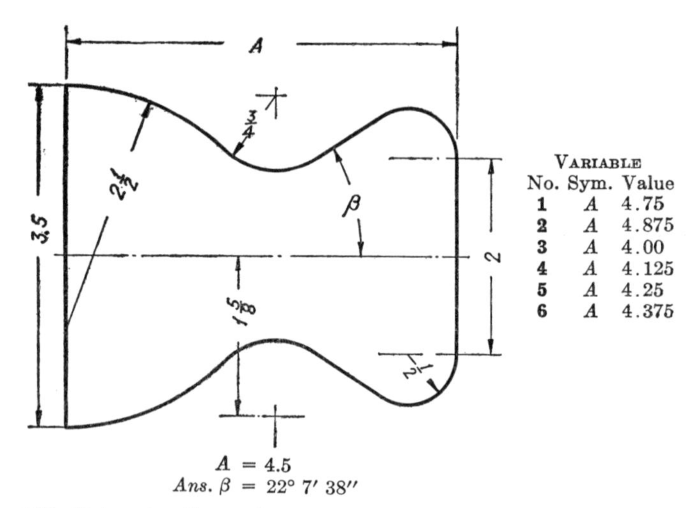

No.	Sym.	Value
	Variable	
1	A	4.75
2	A	4.875
3	A	4.00
4	A	4.125
5	A	4.25
6	A	4.375

$A = 4.5$

Ans. $\beta = 22° \, 7' \, 38''$

146. Determine the angle β.

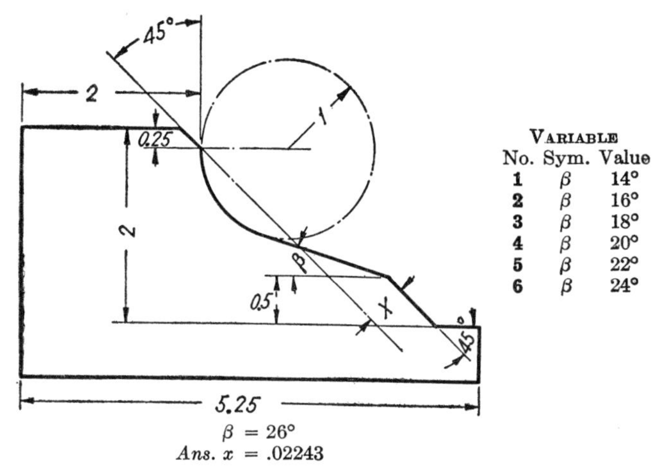

VARIABLE		
No.	Sym.	Value
1	β	14°
2	β	16°
3	β	18°
4	β	20°
5	β	22°
6	β	24°

$\beta = 26°$
Ans. $x = .02243$

147. Determine the distance x.

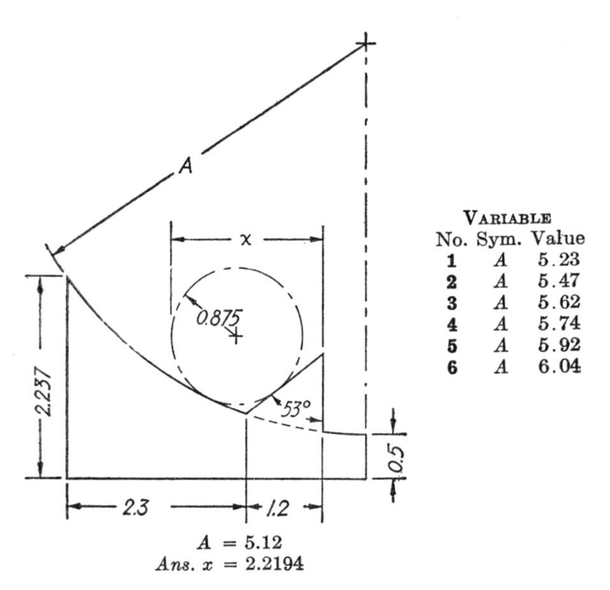

VARIABLE		
No.	Sym.	Value
1	A	5.23
2	A	5.47
3	A	5.62
4	A	5.74
5	A	5.92
6	A	6.04

$A = 5.12$
Ans. $x = 2.2194$

148. Determine the distance x.

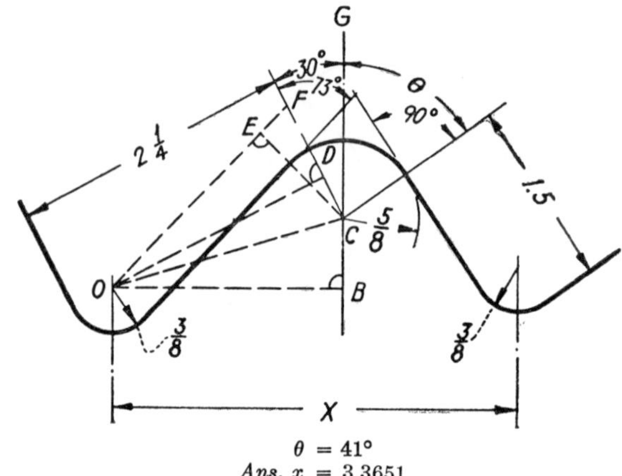

$\theta = 41°$
Ans. $x = 3.3651$

VARIABLE

1. $\theta = 53°$	**2.** $\theta = 51°$	**3.** $\theta = 49°$
4. $\theta = 47°$	**5.** $\theta = 45°$	**6.** $\theta = 43°$

149. Determine the distance x.

Solution:

$\angle CFE = 73°$. Why? $OD = 1.875$
In $\triangle ODF$, solve for OF.
In $\triangle ECF$, solve for EF. $OE = OF - EF$.
$CE = .375 + .625$. In $\triangle OEC$, solve for $\angle EOC$ and OC.
$\angle ECG = 30° + 90° - 73°$. Why?
$\angle EOB = \angle ECG$. Why?
$\angle COB = \angle EOB - \angle EOC$.
In $\triangle OBC$, solve for OB.
The balance of the problem is left to the student.

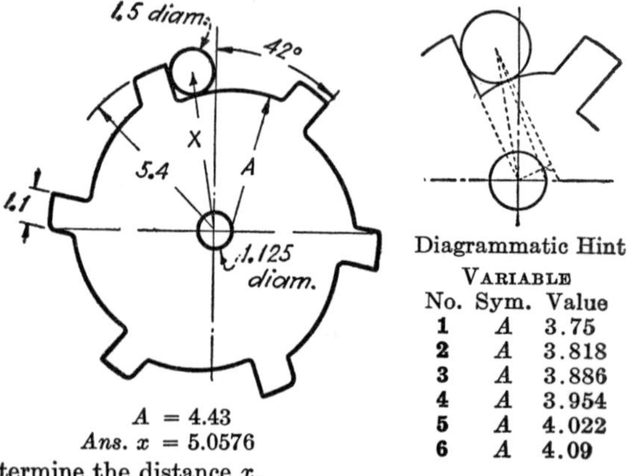

Diagrammatic Hint

$A = 4.43$
Ans. $x = 5.0576$

VARIABLE

No.	Sym.	Value
1	A	3.75
2	A	3.818
3	A	3.886
4	A	3.954
5	A	4.022
6	A	4.09

150. Determine the distance x.

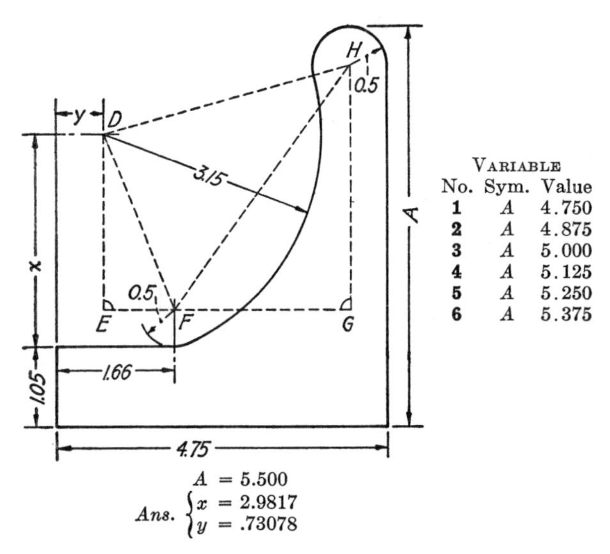

No.	Sym.	Value
1	A	4.750
2	A	4.875
3	A	5.000
4	A	5.125
5	A	5.250
6	A	5.375

$A = 5.500$

Ans. $\begin{cases} x = 2.9817 \\ y = .73078 \end{cases}$

151. Determine the distance x.
152. Determine the distance y.
Solution:

In $\triangle FGH$, solve for $\angle HFG$ and FH.
In $\triangle DFH$, solve for $\angle DFH$.
$\angle EFD = 180° - \angle DFH - \angle HFG$.
In $\triangle EFD$, solve for ED and EF.

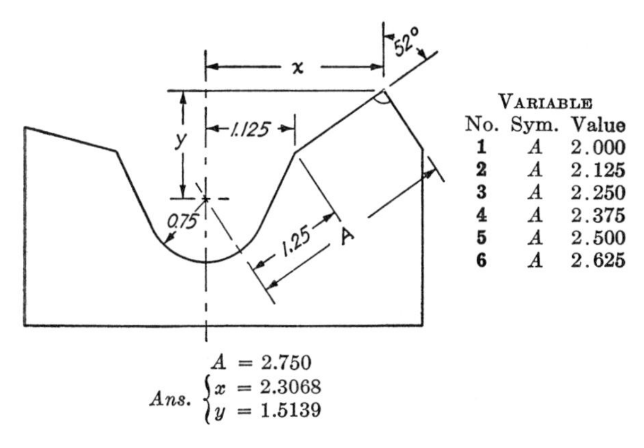

No.	Sym.	Value
1	A	2.000
2	A	2.125
3	A	2.250
4	A	2.375
5	A	2.500
6	A	2.625

$A = 2.750$

Ans. $\begin{cases} x = 2.3068 \\ y = 1.5139 \end{cases}$

153. Determine the distance x.
154. Determine the distance y.

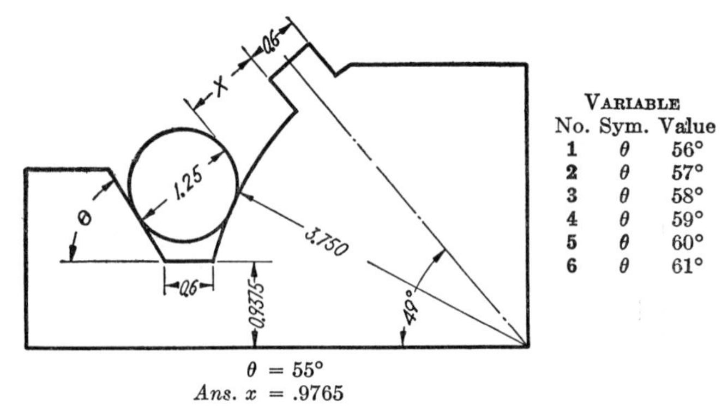

VARIABLE		
No.	Sym.	Value
1	θ	56°
2	θ	57°
3	θ	58°
4	θ	59°
5	θ	60°
6	θ	61°

$\theta = 55°$
Ans. $x = .9765$

155. Determine the distance x.

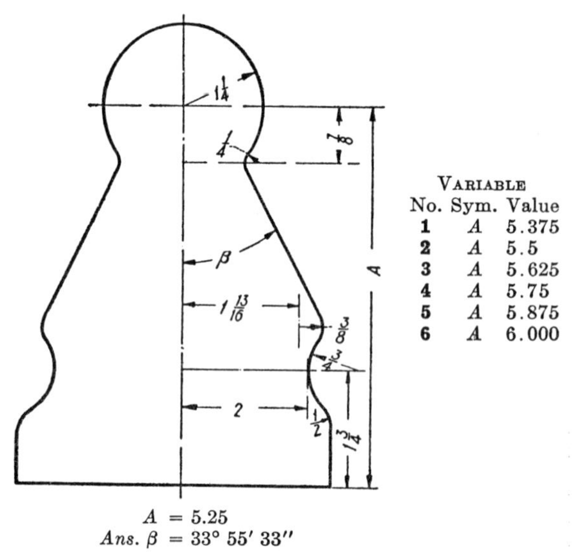

VARIABLE		
No.	Sym.	Value
1	A	5.375
2	A	5.5
3	A	5.625
4	A	5.75
5	A	5.875
6	A	6.000

$A = 5.25$
Ans. $\beta = 33° \ 55' \ 33''$

156. Determine the angle β.

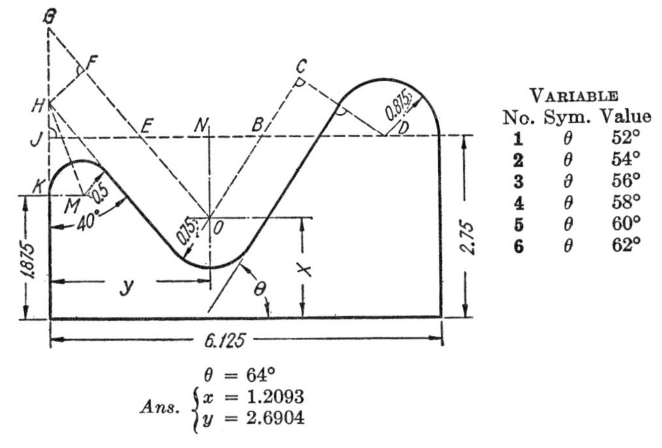

No.	Sym.	Value
1	θ	52°
2	θ	54°
3	θ	56°
4	θ	58°
5	θ	60°
6	θ	62°

$\theta = 64°$

Ans. $\begin{cases} x = 1.2093 \\ y = 2.6904 \end{cases}$

157. Determine the distance x.

158. Determine the distance y.

Solution:

$CD = .875 + .75$. $\angle CBD = \theta$. In $\triangle CBD$, solve for BD.

$\angle KHM = 40° \div 2$. $KM = .5$. In $\triangle KHM$, solve for HK.

$HF = .75$. $\angle HGF = 40°$. In $\triangle HGF$, solve for GH.

$GJ = 1.875 + HK + GH - 2.75$.

In $\triangle JGE$, solve for EJ. $EB = 6.125 - .875 - BD - EJ$.

$\angle NEO = 90° - 40°$. $\angle NBO = \theta$. In $\triangle EON$, solve for NO and NE.

$NE + EJ = y$. $2.75 - NO = x$.

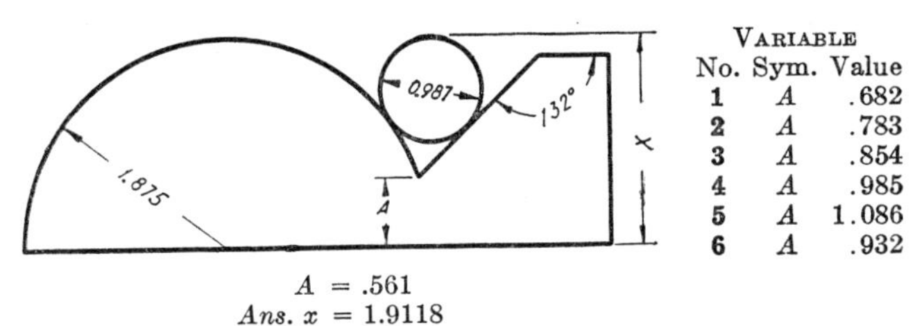

No.	Sym.	Value
1	A	.682
2	A	.783
3	A	.854
4	A	.985
5	A	1.086
6	A	.932

$A = .561$

Ans. $x = 1.9118$

159. Determine the distance x.

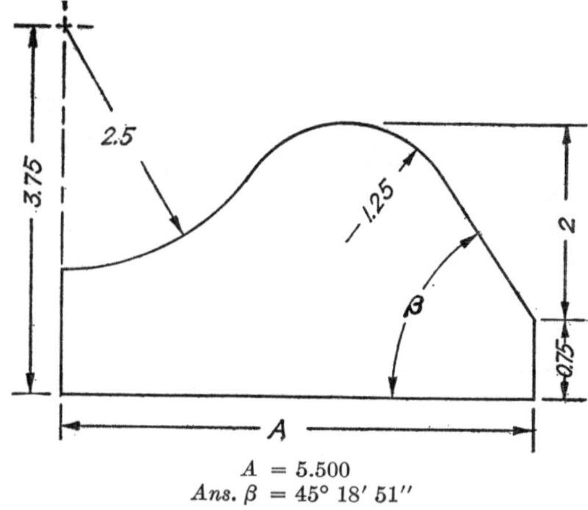

$A = 5.500$
Ans. $\beta = 45° \, 18' \, 51''$

VARIABLE

1. $A = 4.750$	**2.** $A = 4.875$	**3.** $A = 5.000$
4. $A = 5.125$	**5.** $A = 5.250$	**6.** $A = 5.375$

160. Determine the angle β.

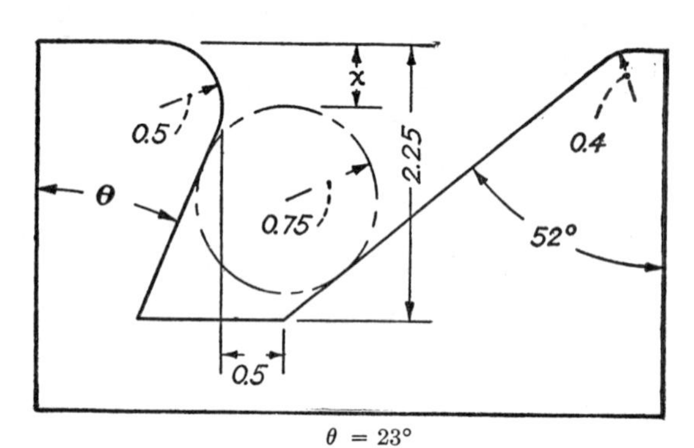

$\theta = 23°$
Ans. $x = .5259$

VARIABLE

1. $\theta = 17°$	**2.** $\theta = 18°$	**3.** $\theta = 19°$
4. $\theta = 20°$	**5.** $\theta = 21°$	**6.** $\theta = 22°$

161. Determine the distance x.

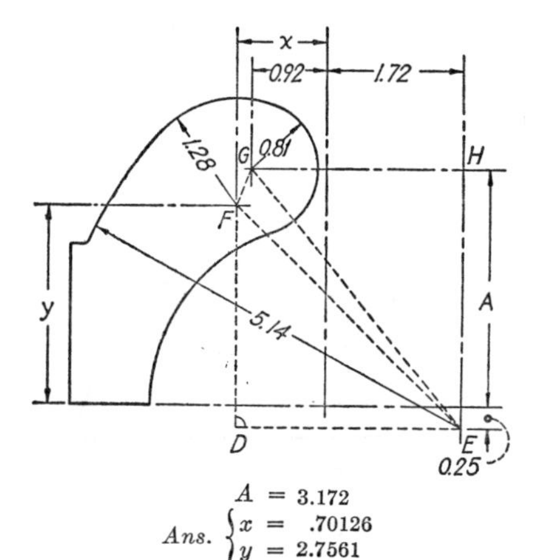

$$Ans. \begin{cases} A = 3.172 \\ x = .70126 \\ y = 2.7561 \end{cases}$$

VARIABLE

1. $A = 2.593$	**2.** $A = 2.691$	**3.** $A = 2.787$
4. $A = 2.884$	**5.** $A = 2.981$	**6.** $A = 3.078$

162. Determine the distance x.
163. Determine the distance y.

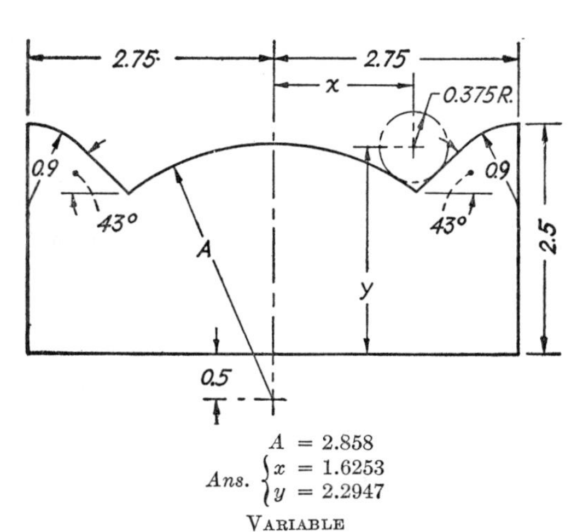

$$Ans. \begin{cases} A = 2.858 \\ x = 1.6253 \\ y = 2.2947 \end{cases}$$

VARIABLE

1. $A = 2.537$	**2.** $A = 2.588$	**3.** $A = 2.642$
4. $A = 2.695$	**5.** $A = 2.750$	**6.** $A = 2.804$

164. Determine the distance x.
165. Determine the distance y.

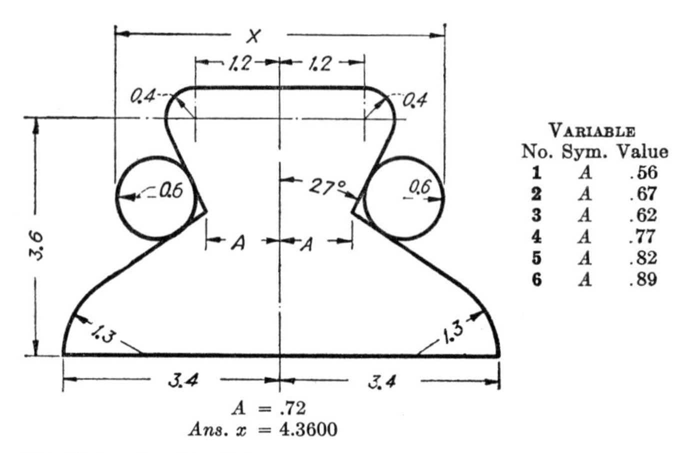

No.	Sym.	Value
1	A	.56
2	A	.67
3	A	.62
4	A	.77
5	A	.82
6	A	.89

$A = .72$
Ans. $x = 4.3600$

166. Determine the distance x.

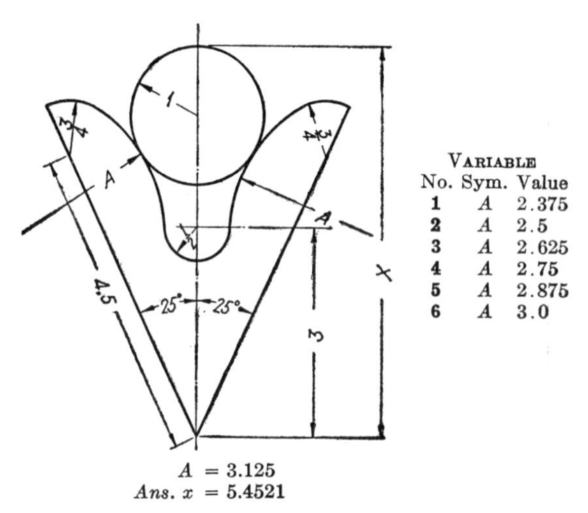

No.	Sym.	Value
1	A	2.375
2	A	2.5
3	A	2.625
4	A	2.75
5	A	2.875
6	A	3.0

$A = 3.125$
Ans. $x = 5.4521$

167. Determine the distance x.

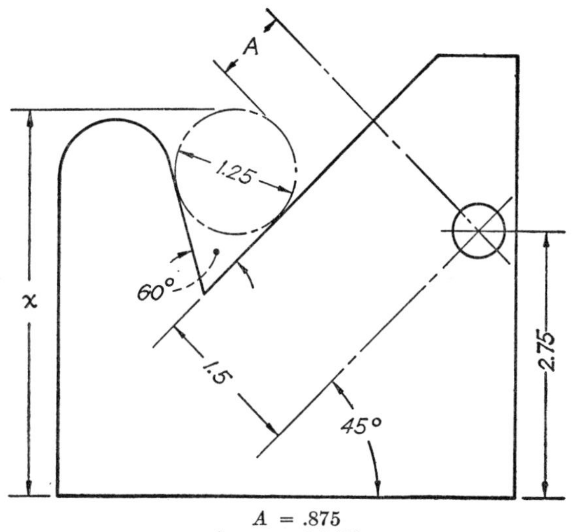

$A = .875$
Ans. $x = 3.8168$

Variable

1. $A = .125$
2. $A = .250$
3. $A = .375$
4. $A = .500$
5. $A = .625$
6. $A = .750$

168. Determine the distance x.

Trip Lever

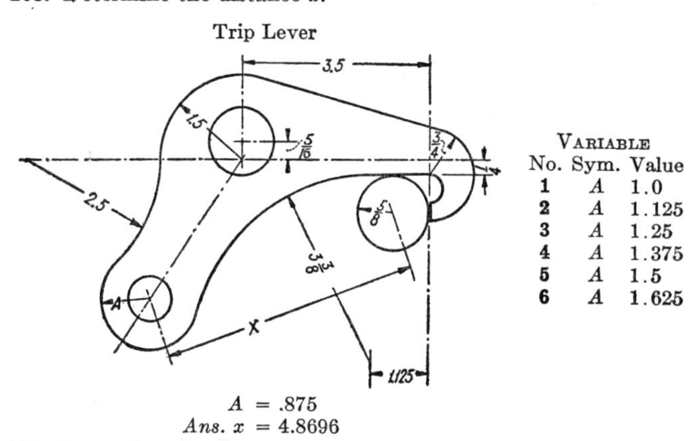

Variable

No.	Sym.	Value
1	A	1.0
2	A	1.125
3	A	1.25
4	A	1.375
5	A	1.5
6	A	1.625

$A = .875$
Ans. $x = 4.8696$

169. Determine the distance x.

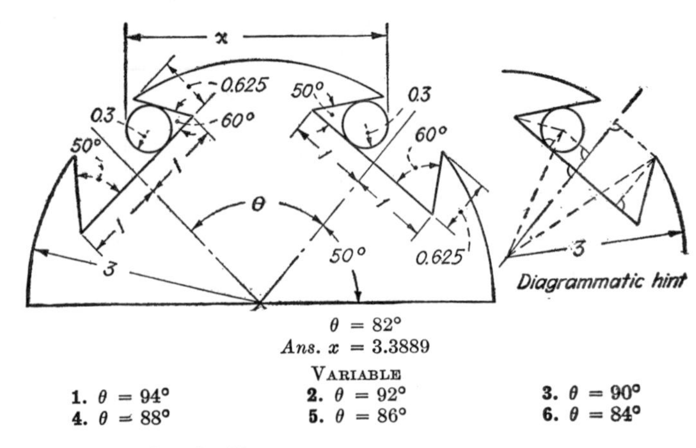

$\theta = 82°$
Ans. $x = 3.3889$

VARIABLE

1. $\theta = 94°$	**2.** $\theta = 92°$	**3.** $\theta = 90°$
4. $\theta = 88°$	**5.** $\theta = 86°$	**6.** $\theta = 84°$

170. Determine the distance x.

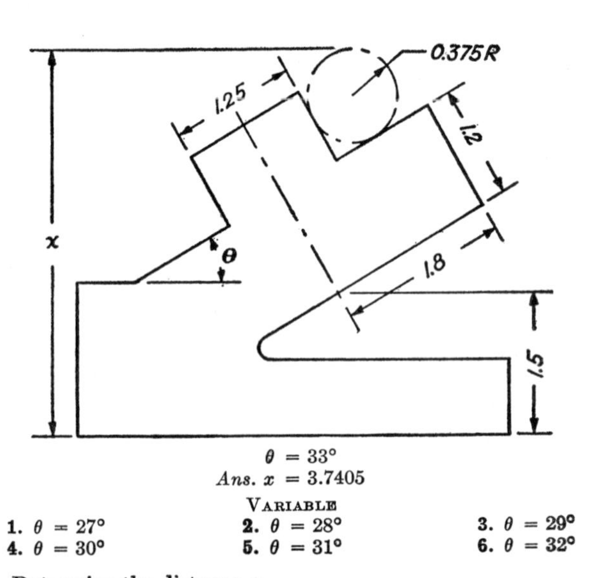

$\theta = 33°$
Ans. $x = 3.7405$

VARIABLE

1. $\theta = 27°$	**2.** $\theta = 28°$	**3.** $\theta = 29°$
4. $\theta = 30°$	**5.** $\theta = 31°$	**6.** $\theta = 32°$

171. Determine the distance x.

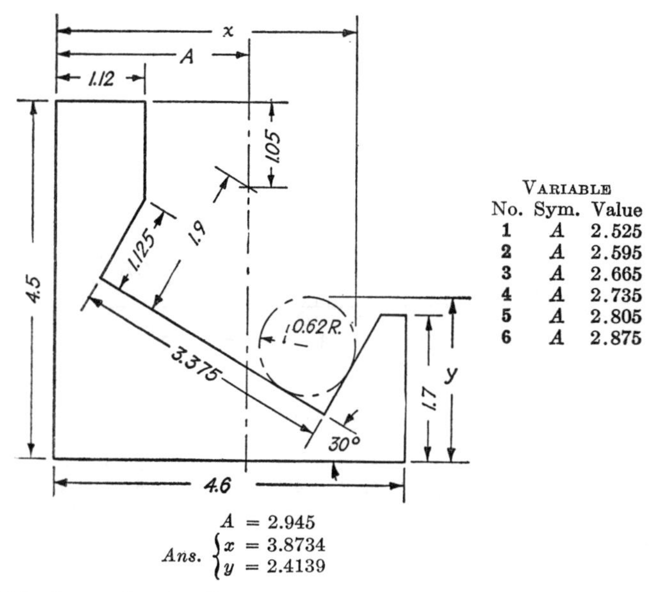

VARIABLE

No.	Sym.	Value
1	A	2.525
2	A	2.595
3	A	2.665
4	A	2.735
5	A	2.805
6	A	2.875

$A = 2.945$

Ans. $\begin{cases} x = 3.8734 \\ y = 2.4139 \end{cases}$

172. Determine the distance x.
173. Determine the distance y.

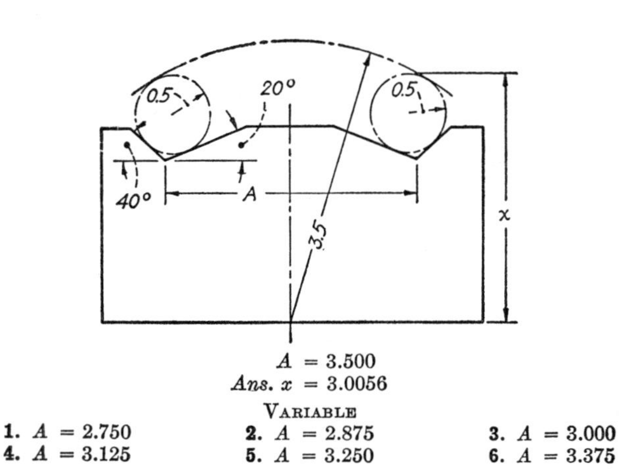

$A = 3.500$

Ans. $x = 3.0056$

VARIABLE

1. $A = 2.750$	**2.** $A = 2.875$	**3.** $A = 3.000$
4. $A = 3.125$	**5.** $A = 3.250$	**6.** $A = 3.375$

174. Determine the distance x.

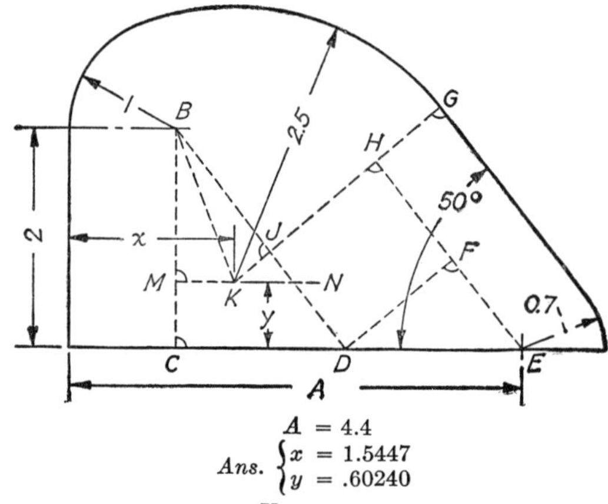

$A = 4.4$

Ans. $\begin{cases} x = 1.5447 \\ y = .60240 \end{cases}$

VARIABLE

1. $A = 3.8$ **2.** $A = 3.9$ **3.** $A = 4.0$
4. $A = 4.1$ **5.** $A = 4.2$ **6.** $A = 4.3$

175. Determine the distance x.
176. Determine the distance y.

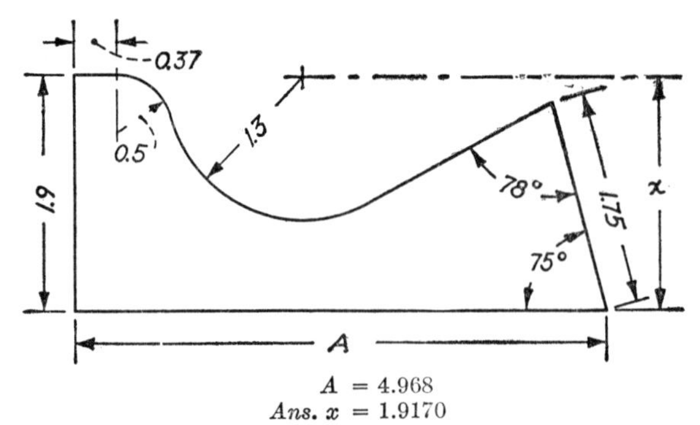

$A = 4.968$
Ans. $x = 1.9170$

VARIABLE

1. $A = 4.560$ **2.** $A = 4.628$ **3.** $A = 4.696$
4. $A = 4.764$ **5.** $A = 4.832$ **6.** $A = 4.900$

177. Determine the distance x.

Cam

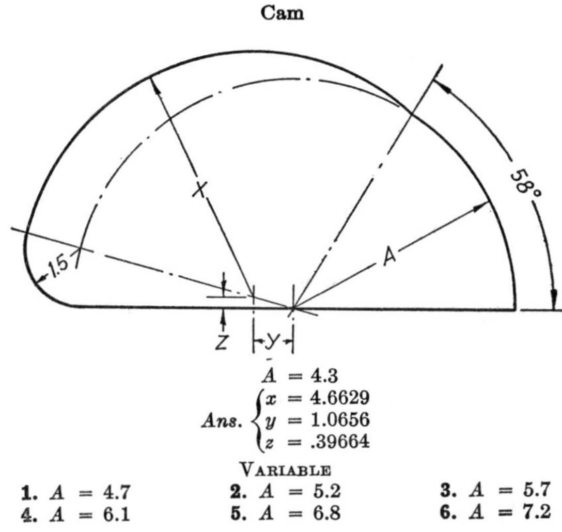

$A = 4.3$

Ans. $\begin{cases} x = 4.6629 \\ y = 1.0656 \\ z = .39664 \end{cases}$

VARIABLE

1. $A = 4.7$	**2.** $A = 5.2$	**3.** $A = 5.7$
4. $A = 6.1$	**5.** $A = 6.8$	**6.** $A = 7.2$

178. Determine the radius x.
179. Determine the distance y.
180. Determine the distance z.

Cam

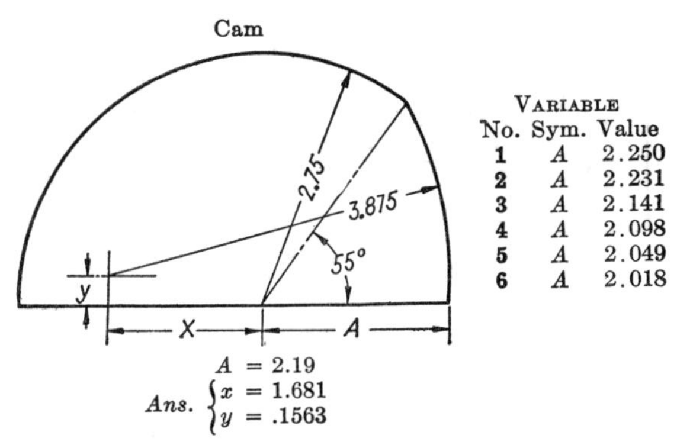

VARIABLE

No.	Sym.	Value
1	A	2.250
2	A	2.231
3	A	2.141
4	A	2.098
5	A	2.049
6	A	2.018

$A = 2.19$

Ans. $\begin{cases} x = 1.681 \\ y = .1563 \end{cases}$

181. Determine the distance x.
182. Determine the distance y.

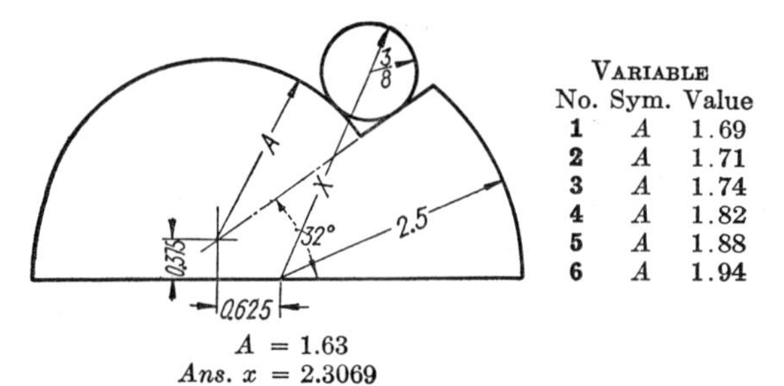

Variable		
No.	Sym.	Value
1	A	1.69
2	A	1.71
3	A	1.74
4	A	1.82
5	A	1.88
6	A	1.94

$A = 1.63$
Ans. $x = 2.3069$

183. Determine the distance x.

Cam

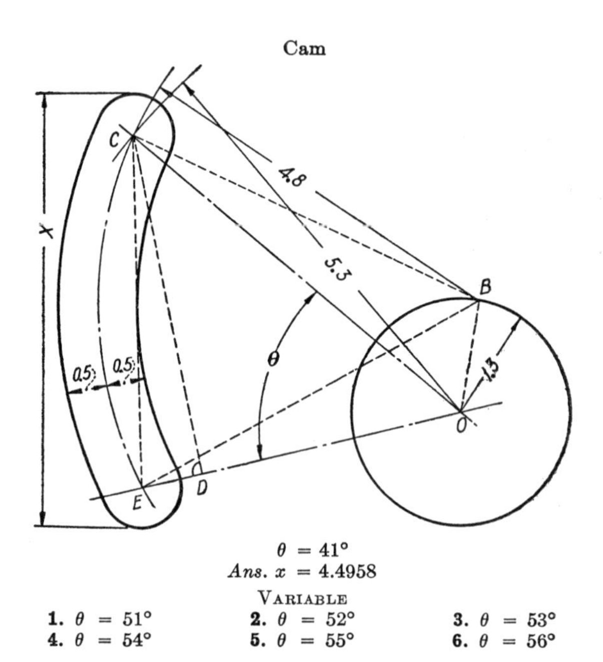

$\theta = 41°$
Ans. $x = 4.4958$

Variable

1. $\theta = 51°$ **2.** $\theta = 52°$ **3.** $\theta = 53°$
4. $\theta = 54°$ **5.** $\theta = 55°$ **6.** $\theta = 56°$

184. Determine the distance x.

Solution:

$CO = 5.3$. $BC = BE = 4.8$. $BO = 1.3$.
In $\triangle BCO$, solve for $\angle COB$. $\angle EOB = \angle COB + \theta$.
In $\triangle EOB$, solve for EO. In $\triangle DCO$, solve for DO and DC.
$DE = EO - DO$. In $\triangle EDC$, solve for CE.

Cam

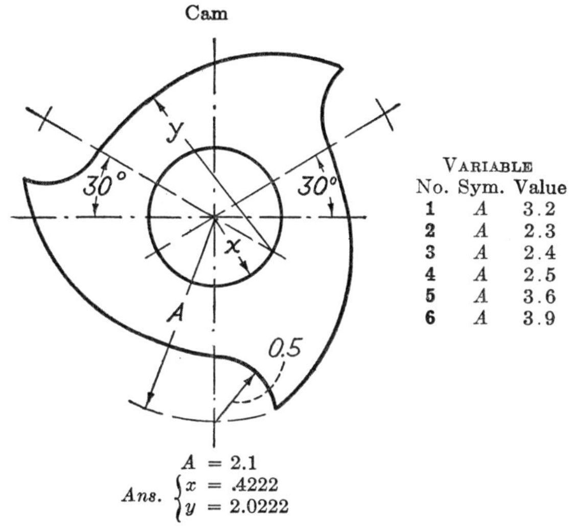

No.	Sym.	Value
	Variable	
1	A	3.2
2	A	2.3
3	A	2.4
4	A	2.5
5	A	3.6
6	A	3.9

$A = 2.1$

Ans. $\begin{cases} x = .4222 \\ y = 2.0222 \end{cases}$

185. Determine the radius x.

186. Determine the radius y.

Cam

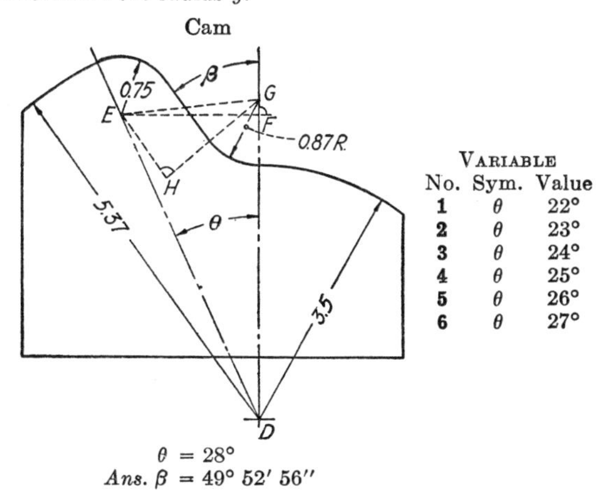

No.	Sym.	Value
	Variable	
1	θ	22°
2	θ	23°
3	θ	24°
4	θ	25°
5	θ	26°
6	θ	27°

$\theta = 28°$

Ans. $\beta = 49° 52' 56''$

187. Determine the angle β.

Solution:

In $\triangle EDF$, solve for EF and DF.

$FG = 3.5 + .87 - DF$.

In $\triangle EFG$, solve for $\angle GEF$ and EG.

In $\triangle EGH$, solve for $\angle GEH$.

$\angle GEH - \angle GEF = \angle FEH$.

$\beta = 90° - \angle FEH$. Why?

Cam

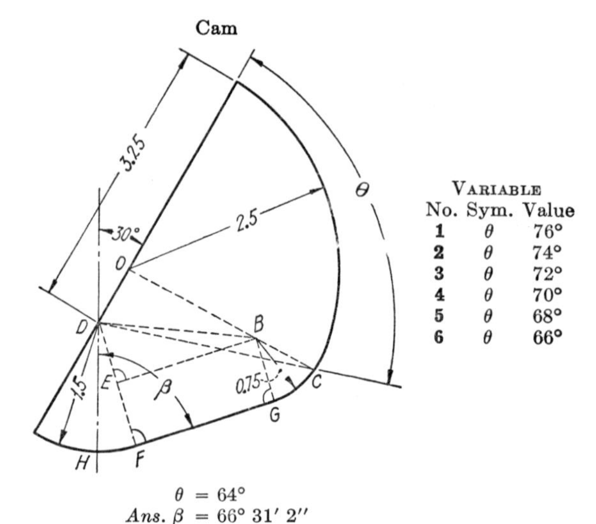

No.	Sym.	Value
1	θ	76°
2	θ	74°
3	θ	72°
4	θ	70°
5	θ	68°
6	θ	66°

$\theta = 64°$
Ans. $\beta = 66° \, 31' \, 2''$

188. Determine the angle β.

Solution:

$CO = 2.5$. $DO = 3.25 - 2.5$. $\angle ODC = \theta$.
In $\triangle ODC$, solve for $\angle OCD$ and CD. $BC = .75$.
In $\triangle BCD$, solve for $\angle BDC$ and BD. $DE = 1.5 - .75$.
In $\triangle DEB$, solve for $\angle BDE$. $\angle CDE = \angle BDE - \angle BDC$.
$\angle FDH = 180° - \theta - \angle CDE - 30°$.
$\beta = 90° - \angle FDH$. Why?

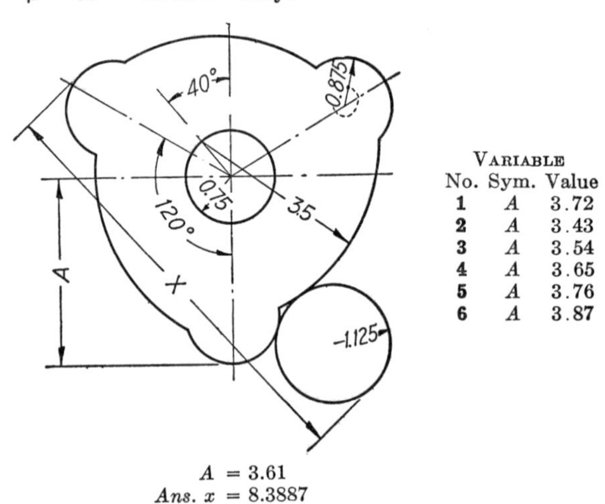

No.	Sym.	Value
1	A	3.72
2	A	3.43
3	A	3.54
4	A	3.65
5	A	3.76
6	A	3.87

$A = 3.61$
Ans. $x = 8.3887$

189. Determine the distance x.

Cam

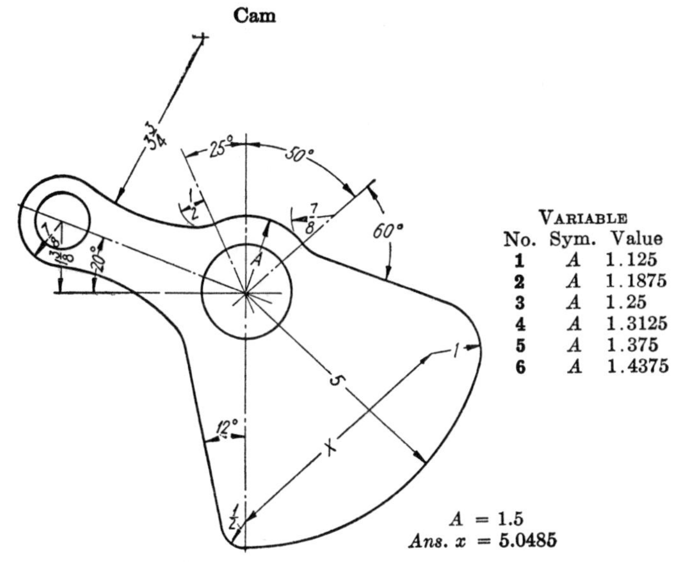

No.	Sym.	Value
1	A	1.125
2	A	1.1875
3	A	1.25
4	A	1.3125
5	A	1.375
6	A	1.4375

(Variable)

A = **1.5**
Ans. x = **5.0485**

190. Determine the distance x.

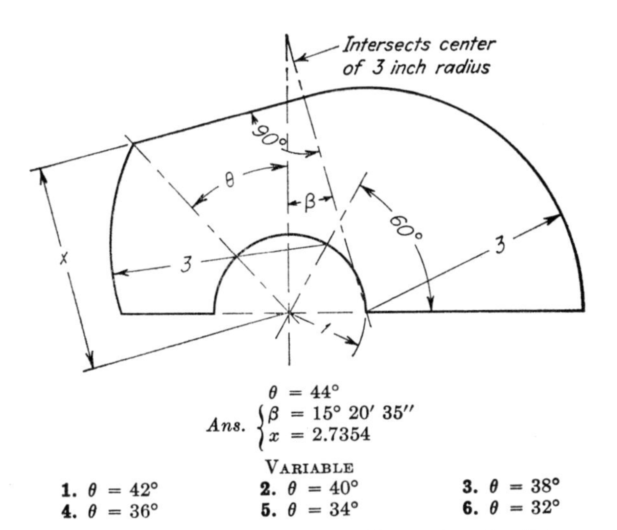

Ans. $\begin{cases} \theta = 44° \\ \beta = 15° \, 20' \, 35'' \\ x = 2.7354 \end{cases}$

Variable

1. $\theta = 42°$ **2.** $\theta = 40°$ **3.** $\theta = 38°$
4. $\theta = 36°$ **5.** $\theta = 34°$ **6.** $\theta = 32°$

191. Determine the angle β. **192.** Determine the distance x.

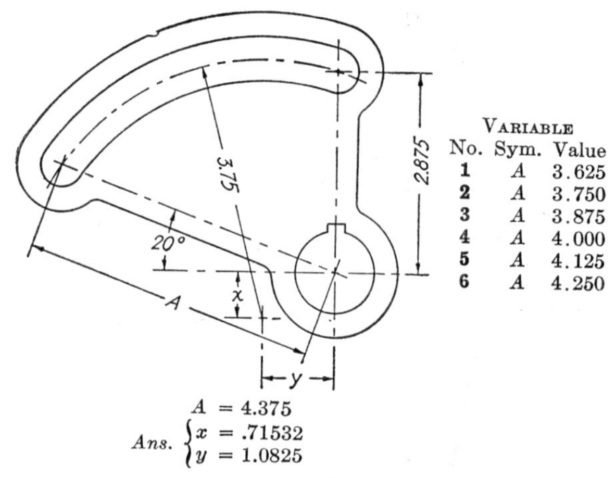

No.	Sym.	Value
1	A	3.625
2	A	3.750
3	A	3.875
4	A	4.000
5	A	4.125
6	A	4.250

VARIABLE

$A = 4.375$

Ans. $\begin{cases} x = .71532 \\ y = 1.0825 \end{cases}$

193. Determine the distance x.
194. Determine the distance y.

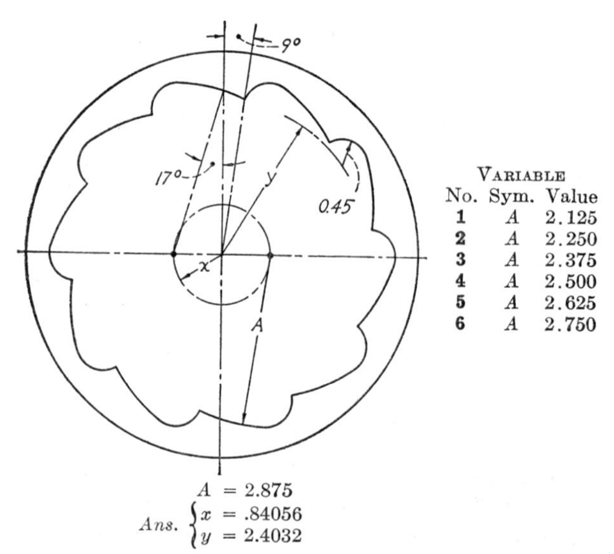

VARIABLE

No.	Sym.	Value
1	A	2.125
2	A	2.250
3	A	2.375
4	A	2.500
5	A	2.625
6	A	2.750

$A = 2.875$

Ans. $\begin{cases} x = .84056 \\ y = 2.4032 \end{cases}$

195. Determine the distance x.
196. Determine the distance y.

Cam

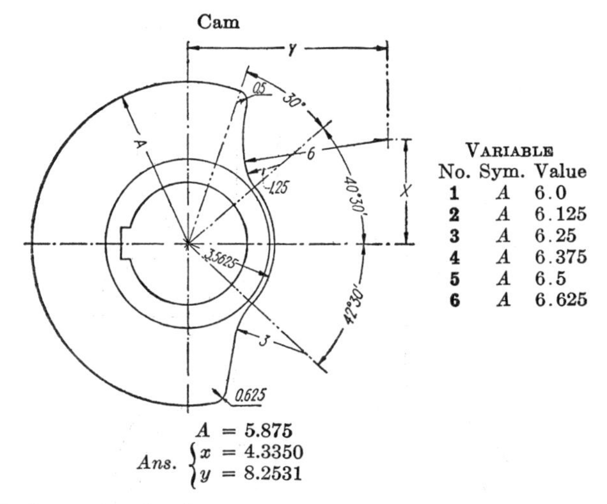

No.	Sym.	Value
1	A	6.0
2	A	6.125
3	A	6.25
4	A	6.375
5	A	6.5
6	A	6.625

$A = 5.875$

Ans. $\begin{cases} x = 4.3350 \\ y = 8.2531 \end{cases}$

197. Determine the distance x.
198. Determine the distance y.

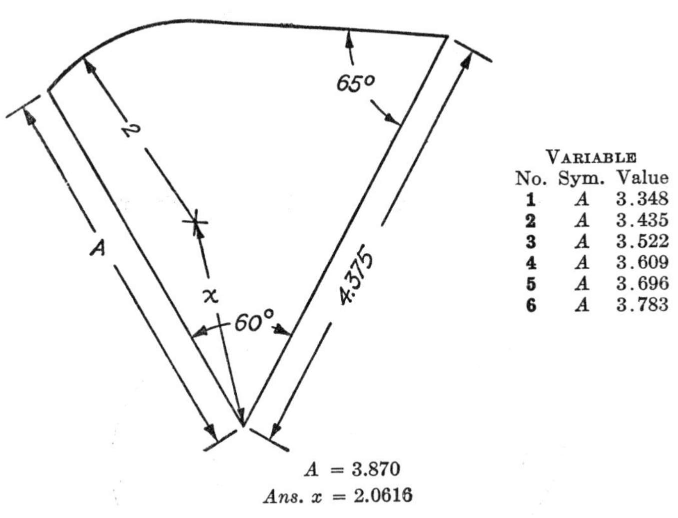

No.	Sym.	Value
1	A	3.348
2	A	3.435
3	A	3.522
4	A	3.609
5	A	3.696
6	A	3.783

$A = 3.870$

Ans. $x = 2.0616$

199. Determine the distance x.

CIRCULAR FORM CUTTERS

Axes of the Cutter and Work Parallel

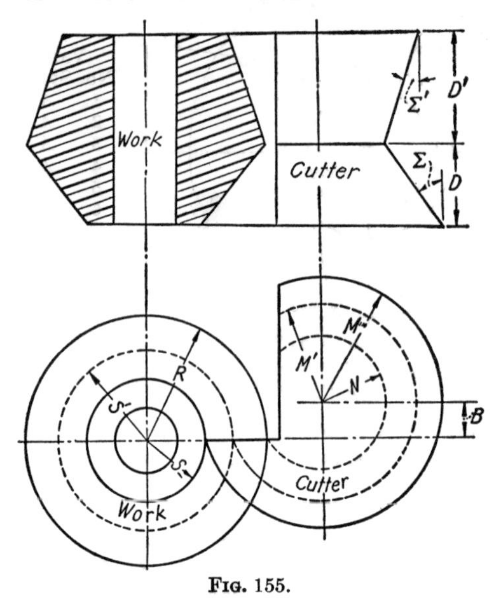

FIG. 155.

Form 1. No Rake on Cutting Face of Cutter

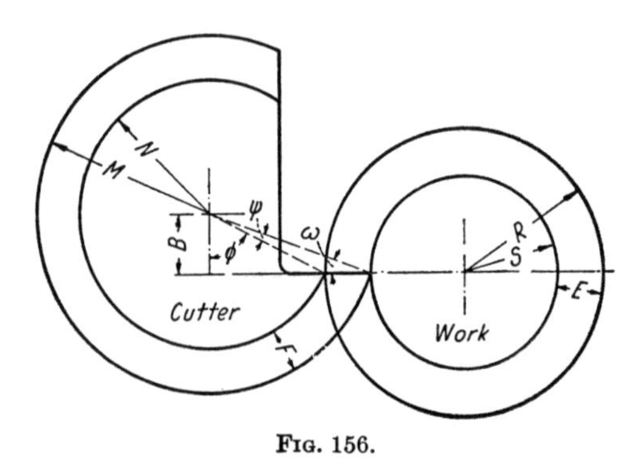

FIG. 156.

Figures 155 and 156 show a circular form cutter in relation to the work. It is required to determine the angle Σ and the depth F of the cutter in order to produce a given angle and a depth E of the work.

From Fig. 156, the following formulas may be derived:

For M given:

$$\sin \omega = \frac{B}{M}. \qquad E = R - S$$

$$\cot \psi = \frac{M}{E \sin \omega} - \cot \omega.$$

$$N = E \sin \omega \csc \psi.$$

$$F = M - N.$$

$$\tan \Sigma = \frac{F}{D}.$$

For N given:

$$\cos \phi = \frac{B}{N}.$$

$$\cot \omega' = \tan \phi + \frac{E'}{B}.$$

$$M' = B \csc \omega'.$$

$$F' = M' - N.$$

$$\tan \Sigma' = \frac{F'}{D'}.$$

D and D' are the distances along the axis between the maximum and minimum radii of the cutter represented by M and N and N and M', respectively. These distances correspond to the distances along the axis between the minimum and maximum radii of the work represented by S and R and R and S'. respectively, as shown in Fig. 155.

Form 2. Rake on Cutting Face of Cutter

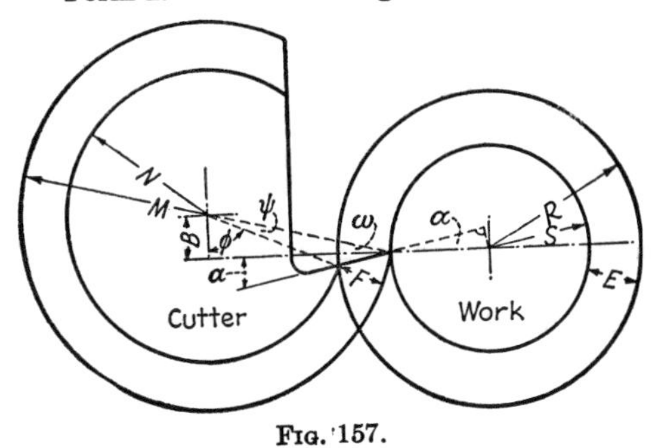

Fig. 157.

From Fig. 157, the following formulas may be derived:
For M given:

$$\sin \omega = \frac{B}{M} \qquad \sin \rho = \frac{S \sin \alpha}{R}$$

$$\cot \psi = \frac{M \csc (\alpha + \omega)}{R \cos \rho - S \cos \alpha} - \cot (\alpha + \omega)$$

$$N = (R \cos \rho - S \cos \alpha) \sin (\alpha + \omega) \csc \psi$$

$$F = M - N \qquad \tan \Sigma = \frac{F}{D}$$

For N given:

$$\sin \rho = \frac{S' \sin \alpha}{R}$$

$$\cos \phi = \frac{(R \cos \rho - S' \cos \alpha) \sin \alpha + B}{N}$$

$$\cot \omega' = \frac{N \sin \phi + (R \cos \rho - S' \cos \alpha) \cos \alpha}{B}$$

$$M' = B \csc \omega' \qquad F' = M' - N \qquad \tan \Sigma' = \frac{F'}{D'}$$

PROBLEMS

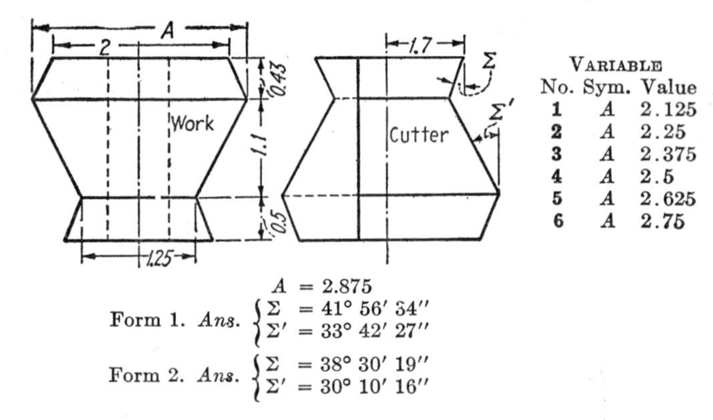

No.	Sym.	Value
1	A	2.125
2	A	2.25
3	A	2.375
4	A	2.5
5	A	2.625
6	A	2.75

$A = 2.875$

Form 1. *Ans.* $\begin{cases} \Sigma = 41° 56' 34'' \\ \Sigma' = 33° 42' 27'' \end{cases}$

Form 2. *Ans.* $\begin{cases} \Sigma = 38° 30' 19'' \\ \Sigma' = 30° 10' 16'' \end{cases}$

200. Determine (*a*) the angle Σ, and (*b*) the angle Σ'. Form 1. $B = 0.7$.

201. Determine (*a*) the angle Σ, and (*b*) the angle Σ'. Form 2. $B = 0.7$ and $\alpha = 10°$.

NATURAL TANGENTS AND COTANGENTS

′	0° tan	0° cotan	1° tan	1° cotan	2° tan	2° cotan	3° tan	3° cotan	′
0	.00000	Infinite.	.01746	57.2900	.03492	28.6363	.05241	19.0811	60
1	.00029	3437.750	.01775	56.3506	.03521	28.3994	.05270	18.9755	59
2	.00058	1718.870	.01804	55.4415	.03550	28.1664	.05299	18.8711	58
3	.00087	1145.920	.01833	54.5613	.03579	27.9372	.05328	18.7678	57
4	.00116	859.436	.01862	53.7086	.03609	27.7117	.05357	18.6656	56
5	.00145	687.549	.01891	52.8821	.03638	27.4899	.05387	18.5645	55
6	.00175	572.957	.01920	52.0807	.03667	27.2715	.05416	18.4645	54
7	.00204	491.106	.01949	51.3032	.03696	27.0566	.05445	18.3655	53
8	.00233	429.718	.01978	50.5485	.03725	26.8450	.05474	18.2677	52
9	.00262	381.971	.02007	49.8157	.03754	26.6367	.05503	18.1708	51
10	.00291	343.774	.02036	49.1039	.03783	26.4316	.05533	18.0750	50
11	.00320	312.521	.02066	48.4121	.03812	26.2296	.05562	17.9802	49
12	.00349	286.478	.02095	47.7395	.03842	26.0307	.05591	17.8863	48
13	.00378	264.441	.02124	47.0853	.03871	25.8348	.05620	17.7934	47
14	.00407	245.552	.02153	46.4489	.03900	25.6418	.05649	17.7015	46
15	.00436	229.182	.02182	45.8294	.03929	25.4517	.05678	17.6106	45
16	.00465	214.858	.02211	45.2261	.03958	25.2644	.05708	17.5205	44
17	.00495	202.219	.02240	44.6386	.03987	25.0798	.05737	17.4314	43
18	.00524	190.984	.02269	44.0661	.04016	24.8978	.05766	17.3432	42
19	.00553	180.932	.02298	43.5081	.04046	24.7185	.05795	17.2558	41
20	.00582	171.885	.02328	42.9641	.04075	24.5418	.05824	17.1693	40
21	.00611	163.700	.02357	42.4335	.04104	24.3675	.05854	17.0837	39
22	.00640	156.259	.02386	41.9158	.04133	24.1957	.05883	16.9990	38
23	.00669	149.465	.02415	41.4106	.04162	24.0263	.05912	16.9150	37
24	.00698	143.237	.02444	40.9174	.04191	23.8593	.05941	16.8319	36
25	.00727	137.507	.02473	40.4358	.04220	23.6945	.05970	16.7496	35
26	.00756	132.219	.02502	39.9655	.04250	23.5321	.05999	16.6681	34
27	.00785	127.321	.02531	39.5059	.04279	23.3718	.06029	16.5874	33
28	.00814	122.774	.02560	39.0568	.04308	23.2137	.06058	16.5075	32
29	.00844	118.540	.02589	38.6177	.04337	23.0577	.06087	16.4283	31
30	.00873	114.589	.02619	38.1885	.04366	22.9038	.06116	16.3499	30
31	.00902	110.892	.02648	37.7686	.04395	22.7519	.06145	16.2722	29
32	.00931	107.426	.02677	37.3579	.04424	22.6020	.06175	16.1952	28
33	00960	104.171	.02706	36.9560	.04454	22.4541	.06204	16.1190	27
34	.00989	101.107	.02735	36.5627	.04483	22.3081	.06233	16.0435	26
35	.01018	98.2179	.02764	36.1776	.04512	22.1640	.06262	15.9687	25
36	.01047	95.4895	.02793	35.8006	.04541	22.0217	.06291	15.8945	24
37	.01076	92.9085	.02822	35.4313	.04570	21.8813	.06321	15.8211	23
38	.01105	90.4633	.02851	35.0695	.04599	21.7426	.06350	15.7483	22
39	.01135	88.1436	.02881	34.7151	.04628	21.6056	.06379	15.6762	21
40	.01164	85.9398	.02910	34.3678	.04658	21.4704	.06408	15.6048	20
41	.01193	83.8435	.02939	34.0273	.04687	21.3369	.06437	15.5340	19
42	.01222	81.8470	.02968	33.6935	.04716	21.2049	.06467	15.4638	18
43	.01251	79.9434	.02997	33.3662	.04745	21.0747	.06496	15.3943	17
44	.01280	78.1263	.03026	33.0452	.04774	20.9460	.06525	15.3254	16
45	.01309	76.3900	.03055	32.7303	.04803	20.8188	.06554	15.2571	15
46	.01338	74.7292	.03084	32.4213	.04832	20.6932	.06584	15.1893	14
47	.01367	73.1390	.03114	32.1181	.04862	20.5691	.06613	15.1222	13
48	.01396	71.6151	.03143	31.8205	.04891	20.4465	.06642	15.0557	12
49	.01425	70.1533	.03172	31.5284	.04920	20.3253	.06671	14.9898	11
50	.01455	68.7501	.03201	31.2416	.04949	20.2056	.06700	14.9244	10
51	.01484	67.4019	.03230	30.9599	.04978	20.0872	.06730	14.8596	9
52	.01513	66.1055	.03259	30.6833	.05007	19.9702	.06759	14.7954	8
53	.01542	64.8580	.03288	30.4116	.05037	19.8546	.06788	14.7317	7
54	.01571	63.6567	.03317	30.1446	.05066	19.7403	.06817	14.6685	6
55	.01600	62.4992	.03346	29.8823	.05095	19.6273	.06847	14.6059	5
56	.01629	61.3829	.03376	29.6245	.05124	19.5156	.06876	14.5438	4
57	.01658	60.3058	.03405	29.3711	.05153	19.4051	.06905	14.4823	3
58	.01687	59.2659	.03434	29.1220	.05182	19.2959	.06934	14.4212	2
59	.01716	58.2612	.03463	28.8771	.05212	19.1879	.06963	14.3607	1
60	.01746	57.2900	.03492	28.6363	.05241	19.0811	.06993	14.3007	0
′	cotan	tan	cotan	tan	cotan	tan	cotan	tan	′
	89°		88°		87°		86°		

′	4°		5°		6°		7°		′
	tan	cotan	tan	cotan	tan	cotan	tan	cotan	
0	.06993	14.3007	.08749	11.4301	.10510	9.51436	.12278	8.14435	60
1	.07022	14.2411	.08778	11.3919	.10540	9.48781	.12308	8.12481	59
2	.07051	14.1821	.08807	11.3540	.10569	9.46141	.12338	8.10536	58
3	.07080	14.1235	.08837	11.3163	.10599	9.43515	.12367	8.08600	57
4	.07110	14.0655	.08866	11.2789	.10628	9.40904	.12397	8.06674	56
5	.07139	14.0079	.08895	11.2417	.10657	9.38307	.12426	8.04756	55
6	.07168	13.9507	.08925	11.2048	.10687	9.35724	.12456	8.02848	54
7	.07197	13.8940	.08954	11.1681	.10716	9.33154	.12485	8.00948	53
8	.07227	13.8378	.08983	11.1316	.10746	9.30599	.12515	7.99058	52
9	.07256	13.7821	.09013	11.0954	.10775	9.28058	.12544	7.97176	51
10	.07285	13.7267	.09042	11.0594	.10805	9.25530	.12574	7.95302	50
11	.07314	13.6719	.09071	11.0237	.10834	9.23016	.12603	7.93438	49
12	.07344	13.6174	.09101	10.9882	.10863	9.20516	.12633	7.91582	48
13	.07373	13.5634	.09130	19.9529	.10893	9.18028	.12662	7.89734	47
14	.07402	13.5098	.09159	10.9178	.10922	9.15554	.12692	7.87895	46
15	.07431	13.4566	.09189	10.8829	.10952	9.13093	.12722	7.86064	45
16	.07461	13.4039	.09218	10.8483	.10981	9.10646	.12751	7.84242	44
17	.07490	13.3515	.09247	10.8139	.11011	9.08211	.12781	7.82428	43
18	.07519	13.2996	.09277	10.7797	.11040	9.05789	.12810	7.80622	42
19	.07548	13.2480	.09306	10.7457	.11070	9.03379	.12840	7.78825	41
20	.07578	13.1969	.09335	10.7119	.11099	9.00983	.12869	7.77035	40
21	.07607	13.1461	.09365	10.6783	.11128	8.98598	.12899	7.75254	39
22	.07636	13.0958	.09394	16.6450	.11158	8.96227	.12929	7.73480	38
23	.07665	13.0458	.09423	10.6118	.11187	8.93867	.12958	7.71715	37
24	.07695	12.9962	.09453	10.5789	.11217	8.91520	.12988	7.69957	36
25	.07724	12.9469	.09482	10.5462	.11246	8.89185	.13017	7.68208	35
26	.07753	12.8981	.09511	10.5136	.11276	8.86862	.13047	7.66466	34
27	.07782	12.8496	.09541	10.4813	.11305	8.84551	.13076	7.64732	33
28	.07812	12.8014	.09570	10.4491	.11335	8.82252	.13106	7.63005	32
29	.07841	12.7536	.09600	10.4172	.11364	8.79964	.13136	7.61287	31
30	.07870	12.7062	.09629	10.3854	.11394	8.77689	.13165	7.59575	30
31	.07899	12.6591	.09658	10.3538	.11423	8.75425	.13195	7.57872	29
32	.07929	12.6124	.09688	10.3224	.11452	8.73172	.13224	7.56176	28
33	.07958	12.5660	.09717	10.2913	.11482	8.70931	.13254	7.54487	27
34	.07987	12.5199	.09746	10.2602	.11511	8.68701	.13284	7.52806	26
35	.08017	12.4742	.09776	10.2294	.11541	8.66482	.13313	7.51132	25
36	.08046	12.4288	.09805	10.1988	.11570	8.64275	.13343	7.49465	24
37	.08075	12.3838	.09834	10.1683	.11600	8.62078	.13372	7.47806	23
38	.08104	12.3390	.09864	10.1381	.11629	8.59893	.13402	7.46154	22
39	.08134	12.2946	.09893	10.1080	.11659	8.57718	.13432	7.44509	21
40	.08163	12.2505	.09923	10.0780	.11688	8.55555	.13461	7.42871	20
41	.08192	12.2067	.09952	10.0483	.11718	8.53402	.13491	7.41240	19
42	.08221	12.1632	.09981	10.0187	.11747	8.51259	.13521	7.39616	18
43	.08251	12.1201	.10011	9.98931	.11777	8.49128	.13550	7.37999	17
44	.08280	12.0772	.10040	9.96007	.11806	8.47007	.13580	7.36389	16
45	.08309	12.0346	.10069	9.93101	.11836	8.44896	.13609	7.34786	15
46	.08339	11.9923	.10099	9.90211	.11865	8.42795	.13639	7.33190	14
47	.08368	11.9504	.10128	9.87338	.11895	8.40705	.13669	7.31600	13
48	.08397	11.9087	.10158	9.84482	.11924	8.38625	.13698	7.30018	12
49	.08427	11.8673	.10187	9.81641	.11954	8.36555	.13728	7.28442	11
50	.08456	11.8262	.10216	9.78817	.11983	8.34496	.13758	7.26873	10
51	.08485	11.7853	.10246	9.76009	.12013	8.32446	.13787	7.25310	9
52	.08514	11.7448	.10275	9.73217	.12042	8.30406	.13817	7.23754	8
53	.08544	11.7045	.10305	9.70441	.12072	8.28376	.13846	7.22204	7
54	.08573	11.6645	.10334	9.67680	.12101	8.26355	.13876	7.20661	6
55	.08602	11.6248	.10363	9.64935	.12131	8.24345	.13906	7.19125	5
56	.08632	11.5853	.10393	9.62205	.12160	8.22344	.13935	7.17594	4
57	.08661	11.5461	.10422	9.59490	.12190	8.20352	.13965	7.16071	3
58	.08690	11.5072	.10452	9.56791	.12219	8.18370	.13995	7.14553	2
59	.08720	11.4685	.10481	9.54106	.12249	8.16398	.14024	7.13042	1
60	.08749	11.4301	.10510	9.51436	.12278	8.14435	.14054	7.11537	0
′	cotan	tan	cotan	tan	cotan	tan	cotan	tan	′
	85°		84°		83°		82°		

′	8° tan	8° cotan	9° tan	9° cotan	10° tan	10° cotan	11° tan	11° cotan	′
0	.14054	7.11537	.15838	6.31375	.17633	5.67128	.19438	5.14455	60
1	.14084	7.10038	.15868	6.30189	.17663	5.66165	.19468	5.13658	59
2	.14113	7.08546	.15898	6.29007	.17693	5.65205	.19498	5.12862	58
3	.14143	7.07059	.15928	6.27829	.17723	5.64248	.19529	5.12069	57
4	.14173	7.05579	.15958	6.26655	.17753	5.63295	.19559	5.11279	56
5	.14202	7.04105	.15988	6.25486	.17783	5.62344	.19589	5.10490	55
6	.14232	7.02637	.16017	6.24321	.17813	5.61397	.19619	5.09704	54
7	.14262	7.01174	.16047	6.23160	.17843	5.60452	.19649	5.08921	53
8	.14291	6.99718	.16077	6.22003	.17873	5.59511	.19680	5.08139	52
9	.14321	6.98268	.16107	6.20851	.17903	5.58573	.19710	5.07360	51
10	.14351	6.96823	.16137	6.19703	.17933	5.57638	.19740	5.06584	50
11	.14381	6.95385	.16167	6.18559	.17963	5.56706	.19770	5.05809	49
12	.14410	6.93952	.16196	6.17419	.17993	5.55777	.19801	5.05037	48
13	.14440	6.92525	.16226	6.16283	.18023	5.54851	.19831	5.04267	47
14	.14470	6.91104	.16256	6.15151	.18053	5.53927	.19861	5.03499	46
15	.14499	6.89688	.16286	6.14023	.18083	5.53007	.19891	5.02734	45
16	.14529	6.88278	.16316	6.12899	.18113	5.52090	.19921	5.01971	44
17	.14559	6.86874	.16346	6.11779	.18143	5.51176	.19952	5.01210	43
18	.14588	6.85475	.16376	6.10664	.18173	5.50264	.19982	5.00451	42
19	.14618	6.84082	.16405	6.09552	.18203	5.49356	.20012	4.99695	41
20	.14648	6.82694	.16435	6.08444	.18233	5.48451	.20042	4.98940	40
21	.14678	6.81312	.16465	6.07340	.18263	5.47548	.20073	4.98188	39
22	.14707	6.79936	.16495	6.06240	.18293	5.46648	.20103	4.97438	38
23	.14737	6.78564	.16525	6.05143	.18323	5.45751	.20133	4.96690	37
24	.14767	6.77199	.16555	6.04051	.18353	5.44857	.20164	4.95945	36
25	.14796	6.75838	.16585	6.02962	.18383	5.43966	.20194	4.95201	35
26	.14826	6.74483	.16615	6.01878	.18414	5.43077	.20224	4.94460	34
27	.14856	6.73133	.16645	6.00797	.18444	5.42192	.20254	4.93721	33
28	.14886	6.71789	.16674	5.99720	.18474	5.41309	.20285	4.92984	32
29	.14915	6.70450	.16704	5.98646	.18504	5.40429	.20315	4.92249	31
30	.14945	6.69116	.16734	5.97576	.18534	5.39552	.20345	4.91516	30
31	.14975	6.67787	.16764	5.96510	.18564	5.38677	.20376	4.90785	29
32	.15005	6.66463	.16794	5.95448	.18594	5.37805	.20406	4.90056	28
33	.15034	6.65144	.16824	5.94390	.18624	5.36936	.20436	4.89330	27
34	.15064	6.63831	.16854	5.93335	.18654	5.36070	.20466	4.88605	26
35	.15094	6.62523	.16884	5.92283	.18684	5.35206	.20497	4.87882	25
36	.15124	6.61219	.16914	5.91235	.18714	5.34345	.20527	4.87162	24
37	.15153	6.59921	.16944	5.90191	.18745	5.33487	.20557	4.86444	23
38	.15183	6.58627	.16974	5.89151	.18775	5.32631	.20588	4.85727	22
39	.15213	6.57339	.17004	5.88114	.18805	5.31778	.20618	4.85013	21
40	.15243	6.56055	.17033	5.87080	.18835	5.30928	.20648	4.84300	20
41	.15272	6.54777	.17063	5.86051	.18865	5.30080	.20679	4.83590	19
42	.15302	6.53503	.17093	5.85024	.18895	5.29235	.20709	4.82882	18
43	.15332	6.52234	.17123	5.84001	.18925	5.28393	.20739	4.82175	17
44	.15362	6.50970	.17153	5.82982	.18955	5.27553	.20770	4.81471	16
45	.15391	6.49710	.17183	5.81966	.18986	5.26715	.20800	4.80769	15
46	.15421	6.48456	.17213	5.80953	.19016	5.25880	.20830	4.80068	14
47	.15451	6.47206	.17243	5.79944	.19046	5.25048	.20861	4.79370	13
48	.15481	6.45961	.17273	5.78938	.19076	5.24218	.20891	4.78673	12
49	.15511	6.44720	.17303	5.77936	.19106	5.23391	.20921	4.77978	11
50	.15540	6.43484	.17333	5.76937	.19136	5.22566	.20952	4.77286	10
51	.15570	6.42253	.17363	5.75941	.19166	5.21744	.20982	4.76595	9
52	.15600	6.41026	.17393	5.74949	.19197	5.20925	.21013	4.75906	8
53	.15630	6.39804	.17423	5.73960	.19227	5.20107	.21043	4.75219	7
54	.15660	6.38587	.17453	5.72974	.19257	5.19293	.21073	4.74534	6
55	.15689	6.37374	.17483	5.71992	.19287	5.18480	.21104	4.73851	5
56	.15719	6.36165	.17513	5.71013	.19317	5.17671	.21134	4.73170	4
57	.15749	6.34961	.17543	5.70037	.19347	5.16863	.21164	4.72490	3
58	.15779	6.33761	.17573	5.69064	.19378	5.16058	.21195	4.71813	2
59	.15809	6.32566	.17603	5.68094	.19408	5.15256	.21225	4.71137	1
60	.15838	6.31375	.17633	5.67128	.19438	5.14455	.21256	4.70463	0
′	cotan	tan	cotan	tan	cotan	tan	cotan	tan	′
	81°		80°		79°		78°		

′	12° tan	12° cotan	13° tan	13° cotan	14° tan	14° cotan	15° tan	15° cotan	′
0	.21256	4.70463	.23087	4.33148	.24933	4.01078	.26795	3.73205	60
1	.21286	4.69791	.23117	4.32573	.24964	4.00582	.26826	3.72771	59
2	.21316	4.69121	.23148	4.32001	.24995	4.00086	.26857	3.72338	58
3	.21347	4.68452	.23179	4.31430	.25026	3.99592	.26888	3.71907	57
4	.21377	4.67786	.23209	4.30860	.25056	3.99099	.26920	3.71476	56
5	.21408	4.67121	.23240	4.30291	.25087	3.98607	.26951	3.71046	55
6	.21438	4.66458	.23271	4.29724	.25118	3.98117	.26982	3.70616	54
7	.21469	4.65797	.23301	4.29159	.25149	3.97627	.27013	3.70188	53
8	.21499	4.65138	.23332	4.28595	.25180	3.97139	.27044	3.69761	52
9	.21529	4.64480	.23363	4.28032	.25211	3.96651	.27076	3.69335	51
10	.21560	4.63825	.23393	4.27471	.25242	3.96165	.27107	3.68909	50
11	.21590	4.63171	.23424	4.26911	.25273	3.95680	.27138	3.68485	49
12	.21621	4.62518	.23455	4.26352	.25304	3.95196	.27169	3.68061	48
13	.21651	4.61868	.23485	4.25795	.25335	3.94713	.27201	3.67638	47
14	.21682	4.61219	.23516	4.25239	.25366	3.94232	.27232	3.67217	46
15	.21712	4.60572	.23547	4.24685	.25397	3.93751	.27263	3.66796	45
16	.21743	4.59927	.23578	4.24132	.25428	3.93271	.27294	3.66376	44
17	.21773	4.59283	.23608	4.23580	.25459	3.92793	.27326	3.65957	43
18	.21804	4.58641	.23639	4.23030	.25490	3.92316	.27357	3.65538	42
19	.21834	4.58001	.23670	4.22481	.25521	3.91839	.27388	3.65121	41
20	.21864	4.57363	.23700	4.21933	.25552	3.91364	.27419	3.64705	40
21	.21895	4.56726	.23731	4.21387	.25583	3.90890	.27451	3.64289	39
22	.21925	4.56091	.23762	4.20842	.25614	3.90417	.27482	3.63874	38
23	.21956	4.55458	.23793	4.20298	.25645	3.89945	.27513	3.63461	37
24	.21986	4.54826	.23823	4.19756	.25676	3.89474	.27545	3.63048	36
25	.22017	4.54196	.23854	4.19215	.25707	3.89004	.27576	3.62636	35
26	.22047	4.53568	.23885	4.18675	.25738	3.88536	.27607	3.62224	34
27	.22078	4.52941	.23916	4.18137	.25769	3.88068	.27638	3.61814	33
28	.22108	4.52316	.23946	4.17600	.25800	3.87601	.27670	3.61405	32
29	.22139	4.51693	.23977	4.17064	.25831	3.87136	.27701	3.60996	31
30	.22169	4.51071	.24008	4.16530	.25862	3.86671	.27732	3.60588	30
31	.22200	4.50451	.24039	4.15997	.25893	3.86208	.27764	3.60181	29
32	.22231	4.49832	.24069	4.15465	.25924	3.85745	.27795	3.59775	28
33	.22261	4.49215	.24100	4.14934	.25955	3.85284	.27826	3.59370	27
34	.22292	4.48600	.24131	4.14405	.25986	3.84824	.27858	3.58966	26
35	.22322	4.47986	.24162	4.13877	.26017	3.84364	.27889	3.58562	25
36	.22353	4.47374	.24193	4.13350	.26048	3.83906	.27920	3.58160	24
37	.22383	4.46764	.24223	4.12825	.26079	3.83449	.27952	3.57758	23
38	.22414	4.46155	.24254	4.12301	.26110	3.82992	.27983	3.57357	22
39	.22444	4.45548	.24285	4.11778	.26141	3.82537	.28015	3.56957	21
40	.22475	4.44942	.24316	4.11256	.26172	3.82083	.28046	3.56557	20
41	.22505	4.44338	.24347	4.10736	.26203	3.81630	.28077	3.56159	19
42	.22536	4.43735	.24377	4.10216	.26235	3.81177	.28109	3.55761	18
43	.22567	4.43134	.24408	4.09699	.26266	3.80726	.28140	3.55364	17
44	.22597	4.42534	.24439	4.09182	.26297	3.80276	.28172	3.54968	16
45	.22628	4.41936	.24470	4.08666	.26328	3.79827	.28203	3.54573	15
46	.22658	4.41340	.24501	4.08152	.26359	3.79378	.28234	3.54179	14
47	.22689	4.40745	.24532	4.07639	.26390	3.78931	.28266	3.53785	13
48	.22719	4.40152	.24562	4.07127	.26421	3.78485	.28297	3.53393	12
49	.22750	4.39560	.24593	4.06616	.26452	3.78040	.28329	3.53001	11
50	.22781	4.38969	.24624	4.06107	.26483	3.77595	.28360	3.52609	10
51	.22811	4.38381	.24655	4.05599	.26515	3.77152	.28391	3.52219	9
52	.22842	4.37793	.24686	4.05092	.26546	3.76709	.28423	3.51829	8
53	.22872	4.37207	.24717	4.04586	.26577	3.76268	.28454	3.51441	7
54	.22903	4.36623	.24747	4.04081	.26608	3.75828	.28486	3.51053	6
55	.22934	4.36040	.24778	4.03578	.26639	3.75388	.28517	3.50666	5
56	.22964	4.35459	.24809	4.03075	.26670	3.74950	.28549	3.50279	4
57	.22995	4.34879	.24840	4.02574	.26701	3.74512	.28580	3.49894	3
58	.23026	4.34300	.24871	4.02074	.26733	3.74075	.28612	3.49509	2
59	.23056	4.33723	.24902	4.01576	.26764	3.73640	.28643	3.49125	1
60	.23087	4.33148	.24933	4.01078	.26795	3.73205	.28675	3.48741	0
′	cotan 77°	tan 77°	cotan 76°	tan 76°	cotan 75°	tan 75°	cotan 74°	tan 74°	′

′	16°		17°		18°		19°		′
	tan	cotan	tan	cotan	tan	cotan	tan	cotan	
0	.28675	3.48741	.30573	3.27085	.32492	3.07768	.34433	2.90421	60
1	.28706	3.48359	.30605	3.26745	.32524	3.07464	.34465	2.90147	59
2	.28738	3.47977	.30637	3.26406	.32556	3.07160	.34498	2.89873	58
3	.28769	3.47596	.30669	3.26067	.32588	3.06857	.34530	2.89600	57
4	.28800	3.47216	.30700	3.25729	.32621	3.06554	.34563	2.89327	56
5	.28832	3.46837	.30732	3.25392	.32653	3.06252	.34596	2.89055	55
6	.28864	3.46458	.30764	3.25055	.32685	3.05950	.34628	2.88783	54
7	.28895	3.46080	.30796	3.24719	.32717	3.05649	.34661	2.88511	53
8	.28927	3.45703	.30828	3.24383	.32749	3.05349	.34693	2.88240	52
9	.28958	3.45327	.30860	3.24049	.32782	3.05049	.34726	2.87970	51
10	.28990	3.44951	.30891	3.23714	.32814	3.04749	.34758	2.87700	50
11	.29021	3.44576	.30923	3.23381	.32846	3.04450	.34791	2.87430	49
12	.29053	3.44202	.30955	3.23048	.32878	3.04152	.34824	2.87161	48
13	.29084	3.43829	.30987	3.22715	.32911	3.03854	.34856	2.86892	47
14	.29116	3.43456	.31019	3.22384	.32943	3.03556	.34889	2.86624	46
15	.29147	3.43084	.31051	3.22053	.32975	3.03260	.34922	2.86356	45
16	.29179	3.42713	.31083	3.21722	.33007	3.02963	.34954	2.86089	44
17	.29210	3.42343	.31115	3.21392	.33040	3.02667	.34987	2.85822	43
18	.29242	3.41973	.31147	3.21063	.33072	3.02372	.35019	2.85555	42
19	.29274	3.41604	.31178	3.20734	.33104	3.02077	.35052	2.85289	41
20	.29305	3.41236	.31210	3.20406	.33136	3.01783	.35085	2.85023	40
21	.29337	3.40869	.31242	3.20079	.33169	3.01489	.35117	2.84758	39
22	.29368	3.40502	.31274	3.19752	.33201	3.01196	.35150	2.84494	38
23	.29400	3.40136	.31306	3.19426	.33233	3.00903	.35183	2.84229	37
24	.29432	3.39771	.31338	3.19100	.33266	3.00611	.35216	2.83965	36
25	.29463	3.39406	.31370	3.18775	.33298	3.00319	.35248	2.83702	35
26	.29495	3.39042	.31402	3.18451	.33330	3.00028	.35281	2.83439	34
27	.29526	3.38679	.31434	3.18127	.33363	2.99738	.35314	2.83176	33
28	.29558	3.38317	.31466	3.17804	.33395	2.99447	.35346	2.82914	32
29	.29590	3.37955	.31498	3.17481	.33427	2.99158	.35379	2.82653	31
30	.29621	3.37594	.31530	3.17159	.33460	2.98868	.35412	2.82391	30
31	.29653	3.37234	.31562	3.16838	.33492	2.98580	.35445	2.82130	29
32	.29685	3.36875	.31594	3.16517	.33524	2.98292	.35477	2.81870	28
33	.29716	3.36516	.31626	3.16197	.33557	2.98004	.35510	2.81610	27
34	.29748	3.36158	.31658	3.15877	.33589	2.97717	.35543	2.81350	26
35	.29780	3.35800	.31690	3.15558	.33621	2.97430	.35576	2.81091	25
36	.29811	3.35443	.31722	3.15240	.33654	2.97144	.35608	2.80833	24
37	.29843	3.35087	.31754	3.14922	.33686	2.96858	.35641	2.80574	23
38	.29875	3.34732	.31786	3.14605	.33718	2.96573	.35674	2.80316	22
39	.29906	3.34377	.31818	3.14288	.33751	2.96288	.35707	2.80059	21
40	.29938	3.34023	.31850	3.13972	.33783	2.96004	.35740	2.79802	20
41	.29970	3.33670	.31882	3.13656	.33816	2.95721	.35772	2.79545	19
42	.30001	3.33317	.31914	3.13341	.33848	2.95437	.35805	2.79289	18
43	.30033	3.32965	.31946	3.13027	.33881	2.95155	.35838	2.79033	17
44	.30065	3.32614	.31978	3.12713	.33913	2.94872	.35871	2.78778	16
45	.30097	3.32264	.32010	3.12400	.33945	2.94590	.35904	2.78523	15
46	.30128	3.31914	.32042	3.12087	.33978	2.94309	.35937	2.78269	14
47	.30160	3.31565	.32074	3.11775	.34010	2.94028	.35969	2.78014	13
48	.30192	3.31216	.32106	3.11464	.34043	2.93748	.36002	2.77761	12
49	.30224	3.30868	.32139	3.11153	.34075	2.93468	.36035	2.77507	11
50	.30255	3.30521	.32171	3.10842	.34108	2.93189	.36068	2.77254	10
51	.30287	3.30174	.32203	3.10532	.34140	2.92910	.36101	2.77002	9
52	.30319	3.29829	.32235	3.10223	.34173	2.92632	.36134	2.76750	8
53	.30351	3.29483	.32267	3.09914	.34205	2.92354	.36167	2.76498	7
54	.30382	3.29139	.32299	3.09606	.34238	2.92076	.36199	2.76247	6
55	.30414	3.28795	.32331	3.09298	.34270	2.91799	.36232	2.75996	5
56	.30446	3.28452	.32363	3.08991	.34303	2.91523	.36265	2.75746	4
57	.30478	3.28109	.32396	3.08685	.34335	2.91246	.36298	2.75496	3
58	.30509	3.27767	.32428	3.08379	.34368	2.90971	.36331	2.75246	2
59	.30541	3.27426	.32460	3.08073	.34400	2.90696	.36364	2.74997	1
60	.30573	3.27085	.32492	3.07768	.34433	2.90421	.36397	2.74748	0
′	cotan	tan	cotan	tan	cotan	tan	cotan	tan	′
	73°		72°		71°		70°		

′	20°		21°		22°		23°		′
	tan	cotan	tan	cotan	tan	cotan	tan	cotan	
0	.36397	2.74748	.38386	2.60509	.40403	2.47509	.42447	2.35585	60
1	.36430	2.74499	.38420	2.60283	.40436	2.47302	.42482	2.35395	59
2	.36463	2.74251	.38453	2.60057	.40470	2.47095	.42516	2.35205	58
3	.36496	2.74004	.38487	2.59831	.40504	2.46888	.42551	2.35015	57
4	.36529	2.73756	.38520	2.59606	.40538	2.46682	.42585	2.34825	56
5	.36562	2.73509	.38553	2.59381	.40572	2.46476	.42619	2.34636	55
6	.36595	2.73263	.38587	2.59156	.40606	2.46270	.42654	2.34447	54
7	.36628	2.73017	.38620	2.58932	.40640	2.46065	.42688	2.34258	53
8	.36661	2.72771	.38654	2.58708	.40674	2.45860	.42722	2.34069	52
9	.36694	2.72526	.38687	2.58484	.40707	2.45655	.42757	2.33881	51
10	.36727	2.72281	.38721	2.58261	.40741	2.45451	.42791	2.33693	50
11	.36760	2.72036	.38754	2.58038	.40775	2.45246	.42826	2.33505	49
12	.36793	2.71792	.38787	2.57815	.40809	2.45043	.42860	2.33317	48
13	.36826	2.71548	.38821	2.57593	.40843	2.44839	.42894	2.33130	47
14	.36859	2.71305	.38854	2.57371	.40877	2.44636	.42929	2.32943	46
15	.36892	2.71062	.38888	2.57150	.40911	2.44433	.42963	2.32756	45
16	.36925	2.70819	.38921	2.56928	.40945	2.44230	.42998	2.32570	44
17	.36958	2.70577	.38955	2.56707	.40979	2.44027	.43032	2.32383	43
18	.36991	2.70335	.38988	2.56487	.41013	2.43825	.43067	2.32197	42
19	.37024	2.70094	.39022	2.56266	.41047	2.43623	.43101	2.32012	41
20	.37057	2.69853	.39055	2.56046	.41081	2.43422	.43136	2.31826	40
21	.37090	2.69612	.39089	2.55827	.41115	2.43220	.43170	2.31641	39
22	.37124	2.69371	.39122	2.55608	.41149	2.43019	.43205	2.31456	38
23	.37157	2.69131	.39156	2.55389	.41183	2.42819	.43239	2.31271	37
24	.37190	2.68892	.39190	2.55170	.41217	2.42618	.43274	2.31086	36
25	.37223	2.68653	.39223	2.54952	.41251	2.42418	.43308	2.30902	35
26	.37256	2.68414	.39257	2.54734	.41285	2.42218	.43343	2.30718	34
27	.37289	2.68175	.39290	2.54516	.41319	2.42019	.43378	2.30534	33
28	.37322	2.67937	.39324	2.54299	.41353	2.41819	.43412	2.30351	32
29	.37355	2.07700	.39357	2.54082	.41387	2.41020	.43447	2.30167	31
30	.37388	2.67462	.39391	2.53865	.41421	2.41421	.43481	2.29984	30
31	.37422	2.67225	.39425	2.53648	.41455	2.41223	.43516	2.29801	29
32	.37455	2.66989	.39458	2.53432	.41490	2.41025	.43550	2.29619	28
33	.37488	2.66752	.39492	2.53217	.41524	2.40827	.43585	2.29437	27
34	.37521	2.66516	.29526	2.53001	.41558	2.40629	.43620	2.29254	26
35	.37554	2.66281	.39559	2.52786	.41592	2.40432	.43654	2.29073	25
36	.37588	2.66046	.39593	2.52571	.41626	2.40235	.43689	2.28891	24
37	.37621	2.65811	.39626	2.52357	.41660	2.40038	.43724	2.28710	23
38	.37654	2.65576	.39660	2.52142	.41694	2.39841	.43758	2.28528	22
39	.37687	2.65342	.39694	2.51929	.41728	2.39645	.43793	2.28348	21
40	.37720	2.65109	.39727	2.51715	.41763	2.39449	.43828	2.28167	20
41	.37754	2.64875	.39761	2.51502	.41797	2.39253	.43862	2.27987	19
42	.37787	2.64642	.39795	2.51289	.41831	2.39058	.43897	2.27806	18
43	.37820	2.64410	.39829	2.51076	.41865	2.38862	.43932	2.27626	17
44	.37853	2.64177	.39862	2.50864	.41899	2.38668	.43966	2.27447	16
45	.37887	2.63945	.39896	2.50652	.41933	2.38473	.44001	2.27267	15
46	.37920	2.63714	.39930	2.50440	.41968	2.38279	.44036	2.27088	14
47	.37953	2.63483	.39963	2.50229	.42002	2.38084	.44071	2.26909	13
48	.37986	2.63252	.39997	2.50018	.42036	2.37891	.44105	2.26730	12
49	.38020	2.63021	.40031	2.49807	.42070	2.37697	.44140	2.26552	11
50	.38053	2.62791	.40065	2.49597	.42105	2.37504	.44175	2.26374	10
51	.38086	2.62561	.40098	2.49386	.42139	2.37311	.44210	2.26196	9
52	.38120	2.62332	.40132	2.49177	.42173	2.37118	.44244	2.26018	8
53	.38153	2.62103	.40166	2.48967	.42207	2.36925	.44279	2.25840	7
54	.38186	2.61874	.40200	2.48758	.42242	2.36733	.44314	2.25663	6
55	.38220	2.61646	.40234	2.48549	.42276	2.36541	.44349	2.25486	5
56	.38253	2.61418	.40267	2.48340	.42310	2.36349	.44384	2.25309	4
57	.38286	2.61190	.40301	2.48132	.42345	2.36158	.44418	2.25132	3
58	.38320	2.60963	.40335	2.47924	.42379	2.35967	.44453	2.24956	2
59	.38353	2.60736	.40369	2.47716	.42413	2.35776	.44488	2.24780	1
60	.38386	2.60509	.40403	2.47509	.42447	2.35585	.44523	2.24604	0
′	cotan	tan	cotan	tan	cotan	tan	cotan	tan	′
	69°		68°		67°		66°		

′	24° tan	24° cotan	25° tan	25° cotan	26° tan	26° cotan	27° tan	27° cotan	′
0	.44523	2.24604	.46631	2.14451	.48773	2.05030	.50953	1.96261	60
1	.44558	2.24428	.46666	2.14288	.48809	2.04879	.50989	1.96120	59
2	.44593	2.24252	.46702	2.14125	.48845	2.04728	.51026	1.95979	58
3	.44627	2.24077	.46737	2.13963	.48881	2.04577	.51063	1.95838	57
4	.44662	2.23902	.46772	2.13801	.48917	2.04426	.51099	1.95698	56
5	.44697	2.23727	.46808	2.13639	.48953	2.04276	.51136	1.95557	55
6	.44732	2.23553	.46843	2.13477	.48989	2.04125	.51173	1.95417	54
7	.44767	2.23378	.46879	2.13316	.49026	2.03975	.51209	1.95277	53
8	.44802	2.23204	.46914	2.13154	.49062	2.03825	.51246	1.95137	52
9	.44837	2.23030	.46950	2.12993	.49098	2.03675	.51283	1.94997	51
10	.44872	2.22857	.46985	2.12832	.49134	2.03526	.51319	1.94858	50
11	.44907	2.22683	.47021	2.12671	.49170	2.03376	.51356	1.94718	49
12	.44942	2.22510	.47056	2.12511	.49206	2.03227	.51393	1.94579	48
13	.44977	2.22337	.47092	2.12350	.49242	2.03078	.51430	1.94440	47
14	.45012	2.22164	.47128	2.12190	.49278	2.02929	.51467	1.94301	46
15	.45047	2.21992	.47163	2.12030	.49315	2.02780	.51503	1.94162	45
16	.45082	2.21819	.47199	2.11871	.49351	2.02631	.51540	1.94023	44
17	.45117	2.21647	.47234	2.11711	.49387	2.02483	.51577	1.93885	43
18	.45152	2.21475	.47270	2.11552	.49423	2.02335	.51614	1.93746	42
19	.45187	2.21304	.47305	2.11392	.49459	2.02187	.51651	1.93608	41
20	.45222	2.21132	.47341	2.11233	.49495	2.02039	.51688	1.93470	40
21	.45257	2.20961	.47377	2.11075	.49532	2.01891	.51724	1.93332	39
22	.45292	2.20790	.47412	2.10916	.49568	2.01743	.51761	1.93195	38
23	.45327	2.20619	.47448	2.10758	.49604	2.01596	.51798	1.93057	37
24	.45362	2.20449	.47483	2.10600	.49640	2.01449	.51835	1.92920	36
25	.45397	2.20278	.47519	2.10442	.49677	2.01302	.51872	1.92782	35
26	.45432	2.20108	.47555	2.10284	.49713	2.01155	.51909	1.92645	34
27	.45467	2.19938	.47590	2.10126	.49749	2.01008	.51946	1.92508	33
28	.45502	2.19769	.47626	2.09969	.49786	2.00862	.51983	1.92371	32
29	.45537	2.19599	.47662	2.09811	.49822	2.00715	.52020	1.92235	31
30	.45573	2.19430	.47698	2.09654	.49858	2.00569	.52057	1.92098	30
31	.45608	2.19261	.47733	2.09498	.49894	2.00423	.52094	1.91962	29
32	.45643	2.19092	.47769	2.09341	.49931	2.00277	.52131	1.91826	28
33	.45678	2.18923	.47805	2.09184	.49967	2.00131	.52168	1.91690	27
34	.45713	2.18755	.47840	2.09028	.50004	1.99986	.52205	1.91554	26
35	.45748	2.18587	.47876	2.08872	.50040	1.99841	.52242	1.91418	25
36	.45784	2.18419	.47912	2.08716	.50076	1.99695	.52279	1.91282	24
37	.45819	2.18251	.47948	2.08560	.50113	1.99550	.52316	1.91147	23
38	.45854	2.18084	.47984	2.08405	.50149	1.99406	.52353	1.91012	22
39	.45889	2.17916	.48019	2.08250	.50185	1.99261	.52390	1.90876	21
40	.45924	2.17749	.48055	2.08094	.50222	1.99116	.52427	1.90741	20
41	.45960	2.17582	.48091	2.07939	.50258	1.98972	.52464	1.90607	19
42	.45995	2.17416	.48127	2.07785	.50295	1.98828	.52501	1.90472	18
43	.46030	2.17249	.48163	2.07630	.50331	1.98684	.52538	1.90337	17
44	.46065	2.17083	.48198	2.07476	.50368	1.98540	.52575	1.90203	16
45	.46101	2.16917	.48234	2.07321	.50404	1.98396	.52613	1.90069	15
46	.46136	2.16751	.48270	2.07167	.50441	1.98253	.52650	1.89935	14
47	.46171	2.16585	.48306	2.07014	.50477	1.98110	.52687	1.89801	13
48	.46206	2.16420	.48342	2.06860	.50514	1.97966	.52724	1.89667	12
49	.46242	2.16255	.48378	2.06706	.50550	1.97823	.52761	1.89533	11
50	.46277	2.16090	.48414	2.06553	.50587	1.97680	.52798	1.89400	10
51	.46312	2.15925	.48450	2.06400	.50623	1.97538	.52836	1.89266	9
52	.46348	2.15760	.48486	2.06247	.50660	1.97395	.52873	1.89133	8
53	.46383	2.15596	.48521	2.06094	.50696	1.97253	.52910	1.89000	7
54	.46418	2.15432	.48557	2.05942	.50733	1.97111	.52947	1.88867	6
55	.46454	2.15268	.48593	2.05790	.50769	1.96969	.52984	1.88734	5
56	.46489	2.15104	.48629	2.05637	.50806	1.96827	.53022	1.88602	4
57	.46525	2.14940	.48665	2.05485	.50843	1.96685	.53059	1.88469	3
58	.46560	2.14777	.48701	2.05333	.50879	1.96544	.53096	1.88337	2
59	.46595	2.14614	.48737	2.05182	.50916	1.96402	.53134	1.88205	1
60	.46631	2.14451	.48773	2.05030	.50953	1.96261	.53171	1.88073	0
′	65° cotan	65° tan	64° cotan	64° tan	63° cotan	63° tan	62° cotan	62° tan	′

	28°		29°		30°		31°		
′	tan	cotan	tan	cotan	tan	cotan	tan	cotan	′
0	.53171	1.88073	.55431	1.80405	.57735	1.73205	.60086	1.66428	60
1	.53208	1.87941	.55469	1.80281	.57774	1.73089	.60126	1.66318	59
2	.53246	1.87809	.55507	1.80158	.57813	1.72973	.60165	1.66209	58
3	.53283	1.87677	.55545	1.80034	.57851	1.72857	.60205	1.66099	57
4	.53320	1.87546	.55583	1.79911	.57890	1.72741	.60245	1.65990	56
5	.53358	1.87415	.55621	1.79788	.57929	1.72625	.60284	1.65881	55
6	.53395	1.87283	.55659	1.79665	.57968	1.72509	.60324	1.65772	54
7	.53432	1.87152	.55697	1.79542	.58007	1.72393	.60364	1.65663	53
8	.53470	1.87021	.55736	1.79419	.58046	1.72278	.60403	1.65554	52
9	.53507	1.86891	.55774	1.79296	.58085	1.72163	.60443	1.65445	51
10	.53545	1.86760	.55812	1.79174	.58124	1.72047	.60483	1.65337	50
11	.53582	1.86630	.55850	1.79051	.58162	1.71932	.60522	1.65228	49
12	.53620	1.86499	.55888	1.78929	.58201	1.71817	.60562	1.65120	48
13	.53657	1.86369	.55926	1.78807	.58240	1.71702	.60602	1.65011	47
14	.53694	1.86239	.55964	1.78685	.58279	1.71588	.60642	1.64903	46
15	.53732	1.86109	.56003	1.78563	.58318	1.71473	.60681	1.64795	45
16	.53769	1.85979	.56041	1.78441	.58357	1.71358	.60721	1.64687	44
17	.53807	1.85850	.56079	1.78319	.58396	1.71244	.60761	1.64579	43
18	.53844	1.85720	.56117	1.78198	.58435	1.71129	.60801	1.64471	42
19	.53882	1.85591	.56156	1.78077	.58474	1.71015	.60841	1.64363	41
20	.53920	1.85462	.56194	1.77955	.58513	1.70901	.60881	1.64256	40
21	.53957	1.85333	.56232	1.77834	.58552	1.70787	.60921	1.64148	39
22	.53995	1.85204	.56270	1.77713	.58591	1.70673	.60960	1.64041	38
23	.54032	1.85075	.56309	1.77592	.58631	1.70560	.61000	1.63934	37
24	.54070	1.84946	.56347	1.77471	.58670	1.70446	.61040	1.63826	36
25	.54107	1.84818	.56385	1.77351	.58709	1.70332	.61080	1.63719	35
26	.54145	1.84689	.56424	1.77230	.58748	1.70219	.61120	1.63612	34
27	.54183	1.84561	.56462	1.77110	.58787	1.70106	.61160	1.63505	33
28	.54220	1.84433	.56500	1.76990	.58826	1.69992	.61200	1.63398	32
29	.54258	1.84305	.56539	1.76869	.58865	1.69879	.61240	1.63292	31
30	.54296	1.84177	.56577	1.76749	.58904	1.69766	.61280	1.63185	30
31	.54333	1.84049	.56616	1.76630	.58944	1.69653	.61320	1.63079	29
32	.54371	1.83922	.56654	1.76510	.58983	1.69541	.61360	1.62972	28
33	.54409	1.83794	.56693	1.76390	.59022	1.69428	.61400	1.62866	27
34	.54446	1.83667	.56731	1.76271	.59061	1.69316	.61440	1.62760	26
35	.54484	1.83540	.56769	1.76151	.59101	1.69203	.61480	1.62654	25
36	.54522	1.83413	.56808	1.76032	.59140	1.69091	.61520	1.62548	24
37	.54560	1.83286	.56846	1.75913	.59179	1.68979	.61561	1.62442	23
38	.54597	1.83159	.56885	1.75794	.59218	1.68866	.61601	1.62336	22
39	.54635	1.83033	.56923	1.75675	.59258	1.68754	.61641	1.62230	21
40	.54673	1.82906	.56962	1.75556	.59297	1.68643	.61681	1.62125	20
41	.54711	1.82780	.57000	1.75437	.59336	1.68531	.61721	1.62019	19
42	.54748	1.82654	.57039	1.75319	.59376	1.68419	.61761	1.61914	18
43	.54786	1.82528	.57078	1.75200	.59415	1.68308	.61801	1.61808	17
44	.54824	1.82402	.57116	1.75082	.59454	1.68196	.61842	1.61703	16
45	.54862	1.82276	.57155	1.74964	.59494	1.68085	.61882	1.61598	15
46	.54900	1.82150	.57193	1.74846	.59533	1.67974	.61922	1.61493	14
47	.54938	1.82025	.57232	1.74728	.59573	1.67863	.61962	1.61388	13
48	.54975	1.81899	.57271	1.74610	.59612	1.67752	.62003	1.61283	12
49	.55013	1.81774	.57309	1.74492	.59651	1.67641	.62043	1.61179	11
50	.55051	1.81649	.57348	1.74375	.59691	1.67530	.62083	1.61074	10
51	.55089	1.81524	.57386	1.74257	.59730	1.67419	.62124	1.60970	9
52	.55127	1.81399	.57425	1.74140	.59770	1.67309	.62164	1.60865	8
53	.55165	1.81274	.57464	1.74022	.59809	1.67198	.62204	1.60761	7
54	.55203	1.81150	.57503	1.73905	.59849	1.67088	.62245	1.60657	6
55	.55241	1.81025	.57541	1.73788	.59888	1.66978	.62285	1.60553	5
56	.55279	1.80901	.57580	1.73671	.59928	1.66867	.62325	1.60449	4
57	.55317	1.80777	.57619	1.73555	.59967	1.66757	.62366	1.60345	3
58	.55355	1.80653	.57657	1.73438	.60007	1.66647	.62406	1.60241	2
59	.55393	1.80529	.57696	1.73321	.60046	1.66538	.62446	1.60137	1
60	.55431	1.80405	.57735	1.73205	.60086	1.66428	.62487	1.60033	0
′	cotan	tan	cotan	tan	cotan	tan	cotan	tan	′
	61°		60°		59°		58°		

′	32°		33°		34°		35°		′
	tan	cotan	tan	cotan	tan	cotan	tan	cotan	
0	.62487	1.60033	.64941	1.53986	.67451	1.48256	.70021	1.42815	60
1	.62527	1.59930	.64982	1.53888	.67493	1.48163	.70064	1.42726	59
2	.62568	1.59826	.65023	1.53791	.67536	1.48070	.70107	1.42638	58
3	.62608	1.59723	.65065	1.53693	.67578	1.47977	.70151	1.42550	57
4	.62649	1.59620	.65106	1.53595	.67620	1.47885	.70194	1.42462	56
5	.62689	1.59517	.65148	1.53497	.67663	1.47792	.70238	1.42374	55
6	.62730	1.59414	.65189	1.53400	.67705	1.47699	.70281	1.42286	54
7	.62770	1.59311	.65231	1.53302	.67748	1.47607	.70325	1.42198	53
8	.62811	1.59208	.65272	1.53205	.67790	1.47514	.70368	1.42110	52
9	.62852	1.59105	.65314	1.53107	.67832	1.47422	.70412	1.42022	51
10	.62892	1.59002	.65355	1.53010	.67875	1.47330	.70455	1.41934	50
11	.62933	1.58900	.65397	1.52913	.67917	1.47238	.70499	1.41847	49
12	.62973	1.58797	.65438	1.52816	.67960	1.47146	.70542	1.41759	48
13	.63014	1.58695	.65480	1.52719	.68002	1.47053	.70586	1.41672	47
14	.63055	1.58593	.65521	1.52622	.68045	1.46962	.70629	1.41584	46
15	.63095	1.58490	.65563	1.52525	.68088	1.46870	.70673	1.41497	45
16	.63136	1.58388	.65604	1.52429	.68130	1.46778	.70717	1.41409	44
17	.63177	1.58286	.65646	1.52332	.68173	1.46686	.70760	1.41322	43
18	.63217	1.58184	.65688	1.52235	.68215	1.46595	.70804	1.41235	42
19	.63258	1.58083	.65729	1.52139	.68258	1.46503	.70848	1.41148	41
20	.63299	1.57981	.65771	1.52043	.68301	1.46411	.70891	1.41061	40
21	.63340	1.57879	.65813	1.51946	.68343	1.46320	.70935	1.40974	39
22	.63380	1.57778	.65854	1.51850	.68386	1.46229	.70979	1.40887	38
23	.63421	1.57676	.65896	1.51754	.68429	1.46137	.71023	1.40800	37
24	.63462	1.57575	.65938	1.51658	.68471	1.46046	.71066	1.40714	36
25	.63503	1.57474	.65980	1.51562	.68514	1.45955	.71110	1.40627	35
26	.63544	1.57372	.66021	1.51466	.68557	1.45864	.71154	1.40540	34
27	.63584	1.57271	.66063	1.51370	.68600	1.45773	.71198	1.40454	33
28	.63625	1.57170	.66105	1.51275	.68642	1.45682	.71242	1.40367	32
29	.63666	1.57069	.66147	1.51179	.68685	1.45592	.71285	1.40281	31
30	.63707	1.56969	.66189	1.51084	.68728	1.45501	.71329	1.40195	30
31	.63748	1.56868	.66230	1.50988	.68771	1.45410	.71373	1.40109	29
32	.63789	1.56767	.66272	1.50893	.68814	1.45320	.71417	1.40022	28
33	.63830	1.56667	.66314	1.50799	.68857	1.45229	.71461	1.39936	27
34	.63871	1.56566	.66356	1.50702	.68900	1.45138	.71505	1.39850	26
35	.63912	1.56466	.66398	1.50607	.68942	1.45049	.71549	1.39764	25
36	.63953	1.56366	.66440	1.50512	.68985	1.44958	.71593	1.39679	24
37	.63994	1.56265	.66482	1.50417	.69028	1.44868	.71637	1.39593	23
38	.64035	1.56165	.66524	1.50322	.69071	1.44778	.71681	1.39507	22
39	.64076	1.56065	.66566	1.50228	.69114	1.44688	.71725	1.39421	21
40	.64117	1.55966	.66608	1.50133	.69157	1.44598	.71769	1.39336	20
41	.64158	1.55866	.66650	1.50038	.69200	1.44508	.71813	1.39250	19
42	.64199	1.55766	.66692	1.49944	.69243	1.44418	.71857	1.39165	18
43	.64240	1.55666	.66734	1.49849	.69286	1.44329	.71901	1.39079	17
44	.64281	1.55567	.66776	1.49755	.69329	1.44239	.71946	1.38994	16
45	.64322	1.55467	.66818	1.49661	.69372	1.44149	.71990	1.38909	15
46	.64363	1.55368	.66860	1.49566	.69416	1.44060	.72034	1.38824	14
47	.64404	1.55269	.66902	1.49472	.69459	1.43970	.72078	1.38738	13
48	.64446	1.55170	.66944	1.49378	.69502	1.43881	.72122	1.38653	12
49	.64487	1.55071	.66986	1.49284	.69545	1.43792	.72166	1.38568	11
50	.64528	1.54972	.67028	1.49190	.69588	1.43703	.72211	1.38484	10
51	.64569	1.54873	.67071	1.49097	.69631	1.43614	.72255	1.38399	9
52	.64610	1.54774	.67113	1.49003	.69675	1.43525	.72299	1.38314	8
53	.64652	1.54675	.67155	1.48909	.69718	1.43436	.72344	1.38229	7
54	.64693	1.54576	.67197	1.48816	.69761	1.43347	.72388	1.38145	6
55	.64734	1.54478	.67239	1.48722	.69804	1.43258	.72432	1.38060	5
56	.64775	1.54379	.67282	1.48629	.69847	1.43169	.72477	1.37976	4
57	.64817	1.54281	.67324	1.48536	.69891	1.43080	.72521	1.37891	3
58	.64858	1.54183	.67366	1.48442	.69934	1.42992	.72565	1.37807	2
59	.64899	1.54085	.67409	1.48349	.69966	1.42903	.72610	1.37722	1
60	.64941	1.53986	.67451	1.48256	.70021	1.42815	.72654	1.37638	0
′	cotan	tan	cotan	tan	cotan	tan	cotan	tan	′
	57°		56°		55°		54°		

′	36° tan	36° cotan	37° tan	37° cotan	38° tan	38° cotan	39° tan	39° cotan	′
0	.72654	1.37638	.75355	1.32704	.78129	1.27994	.80978	1.23490	60
1	.72699	1.37554	.75401	1.32624	.78175	1.27917	.81027	1.23416	59
2	.72743	1.37470	.75447	1.32544	.78222	1.27841	.81075	1.23343	58
3	.72788	1.37386	.75492	1.32464	.78269	1.27764	.81123	1.23270	57
4	.72832	1.37302	.75538	1.32384	.78316	1.27688	.81171	1.23196	56
5	.72877	1.37218	.75584	1.32304	.78363	1.27611	.81220	1.23123	55
6	.72921	1.37134	.75629	1.32224	.78410	1.27535	.81268	1.23050	54
7	.72966	1.37050	.75675	1.32144	.78457	1.27458	.81316	1.22977	53
8	.73010	1.36967	.75721	1.32064	.78504	1.27382	.81364	1.22904	52
9	.73055	1.36883	.75767	1.31984	.78551	1.27306	.81413	1.22831	51
10	.73100	1.36800	.75812	1.31904	.78598	1.27230	.81461	1.22758	50
11	.73144	1.36716	.75858	1.31825	.78645	1.27153	.81510	1.22685	49
12	.73189	1.36633	.75904	1.31745	.78692	1.27077	.81558	1.22612	48
13	.73234	1.36549	.75950	1.31666	.78739	1.27001	.81606	1.22539	47
14	.73278	1.36466	.75996	1.31586	.78786	1.26925	.81655	1.22467	46
15	.73323	1.36383	.76042	1.31507	.78834	1.26849	.81703	1.22394	45
16	.73368	1.36300	.76088	1.31427	.78881	1.26774	.81752	1.22321	44
17	.73413	1.36217	.76134	1.31348	.78928	1.26698	.81800	1.22249	43
18	.73457	1.36133	.76180	1.31269	.78975	1.26622	.81849	1.22176	42
19	.73502	1.36051	.76226	1.31190	.79022	1.26546	.81898	1.22104	41
20	.73547	1.35968	.76272	1.31110	.79070	1.26471	.81946	1.22031	40
21	.73592	1.35885	.76318	1.31031	.79117	1.26395	.81995	1.21959	39
22	.73637	1.35802	.76364	1.30952	.79164	1.26319	.82044	1.21886	38
23	.73681	1.35719	.76410	1.30873	.79212	1.26244	.82092	1.21814	37
24	.73726	1.35637	.76456	1.30795	.79259	1.26169	.82141	1.21742	36
25	.73771	1.35554	.76502	1.30716	.79306	1.26093	.82190	1.21670	35
26	.73816	1.35472	.76548	1.30637	.79354	1.26018	.82238	1.21598	34
27	.73861	1.35389	.76594	1.30558	.79401	1.25943	.82287	1.21526	33
28	.73906	1.35307	.76640	1.30480	.79449	1.25867	.82336	1.21454	32
29	.73951	1.35224	.76686	1.30401	.79496	1.25792	.82385	1.21382	31
30	.73996	1.35142	.76733	1.30323	.79544	1.25717	.82434	1.21310	30
31	.74041	1.35060	.76779	1.30244	.79591	1.25642	.82483	1.21238	29
32	.74086	1.34978	.76825	1.30166	.79639	1.25567	.82531	1.21166	28
33	.74131	1.34896	.76871	1.30087	.79686	1.25492	.82580	1.21094	27
34	.74176	1.34814	.76918	1.30009	.79734	1.25417	.82629	1.21023	26
35	.74221	1.34732	.76964	1.29931	.79781	1.25343	.82678	1.20951	25
36	.74267	1.34650	.77010	1.29853	.79829	1.25268	.82727	1.20879	24
37	.74312	1.34568	.77057	1.29775	.79877	1.25193	.82776	1.20808	23
38	.74357	1.34487	.77103	1.29696	.79924	1.25118	.82825	1.20736	22
39	.74402	1.34405	.77149	1.29618	.79972	1.25044	.82874	1.20665	21
40	.74447	1.34323	.77196	1.29541	.80020	1.24969	.82923	1.20593	20
41	.74492	1.34242	.77242	1.29463	.80067	1.24895	.82972	1.20522	19
42	.74538	1.34160	.77289	1.29385	.80115	1.24820	.83022	1.20451	18
43	.74583	1.34070	.77335	1.29307	.80163	1.24746	.83071	1.20379	17
44	.74628	1.33998	.77382	1.29229	.80211	1.24672	.83120	1.20308	16
45	.74674	1.33916	.77428	1.29152	.80258	1.24597	.83169	1.20237	15
46	.74719	1.33835	.77475	1.29074	.80306	1.24523	.83218	1.20166	14
47	.74764	1.33754	.77521	1.28997	.80354	1.24449	.83268	1.20095	13
48	.74810	1.33673	.77568	1.28919	.80402	1.24375	.83317	1.20024	12
49	.74855	1.33592	.77615	1.28842	.80450	1.24301	.83366	1.19953	11
50	.74900	1.33511	.77661	1.28764	.80498	1.24227	.83415	1.19882	10
51	.74946	1.33430	.77708	1.28687	.80546	1.24153	.83465	1.19811	9
52	.74991	1.33349	.77754	1.28610	.80594	1.24079	.83514	1.19740	8
53	.75037	1.33268	.77801	1.28533	.80642	1.24005	.83564	1.19669	7
54	.75082	1.33187	.77848	1.28456	.80690	1.23931	.83613	1.19599	6
55	.75128	1.33107	.77895	1.28379	.80738	1.23858	.83662	1.19528	5
56	.75173	1.33026	.77941	1.28302	.80786	1.23784	.83712	1.19457	4
57	.75219	1.32946	.77988	1.28225	.80834	1.23710	.83761	1.19387	3
58	.75264	1.32865	.78035	1.28148	.80882	1.23637	.83811	1.19316	2
59	.75310	1.32785	.78082	1.28071	.80930	1.23563	.83860	1.19246	1
60	.75355	1.32704	.78129	1.27994	.80978	1.23490	.83910	1.19175	0
′	cotan 53°	tan 53°	cotan 52°	tan 52°	cotan 51°	tan 51°	cotan 50°	tan 50°	′

′	40° tan	40° cotan	41° tan	41° cotan	42° tan	42° cotan	43° tan	43° cotan	′
0	.83910	1.19175	.86929	1.15037	.90040	1.11061	.93252	1.07237	60
1	.83960	1.19105	.86980	1.14969	.90093	1.10996	.93306	1.07174	59
2	.84009	1.19035	.87031	1.14902	.90146	1.10931	.93360	1.07112	58
3	.84059	1.18964	.87082	1.14834	.90199	1.10867	.93415	1.07049	57
4	.84108	1.18894	.87133	1.14767	.90251	1.10802	.93469	1.06987	56
5	.84158	1.18824	.87184	1.14699	.90304	1.10737	.93524	1.06925	55
6	.84208	1.18754	.87236	1.14632	.90357	1.10672	.93578	1.06862	54
7	.84258	1.18684	.87287	1.14565	.90410	1.10607	.93633	1.06800	53
8	.84307	1.18614	.87338	1.14498	.90463	1.10543	.93688	1.06738	52
9	.84357	1.18544	.87389	1.14430	.90516	1.10478	.93742	1.06676	51
10	.84407	1.18474	.87441	1.14363	.90569	1.10414	.93797	1.06613	50
11	.84457	1.18404	.87492	1.14296	.90621	1.10349	.93852	1.06551	49
12	.84507	1.18334	.87543	1.14229	.90674	1.10285	.93906	1.06489	48
13	.84556	1.18264	.87595	1.14162	.90727	1.10220	.93961	1.06427	47
14	.84606	1.18194	.87646	1.14095	.90781	1.10156	.94016	1.06365	46
15	.84656	1.18125	.87698	1.14028	.90834	1.10091	.94071	1.06303	45
16	.84706	1.18055	.87749	1.13961	.90887	1.10027	.94125	1.06241	44
17	.84756	1.17986	.87801	1.13894	.90940	1.09963	.94180	1.06179	43
18	.84806	1.17916	.87852	1.13828	.90993	1.09899	.94235	1.06117	42
19	.84856	1.17846	.87904	1.13761	.91046	1.09834	.94290	1.06056	41
20	.84906	1.17777	.87955	1.13694	.91099	1.09770	.94345	1.05994	40
21	.84956	1.17708	.88007	1.13627	.91153	1.09706	.94400	1.05932	39
22	.85006	1.17638	.88059	1.13561	.91206	1.09642	.94455	1.05870	38
23	.85057	1.17569	.88110	1.13494	.91259	1.09578	.94510	1.05809	37
24	.85107	1.17500	.88162	1.13428	.91313	1.09514	.94565	1.05747	36
25	.85157	1.17430	.88214	1.13361	.91366	1.09450	.94620	1.05685	35
26	.85207	1.17361	.88265	1.13295	.91419	1.09386	.94676	1.05624	34
27	.85257	1.17292	.88317	1.13228	.91473	1.09322	.94731	1.05562	33
28	.85307	1.17223	.88369	1.13162	.91526	1.09258	.94786	1.05501	32
29	.85358	1.17154	.88421	1.13096	.91580	1.09195	.94841	1.05439	31
30	.85408	1.17085	.88473	1.13029	.91633	1.09131	.94896	1.05378	30
31	.85458	1.17016	.88524	1.12963	.91687	1.09067	.94952	1.05317	29
32	.85509	1.16947	.88576	1.12897	.91740	1.09003	.95007	1.05255	28
33	.85559	1.16878	.88628	1.12831	.91794	1.08940	.95062	1.05194	27
34	.85609	1.16809	.88680	1.12765	.91847	1.08876	.95118	1.05133	26
35	.85660	1.16741	.88732	1.12699	.91901	1.08813	.95173	1.05072	25
36	.85710	1.16672	.88784	1.12633	.91955	1.08749	.95229	1.05010	24
37	.85761	1.16603	.88836	1.12567	.92008	1.08686	.95284	1.04949	23
38	.85811	1.16535	.88888	1.12501	.92062	1.08622	.95340	1.04888	22
39	.85862	1.16466	.88940	1.12435	.92116	1.08559	.95395	1.04827	21
40	.85912	1.16398	.88992	1.12369	.92170	1.08496	.95451	1.04766	20
41	.85963	1.16329	.89045	1.12303	.92224	1.08432	.95506	1.04705	19
42	.86014	1.16261	.89097	1.12238	.92277	1.08369	.95562	1.04644	18
43	.86064	1.16192	.89149	1.12172	.92331	1.08306	.95618	1.04583	17
44	.86115	1.16124	.89201	1.12106	.92385	1.08243	.95673	1.04522	16
45	.86166	1.16056	.89253	1.12041	.92439	1.08179	.95729	1.04461	15
46	.86216	1.15987	.89306	1.11975	.92493	1.08116	.95785	1.04401	14
47	.86267	1.15919	.89358	1.11909	.92547	1.08053	.95841	1.04340	13
48	.86318	1.15851	.89410	1.11844	.92601	1.07990	.95897	1.04279	12
49	.86368	1.15783	.89463	1.11778	.92655	1.07927	.95952	1.04218	11
50	.86419	1.15715	.89515	1.11713	.92709	1.07864	.96008	1.04158	10
51	.86470	1.15647	.89567	1.11648	.92763	1.07801	.96064	1.04097	9
52	.86521	1.15579	.89620	1.11582	.92817	1.07738	.96120	1.04036	8
53	.86572	1.15511	.89672	1.11517	.92872	1.07676	.96176	1.03976	7
54	.86623	1.15443	.89725	1.11452	.92926	1.07613	.96232	1.03915	6
55	.86674	1.15375	.89777	1.11387	.92980	1.07550	.96288	1.03855	5
56	.86725	1.15308	.89830	1.11321	.93034	1.07487	.96344	1.03794	4
57	.86776	1.15240	.89883	1.11256	.93088	1.07425	.96400	1.03734	3
58	.86827	1.15172	.89935	1.11191	.93143	1.07362	.96457	1.03674	2
59	.86878	1.15104	.89988	1.11126	.93197	1.07299	.96513	1.03613	1
60	.86929	1.15037	.90040	1.11061	.93252	1.07237	.96569	1.03553	0
′	49° cotan	49° tan	48° cotan	48° tan	47° cotan	47° tan	46° cotan	46° tan	′

′	44° tan	cotan	′	′	44° tan	cotan	′	′	44° tan	cotan	′
0	.96569	1.03553	60	21	.97756	1.02295	39	41	.98901	1.01112	19
1	.96625	1.03493	59	22	.97813	1.02236	38	42	.98958	1.01053	18
2	.96681	1.03433	58	23	.97870	1.02176	37	43	.99016	1.00994	17
3	.96738	1.03372	57	24	.97927	1.02117	36	44	.99073	1.00935	16
4	.96794	1.03312	56	25	.97984	1.02057	35	45	.99131	1.00876	15
5	.96850	1.03252	55	26	.98041	1.01998	34	46	.99189	1.00818	14
6	.96907	1.03192	54	27	.98098	1.01939	33	47	.99247	1.00759	13
7	.96963	1.03132	53	28	.98155	1.01879	32	48	.99304	1.00701	12
8	.97020	1.03072	52	29	.98213	1.01820	31	49	.99362	1.00642	11
9	.97076	1.03012	51	30	.98270	1.01761	30	50	.99420	1.00583	10
10	.97133	1.02952	50								
				31	.98327	1.01702	29	51	.99478	1.00525	9
11	.97189	1.02892	49	32	.98384	1.01642	28	52	.99536	1.00467	8
12	.97246	1.02832	48	33	.98441	1.01583	27	53	.99594	1.00408	7
13	.97302	1.02772	47	34	.98499	1.01524	26	54	.99652	1.00350	6
14	.97359	1.02713	46	35	.98556	1.01465	25	55	.99710	1.00291	5
15	.97416	1.02653	45	36	.98613	1.01406	24	56	.99768	1.00233	4
16	.97472	1.02593	44	37	.98671	1.01347	23	57	.99826	1.00175	3
17	.97529	1.02533	43	38	.98728	1.01288	22	58	.99884	1.00116	2
18	.97586	1.02474	42	39	.98786	1.01229	21	59	.99942	1.00058	1
19	.97643	1.02414	41	40	.98843	1.01170	20	60	1	1	0
20	.97700	1.02355	40								
′	cotan 45°	tan	′	′	cotan 45°	tan	′	′	cotan 45°	tan	′

NATURAL SINES AND COSINES

′	0° sine	cosine	′	′	0° sine	cosine	′	′	0° sine	cosine	′
0	.00000	1	60	21	.00611	.99998	39	41	.01193	.99993	19
1	.00029	1	59	22	.00640	.99998	38	42	.01222	.99993	18
2	.00058	1	58	23	.00669	.99998	37	43	.01251	.99992	17
3	.00087	1	57	24	.00698	.99998	36	44	.01280	.99992	16
4	.00116	1	56	25	.00727	.99997	35	45	.01309	.99991	15
5	.00145	1	55	26	.00756	.99997	34	46	.01338	.99991	14
6	.00175	1	54	27	.00785	.99997	33	47	.01367	.99991	13
7	.00204	1	53	28	.00814	.99997	32	48	.01396	.99990	12
8	.00233	1	52	29	.00844	.99996	31	49	.01425	.99990	11
9	.00262	1	51	30	.00873	.99996	30	50	.01454	.99989	10
10	.00291	1	50								
				31	.00902	.99996	29	51	.01483	.99989	9
11	.00320	.99999	49	32	.00931	.99996	28	52	.01513	.99989	8
12	.00349	.99999	48	33	.00960	.99995	27	53	.01542	.99988	7
13	.00378	.99999	47	34	.00989	.99995	26	54	.01571	.99988	6
14	.00407	.99999	46	35	.01018	.99995	25	55	.01600	.99987	5
15	.00436	.99999	45	36	.01047	.99995	24	56	.01629	.99987	4
16	.00465	.99999	44	37	.01076	.99994	23	57	.01658	.99986	3
17	.00495	.99999	43	38	.01105	.99994	22	58	.01687	.99986	2
18	.00524	.99999	42	39	.01134	.99994	21	59	.01716	.99985	1
19	.00553	.99998	41	40	.01164	.99993	20	60	.01745	.99985	0
20	.00582	.99998	40								
′	cosine 89°	sine	′	′	cosine 89°	sine	′	′	cosine 89°	sine	′

′	1° sine	1° cosine	2° sine	2° cosine	3° sine	3° cosine	4° sine	4° cosine	′
0	.01745	.99985	.03490	.99939	.05234	.99863	.06976	.99756	60
1	.01774	.99984	.03519	.99938	.05263	.99861	.07005	.99754	59
2	.01803	.99984	.03548	.99937	.05292	.99860	.07034	.99752	58
3	.01832	.99983	.03577	.99936	.05321	.99858	.07063	.99750	57
4	.01862	.99983	.03606	.99935	.05350	.99857	.07092	.99748	56
5	.01891	.99982	.03635	.99934	.05379	.99855	.07121	.99746	55
6	.01920	.99982	.03664	.99933	.05408	.99854	.07150	.99744	54
7	.01949	.99981	.03693	.99932	.05437	.99852	.07179	.99742	53
8	.01978	.99980	.03723	.99931	.05466	.99851	.07208	.99740	52
9	.02007	.99980	.03752	.99930	.05495	.99849	.07237	.99738	51
10	.02036	.99979	.03781	.99929	.05524	.99847	.07266	.99736	50
11	.02065	.99979	.03810	.99927	.05553	.99846	.07295	.99734	49
12	.02094	.99978	.03839	.99926	.05582	.99844	.07324	.99731	48
13	.02123	.99977	.03868	.99925	.05611	.99842	.07353	.99729	47
14	.02152	.99977	.03897	.99924	.05640	.99841	.07382	.99727	46
15	.02181	.99976	.03926	.99923	.05669	.99839	.07411	.99725	45
16	.02211	.99976	.03955	.99922	.05698	.99838	.07440	.99723	44
17	.02240	.99975	.03984	.99921	.05727	.99836	.07469	.99721	43
18	.02269	.99974	.04013	.99919	.05756	.99834	.07498	.99719	42
19	.02298	.99974	.04042	.99918	.05785	.99833	.07527	.99716	41
20	.02327	.99973	.04071	.99917	.05814	.99831	.07556	.99714	40
21	.02356	.99972	.04100	.99916	.05844	.99829	.07585	.99712	39
22	.02385	.99972	.04129	.99915	.05873	.99827	.07614	.99710	38
23	.02414	.99971	.04159	.99913	.05902	.99826	.07643	.99708	37
24	.02443	.99970	.04188	.99912	.05931	.99824	.07672	.99705	36
25	.02472	.99969	.04217	.99911	.05960	.99822	.07701	.99703	35
26	.02501	.99969	.04246	.99910	.05989	.99821	.07730	.99701	34
27	.02530	.99968	.04275	.99909	.06018	.99819	.07759	.99699	33
28	.02560	.99967	.04304	.99907	.06047	.99817	.07788	.99696	32
29	.02589	.99966	.04333	.99906	.06076	.99815	.07817	.99694	31
30	.02618	.99966	.04362	.99905	.06105	.99813	.07846	.99692	30
31	.02647	.99965	.04391	.99904	.06134	.99812	.07875	.99689	29
32	.02676	.99964	.04420	.99902	.06163	.99810	.07904	.99687	28
33	.02705	.99963	.04449	.99901	.06192	.99808	.07933	.99685	27
34	.02734	.99963	.04478	.99900	.06221	.99806	.07962	.99683	26
35	.02763	.99962	.04507	.99898	.06250	.99804	.07991	.99680	25
36	.02792	.99961	.04536	.99897	.06279	.99803	.08020	.99678	24
37	.02821	.99960	.04565	.99896	.06308	.99801	.08049	.99676	23
38	.02850	.99959	.04594	.99894	.06337	.99799	.08078	.99673	22
39	.02879	.99959	.04623	.99893	.06366	.99797	.08107	.99671	21
40	.02908	.99958	.04653	.99892	.06395	.99795	.08136	.99668	20
41	.02938	.99957	.04682	.99890	.06424	.99793	.08165	.99666	19
42	.02967	.99956	.04711	.99889	.06453	.99792	.08194	.99664	18
43	.02996	.99955	.04740	.99888	.06482	.99790	.08223	.99661	17
44	.03025	.99954	.04769	.99886	.06511	.99788	.08252	.99659	16
45	.03054	.99953	.04798	.99885	.06540	.99786	.08281	.99657	15
46	.03083	.99952	.04827	.99883	.06569	.99784	.08310	.99654	14
47	.03112	.99952	.04856	.99882	.06598	.99782	.08339	.99652	13
48	.03141	.99951	.04885	.99881	.06627	.99780	.08368	.99649	12
49	.03170	.99950	.04914	.99879	.06656	.99778	.08397	.99647	11
50	.03199	.99949	.04943	.99878	.06685	.99776	.08426	.99644	10
51	.03228	.99948	.04972	.99876	.06714	.99774	.08455	.99642	9
52	.03257	.99947	.05001	.99875	.06743	.99772	.08484	.99639	8
53	.03286	.99946	.05030	.99873	.06773	.99770	.08513	.99637	7
54	.03316	.99945	.05059	.99872	.06802	.99768	.08542	.99635	6
55	.03345	.99944	.05088	.99870	.06831	.99766	.08571	.99632	5
56	.03374	.99943	.05117	.99869	.06860	.99764	.08600	.99630	4
57	.03403	.99942	.05146	.99867	.06889	.99762	.08629	.99627	3
58	.03432	.99941	.05175	.99866	.06918	.99760	.08658	.99625	2
59	.03461	.99940	.05205	.99864	.06947	.99758	.08687	.99622	1
60	.03490	.99939	.05234	.99863	.06976	.99756	.08716	.99619	0
′	88° cosine	88° sine	87° cosine	87° sine	86° cosine	86° sine	85° cosine	85° sine	′

′	5° sine	5° cosine	6° sine	6° cosine	7° sine	7° cosine	8° sine	8° cosine	′
0	.08716	.99619	.10453	.99452	.12187	.99255	.13917	.99027	60
1	.08745	.99617	.10482	.99449	.12216	.99251	.13946	.99023	59
2	.08774	.99614	.10511	.99446	.12245	.99248	.13975	.99019	58
3	.08803	.99612	.10540	.99443	.12274	.99244	.14004	.99015	57
4	.08831	.99609	.10569	.99440	.12302	.99240	.14033	.99011	56
5	.08860	.99607	.10597	.99437	.12331	.99237	.14061	.99006	55
6	.08889	.99604	.10626	.99434	.12360	.99233	.14090	.99002	54
7	.08918	.99602	.10655	.99431	.12389	.99230	.14119	.98998	53
8	.08947	.99599	.10684	.99428	.12418	.99226	.14148	.98994	52
9	.08976	.99596	.10713	.99424	.12447	.99222	.14177	.98990	51
10	.09005	.99594	.10742	.99421	.12476	.99219	.14205	.98986	50
11	.09034	.99591	.10771	.99418	.12504	.99215	.14234	.98982	49
12	.09063	.99588	.10800	.99415	.12533	.99211	.14263	.98978	48
13	.09092	.99586	.10829	.99412	.12562	.99208	.14292	.98973	47
14	.09121	.99583	.10858	.99409	.12591	.99204	.14320	.98969	46
15	.09150	.99580	.10887	.99406	.12620	.99200	.14349	.98965	45
16	.09179	.99578	.10916	.99402	.12649	.99197	.14378	.98961	44
17	.09208	.99575	.10945	.99399	.12678	.99193	.14407	.98957	43
18	.09237	.99572	.10973	.99396	.12706	.99189	.14436	.98953	42
19	.09266	.99570	.11002	.99393	.12735	.99186	.14464	.98948	41
20	.09295	.99567	.11031	.99390	.12764	.99182	.14493	.98944	40
21	.09324	.99564	.11060	.99386	.12793	.99178	.14522	.98940	39
22	.09353	.99562	.11089	.99383	.12822	.99175	.14551	.98936	38
23	.09382	.99559	.11118	.99380	.12851	.99171	.14580	.98931	37
24	.09411	.99556	.11147	.99377	.12880	.99167	.14608	.98927	36
25	.09440	.99553	.11176	.99374	.12908	.99163	.14637	.98923	35
26	.09469	.99551	.11205	.99370	.12937	.99160	.14666	.98919	34
27	.09498	.99548	.11234	.99367	.12966	.99156	.14695	.98914	33
28	.00527	.99545	.11263	.99364	.12995	.99152	.14723	.98910	32
29	.09556	.99542	.11291	.99360	.13024	.99148	.14752	.98906	31
30	.09585	.99540	.11320	.99357	.13053	.99144	.14781	.98902	30
31	.09614	.99537	.11349	.99354	.13081	.99141	.14810	.98897	29
32	.09642	.99534	.11378	.99351	.13110	.99137	.14838	.98893	28
33	.09671	.99531	.11407	.99347	.13139	.99133	.14867	.98889	27
34	.09700	.99528	.11436	.99344	.13168	.99129	.14896	.98884	26
35	.09729	.99526	.11465	.99341	.13197	.99125	.14925	.98880	25
36	.09758	.99523	.11494	.99337	.13226	.99122	.14954	.98876	24
37	.09787	.99520	.11523	.99334	.13254	.99118	.14982	.98871	23
38	.09816	.99517	.11552	.99331	.13283	.99114	.15011	.98867	22
39	.09845	.99514	.11580	.99327	.13312	.99110	.15040	.98863	21
40	.09874	.99511	.11609	.99324	.13341	.99106	.15069	.98858	20
41	.09903	.99508	.11638	.99320	.13370	.99102	.15097	.98854	19
42	.09932	.99506	.11667	.99317	.13399	.99098	.15126	.98849	18
43	.09961	.99503	.11696	.99314	.13427	.99094	.15155	.98845	17
44	.09990	.99500	.11725	.99310	.13456	.99091	.15184	.98841	16
45	.10019	.99497	.11754	.99307	.13485	.99087	.15212	.98836	15
46	.10048	.99494	.11783	.99303	.13514	.99083	.15241	.98832	14
47	.10077	.99491	.11812	.99300	.13543	.99079	.15270	.98827	13
48	.10106	.99488	.11840	.99297	.13572	.99075	.15299	.98823	12
49	.10135	.99485	.11869	.99293	.13600	.99071	.15327	.98818	11
50	.10164	.99482	.11898	.99290	.13629	.99067	.15356	.98814	10
51	.10192	.99479	.11927	.99286	.13658	.99063	.15385	.98809	9
52	.10221	.99476	.11956	.99283	.13687	.99059	.15414	.98805	8
53	.10250	.99473	.11985	.99279	.13716	.99055	.15442	.98800	7
54	.10279	.99470	.12014	.99276	.13744	.99051	.15471	.98796	6
55	.10308	.99467	.12043	.99272	.13773	.99047	.15500	.98791	5
56	.10337	.99464	.12071	.99269	.13802	.99043	.15529	.98787	4
57	.10366	.99461	.12100	.99265	.13831	.99039	.15557	.98782	3
58	.10395	.99458	.12129	.99262	.13860	.99035	.15586	.98778	2
59	.10424	.99455	.12158	.99258	.13889	.99031	.15615	.98773	1
60	.10453	.99452	.12187	.99255	.13917	.99027	.15643	.98769	0
′	84° cosine	84° sine	83° cosine	83° sine	82° cosine	82° sine	81° cosine	81° sine	′

′	9° sine	9° cosine	10° sine	10° cosine	11° sine	11° cosine	12° sine	12° cosine	′
0	.15643	.98769	.17365	.98481	.19081	.98163	.20791	.97815	60
1	.15672	.98764	.17393	.98476	.19109	.98157	.20820	.97809	59
2	.15701	.98760	.17422	.98471	.19138	.98152	.20848	.97803	58
3	.15730	.98755	.17451	.98466	.19167	.98146	.20877	.97797	57
4	.15758	.98751	.17479	.98461	.19195	.98140	.20905	.97791	56
5	.15787	.98746	.17508	.98455	.19224	.98135	.20933	.97784	55
6	.15816	.98741	.17537	.98450	.19252	.98129	.20962	.97778	54
7	.15845	.98737	.17565	.98445	.19281	.98124	.20990	.97772	53
8	.15873	.98732	.17594	.98440	.19309	.98118	.21019	.97766	52
9	.15902	.98728	.17623	.98435	.19338	.98112	.21047	.97760	51
10	.15931	.98723	.17651	.98430	.19366	.98107	.21076	.97754	50
11	.15959	.98718	.17680	.98425	.19395	.98101	.21104	.97748	49
12	.15988	.98714	.17708	.98420	.19423	.98096	.21132	.97742	48
13	.16017	.98709	.17737	.98414	.19452	.98090	.21161	.97735	47
14	.16046	.98704	.17766	.98409	.19481	.98084	.21189	.97729	46
15	.16074	.98700	.17794	.98404	.19509	.98079	.21218	.97723	45
16	.16103	.98695	.17823	.98399	.19538	.98073	.21246	.97717	44
17	.16132	.98690	.17852	.98394	.19566	.98067	.21275	.97711	43
18	.16160	.98686	.17880	.98389	.19595	.98061	.21303	.97705	42
19	.16189	.98681	.17909	.98383	.19623	.98056	.21331	.97698	41
20	.16218	.98676	.17937	.98378	.19652	.98050	.21360	.97692	40
21	.16246	.98671	.17966	.98373	.19680	.98044	.21388	.97686	39
22	.16275	.98667	.17995	.98368	.19709	.98039	.21417	.97680	38
23	.16304	.98662	.18023	.98362	.19737	.98033	.21445	.97673	37
24	.16333	.98657	.18052	.98357	.19766	.98027	.21474	.97667	36
25	.16361	.98652	.18081	.98352	.19794	.98021	.21502	.97661	35
26	.16390	.98648	.18109	.98347	.19823	.98016	.21530	.97655	34
27	.16419	.98643	.18138	.98341	.19851	.98010	.21559	.97648	33
28	.16447	.98638	.18166	.98336	.19880	.98004	.21587	.97642	32
29	.16476	.98633	.18195	.98331	.19908	.97997	.21616	.97636	31
30	.16505	.98629	.18224	.98325	.19937	.97992	.21644	.97630	30
31	.16533	.98624	.18252	.98320	.19965	.97987	.21672	.97623	29
32	.16562	.98619	.18281	.98315	.19994	.97981	.21701	.97617	28
33	.16591	.98614	.18309	.98310	.20022	.97975	.21729	.97611	27
34	.16620	.98609	.18338	.98304	.20051	.97969	.21758	.97604	26
35	.16648	.98604	.18367	.98299	.20079	.97963	.21786	.97598	25
36	.16677	.98600	.18395	.98294	.20108	.97958	.21814	.97592	24
37	.16706	.98595	.18424	.98288	.20136	.97952	.21843	.97585	23
38	.16734	.98590	.18452	.98283	.20165	.97946	.21871	.97579	22
39	.16763	.98585	.18481	.98277	.20193	.97940	.21899	.97573	21
40	.16792	.98580	.18509	.98272	.20222	.97934	.21928	.97566	20
41	.16820	.98575	.18538	.98267	.20250	.97928	.21956	.97560	19
42	.16849	.98570	.18567	.98261	.20279	.97922	.21985	.97553	18
43	.16878	.98565	.18595	.98256	.20307	.97916	.22013	.97547	17
44	.16906	.98561	.18624	.98250	.20336	.97910	.22041	.97541	16
45	.16935	.98556	.18652	.98245	.20364	.97905	.22070	.97534	15
46	.16964	.98551	.18681	.98240	.20393	.97899	.22098	.97528	14
47	.16992	.98546	.18710	.98234	.20421	.97893	.22126	.97521	13
48	.17021	.98541	.18738	.98229	.20450	.97887	.22155	.97515	12
49	.17050	.98536	.18767	.98223	.20478	.97881	.22183	.97508	11
50	.17078	.98531	.18795	.98218	.20507	.97875	.22212	.97502	10
51	.17107	.98526	.18824	.98212	.20535	.97869	.22240	.97496	9
52	.17136	.98521	.18852	.98207	.20563	.97863	.22268	.97489	8
53	.17164	.98516	.18881	.98201	.20592	.97857	.22297	.97483	7
54	.17193	.98511	.18910	.98196	.20620	.97851	.22325	.97476	6
55	.17222	.98506	.18938	.98190	.20649	.97845	.22353	.97470	5
56	.17250	.98501	.18967	.98185	.20677	.97839	.22382	.97463	4
57	.17279	.98496	.18995	.98179	.20706	.97833	.22410	.97457	3
58	.17308	.98491	.19024	.98174	.20734	.97827	.22438	.97450	2
59	.17336	.98486	.19052	.98168	.20763	.97821	.22467	.97444	1
60	.17365	.98481	.19081	.98163	.20791	.97815	.22495	.97437	0
′	80° cosine	80° sine	79° cosine	79° sine	78° cosine	78° sine	77° cosine	77° sine	′

′	13° sine	13° cosine	14° sine	14° cosine	15° sine	15° cosine	16° sine	16° cosine	′
0	.22495	.97437	.24192	.97030	.25882	.96593	.27564	.96126	60
1	.22523	.97430	.24220	.97023	.25910	.96585	.27592	.96118	59
2	.22552	.97424	.24249	.97015	.25938	.96578	.27620	.86110	58
3	.22580	.97417	.24277	.97008	.25966	.96570	.27648	.96102	57
4	.22608	.97411	.24305	.97001	.25994	.96562	.27676	.96094	56
5	.22637	.97404	.24333	.96994	.26022	.96555	.27704	.96086	55
6	.22665	.97398	.24362	.96987	.26050	.96547	.27731	.96078	54
7	.22693	.97391	.24390	.96980	.26079	.96540	.27759	.96070	53
8	.22722	.97384	.24418	.96973	.26107	.96532	.27787	.96062	52
9	.22750	.97378	.24446	.96966	.26135	.96524	.27815	.96054	51
10	.22778	.97371	.24474	.96959	.26163	.96517	.27843	.96046	50
11	.22807	.97365	.24503	.96952	.26191	.96509	.27871	.96037	49
12	.22835	.97358	.24531	.96945	.26219	.96502	.27899	.96029	48
13	.22863	.97351	.24559	.96937	.26247	.96494	.27927	.96021	47
14	.22892	.97345	.24587	.96930	.26275	.96486	.27955	.96013	46
15	.22920	.97338	.24615	.96923	.26303	.96479	.27983	.96005	45
16	.22948	.97331	.24644	.96916	.26331	.96471	.28011	.95997	44
17	.22977	.97325	.24672	.96909	.26359	.96463	.28039	.95989	43
18	.23005	.97318	.24700	.96902	.26387	.96456	.28067	.95981	42
19	.23033	.97311	.24728	.96894	.26415	.96448	.28095	.95972	41
20	.23062	.97304	.24756	.96887	.26443	.96440	.28123	.95964	40
21	.23090	.97298	.24784	.96880	.26471	.96433	.28150	.95956	39
22	.23118	.97291	.24813	.96873	.26500	.96425	.28178	.95948	38
23	.23146	.97284	.24841	.96866	.26528	.96417	.28206	.95940	37
24	.23175	.97278	.24869	.96858	.26556	.96410	.28234	.95931	36
25	.23203	.97271	.24897	.96851	.26584	.96402	.28262	.95923	35
26	.23231	.97264	.24925	.96844	.26612	.96394	.28290	.95915	34
27	.23260	.97257	.24954	.96837	.26640	.96386	.28318	.95907	33
28	.23288	.97251	.24982	.96829	.26668	.96379	.28346	.95898	32
29	.23316	.97244	.25010	.96822	.26696	.96371	.28374	.95890	31
30	.23345	.97237	.25038	.96815	.26724	.96363	.28402	.95882	30
31	.23373	.97230	.25066	.96807	.26752	.96355	.28429	.95874	29
32	.23401	.97223	.25094	.96800	.26780	.96347	.28457	.95865	28
33	.23429	.97217	.25122	.96793	.26808	.96340	.28485	.95857	27
34	.23458	.97210	.25151	.96786	.26836	.96332	.28513	.95849	26
35	.23486	.97203	.25179	.96778	.26864	.96324	.28541	.95841	25
36	.23514	.97196	.25207	.96771	.26892	.96316	.28569	.95832	24
37	.23542	.97189	.25235	.96764	.26920	.96308	.28597	.95824	23
38	.23571	.97182	.25263	.96756	.26948	.96301	.28625	.95816	22
39	.23599	.97176	.25291	.96749	.26976	.96293	.28652	.95807	21
40	.23627	.97169	.25320	.96742	.27004	.96285	.28680	.95799	20
41	.23656	.97162	.25348	.96734	.27032	.96277	.28708	.95791	19
42	.23684	.97155	.25376	.96727	.27060	.96269	.28736	.95782	18
43	.23712	.97148	.25404	.96719	.27088	.96261	.28764	.95774	17
44	.23740	.97141	.25432	.96712	.27116	.96253	.28792	.95766	16
45	.23769	.97134	.25460	.96705	.27144	.96246	.28820	.95757	15
46	.23797	.97127	.25488	.96697	.27172	.96238	.28847	.95749	14
47	.23825	.97120	.25516	.96690	.27200	.96230	.28875	.95740	13
48	.23853	.97113	.25545	.96682	.27228	.96222	.28903	.95732	12
49	.23882	.97106	.25573	.96675	.27256	.96214	.28931	.95724	11
50	.23910	.97100	.25601	.96667	.27284	.96206	.28959	.95715	10
51	.23938	.97093	.25629	.96660	.27312	.96198	.28987	.95707	9
52	.23966	.97086	.25657	.96653	.27340	.96190	.29015	.95698	8
53	.23995	.97079	.25685	.96645	.27368	.96182	.29042	.95690	7
54	.24023	.97072	.25713	.96638	.27396	.96174	.29070	.95681	6
55	.24051	.97065	.25741	.96630	.27424	.96166	.29098	.95673	5
56	.24079	.97058	.25769	.96623	.27452	.96158	.29126	.95664	4
57	.24108	.97051	.25798	.96615	.27480	.96150	.29154	.95656	3
58	.24136	.97044	.25826	.96608	.27508	.96142	.29182	.95647	2
59	.24164	.97037	.25854	.96600	.27536	.96134	.29209	.95639	1
60	.24192	.97030	.25882	.96593	.27564	.96126	.29237	.95630	0
′	76° cosine	76° sine	75° cosine	75° sine	74° cosine	74° sine	73° cosine	73° sine	′

	17°		18°		19°		20°		
′	sine	cosine	sine	cosine	sine	cosine	sine	cosine	′
0	.29237	.95630	.30902	.95106	.32557	.94552	.34202	.93969	60
1	.29265	.95622	.30929	.95097	.32584	.94542	.34229	.93959	59
2	.29293	.95613	.30957	.95088	.32612	.94533	.34257	.93949	58
3	.29321	.95605	.30985	.95079	.32639	.94523	.34284	.93939	57
4	.29348	.95596	.31012	.95070	.32667	.94514	.34311	.93929	56
5	.29376	.95588	.31040	.95061	.32694	.94504	.34339	.93919	55
6	.29404	.95579	.31068	.95052	.32722	.94495	.34366	.93909	54
7	.29432	.95571	.31095	.95043	.32749	.94485	.34393	.93899	53
8	.29460	.95562	.31123	.95033	.32777	.94476	.34421	.93889	52
9	.29487	.95554	.31151	.95024	.32804	.94466	.34448	.93879	51
10	.29515	.95545	.31178	.95015	.32832	.94457	.34475	.93869	50
11	.29543	.95536	.31206	.95006	.32859	.94447	.34503	.93859	49
12	.29571	.95528	.31233	.94997	.32887	.94438	.34530	.93849	48
13	.29599	.95519	.31261	.94988	.32914	.94428	.34557	.93839	47
14	.29626	.95511	.31289	.94979	.32942	.94418	.34584	.93829	46
15	.29654	.95502	.31316	.94970	.32969	.94409	.34612	.93819	45
16	.29682	.95493	.31344	.94961	.32997	.94399	.34639	.93809	44
17	.29710	.95485	.31372	.94952	.33024	.94390	.34666	.93799	43
18	.29737	.95476	.31399	.94943	.33051	.94380	.34694	.93789	42
19	.29765	.95467	.31427	.94933	.33079	.94370	.34721	.93779	41
20	.29793	.95459	.31454	.94924	.33106	.94361	.34748	.93769	40
21	.29821	.95450	.31482	.94915	.33134	.94351	.34775	.93759	39
22	.29849	.95441	.31510	.94906	.33161	.94342	.34803	.93748	38
23	.29876	.95433	.31537	.94897	.33189	.94332	.34830	.93738	37
24	.29904	.95424	.31565	.94888	.33216	.94322	.34857	.93728	36
25	.29932	.95415	.31593	.94878	.33244	.94313	.34884	.93718	35
26	.29960	.95407	.31620	.94869	.33271	.94303	.34912	.93708	34
27	.29987	.95398	.31648	.94860	.33298	.94293	.34939	.93698	33
28	.30015	.95389	.31675	.94851	.33326	.94284	.34966	.93688	32
29	.30043	.95380	.31703	.94842	.33353	.94274	.34993	.93677	31
30	.30071	.95372	.31730	.94832	.33381	.94264	.35021	.93667	30
31	.30098	.95363	.31758	.94823	.33408	.94254	.35048	.93657	29
32	.30126	.95354	.31786	.94814	.33436	.94245	.35075	.93647	28
33	.30154	.95345	.31813	.94805	.33463	.94235	.35102	.93637	27
34	.30182	.95337	.31841	.94795	.33490	.94225	.35130	.93626	26
35	.30209	.95328	.31868	.94786	.33518	.94215	.35157	.93616	25
36	.30237	.95319	.31896	.94777	.33545	.94206	.35184	.93606	24
37	.30265	.95310	.31923	.94768	.33573	.94196	.35211	.93596	23
38	.30292	.95301	.31951	.94758	.33600	.94186	.35239	.93585	22
39	.30320	.95293	.31979	.94749	.33627	.94176	.35266	.93575	21
40	.30348	.95284	.32006	.94740	.33655	.94167	.35293	.93565	20
41	.30376	.95275	.32034	.94730	.33682	.94157	.35320	.93555	19
42	.30403	.95266	.32061	.94721	.33710	.94147	.35347	.93544	18
43	.30431	.95257	.32089	.94712	.33737	.94137	.35375	.93534	17
44	.30459	.95248	.32116	.94702	.33764	.94127	.35402	.93524	16
45	.30486	.95240	.32144	.94693	.33792	.94118	.35429	.93514	15
46	.30514	.95231	.32171	.94684	.33819	.94108	.35456	.93503	14
47	.30542	.95222	.32199	.94674	.33846	.94098	.35484	.93493	13
48	.30570	.95213	.32227	.94665	.33874	.94088	.35511	.93483	12
49	.30597	.95204	.32254	.94656	.33901	.94078	.35538	.93472	11
50	.30625	.95195	.32282	.94646	.33929	.94068	.35565	.93462	10
51	.30653	.95186	.32309	.94637	.33956	.94058	.35592	.93452	9
52	.30680	.95177	.32337	.94627	.33983	.94049	.35619	.93441	8
53	.30708	.95168	.32364	.94618	.34011	.94039	.35647	.93431	7
54	.30736	.95159	.32392	.94609	.34038	.94029	.35674	.93420	6
55	.30763	.95150	.32419	.94599	.34065	.94019	.35701	.93410	5
56	.30791	.95142	.32447	.94590	.34093	.94009	.35728	.93400	4
57	.30819	.95133	.32474	.94580	.34120	.93999	.35755	.93389	3
58	.30846	.95124	.32502	.94571	.34147	.93989	.35782	.93379	2
59	.30874	.95115	.32529	.94561	.34175	.93979	.35810	.93368	1
60	.30902	.95106	.32557	.94552	.34202	.93969	.35837	.93358	0
′	cosine	sine	cosine	sine	cosine	sine	cosine	sine	′
	72°		71°		70°		69°		

′	21° sine	21° cosine	22° sine	22° cosine	23° sine	23° cosine	24° sine	24° cosine	′
0	.35837	.93358	.37461	.92718	.39073	.92050	.40674	.91355	60
1	.35864	.93348	.37488	.92707	.39100	.92039	.40700	.91343	59
2	.35891	.93337	.37515	.92697	.39127	.92028	.40727	.91331	58
3	.35918	.93327	.37542	.92686	.39153	.92016	.40753	.91319	57
4	.35945	.93316	.37569	.92675	.39180	.92005	.40780	.91307	56
5	.35973	.93306	.37595	.92664	.39207	.91994	.40806	.91295	55
6	.36000	.93295	.37622	.92653	.39234	.91982	.40833	.91283	54
7	.36027	.93285	.37649	.92642	.39260	.91971	.40860	.91272	53
8	.36054	.93274	.37676	.92631	.39287	.91959	.40886	.91260	52
9	.36081	.93264	.37703	.92620	.39314	.91948	.40913	.91248	51
10	.36108	.93253	.37730	.92609	.39341	.91936	.40939	.91236	50
11	.36135	.93243	.37757	.92598	.39367	.91925	.40966	.91224	49
12	.36162	.93232	.37784	.92587	.39394	.91914	.40992	.91212	48
13	.36190	.93222	.37811	.92576	.39421	.91902	.41019	.91200	47
14	.36217	.93211	.37838	.92565	.39448	.91891	.41045	.91188	46
15	.36244	.93201	.37865	.92554	.39474	.91879	.41072	.91176	45
16	.36271	.93190	.37892	.92543	.39501	.91868	.41098	.91164	44
17	.36298	.93180	.37919	.92532	.39528	.91856	.41125	.91152	43
18	.36325	.93169	.37946	.92521	.39555	.91845	.41151	.91140	42
19	.36352	.93159	.37973	.92510	.39581	.91833	.41178	.91128	41
20	.36379	.93148	.37999	.92499	.39608	.91822	.41204	.91116	40
21	.36406	.93137	.38026	.92488	.39635	.91810	.41231	.91104	39
22	.36434	.93127	.38053	.92477	.39661	.91799	.41257	.91092	38
23	.36461	.93116	.38080	.92466	.39688	.91787	.41284	.91080	37
24	.36488	.93106	.38107	.92455	.39715	.91775	.41310	.91068	36
25	.36515	.93095	.38134	.92444	.39741	.91764	.41337	.91056	35
26	.36542	.93084	.38161	.92432	.39768	.91752	.41363	.91044	34
27	.36569	.93074	.38188	.92421	.39795	.91741	.41390	.91032	33
28	.36596	.93063	.38215	.92410	.39822	.91729	.41416	.91020	32
29	.36623	.93052	.38241	.92399	.39848	.91718	.41443	.91008	31
30	.36650	.93042	.38268	.92388	.39875	.91706	.41469	.90996	30
31	.36677	.93031	.38295	.92377	.39902	.91694	.41496	.90984	29
32	.36704	.93020	.38322	.92366	.39928	.91683	.41522	.90972	28
33	.36731	.93010	.38349	.92355	.39955	.91671	.41549	.90960	27
34	.36758	.92999	.38376	.92343	.39982	.91660	.41575	.90948	26
35	.36785	.92988	.38403	.92332	.40008	.91648	.41602	.90936	25
36	.36812	.92978	.38430	.92321	.40035	.91636	.41628	.90924	24
37	.36839	.92967	.38456	.92310	.40062	.91625	.41655	.90911	23
38	.36867	.92956	.38483	.92299	.40088	.91613	.41681	.90899	22
39	.36894	.92945	.38510	.92287	.40115	.91601	.41707	.90887	21
40	.36921	.92935	.38537	.92276	.40141	.91590	.41734	.90875	20
41	.36948	.92924	.38564	.92265	.40168	.91578	.41760	.90863	19
42	.36975	.92913	.38591	.92254	.40195	.91566	.41787	.90851	18
43	.37002	.92902	.38617	.92243	.40221	.91555	.41813	.90839	17
44	.37029	.92892	.38644	.92231	.40248	.91543	.41840	.90826	16
45	.37056	.92881	.38671	.92220	.40275	.91531	.41866	.90814	15
46	.37083	.92870	.38698	.92209	.40301	.91519	.41892	.90802	14
47	.37110	.92859	.38725	.92198	.40328	.91508	.41919	.90790	13
48	.37137	.92849	.38752	.92186	.40355	.91496	.41945	.90778	12
49	.37164	.92838	.38778	.92175	.40381	.91484	.41972	.90766	11
50	.37191	.92827	.38805	.92164	.40408	.91472	.41998	.90753	10
51	.37218	.92816	.38832	.92152	.40434	.91461	.42024	.90741	9
52	.37245	.92805	.38859	.92141	.40461	.91449	.42051	.90729	8
53	.37272	.92794	.38886	.92130	.40488	.91437	.42077	.90717	7
54	.37299	.92784	.38912	.92119	.40514	.91425	.42104	.90704	6
55	.37326	.92773	.38939	.92107	.40541	.91414	.42130	.90692	5
56	.37353	.92762	.38966	.92096	.40567	.91402	.42156	.90680	4
57	.37380	.92751	.38993	.92085	.40594	.91390	.42183	.90668	3
58	.37407	.92740	.39020	.92073	.40621	.91378	.42209	.90655	2
59	.37434	.92729	.39046	.92062	.40647	.91366	.42235	.90643	1
60	.37461	.92718	.39073	.92050	.40674	.91355	.42262	.90631	0
′	cosine 68°	sine 68°	cosine 67°	sine 67°	cosine 66°	sine 66°	cosine 65°	sine 65°	′

′	25° sine	25° cosine	26° sine	26° cosine	27° sine	27° cosine	28° sine	28° cosine	′
0	.42262	.90631	.43837	.89879	.45399	.89101	.46947	.88295	60
1	.42288	.90618	.43863	.89867	.45425	.89087	.46973	.88281	59
2	.42315	.90606	.43889	.89854	.45451	.89074	.46999	.88267	58
3	.42341	.90594	.43916	.89841	.45477	.89061	.47024	.88254	57
4	.42367	.90582	.43942	.89828	.45503	.89048	.47050	.88240	56
5	.42394	.90569	.43968	.89816	.45529	.89035	.47076	.88226	55
6	.42420	.90557	.43994	.89803	.45554	.89021	.47101	.88213	54
7	.42446	.90545	.44020	.89790	.45580	.89008	.47127	.88199	53
8	.42473	.90532	.44046	.89777	.45606	.88995	.47153	.88185	52
9	.42499	.90520	.44072	.89764	.45632	.88981	.47178	.88172	51
10	.42525	.90507	.44098	.89752	.45658	.88968	.47204	.88158	50
11	.42552	.90495	.44124	.89739	.45684	.88955	.47229	.88144	49
12	.42578	.90483	.44151	.89726	.45710	.88942	.47255	.88130	48
13	.42604	.90470	.44177	.89713	.45736	.88928	.47281	.88117	47
14	.42631	.90458	.44203	.89700	.45762	.88915	.47306	.88103	46
15	.42657	.90446	.44229	.89687	.45787	.88902	.47332	.88089	45
16	.42683	.90433	.44255	.89674	.45813	.88888	.47358	.88075	44
17	.42709	.90421	.44281	.89662	.45839	.88875	.47383	.88062	43
18	.42736	.90408	.44307	.89649	.45865	.88862	.47409	.88048	42
19	.42762	.90396	.44333	.89636	.45891	.88848	.47434	.88034	41
20	.42788	.90383	.44359	.89623	.45917	.88835	.47460	.88020	40
21	.42815	.90371	.44385	.89610	.45942	.88822	.47486	.88006	39
22	.42841	.90358	.44411	.89597	.45968	.88808	.47511	.87993	38
23	.42867	.90346	.44437	.89584	.45994	.88795	.47537	.87979	37
24	.42894	.90334	.44464	.89571	.46020	.88782	.47562	.87965	36
25	.42920	.90321	.44490	.89558	.46046	.88768	.47588	.87951	35
26	.42946	.90309	.44516	.89545	.46072	.88755	.47614	.87937	34
27	.42972	.90296	.44542	.89532	.46097	.88741	.47639	.87923	33
28	.42999	.90284	.44568	.89519	.46123	.88728	.47665	.87909	32
29	.43025	.90271	.44594	.89506	.46149	.88715	.47690	.87896	31
30	.43051	.90259	.44620	.89493	.46175	.88701	.47716	.87882	30
31	.43077	.90246	.44646	.89480	.46201	.88688	.47741	.87868	29
32	.43104	.90233	.44672	.89467	.46226	.88674	.47767	.87854	28
33	.43130	.90221	.44698	.89454	.46252	.8866[illegible]	.47793	.87840	27
34	.43156	.90208	.44724	.89441	.46278	.88647	.47818	.87826	26
35	.43182	.90196	.44750	.89428	.46304	.88634	.47844	.87812	25
36	.43209	.90183	.44776	.89415	.46330	.88620	.47869	.87798	24
37	.43235	.90171	.44802	.89402	.46355	.88607	.47895	.87784	23
38	.43261	.90158	.44828	.89389	.46381	.88593	.47920	.87770	22
39	.43287	.90146	.44854	.89376	.46407	.88580	.47946	.87756	21
40	.43313	.90133	.44880	.89363	.46433	.88566	.47971	.87743	20
41	.43340	.90120	.44906	.89350	.46458	.88553	.47997	.87729	19
42	.43366	.90108	.44932	.89337	.46484	.88539	.48022	.87715	18
43	.43392	.90095	.44958	.89324	.46510	.88526	.48048	.87701	17
44	.43418	.90082	.44984	.89311	.46536	.88512	.48073	.87687	16
45	.43445	.90070	.45010	.89298	.46561	.88499	.48099	.87673	15
46	.43471	.90057	.45036	.89285	.46587	.88485	.48124	.87659	14
47	.43497	.90045	.45062	.89272	.46613	.88472	.48150	.87645	13
48	.43523	.90032	.45088	.89259	.46639	.88458	.48175	.87631	12
49	.43549	.90019	.45114	.89245	.46664	.88445	.48201	.87617	11
50	.43575	.90007	.45140	.89232	.46690	.88431	.48226	.87603	10
51	.43602	.89994	.45166	.89219	.46716	.88417	.48252	.87589	9
52	.43628	.89981	.45192	.89206	.46742	.88404	.48277	.87575	8
53	.43654	.89968	.45218	.89193	.46767	.88390	.48303	.87561	7
54	.43680	.89956	.45243	.89180	.46793	.88377	.48328	.87546	6
55	.43706	.89943	.45269	.89167	.46819	.88363	.48354	.87532	5
56	.43733	.89930	.45295	.89153	.46844	.88349	.48379	.87518	4
57	.43759	.89918	.45321	.89140	.46870	.88336	.48405	.87504	3
58	.43785	.89905	.45347	.89127	.46896	.88322	.48430	.87490	2
59	.43811	.89892	.45373	.89114	.46921	.88308	.48456	.87476	1
60	.43837	.89879	.45399	.89101	.46947	.88295	.48481	.87462	0
′	cosine 64°	sine 64°	cosine 63°	sine 63°	cosine 62°	sine 62°	cosine 61°	sine 61°	′

′	29° sine	29° cosine	30° sine	30° cosine	31° sine	31° cosine	32° sine	32° cosine	′
0	.48481	.87462	.50000	.86603	.51504	.85717	.52992	.84805	60
1	.48506	.87448	.50025	.86588	.51529	.85702	.53017	.84789	59
2	.48532	.87434	.50050	.86573	.51554	.85687	.53041	.84774	58
3	.48557	.87420	.50076	.86559	.51579	.85672	.53066	.84759	57
4	.48583	.87406	.50101	.86544	.51604	.85657	.53091	.84743	56
5	.48608	.87391	.50126	.86530	.51628	.85642	.53115	.84728	55
6	.48634	.87377	.50151	.86515	.51653	.85627	.53140	.84712	54
7	.48659	.87363	.50176	.86501	.51678	.85612	.53164	.84697	53
8	.48684	.87349	.50201	.86486	.51703	.85597	.53189	.84681	52
9	.48710	.87335	.50227	.86471	.51728	.85582	.53214	.84666	51
10	.48735	.87321	.50252	.86457	.51753	.85567	.53238	.84650	50
11	.48761	.87306	.50277	.86442	.51778	.85551	.53263	.84635	49
12	.48786	.87292	.50302	.86427	.51803	.85536	.53288	.84619	48
13	.48811	.87278	.50327	.86413	.51828	.85521	.53312	.84604	47
14	.48837	.87264	.50352	.86398	.51852	.85506	.53337	.84588	46
15	.48862	.87250	.50377	.86384	.51877	.85491	.53361	.84573	45
16	.48888	.87235	.50403	.86369	.51902	.85476	.53386	.84557	44
17	.48913	.87221	.50428	.86354	.51927	.85461	.53411	.84542	43
18	.48938	.87207	.50453	.86340	.51952	.85446	.53435	.84526	42
19	.48964	.87193	.50478	.86325	.51977	.85431	.53460	.84511	41
20	.48989	.87178	.50503	.86310	.52002	.85416	.53484	.84495	40
21	.49014	.87164	.50528	.86295	.52026	.85401	.53509	.84480	39
22	.49040	.87150	.50553	.86281	.52051	.85385	.53534	.84464	38
23	.49065	.87136	.50578	.86266	.52076	.85370	.53558	.84448	37
24	.49090	.87121	.50603	.86251	.52101	.85355	.53583	.84433	36
25	.49116	.87107	.50628	.86237	.52126	.85340	.53607	.84417	35
26	.49141	.87093	.50654	.86222	.52151	.85325	.53632	.84402	34
27	.49166	.87079	.50679	.86207	.52175	.85310	.53656	.84386	33
28	.49192	.87064	.50704	.86192	.52200	.85294	.53681	.84370	32
29	.49217	.87050	.50729	.86178	.52225	.85279	.53705	.84355	31
30	.49242	.87036	.50754	.86163	.52250	.85264	.53730	.84339	30
31	.49268	.87021	.50779	.86148	.52275	.85249	.53754	.84324	29
32	.49293	.87007	.50804	.86133	.52299	.85234	.53779	.84308	28
33	.49318	.86993	.50829	.86119	.52324	.85218	.53804	.84292	27
34	.49344	.86978	.50854	.86104	.52349	.85203	.53828	.84277	26
35	.49369	.86964	.50879	.86089	.52374	.85188	.53853	.84261	25
36	.49394	.86949	.50904	.86074	.52399	.85173	.53877	.84245	24
37	.49419	.86935	.50929	.86059	.52423	.85157	.53902	.84230	23
38	.49445	.86921	.50954	.86045	.52448	.85142	.53926	.84214	22
39	.49470	.86906	.50979	.86030	.52473	.85127	.53951	.84198	21
40	.49495	.86892	.51004	.86015	.52498	.85112	.53975	.84182	20
41	.49521	.86878	.51029	.86000	.52522	.85096	.54000	.84167	19
42	.49546	.86863	.51054	.85985	.52547	.85081	.54024	.84151	18
43	.49571	.86849	.51079	.85970	.52572	.85066	.54049	.84135	17
44	.49596	.86834	.51104	.85956	.52597	.85051	.54073	.84120	16
45	.49622	.86820	.51129	.85941	.52621	.85035	.54097	.84104	15
46	.49647	.86805	.51154	.85926	.52646	.85020	.54122	.84088	14
47	.49672	.86791	.51179	.85911	.52671	.85005	.54146	.84072	13
48	.49697	.86777	.51204	.85896	.52696	.84989	.54171	.84057	12
49	.49723	.86762	.51229	.85881	.52720	.84974	.54195	.84041	11
50	.49748	.86748	.51254	.85866	.52745	.84959	.54220	.84025	10
51	.49773	.86733	.51279	.85851	.52770	.84943	.54244	.84009	9
52	.49798	.86719	.51304	.85836	.52794	.84928	.54269	.83994	8
53	.49824	.86704	.51329	.85821	.52819	.84913	.54293	.83978	7
54	.49849	.86690	.51354	.85806	.52844	.84897	.54317	.83962	6
55	.49874	.86675	.51379	.85792	.52869	.84882	.54342	.83946	5
56	.49899	.86661	.51404	.85777	.52893	.84866	.54366	.83930	4
57	.49924	.86646	.51429	.85762	.52918	.84851	.54391	.83915	3
58	.49950	.86632	.51454	.85747	.52943	.84836	.54415	.83899	2
59	.49975	.86617	.51479	.85732	.52967	.84820	.54440	.83883	1
60	.50000	.86603	.51504	.85717	.52992	.84805	.54464	.83867	0
′	cosine	sine	cosine	sine	cosine	sine	cosine	sine	′
	60°		59°		58°		57°		

′	33° sine	33° cosine	34° sine	34° cosine	35° sine	35° cosine	36° sine	36° cosine	′
0	.54464	.83867	.55919	.82904	.57358	.81915	.58779	.80902	60
1	.54488	.83851	.55943	.82887	.57381	.81899	.58802	.80885	59
2	.54513	.83835	.55968	.82871	.57405	.81882	.58826	.80867	58
3	.54537	.83819	.55992	.82855	.57429	.81865	.58849	.80850	57
4	.54561	.83804	.56016	.82839	.57453	.81848	.58873	.80833	56
5	.54586	.83788	.56040	.82822	.57477	.81832	.58896	.80816	55
6	.54610	.83772	.56064	.82806	.57501	.81815	.58920	.80799	54
7	.54635	.83756	.56088	.82790	.57524	.81798	.58943	.80782	53
8	.54659	.83740	.56112	.82773	.57548	.81782	.58967	.80765	52
9	.54683	.83724	.56136	.82757	.57572	.81765	.58990	.80748	51
10	.54708	.83708	.56160	.82741	.57596	.81748	.59014	.80730	50
11	.54732	.83692	.56184	.82724	.57619	.81731	.59037	.80713	49
12	.54756	.83676	.56208	.82708	.57643	.81714	.59061	.80696	48
13	.54781	.83660	.56232	.82692	.57667	.81698	.59084	.80679	47
14	.54805	.83645	.56256	.82675	.57691	.81681	.59108	.80662	46
15	.54829	.83629	.56280	.82659	.57715	.81664	.59131	.80644	45
16	.54854	.83613	.56305	.82643	.57738	.81647	.59154	.80627	44
17	.54878	.83597	.56329	.82626	.57762	.81631	.59178	.80610	43
18	.54902	.83581	.56353	.82610	.57786	.81614	.59201	.80593	42
19	.54927	.83565	.56377	.82593	.57810	.81597	.59225	.80576	41
20	.54951	.83549	.56401	.82577	.57833	.81580	.59248	.80558	40
21	.54975	.83533	.56425	.82561	.57857	.81563	.59272	.80541	39
22	.54999	.83517	.56449	.82544	.57881	.81546	.59295	.80524	38
23	.55024	.83501	.56473	.82528	.57904	.81530	.59318	.80507	37
24	.55048	.83485	.56497	.82511	.57928	.81513	.59342	.80489	36
25	.55072	.83469	.56521	.82495	.57952	.81496	.59365	.80472	35
26	.55097	.83453	.56545	.82478	.57976	.81479	.59389	.80455	34
27	.55121	.83437	.56569	.82462	.57999	.81462	.59412	.80438	33
28	.55145	.83421	.56593	.82446	.58023	.81445	.59436	.80420	32
29	.55169	.83405	.56617	.82429	.58047	.81428	.59459	.80403	31
30	.55194	.83389	.56641	.82413	.58070	.81412	.59482	.80386	30
31	.55218	.83373	.56665	.82396	.58094	.81395	.59506	.80368	29
32	.55242	.83356	.56689	.82380	.58118	.81378	.59529	.80351	28
33	.55266	.83340	.56713	.82363	.58141	.81361	.59552	.80334	27
34	.55291	.83324	.56736	.82347	.58165	.81344	.59576	.80316	26
35	.55315	.83308	.56760	.82330	.58189	.81327	.59599	.80299	25
36	.55339	.83292	.56784	.82314	.58212	.81310	.59622	.80282	24
37	.55363	.83276	.56808	.82297	.58236	.81293	.59646	.80264	23
38	.55388	.83260	.56832	.82281	.58260	.81276	.59669	.80247	22
39	.55412	.83244	.56856	.82264	.58283	.81259	.59693	.80230	21
40	.55436	.83228	.56880	.82248	.58307	.81242	.59716	.80212	20
41	.55460	.83212	.56904	.82231	.58330	.81225	.59739	.80195	19
42	.55484	.83195	.56928	.82214	.58354	.81208	.59763	.80178	18
43	.55509	.83179	.56952	.82198	.58378	.81191	.59786	.80160	17
44	.55533	.83163	.56976	.82181	.58401	.81174	.59809	.80143	16
45	.55557	.83147	.57000	.82165	.58425	.81157	.59832	.80125	15
46	.55581	.83131	.57024	.82148	.58449	.81140	.59856	.80108	14
47	.55605	.83115	.57047	.82132	.58472	.81123	.59879	.80091	13
48	.55630	.83098	.57071	.82115	.58496	.81106	.59902	.80073	12
49	.55654	.83082	.57095	.82098	.58519	.81089	.59926	.80056	11
50	.55678	.83066	.57119	.82082	.58543	.81072	.59949	.80038	10
51	.55702	.83050	.57143	.82065	.58567	.81055	.59972	.80021	9
52	.55726	.83034	.57167	.82048	.58590	.81038	.59995	.80003	8
53	.55750	.83017	.57191	.82032	.58614	.81021	.60019	.79986	7
54	.55775	.83001	.57215	.82015	.58637	.81004	.60042	.79968	6
55	.55799	.82985	.57238	.81999	.58661	.80987	.60065	.79951	5
56	.55823	.82969	.57262	.81982	.58684	.80970	.60089	.79934	4
57	.55847	.82953	.57286	.81965	.58708	.80953	.60112	.79916	3
58	.55871	.82936	.57310	.81949	.58731	.80936	.60135	.79899	2
59	.55895	.82920	.57334	.81932	.58755	.80919	.60158	.79881	1
60	.55919	.82904	.57358	.81915	.58779	.80902	.60182	.79864	0
′	cosine 56°	sine 56°	cosine 55°	sine 55°	cosine 54°	sine 54°	cosine 53°	sine 53°	′

′	37° sine	37° cosine	38° sine	38° cosine	39° sine	39° cosine	40° sine	40° cosine	′
0	.60182	.79864	.61566	.78801	.62932	.77715	.64279	.76604	60
1	.60205	.79846	.61589	.78783	.62955	.77696	.64301	.76586	59
2	.60228	.79829	.61612	.78765	.62977	.77678	.64323	.76567	58
3	.60251	.79811	.61635	.78747	.63000	.77660	.64346	.76548	57
4	.60274	.79793	.61658	.78729	.63022	.77641	.64368	.76530	56
5	.60298	.79776	.61681	.78711	.63045	.77623	.64390	.76511	55
6	.60321	.79758	.61704	.78694	.63068	.77605	.64412	.76492	54
7	.60344	.79741	.61726	.78676	.63090	.77586	.64435	.76473	53
8	.60367	.79723	.61749	.78658	.63113	.77568	.64457	.76455	52
9	.60390	.79706	.61772	.78640	.63135	.77550	.64479	.76436	51
10	.60414	.79688	.61795	.78622	.63158	.77531	.64501	.76417	50
11	.60437	.79671	.61818	.78604	.63180	.77513	.64524	.76398	49
12	.60460	.79653	.61841	.78586	.63203	.77494	.64546	.76380	48
13	.60483	.79635	.61864	.78568	.63225	.77476	.64568	.76361	47
14	.60506	.79618	.61887	.78550	.63248	.77458	.64590	.76342	46
15	.60529	.79600	.61909	.78532	.63271	.77439	.64612	.76323	45
16	.60553	.79583	.61932	.78514	.63293	.77421	.64635	76304	44
17	.60576	.79565	.61955	.78496	.63316	.77402	.64657	.76286	43
18	.60599	.79547	.61978	.78478	.63338	.77384	.64679	.76267	42
19	.60622	.79530	.62001	.78460	.63361	.77366	.64701	.76248	41
20	.60645	.79512	.62024	.78442	.63383	.77347	.64723	.76229	40
21	.60668	.79494	.62046	.78424	.63406	.77329	.64746	.76210	39
22	.60691	.79477	.62069	.78405	.63428	.77310	.64768	.76192	38
23	.60714	.79459	.62092	.78387	.63451	.77292	.64790	.76173	37
24	.60738	.79441	.62115	.78369	.63473	.77273	.64812	.76154	36
25	.60761	.79424	.62138	.78351	.63496	.77255	.64834	.76135	35
26	.60784	.79406	.62160	.78333	.63518	.77236	.64856	.76116	34
27	.60807	.79388	.62183	.78315	.63540	.77218	.64878	.76097	33
28	.60830	.79371	.62206	.78297	.63563	.77199	.64901	.76078	32
29	.60853	.79353	.62229	.78279	.63585	.77181	.64923	.76059	31
30	.60876	.79335	.62251	.78261	.63608	.77162	.64945	.76041	30
31	.60899	.79318	.62274	.78243	.63630	.77144	.64967	.76022	29
32	.60922	.79300	.62297	.78225	.63653	.77125	.64989	.76003	28
33	.60945	.79282	.62320	.78206	.63675	.77107	.65011	.75984	27
34	.60968	.79264	.62342	.78188	.63698	.77088	.65033	.75965	26
35	.60991	.79247	.62365	.78170	.63720	.77070	.65055	.75946	25
36	.61015	.79229	.62388	.78152	.63742	.77051	.65077	.75927	24
37	.61038	.79211	.62411	.78134	.63765	.77033	.65100	.75908	23
38	.61061	.79193	.62433	.78116	.63787	.77014	.65122	.75889	22
39	.61084	.79176	.62456	.78098	.63810	.76996	.65144	.75870	21
40	.61107	.79158	.62479	.78079	.63832	.76977	.65166	.75851	20
41	.61130	.79140	.62502	.78061	.63854	.76959	.65188	.75832	19
42	.61153	.79122	.62524	.78043	.63877	.76940	.65210	.75813	18
43	.61176	.79105	.62547	.78025	.63899	.76921	.65232	.75794	17
44	.61199	.79087	.62570	.78007	.63922	.76903	.65254	.75775	16
45	.61222	.79069	.62592	.77988	.63944	.76884	.65276	.75756	15
46	.61245	.79051	.62615	.77970	.63966	.76866	.65298	.75738	14
47	.61268	.79033	.62638	.77952	.63989	.76847	.65320	.75719	13
48	.61291	.79016	.62660	.77934	.64011	.76828	.65342	.75700	12
49	.61314	.78998	.62683	.77916	.64033	.76810	.65364	.75680	11
50	.61337	.78980	.62706	.77897	.64056	.76791	.65386	.75661	10
51	.61360	.78962	.62728	.77879	.64078	.76772	.65408	.75642	9
52	.61383	.78944	.62751	.77861	.64100	.76754	.65430	.75623	8
53	.61406	.78926	.62774	.77843	.64123	.76735	.65452	.75604	7
54	.61429	.78908	.62796	.77824	.64145	.76717	.65474	.75585	6
55	.61451	.78891	.62819	.77806	.64167	.76698	.65496	.75566	5
56	.61474	.78873	.62842	.77788	.64190	.76679	.65518	.75547	4
57	.61497	.78855	.62864	.77769	.64212	.76661	.65540	.75528	3
58	.61520	.78837	.62887	.77751	.64234	.76642	.65562	.75509	2
59	.61543	.78819	.62909	.77733	.64256	.76623	.65584	.75490	1
60	.61566	.78801	.62932	.77715	.64279	.76604	.65606	.75471	0
′	cosine 52°	sine 52°	cosine 51°	sine 51°	cosine 50°	sine 50°	cosine 49°	sine 49°	′

′	41° sine	41° cosine	42° sine	42° cosine	43° sine	43° cosine	44° sine	44° cosine	′
0	.65606	.75471	.66913	.74314	.68200	.73135	.69466	.71934	60
1	.65628	.75452	.66935	.74295	.68221	.73116	.69487	.71914	59
2	.65650	.75433	.66956	.74276	.68242	.73096	.69508	.71894	58
3	.65672	.75414	.66978	.74256	.68264	.73076	.69529	.71873	57
4	.65694	.75395	.66999	.74237	.68285	.73056	.69549	.71853	56
5	.65716	.75375	.67021	.74217	.68306	.73036	.69570	.71833	55
6	.65738	.75356	.67043	.74198	.68327	.73016	.69591	.71813	54
7	.65759	.75337	.67064	.74178	.68349	.72996	.69612	.71792	53
8	.65781	.75318	.67086	.74159	.68370	.72976	.69633	.71772	52
9	.65803	.75299	.67107	.74139	.68391	.72957	.69654	.71752	51
10	.65825	.75280	.67129	.74120	.68412	.72937	.69675	.71732	50
11	.65847	.75261	.67151	.74100	.68434	.72917	.69696	.71711	49
12	65869	.75241	.67172	.74080	.68455	.72897	.69717	.71691	48
13	.65891	.75222	.67194	.74061	.68476	.72877	.69737	.71671	47
14	.65913	.75203	.67215	.74041	.68497	.72857	.69758	.71650	46
15	.65935	.75184	.67237	.74022	.68518	.72837	.69779	.71630	45
16	.65956	.75165	.67258	.74002	.68539	.72817	.69800	.71610	44
17	.65978	.75146	.67280	.73983	.68561	.72797	.69821	.71590	43
18	.66000	.75126	.67301	.73963	.68582	.72777	.69842	.71569	42
19	.66022	.75107	.67323	.73944	.68603	.72757	.69862	.71549	41
20	.66044	.75088	.67344	.73924	.68624	.72737	.69883	.71529	40
21	.66066	.75069	.67366	.73904	.68645	.72717	.69904	.71508	39
22	.66088	.75050	.67387	.73885	.68666	.72697	.69925	.71488	38
23	.66109	.75030	.67409	.73865	.68688	.72677	.69946	.71468	37
24	.66131	.75011	.67430	.73846	.68709	.72657	.69966	.71447	36
25	.66153	.74992	.67452	.73826	.68730	.72637	.69987	.71427	35
26	.66175	.74973	.67473	.73806	.68751	.72617	.70008	.71407	34
27	.66197	.74953	.67495	.73787	.68772	.72597	.70029	.71386	33
28	.66218	.74934	.67516	.73767	.68793	.72577	.70049	.71366	32
29	.66240	.74915	.67538	.73747	.68814	.72557	.70070	.71345	31
30	.66262	.74896	.67559	.73728	.68835	.72537	.70091	.71325	30
31	.66284	.74876	.67580	.73708	.68857	.72517	.70112	.71305	29
32	.66306	.74857	.67602	.73688	.68878	.72497	.70132	.71284	28
33	.66327	.74838	.67623	.73669	.68899	.72477	.70153	.71264	27
34	.66349	.74818	.67645	.73649	.68920	.72457	.70174	.71243	26
35	.66371	.74799	.67666	.73629	.68941	.72437	.70195	.71223	25
36	.66393	.74780	.67688	.73610	.68962	.72417	.70215	.71203	24
37	.66414	.74760	.67709	.73590	.68983	.72397	.70236	.71182	23
38	.66436	.74741	.67730	.73570	.69004	.72377	.70257	.71162	22
39	.66458	.74722	.67752	.73551	.69025	.72357	.70277	.71141	21
40	.66480	.74703	.67773	.73531	.69046	.72337	.70298	.71121	20
41	.66501	.74683	.67795	.73511	.69067	.72317	.70319	.71100	19
42	.66523	.74664	.67816	.73491	.69088	.72297	.70339	.71080	18
43	.66545	.74644	.67837	.73472	.69109	.72277	.70360	.71059	17
44	.66566	.74625	.67859	.73452	.69130	.72257	.70381	.71039	16
45	.66588	.74606	.67880	.73432	.69151	.72236	.70401	.71019	15
46	.66610	.74586	.67901	.73413	.69172	.72216	.70422	.70998	14
47	.66632	.74567	.67923	.73393	.69193	.72196	.70443	.70978	13
48	.66653	.74548	.67944	.73373	.69214	.72176	.70463	.70957	12
49	.66675	.74528	.67965	.73353	.69235	.72156	.70484	.70937	11
50	.66697	.74509	.67987	.73333	.69256	.72136	.70505	.70916	10
51	.66718	.74489	.68008	.73314	.69277	.72116	.70525	.70896	9
52	.66740	.74470	.68029	.73294	.69298	.72095	.70546	.70875	8
53	.66762	.74451	.68051	.73274	.69319	.72075	.70567	.70855	7
54	.66783	.74431	.68072	.73254	.69340	.72055	.70587	.70834	6
55	.66805	.74412	.68093	.73234	.69361	.72035	.70608	.70813	5
56	.66827	.74392	.68115	.73215	.69382	.72015	.70628	.70793	4
57	.66848	.74373	.68136	.73195	.69403	.71995	.70649	.70772	3
58	.66870	.74353	.68157	.73175	.69424	.71974	.70670	.70752	2
59	.66891	.74334	.68179	.73155	.69445	.71954	.70690	.70731	1
60	.66913	.74314	.68200	.73135	.69466	.71934	.70711	.70711	0
′	cosine 48°	sine 48°	cosine 47°	sine 47°	cosine 46°	sine 46°	cosine 45°	sine 45°	′

SECANTS AND COSECANTS

′	0° sec	0° cosec	1° sec	1° cosec	2° sec	2° cosec	3° sec	3° cosec	′
0	1	Infinite.	1.0001	57.299	1.0006	28.654	1.0014	19.107	60
1	1	3437.70	1.0001	56.359	1.0006	28.417	1.0014	19.002	59
2	1	1718.90	1.0002	55.450	1.0006	28.184	1.0014	18.897	58
3	1	1145.90	1.0002	54.570	1.0006	27.955	1.0014	18.794	57
4	1	859.44	1.0002	53.718	1.0006	27.730	1.0014	18.692	56
5	1	687.55	1.0002	52.891	1.0007	27.508	1.0014	18.591	55
6	1	572.96	1.0002	52.090	1.0007	27.290	1.0015	18.491	54
7	1	491.11	1.0002	51.313	1.0007	27.075	1.0015	18.393	53
8	1	429.72	1.0002	50.558	1.0007	26.864	1.0015	18.295	52
9	1	381.97	1.0002	49.826	1.0007	26.655	1.0015	18.198	51
10	1	343.77	1.0002	49.114	1.0007	26.450	1.0015	18.103	50
11	1	312.52	1.0002	48.422	1.0007	26.249	1.0015	18.008	49
12	1	286.48	1.0002	47.750	1.0007	26.050	1.0016	17.914	48
13	1	264.44	1.0002	47.096	1.0007	25.854	1.0016	17.821	47
14	1	245.55	1.0002	46.460	1.0008	25.661	1.0016	17.730	46
15	1	229.18	1.0002	45.840	1.0008	25.471	1.0016	17.639	45
16	1	214.86	1.0002	45.237	1.0008	25.284	1.0016	17.549	44
17	1	202.22	1.0002	44.650	1.0008	25.100	1.0016	17.460	43
18	1	190.99	1.0002	44.077	1.0008	24.918	1.0017	17.372	42
19	1	180.73	1.0003	43.520	1.0008	24.739	1.0017	17.285	41
20	1	171.89	1.0003	42.976	1.0008	24.562	1.0017	17.198	40
21	1	163.70	1.0003	42.445	1.0008	24.388	1.0017	17.113	39
22	1	156.26	1.0003	41.928	1.0008	24.216	1.0017	17.028	38
23	1	149.47	1.0003	41.423	1.0009	24.047	1.0017	16.944	37
24	1	143.24	1.0003	40.930	1.0009	23.880	1.0018	16.861	36
25	1	137.51	1.0003	40.448	1.0009	23.716	1.0018	16.779	35
26	1	132.22	1.0003	39.978	1.0009	23.553	1.0018	16.698	34
27	1	127.32	1.0003	39.518	1.0009	23.393	1.0018	16.617	33
28	1	122.78	1.0003	39.069	1.0009	23.235	1.0018	16.538	32
29	1	118.54	1.0003	38.631	1.0009	23.079	1.0018	16.459	31
30	1	114.59	1.0003	38.201	1.0009	22.925	1.0019	16.380	30
31	1	110.90	1.0003	37.782	1.0010	22.774	1.0019	16.303	29
32	1	107.43	1.0003	37.371	1.0010	22.624	1.0019	16.226	28
33	1	104.17	1.0004	36.969	1.0010	22.476	1.0019	16.150	27
34	1	101.11	1.0004	36.576	1.0010	22.330	1.0019	16.075	26
35	1	98.223	1.0004	36.191	1.0010	22.186	1.0019	16.000	25
36	1	95.495	1.0004	35.814	1.0010	22.044	1.0020	15.926	24
37	1	92.914	1.0004	35.445	1.0010	21.904	1.0020	15.853	23
38	1.0001	90.469	1.0004	35.084	1.0010	21.765	1.0020	15.780	22
39	1.0001	88.149	1.0004	34.729	1.0011	21.629	1.0020	15.708	21
40	1.0001	85.946	1.0004	34.382	1.0011	21.494	1.0020	15.637	20
41	1.0001	83.849	1.0004	34.042	1.0011	21.360	1.0021	15.566	19
42	1.0001	81.853	1.0004	33.708	1.0011	21.228	1.0021	15.496	18
43	1.0001	79.950	1.0004	33.381	1.0011	21.098	1.0021	15.427	17
44	1.0001	78.133	1.0004	33.060	1.0011	20.970	1.0021	15.358	16
45	1.0001	76.396	1.0005	32.745	1.0011	20.843	1.0021	15.290	15
46	1.0001	74.736	1.0005	32.437	1.0012	20.717	1.0022	15.222	14
47	1.0001	73.146	1.0005	32.134	1.0012	20.593	1.0022	15.155	13
48	1.0001	71.622	1.0005	31.836	1.0012	20.471	1.0022	15.089	12
49	1.0001	70.160	1.0005	31.544	1.0012	20.350	1.0022	15.023	11
50	1.0001	68.757	1.0005	31.257	1.0012	20.230	1.0022	14.958	10
51	1.0001	67.409	1.0005	30.976	1.0012	20.112	1.0023	14.893	9
52	1.0001	66.113	1.0005	30.699	1.0012	19.995	1.0023	14.829	8
53	1.0001	64.866	1.0005	30.428	1.0013	19.880	1.0023	14.765	7
54	1.0001	63.664	1.0005	30.161	1.0013	19.766	1.0023	14.702	6
55	1.0001	62.507	1.0005	29.899	1.0013	19.653	1.0023	14.640	5
56	1.0001	61.391	1.0006	29.641	1.0013	19.541	1.0024	14.578	4
57	1.0001	60.314	1.0006	29.388	1.0013	19.431	1.0024	14.517	3
58	1.0001	59.274	1.0006	29.139	1.0013	19.322	1.0024	14.456	2
59	1.0001	58.270	1.0006	28.894	1.0013	19.214	1.0024	14.395	1
60	1.0001	57.299	1.0006	28.654	1.0014	19.107	1.0024	14.335	0
′	89° cosec	89° sec	88° cosec	88° sec	87° cosec	87° sec	86° cosec	86° sec	′

′	4°		5°		6°		7°		′
	sec	cosec	sec	cosec	sec	cosec	sec	cosec	
0	1.0024	14.335	1.0038	11.474	1.0055	9.5668	1.0075	8.2055	60
1	1.0025	14.276	1.0038	11.436	1.0055	9.5404	1.0075	8.1861	59
2	1.0025	14.217	1.0039	11.398	1.0056	9.5141	1.0076	8.1668	58
3	1.0025	14.159	1.0039	11.360	1.0056	9.4880	1.0076	8.1476	57
4	1.0025	14.101	1.0039	11.323	1.0056	9.4620	1.0076	8.1285	56
5	1.0025	14.043	1.0039	11.286	1.0057	9.4362	1.0077	8.1094	55
6	1.0026	13.986	1.0040	11.249	1.0057	9.4105	1.0077	8.0905	54
7	1.0026	13.930	1.0040	11.213	1.0057	9.3850	1.0078	8.0717	53
8	1.0026	13.874	1.0040	11.176	1.0057	9.3596	1.0078	8.0529	52
9	1.0026	13.818	1.0040	11.140	1.0058	9.3343	1.0078	8.0342	51
10	1.0026	13.763	1.0041	11.104	1.0058	9.3092	1.0079	8.0156	50
11	1.0027	13.708	1.0041	11.069	1.0058	9.2842	1.0079	7.9971	49
12	1.0027	13.654	1.0041	11.033	1.0059	9.2593	1.0079	7.9787	48
13	1.0027	13.600	1.0041	10.988	1.0059	9.2346	1.0080	7.9604	47
14	1.0027	13.547	1.0042	10.963	1.0059	9.2100	1.0080	7.9421	46
15	1.0027	13.494	1.0042	10.929	1.0060	9.1855	1.0080	7.9240	45
16	1.0028	13.441	1.0042	10.894	1.0060	9.1612	1.0081	7.9059	44
17	1.0028	13.389	1.0043	10.860	1.0060	9.1370	1.0081	7.8879	43
18	1.0028	13.337	1.0043	10.826	1.0061	9.1129	1.0082	7.8700	42
19	1.0028	13.286	1.0043	10.792	1.0061	9.0890	1.0082	7.8522	41
20	1.0029	13.235	1.0043	10.758	1.0061	9.0651	1.0082	7.8344	40
21	1.0029	13.184	1.0044	10.725	1.0062	9.0414	1.0083	7.8168	39
22	1.0029	13.134	1.0044	10.692	1.0062	9.0179	1.0083	7.7992	38
23	1.0029	13.084	1.0044	10.659	1.0062	8.9944	1.0084	7.7817	37
24	1.0029	13.034	1.0044	10.626	1.0063	8.9711	1.0084	7.7642	36
25	1.0030	12.985	1.0045	10.593	1.0063	8.9479	1.0084	7.7469	35
26	1.0030	12.937	1.0045	10.561	1.0063	8.9248	1.0085	7.7296	34
27	1.0030	12.888	1.0045	10.529	1.0064	8.9018	1.0085	7.7124	33
28	1.0030	12.840	1.0046	10.497	1.0064	8.8790	1.0085	7.6953	32
29	1.0031	12.793	1.0046	10.465	1.0064	8.8563	1.0086	7.6783	31
30	1.0031	12.745	1.0046	10.433	1.0065	8.8337	1.0086	7.6613	30
31	1.0031	12.698	1.0046	10.402	1.0065	8.8112	1.0087	7.6444	29
32	1.0031	12.652	1.0047	10.371	1.0065	8.7888	1.0087	7.6276	28
33	1.0032	12.606	1.0047	10.340	1.0066	8.7665	1.0087	7.6108	27
34	1.0032	12.560	1.0047	10.309	1.0066	8.7444	1.0088	7.5942	26
35	1.0032	12.514	1.0048	10.278	1.0066	8.7223	1.0088	7.5776	25
36	1.0032	12.469	1.0048	10.248	1.0067	8.7004	1.0089	7.5611	24
37	1.0032	12.424	1.0048	10.217	1.0067	8.6786	1.0089	7.5446	23
38	1.0033	12.379	1.0048	10.187	1.0067	8.6569	1.0089	7.5282	22
39	1.0033	12.335	1.0049	10.157	1.0068	8.6353	1.0090	7.5119	21
40	1.0033	12.291	1.0049	10.127	1.0068	8.6138	1.0090	7.4957	20
41	1.0033	12.248	1.0049	10.098	1.0068	8.5924	1.0090	7.4795	19
42	1.0034	12.204	1.0050	10.068	1.0069	8.5711	1.0091	7.4634	18
43	1.0034	12.161	1.0050	10.039	1.0069	8.5499	1.0091	7.4474	17
44	1.0034	12.118	1.0050	10.010	1.0069	8.5289	1.0092	7.4315	16
45	1.0034	12.076	1.0050	9.9812	1.0070	8.5079	1.0092	7.4156	15
46	1.0035	12.034	1.0051	9.9525	1.0070	8.4871	1.0092	7.3998	14
47	1.0035	11.992	1.0051	9.9239	1.0070	8.4663	1.0093	7.3840	13
48	1.0035	11.950	1.0051	9.8955	1.0071	8.4457	1.0093	7.3683	12
49	1.0035	11.909	1.0052	9.8672	1.0071	8.4251	1.0094	7.3527	11
50	1.0036	11.868	1.0052	9.8391	1.0071	8.4046	1.0094	7.3372	10
51	1.0036	11.828	1.0052	9.8112	1.0072	8.3843	1.0094	7.3217	9
52	1.0036	11.787	1.0053	9.7834	1.0072	8.3640	1.0095	7.3063	8
53	1.0036	11.747	1.0053	9.7558	1.0073	8.3439	1.0095	7.2909	7
54	1.0037	11.707	1.0053	9.7283	1.0073	8.3238	1.0096	7.2757	6
55	1.0037	11.668	1.0053	9.7010	1.0073	8.3039	1.0096	7.2604	5
56	1.0037	11.628	1.0054	9.6739	1.0074	8.2840	1.0097	7.2453	4
57	1.0037	11.589	1.0054	9.6469	1.0074	8.2642	1.0097	7.2302	3
58	1.0038	11.550	1.0054	9.6200	1.0074	8.2446	1.0097	7.2152	2
59	1.0038	11.512	1.0055	9.5933	1.0075	8.2250	1.0098	7.2002	1
60	1.0038	11.474	1.0055	9.5668	1.0075	8.2055	1.0098	7.1853	0
′	cosec	sec	cosec	sec	cosec	sec	cosec	sec	′
	85°		84°		83°		82°		

′	8° sec	8° cosec	9° sec	9° cosec	10° sec	10° cosec	11° sec	11° cosec	′
0	1.0098	7.1853	1.0125	6.3924	1.0154	5.7588	1.0187	5.2408	60
1	1.0099	7.1704	1.0125	6.3807	1.0155	5.7493	1.0188	5.2330	59
2	1.0099	7.1557	1.0125	6.3690	1.0155	5.7398	1.0188	5.2252	58
3	1.0099	7.1409	1.0126	6.3574	1.0156	5.7304	1.0189	5.2174	57
4	1.0100	7.1263	1.0126	6.3458	1.0156	5.7210	1.0189	5.2097	56
5	1.0100	7.1117	1.0127	6.3343	1.0157	5.7117	1.0190	5.2019	55
6	1.0101	7.0972	1.0127	6.3228	1.0157	5.7023	1.0191	5.1942	54
7	1.0101	7.0827	1.0128	6.3113	1.0158	5.6930	1.0191	5.1865	53
8	1.0102	7.0683	1.0128	6.2999	1.0158	5.6838	1.0192	5.1788	52
9	1.0102	7.0539	1.0129	6.2885	1.0159	5.6745	1.0192	5.1712	51
10	1.0102	7.0396	1.0129	6.2772	1.0159	5.6653	1.0193	5.1636	50
11	1.0103	7.0254	1.0130	6.2659	1.0160	5.6561	1.0193	5.1560	49
12	1.0103	7.0112	1.0130	6.2546	1.0160	5.6470	1.0194	5.1484	48
13	1.0104	6.9971	1.0131	6.2434	1.0161	5.6379	1.0195	5.1409	47
14	1.0104	6.9830	1.0131	6.2322	1.0162	5.6288	1.0195	5.1333	46
15	1.0104	6.9690	1.0132	6.2211	1.0162	5.6197	1.0196	5.1258	45
16	1.0105	6.9550	1.0132	6.2100	1.0163	5.6107	1.0196	5.1183	44
17	1.0105	6.9411	1.0133	6.1990	1.0163	5.6017	1.0197	5.1109	43
18	1.0106	6.9273	1.0133	6.1880	1.0164	5.5928	1.0198	5.1034	42
19	1.0106	6.9135	1.0134	6.1770	1.0164	5.5838	1.0198	5.0960	41
20	1.0107	6.8998	1.0134	6.1661	1.0165	5.5749	1.0199	5.0886	40
21	1.0107	6.8861	1.0135	6.1552	1.0165	5.5660	1.0199	5.0812	39
22	1.0107	6.8725	1.0135	6.1443	1.0166	5.5572	1.0200	5.0739	38
23	1.0108	6.8589	1.0136	6.1335	1.0166	5.5484	1.0201	5.0666	37
24	1.0108	6.8454	1.0136	6.1227	1.0167	5.5396	1.0201	5.0593	36
25	1.0109	6.8320	1.0136	6.1120	1.0167	5.5308	1.0202	5.0520	35
26	1.0109	6.8185	1.0137	6.1013	1.0168	5.5221	1.0202	5.0447	34
27	1.0110	6.8052	1.0137	6.0906	1.0169	5.5134	1.0203	5.0375	33
28	1.0110	6.7919	1.0138	6.0800	1.0169	5.5047	1.0204	5.0302	32
29	1.0111	6.7787	1.0138	6.0694	1.0170	5.4960	1.0204	5.0230	31
30	1.0111	6.7655	1.0139	6.0588	1.0170	5.4874	1.0205	5.0158	30
31	1.0111	6.7523	1.0139	6.0483	1.0171	5.4788	1.0205	5.0087	29
32	1.0112	6.7392	1.0140	6.0379	1.0171	5.4702	1.0206	5.0015	28
33	1.0112	6.7262	1.0140	6.0274	1.0172	5.4617	1.0207	4.9944	27
34	1.0113	6.7132	1.0141	6.0170	1.0172	5.4532	1.0207	4.9873	26
35	1.0113	6.7003	1.0141	6.0066	1.0173	5.4447	1.0208	4.9802	25
36	1.0114	6.6874	1.0142	5.9963	1.0174	5.4362	1.0208	4.9732	24
37	1.0114	6.6745	1.0142	5.9860	1.0174	5.4278	1.0209	4.9661	23
38	1.0115	6.6617	1.0143	5.9758	1.0175	5.4194	1.0210	4.9591	22
39	1.0115	6.6490	1.0143	5.9655	1.0175	5.4110	1.0210	4.9521	21
40	1.0015	6.6363	1.0144	5.9554	1.0176	5.4026	1.0211	4.9452	20
41	1.0116	6.6237	1.0144	5.9452	1.0176	5.3943	1.0211	4.9382	19
42	1.0116	6.6111	1.0145	5.9351	1.0177	5.3860	1.0212	4.9313	18
43	1.0117	6.5985	1.0145	5.9250	1.0177	5.3777	1.0213	4.9243	17
44	1.0117	6.5860	1.0146	5.9150	1.0178	5.3695	1.0213	4.9175	16
45	1.0118	6.5736	1.0146	5.9049	1.0179	5.3612	1.0214	4.9106	15
46	1.0118	6.5612	1.0147	5.8950	1.0179	5.3530	1.0215	4.9037	14
47	1.0119	6.5488	1.0147	5.8850	1.0180	5.3449	1.0215	4.8969	13
48	1.0119	6.5365	1.0148	5.8751	1.0180	5.3367	1.0216	4.8901	12
49	1.0119	6.5243	1.0148	5.8652	1.0181	5.3286	1.0216	4.8833	11
50	1.0120	6.5121	1.0149	5.8554	1.0181	5.3205	1.0217	4.8765	10
51	1.0120	6.4999	1.0150	5.8456	1.0182	5.3124	1.0218	4.8697	9
52	1.0121	6.4878	1.0150	5.8358	1.0182	5.3044	1.0218	4.8630	8
53	1.0121	6.4757	1.0151	5.8261	1.0183	5.2963	1.0219	4.8563	7
54	1.0122	6.4637	1.0151	5.8163	1.0184	5.2883	1.0220	4.8496	6
55	1.0122	6.4517	1.0152	5.8067	1.0184	5.2803	1.0220	4.8429	5
56	1.0123	6.4398	1.0152	5.7970	1.0185	5.2724	1.0221	4.8362	4
57	1.0123	6.4279	1.0153	5.7874	1.0185	5.2645	1.0221	4.8296	3
58	1.0124	6.4160	1.0153	5.7778	1.0186	5.2566	1.0222	4.8229	2
59	1.0124	6.4042	1.0154	5.7683	1.0186	5.2487	1.0223	4.8163	1
60	1.0125	6.3924	1.0154	5.7588	1.0187	5.2408	1.0223	4.8097	0
′	81° cosec	81° sec	80° cosec	80° sec	79° cosec	79° sec	78° cosec	78° sec	′

′	12°		13°		14°		15°		′
	sec	cosec	sec	cosec	sec	cosec	sec	cosec	
0	1.0223	4.8097	1.0263	4.4454	1.0306	4.1336	1.0353	3.8637	60
1	1.0224	4.8032	1.0264	4.4398	1.0307	4.1287	1.0353	3.8595	59
2	1.0225	4.7966	1.0264	4.4342	1.0308	4.1239	1.0354	3.8553	58
3	1.0225	4.7901	1.0265	4.4287	1.0308	4.1191	1.0355	3.8512	57
4	1.0226	4.7835	1.0266	4.4231	1.0309	4.1144	1.0356	3.8470	56
5	1.0226	4.7770	1.0266	4.4176	1.0310	4.1096	1.0357	3.8428	55
6	1.0227	4.7706	1.0267	4.4121	1.0311	4.1048	1.0358	3.8387	54
7	1.0228	4.7641	1.0268	4.4065	1.0311	4.1001	1.0358	3.8346	53
8	1.0228	4.7576	1.0268	4.4011	1.0312	4.0953	1.0359	3.8304	52
9	1.0229	4.7512	1.0269	4.3956	1.0313	4.0906	1.0360	3.8263	51
10	1.0230	4.7448	1.0270	4.3901	1.0314	4.0859	1.0361	3.8222	50
11	1.0230	4.7384	1.0271	4.3847	1.0314	4.0812	1.0362	3.8181	49
12	1.0231	4.7320	1.0271	4.3792	1.0315	4.0765	1.0362	3.8140	48
13	1.0232	4.7257	1.0272	4.3738	1.0316	4.0718	1.0363	3.8100	47
14	1.0232	4.7193	1.0273	4.3684	1.0317	4.0672	1.0364	3.8059	46
15	1.0233	4.7130	1.0273	4.3630	1.0317	4.0625	1.0365	3.8018	45
16	1.0234	4.7067	1.0274	4.3576	1.0318	4.0579	1.0366	3.7978	44
17	1.0234	4.7004	1.0275	4.3522	1.0319	4.0532	1.0367	3.7937	43
18	1.0235	4.6942	1.0276	4.3469	1.0320	4.0486	1.0367	3.7897	42
19	1.0235	4.6879	1.0276	4.3415	1.0320	4.0440	1.0368	3.7857	41
20	1.0236	4.6817	1.0277	4.3362	1.0321	4.0394	1.0369	3.7816	40
21	1.0237	4.6754	1.0278	4.3309	1.0322	4.0348	1.0370	3.7776	39
22	1.0237	4.6692	1.0278	4.3256	1.0323	4.0302	1.0371	3.7736	38
23	1.0238	4.6631	1.0279	4.3203	1.0323	4.0256	1.0371	3.7697	37
24	1.0239	4.6569	1.0280	4.3150	1.0324	4.0211	1.0372	3.7657	36
25	1.0239	4.6507	1.0280	4.3098	1.0325	4.0165	1.0373	3.7617	35
26	1.0240	4.6446	1.0281	4.3045	1.0326	4.0120	1.0374	3.7577	34
27	1.0241	4.6385	1.0282	4.2993	1.0327	4.0074	1.0375	3.7538	33
28	1.0241	4.6324	1.0283	4.2941	1.0327	4.0029	1.0376	3.7498	32
29	1.0242	4.6263	1.0283	4.2888	1.0328	3.9984	1.0376	3.7459	31
30	1.0243	4.6202	1.0284	4.2836	1.0329	3.9939	1.0377	3.7420	30
31	1.0243	4.6142	1.0285	4.2785	1.0330	3.9894	1.0378	3.7380	29
32	1.0244	4.6081	1.0285	4.2733	1.0330	3.9850	1.0379	3.7341	28
33	1.0245	4.6021	1.0286	4.2681	1.0331	3.9805	1.0380	3.7302	27
34	1.0245	4.5961	1.0287	4.2630	1.0332	3.9760	1.0381	3.7263	26
35	1.0246	4.5901	1.0288	4.2579	1.0333	3.9716	1.0382	3.7224	25
36	1.0247	4.5841	1.0288	4.2527	1.0334	3.9672	1.0382	3.7186	24
37	1.0247	4.5782	1.0289	4.2476	1.0334	3.9627	1.0383	3.7147	23
38	1.0248	4.5722	1.0290	4.2425	1.0335	3.9583	1.0384	3.7108	22
39	1.0249	4.5663	1.0291	4.2375	1.0336	3.9539	1.0385	3.7070	21
40	1.0249	4.5604	1.0291	4.2324	1.0337	3.9495	1.0386	3.7031	20
41	1.0250	4.5545	1.0292	4.2273	1.0338	3.9451	1.0387	3.6993	19
42	1.0251	4.5486	1.0293	4.2223	1.0338	3.9408	1.0387	3.6955	18
43	1.0251	4.5428	1.0293	4.2173	1.0339	3.9364	1.0388	3.6917	17
44	1.0252	4.5369	1.0294	4.2122	1.0340	3.9320	1.0389	3.6878	16
45	1.0253	4.5311	1.0295	4.2072	1.0341	3.9277	1.0390	3.6840	15
46	1.0253	4.5253	1.0296	4.2022	1.0341	3.9234	1.0391	3.6802	14
47	1.0254	4.5195	1.0296	4.1972	1.0342	3.9190	1.0392	3.6765	13
48	1.0255	4.5137	1.0297	4.1923	1.0343	3.9147	1.0393	3.6727	12
49	1.0255	4.5079	1.0298	4.1873	1.0344	3.9104	1.0393	3.6689	11
50	1.0256	4.5021	1.0299	4.1824	1.0345	3.9061	1.0394	3.6651	10
51	1.0257	4.4964	1.0299	4.1774	1.0345	3.9018	1.0395	3.6614	9
52	1.0257	4.4907	1.0300	4.1725	1.0346	3.8976	1.0396	3.6576	8
53	1.0258	4.4850	1.0301	4.1676	1.0347	3.8933	1.0397	3.6539	7
54	1.0259	4.4793	1.0302	4.1627	1.0348	3.8890	1.0398	3.6502	6
55	1.0260	4.4736	1.0302	4.1578	1.0349	3.8848	1.0399	3.6464	5
56	1.0260	4.4679	1.0303	4.1529	1.0349	3.8805	1.0399	3.6427	4
57	1.0261	4.4623	1.0304	4.1481	1.0350	3.8763	1.0400	3.6390	3
58	1.0262	4.4566	1.0305	4.1432	1.0351	3.8721	1.0401	3.6353	2
59	1.0262	4.4510	1.0305	4.1384	1.0352	3.8679	1.0402	3.6316	1
60	1.0263	4.4454	1.0306	4.1336	1.0353	3.8637	1.0403	3.6279	0
′	cosec	sec	cosec	sec	cosec	sec	cosec	sec	′
	77°		76°		75°		74°		

′	16° sec	cosec	17° sec	cosec	18° sec	cosec	19° sec	cosec	′
0	1.0403	3.6279	1.0457	3.4203	1.0515	3.2361	1.0576	3.0715	60
1	1.0404	3.6243	1.0458	3.4170	1.0516	3.2332	1.0577	3.0690	59
2	1.0405	3.6206	1.0459	3.4138	1.0517	3.2303	1.0578	3.0664	58
3	1.0406	3.6169	1.0460	3.4106	1.0518	3.2274	1.0579	3.0638	57
4	1.0406	3.6133	1.0461	3.4073	1.0519	3.2245	1.0580	3.0612	56
5	1.0407	3.6096	1.0461	3.4041	1.0520	3.2216	1.0581	3.0586	55
6	1.0408	3.6060	1.0462	3.4009	1.0521	3.2188	1.0582	3.0561	54
7	1.0409	3.6024	1.0463	3.3977	1.0522	3.2159	1.0584	3.0535	53
8	1.0410	3.5987	1.0464	3.3945	1.0523	3.2131	1.0585	3.0509	52
9	1.0411	3.5951	1.0465	3.3913	1.0524	3.2102	1.0586	3.0484	51
10	1.0412	3.5915	1.0466	3.3881	1.0525	3.2074	1.0587	3.0458	50
11	1.0413	3.5879	1.0467	3.3849	1.0526	3.2045	1.0588	3.0433	49
12	1.0413	3.5843	1.0468	3.3817	1.0527	3.2017	1.0589	3.0407	48
13	1.0414	3.5807	1.0469	3.3785	1.0528	3.1989	1.0590	3.0382	47
14	1.0415	3.5772	1.0470	3.3754	1.0529	3.1960	1.0591	3.0357	46
15	1.0416	3.5736	1.0471	3.3722	1.0530	3.1932	1.0592	3.0331	45
16	1.0417	3.5700	1.0472	3.3690	1.0531	3.1904	1.0593	3.0306	44
17	1.0418	3.5665	1.0473	3.3659	1.0532	3.1876	1.0594	3.0281	43
18	1.0419	3.5629	1.0474	3.3627	1.0533	3.1848	1.0595	3.0256	42
19	1.0420	3.5594	1.0475	3.3596	1.0534	3.1820	1.0596	3.0231	41
20	1.0420	3.5559	1.0476	3.3565	1.0535	3.1792	1.0598	3.0206	40
21	1.0421	3.5523	1.0477	3.3534	1.0536	3.1764	1.0599	3.0181	39
22	1.0422	3.5488	1.0478	3.3502	1.0537	3.1736	1.0600	3.0156	38
23	1.0423	3.5453	1.0478	3.3471	1.0538	3.1708	1.0601	3.0131	37
24	1.0424	3.5418	1.0479	3.3440	1.0539	3.1681	1.0602	3.0106	36
25	1.0425	3.5383	1.0480	3.3409	1.0540	3.1653	1.0603	3.0081	35
26	1.0426	3.5348	1.0481	3.3378	1.0541	3.1625	1.0604	3.0056	34
27	1.0427	3.5313	1.0482	3.3347	1.0542	3.1598	1.0605	3.0031	33
28	1.0428	3.5279	1.0483	3.3316	1.0543	3.1570	1.0606	3.0007	32
29	1.0428	3.5244	1.0484	3.3286	1.0544	3.1543	1.0607	2.9982	31
30	1.0429	3.5209	1.0485	3.3255	1.0545	3.1515	1.0608	2.9957	30
31	1.0430	3.5175	1.0486	3.3224	1.0546	3.1488	1.0609	2.9933	29
32	1.0431	3.5140	1.0487	3.3194	1.0547	3.1461	1.0611	2.9908	28
33	1.0432	3.5106	1.0488	3.3163	1.0548	3.1433	1.0612	2.9884	27
34	1.0433	3.5072	1.0489	3.3133	1.0549	3.1406	1.0613	2.9859	26
35	1.0434	3.5037	1.0490	3.3102	1.0550	3.1379	1.0614	2.9835	25
36	1.0435	3.5003	1.0491	3.3072	1.0551	3.1352	1.0615	2.9810	24
37	1.0436	3.4969	1.0492	3.3042	1.0552	3.1325	1.0616	2.9786	23
38	1.0437	3.4935	1.0493	3.3011	1.0553	3.1298	1.0617	2.9762	22
39	1.0438	3.4901	1.0494	3.2981	1.0554	3.1271	1.0618	2.9738	21
40	1.0438	3.4867	1.0495	3.2951	1.0555	3.1244	1.0619	2.9713	20
41	1.0439	3.4833	1.0496	3.2921	1.0556	3.1217	1.0620	2.9689	19
42	1.0440	3.4799	1.0497	3.2891	1.0557	3.1190	1.0622	2.9665	18
43	1.0441	3.4766	1.0498	3.2861	1.0558	3.1163	1.0623	2.9641	17
44	1.0442	3.4732	1.0499	3.2831	1.0559	3.1137	1.0624	2.9617	16
45	1.0443	3.4698	1.0500	3.2801	1.0560	3.1110	1.0625	2.9593	15
46	1.0444	3.4665	1.0501	3.2772	1.0561	3.1083	1.0626	2.9569	14
47	1.0445	3.4632	1.0502	3.2742	1.0562	3.1057	1.0627	2.9545	13
48	1.0446	3.4598	1.0503	3.2712	1.0563	3.1030	1.0628	2.9521	12
49	1.0447	3.4565	1.0504	3.2683	1.0565	3.1004	1.0629	2.9497	11
50	1.0448	3.4532	1.0505	3.2653	1.0566	3.0977	1.0630	2.9474	10
51	1.0448	3.4498	1.0506	3.2624	1.0567	3.0951	1.0632	2.9450	9
52	1.0449	3.4465	1.0507	3.2594	1.0568	3.0925	1.0633	2.9426	8
53	1.0450	3.4432	1.0508	3.2565	1.0569	3.0898	1.0634	2.9402	7
54	1.0451	3.4399	1.0509	3.2535	1.0570	3.0872	1.0635	2.9379	6
55	1.0452	3.4366	1.0510	3.2506	1.0571	3.0846	1.0636	2.9355	5
56	1.0453	3.4334	1.0511	3.2477	1.0572	3.0820	1.0637	2.9332	4
57	1.0454	3.4301	1.0512	3.2448	1.0573	3.0793	1.0638	2.9308	3
58	1.0455	3.4268	1.0513	3.2419	1.0574	3.0767	1.0639	2.9285	2
59	1.0456	3.4236	1.0514	3.2390	1.0575	3.0741	1.0641	2.9261	1
60	1.0457	3.4203	1.0515	3.2361	1.0576	3.0715	1.0642	2.9238	0
′	cosec 73°	sec	cosec 72°	sec	cosec 71°	sec	cosec 70°	sec	′

′	20° sec	20° cosec	21° sec	21° cosec	22° sec	22° cosec	23° sec	23° cosec	′
0	1.0642	2.9238	1.0711	2.7904	1.0785	2.6695	1.0864	2.5593	60
1	1.0643	2.9215	1.0713	2.7883	1.0787	2.6675	1.0865	2.5575	59
2	1.0644	2.9191	1.0714	2.7862	1.0788	2.6656	1.0866	2.5558	58
3	1.0645	2.9168	1.0715	2.7841	1.0789	2.6637	1.0868	2.5540	57
4	1.0646	2.9145	1.0716	2.7820	1.0790	2.6618	1.0869	2.5523	56
5	1.0647	2.9122	1.0717	2.7799	1.0792	2.6599	1.0870	2.5506	55
6	1.0648	2.9098	1.0719	2.7778	1.0793	2.6580	1.0872	2.5488	54
7	1.0650	2.9075	1.0720	2.7757	1.0794	2.6561	1.0873	2.5471	53
8	1.0651	2.9052	1.0721	2.7736	1.0795	2.6542	1.0874	2.5453	52
9	1.0652	2.9029	1.0722	2.7715	1.0797	2.6523	1.0876	2.5436	51
10	1.0653	2.9006	1.0723	2.7694	1.0798	2.6504	1.0877	2.5419	50
11	1.0654	2.8983	1.0725	2.7674	1.0799	2.6485	1.0878	2.5402	49
12	1.0655	2.8960	1.0726	2.7653	1.0801	2.6466	1.0880	2.5384	48
13	1.0656	2.8937	1.0727	2.7632	1.0802	2.6447	1.0881	2.5367	47
14	1.0658	2.8915	1.0728	2.7611	1.0803	2.6428	1.0882	2.5350	46
15	1.0659	2.8892	1.0729	2.7591	1.0804	2.6410	1.0884	2.5333	45
16	1.0660	2.8869	1.0731	2.7570	1.0806	2.6391	1.0885	2.5316	44
17	1.0661	2.8846	1.0732	2.7550	1.0807	2.6372	1.0886	2.5299	43
18	1.0662	2.8824	1.0733	2.7529	1.0808	2.6353	1.0888	2.5281	42
19	1.0663	2.8801	1.0734	2.7509	1.0810	2.6335	1.0889	2.5264	41
20	1.0664	2.8778	1.0736	2.7488	1.0811	2.6316	1.0891	2.5247	40
21	1.0666	2.8756	1.0737	2.7468	1.0812	2.6297	1.0892	2.5230	39
22	1.0667	2.8733	1.0738	2.7447	1.0813	2.6279	1.0893	2.5213	38
23	1.0668	2.8711	1.0739	2.7427	1.0815	2.6260	1.0895	2.5196	37
24	1.0669	2.8688	1.0740	2.7406	1.0816	2.6242	1.0896	2.5179	36
25	1.0670	2.8666	1.0742	2.7386	1.0817	2.6223	1.0897	2.5163	35
26	1.0671	2.8644	1.0743	2.7366	1.0819	2.6205	1.0899	2.5146	34
27	1.0673	2.8621	1.0744	2.7346	1.0820	2.6186	1.0900	2.5129	33
28	1.0674	2.8599	1.0745	2.7325	1.0821	2.6168	1.0902	2.5112	32
29	1.0675	2.8577	1.0747	2.7305	1.0823	2.6150	1.0903	2.5095	31
30	1.0676	2.8554	1.0748	2.7285	1.0824	2.6131	1.0904	2.5078	30
31	1.0677	2.8532	1.0749	2.7265	1.0825	2.6113	1.0906	2.5062	29
32	1.0678	2.8510	1.0750	2.7245	1.0826	2.6095	1.0907	2.5045	28
33	1.0679	2.8488	1.0751	2.7225	1.0828	2.6076	1.0908	2.5028	27
34	1.0681	2.8466	1.0753	2.7205	1.0829	2.6058	1.0910	2.5011	26
35	1.0682	2.8444	1.0754	2.7185	1.0830	2.6040	1.0911	2.4995	25
36	1.0683	2.8422	1.0755	2.7165	1.0832	2.6022	1.0913	2.4978	24
37	1.0684	2.8400	1.0756	2.7145	1.0833	2.6003	1.0914	2.4961	23
38	1.0685	2.8378	1.0758	2.7125	1.0834	2.5985	1.0915	2.4945	22
39	1.0686	2.8356	1.0759	2.7105	1.0836	2.5967	1.0917	2.4928	21
40	1.0688	2.8334	1.0760	2.7085	1.0837	2.5949	1.0918	2.4912	20
41	1.0689	2.8312	1.0761	2.7065	1.0838	2.5931	1.0920	2.4895	19
42	1.0690	2.8290	1.0763	2.7045	1.0840	2.5913	1.0921	2.4879	18
43	1.0691	2.8269	1.0764	2.7026	1.0841	2.5895	1.0922	2.4862	17
44	1.0692	2.8247	1.0765	2.7006	1.0842	2.5877	1.0924	2.4846	16
45	1.0694	2.8225	1.0766	2.6986	1.0844	2.5859	1.0925	2.4829	15
46	1.0695	2.8204	1.0768	2.6967	1.0845	2.5841	1.0927	2.4813	14
47	1.0696	2.8182	1.0769	2.6947	1.0846	2.5823	1.0928	2.4797	13
48	1.0697	2.8160	1.0770	2.6927	1.0847	2.5805	1.0929	2.4780	12
49	1.0698	2.8139	1.0771	2.6908	1.0849	2.5787	1.0931	2.4764	11
50	1.0699	2.8117	1.0773	2.6888	1.0850	2.5770	1.0932	2.4748	10
51	1.0701	2.8096	1.0774	2.6869	1.0851	2.5752	1.0934	2.4731	9
52	1.0702	2.8074	1.0775	2.6849	1.0853	2.5734	1.0935	2.4715	8
53	1.0703	2.8053	1.0776	2.6830	1.0854	2.5716	1.0936	2.4699	7
54	1.0704	2.8032	1.0778	2.6810	1.0855	2.5699	1.0938	2.4683	6
55	1.0705	2.8010	1.0779	2.6791	1.0857	2.5681	1.0939	2.4666	5
56	1.0707	2.7989	1.0780	2.6772	1.0858	2.5663	1.0941	2.4650	4
57	1.0708	2.7968	1.0781	2.6752	1.0859	2.5646	1.0942	2.4634	3
58	1.0709	2.7947	1.0783	2.6733	1.0861	2.5628	1.0943	2.4618	2
59	1.0710	2.7925	1.0784	2.6714	1.0862	2.5610	1.0945	2.4602	1
60	1.0711	2.7904	1.0785	2.6695	1.0864	2.5593	1.0946	2.4586	0
′	69° cosec	69° sec	68° cosec	68° sec	67° cosec	67° sec	66° cosec	66° sec	′

′	24°		25°		26°		27°		′
	sec	cosec	sec	cosec	sec	cosec	sec	cosec	
0	1.0946	2.4586	1.1034	2.3662	1.1126	2.2812	1.1223	2.2027	60
1	1.0948	2.4570	1.1035	2.3647	1.1127	2.2798	1.1225	2.2014	59
2	1.0949	2.4554	1.1037	2.3632	1.1129	2.2784	1.1226	2.2002	58
3	1.0951	2.4538	1.1038	2.3618	1.1131	2.2771	1.1228	2.1989	57
4	1.0952	2.4522	1.1040	2.3603	1.1132	2.2757	1.1230	2.1977	56
5	1.0953	2.4506	1.1041	2.3588	1.1134	2.2744	1.1231	2.1964	55
6	1.0955	2.4490	1.1043	2.3574	1.1135	2.2730	1.1233	2.1952	54
7	1.0956	2.4474	1.1044	2.3559	1.1137	2.2717	1.1235	2.1939	53
8	1.0958	2.4458	1.1046	2.3544	1.1139	2.2703	1.1237	2.1927	52
9	1.0959	2.4442	1.1047	2.3530	1.1140	2.2690	1.1238	2.1914	51
10	1.0961	2.4426	1.1049	2.3515	1.1142	2.2676	1.1240	2.1902	50
11	1.0962	2.4411	1.1050	2.3501	1.1143	2.2663	1.1242	2.1889	49
12	1.0963	2.4395	1.1052	2.3486	1.1145	2.2650	1.1243	2.1877	48
13	1.0965	2.4379	1.1053	2.3472	1.1147	2.2636	1.1245	2.1865	47
14	1.0966	2.4363	1.1055	2.3457	1.1148	2.2623	1.1247	2.1852	46
15	1.0968	2.4347	1.1056	2.3443	1.1150	2.2610	1.1248	2.1840	45
16	1.0969	2.4332	1.1058	2.3428	1.1151	2.2596	1.1250	2.1828	44
17	1.0971	2.4316	1.1059	2.3414	1.1153	2.2583	1.1252	2.1815	43
18	1.0972	2.4300	1.1061	2.3399	1.1155	2.2570	1.1253	2.1803	42
19	1.0973	2.4285	1.1062	2.3385	1.1156	2.2556	1.1255	2.1791	41
20	1.0975	2.4269	1.1064	2.3371	1.1158	2.2543	1.1257	2.1778	40
21	1.0976	2.4254	1.1065	2.3356	1.1159	2.2530	1.1258	2.1766	39
22	1.0978	2.4238	1.1067	2.3342	1.1161	2.2517	1.1260	2.1754	38
23	1.0979	2.4222	1.1068	2.3328	1.1163	2.2503	1.1262	2.1742	37
24	1.0981	2.4207	1.1070	2.3313	1.1164	2.2490	1.1264	2.1730	36
25	1.0982	2.4191	1.1072	2.3299	1.1166	2.2477	1.1265	2.1717	35
26	1.0984	2.4176	1.1073	2.3285	1.1167	2.2464	1.1267	2.1705	34
27	1.0985	2.4160	1.1075	2.3271	1.1169	2.2451	1.1269	2.1693	33
28	1.0986	2.4145	1.1076	2.3256	1.1171	2.2438	1.1270	2.1681	32
29	1.0988	2.4130	1.1078	2.3242	1.1172	2.2425	1.1272	2.1669	31
30	1.0989	2.4114	1.1079	2.3228	1.1174	2.2411	1.1274	2.1657	30
31	1.0991	2.4099	1.1081	2.3214	1.1176	2.2398	1.1275	2.1645	29
32	1.0992	2.4083	1.1082	2.3200	1.1177	2.2385	1.1277	2.1633	28
33	1.0994	2.4068	1.1084	2.3186	1.1179	2.2372	1.1279	2.1620	27
34	1.0995	2.4053	1.1085	2.3172	1.1180	2.2359	1.1281	2.1608	26
35	1.0997	2.4037	1.1087	2.3158	1.1182	2.2346	1.1282	2.1596	25
36	1.0998	2.4022	1.1088	2.3143	1.1184	2.2333	1.1284	2.1584	24
37	1.1000	2.4007	1.1090	2.3129	1.1185	2.2320	1.1286	2.1572	23
38	1.1001	2.3992	1.1092	2.3115	1.1187	2.2307	1.1287	2.1560	22
39	1.1003	2.3976	1.1093	2.3101	1.1189	2.2294	1.1289	2.1548	21
40	1.1004	2.3961	1.1095	2.3087	1.1190	2.2282	1.1291	2.1536	20
41	1.1005	2.3946	1.1096	2.3073	1.1192	2.2269	1.1293	2.1525	19
42	1.1007	2.3931	1.1098	2.3059	1.1193	2.2256	1.1294	2.1513	18
43	1.1008	2.3916	1.1099	2.3046	1.1195	2.2243	1.1296	2.1501	17
44	1.1010	2.3901	1.1101	2.3032	1.1197	2.2230	1.1298	2.1489	16
45	1.1011	2.3886	1.1102	2.3018	1.1198	2.2217	1.1299	2.1477	15
46	1.1013	2.3871	1.1104	2.3004	1.1200	2.2204	1.1301	2.1465	14
47	1.1014	2.3856	1.1106	2.2990	1.1202	2.2192	1.1303	2.1453	13
48	1.1016	2.3841	1.1107	2.2976	1.1203	2.2179	1.1305	2.1441	12
49	1.1017	2.3826	1.1109	2.2962	1.1205	2.2166	1.1306	2.1430	11
50	1.1019	2.3811	1.1110	2.2949	1.1207	2.2153	1.1308	2.1418	10
51	1.1020	2.3796	1.1112	2.2935	1.1208	2.2141	1.1310	2.1406	9
52	1.1022	2.3781	1.1113	2.2921	1.1210	2.2128	1.1312	2.1394	8
53	1.1023	2.3766	1.1115	2.2907	1.1212	2.2115	1.1313	2.1382	7
54	1.1025	2.3751	1.1116	2.2894	1.1213	2.2103	1.1315	2.1371	6
55	1.1026	2.3736	1.1118	2.2880	1.1215	2.2090	1.1317	2.1359	5
56	1.1028	2.3721	1.1120	2.2866	1.1217	2.2077	1.1319	2.1347	4
57	1.1029	2.3706	1.1121	2.2853	1.1218	2.2065	1.1320	2.1335	3
58	1.1031	2.3691	1.1123	2.2839	1.1220	2.2052	1.1322	2.1324	2
59	1.1032	2.3677	1.1124	2.2825	1.1222	2.2039	1.1324	2.1312	1
60	1.1034	2.3662	1.1126	2.2812	1.1223	2.2027	1.1326	2.1300	0
′	cosec	sec	cosec	sec	cosec	sec	cosec	sec	′
	65°		64°		63°		62°		

′	28°		29°		30°		31°		′
	sec	cosec	sec	cosec	sec	cosec	sec	cosec	
0	1.1326	2.1300	1.1433	2.0627	1.1547	2.0000	1.1666	1.9416	60
1	1.1327	2.1289	1.1435	2.0616	1.1549	1.9990	1.1668	1.9407	59
2	1.1329	2.1277	1.1437	2.0605	1.1551	1.9980	1.1670	1.9397	58
3	1.1331	2.1266	1.1439	2.0594	1.1553	1.9970	1.1672	1.9388	57
4	1.1333	2.1254	1.1441	2.0583	1.1555	1.9960	1.1674	1.9378	56
5	1.1334	2.1242	1.1443	2.0573	1.1557	1.9950	1.1676	1.9369	55
6	1.1336	2.1231	1.1445	2.0562	1.1559	1.9940	1.1678	1.9360	54
7	1.1338	2.1219	1.1446	2.0551	1.1561	1.9930	1.1681	1.9350	53
8	1.1340	2.1208	1.1448	2.0540	1.1562	1.9920	1.1683	1.9341	52
9	1.1341	2.1196	1.1450	2.0530	1.1564	1.9910	1.1685	1.9332	51
10	1.1343	2.1185	1.1452	2.0519	1.1566	1.9900	1.1687	1.9322	50
11	1.1345	2.1173	1.1454	2.0508	1.1568	1.9890	1.1689	1.9313	49
12	1.1347	2.1162	1.1456	2.0498	1.1570	1.9880	1.1691	1.9304	48
13	1.1349	2.1150	1.1458	2.0487	1.1572	1.9870	1.1693	1.9295	47
14	1.1350	2.1139	1.1459	2.0476	1.1574	1.9860	1.1695	1.9285	46
15	1.1352	2.1127	1.1461	2.0466	1.1576	1.9850	1.1697	1.9276	45
16	1.1354	2.1116	1.1463	2.0455	1.1578	1.9840	1.1699	1.9267	44
17	1.1356	2.1104	1.1465	2.0444	1.1580	1.9830	1.1701	1.9258	43
18	1.1357	2.1093	1.1467	2.0434	1.1582	1.9820	1.1703	1.9248	42
19	1.1359	2.1082	1.1469	2.0423	1.1584	1.9811	1.1705	1.9239	41
20	1.1361	2.1070	1.1471	2.0413	1.1586	1.9801	1.1707	1.9230	40
21	1.1363	2.1059	1.1473	2.0402	1.1588	1.9791	1.1709	1.9221	39
22	1.1365	2.1048	1.1474	2.0392	1.1590	1.9781	1.1712	1.9212	38
23	1.1366	2.1036	1.1476	2.0381	1.1592	1.9771	1.1714	1.9203	37
24	1.1368	2.1025	1.1478	2.0370	1.1594	1.9761	1.1716	1.9193	36
25	1.1370	2.1014	1.1480	2.0360	1.1596	1.9752	1.1718	1.9184	35
26	1.1372	2.1002	1.1482	2.0349	1.1598	1.9742	1.1720	1.9175	34
27	1.1373	2.0991	1.1484	2.0339	1.1600	1.9732	1.1722	1.9166	33
28	1.1375	2.0980	1.1486	2.0329	1.1602	1.9722	1.1724	1.9157	32
29	1.1377	2.0969	1.1488	2.0318	1.1604	1.9713	1.1726	1.9148	31
30	1.1379	2.0957	1.1489	2.0308	1.1606	1.9703	1.1728	1.9139	30
31	1.1381	2.0946	1.1491	2.0297	1.1608	1.9693	1.1730	1.9130	29
32	1.1382	2.0935	1.1493	2.0287	1.1610	1.9683	1.1732	1.9121	28
33	1.1384	2.0924	1.1495	2.0276	1.1612	1.9674	1.1734	1.9112	27
34	1.1386	2.0912	1.1497	2.0266	1.1614	1.9664	1.1737	1.9102	26
35	1.1388	2.0901	1.1499	2.0256	1.1616	1.9654	1.1739	1.9093	25
36	1.1390	2.0890	1.1501	2.0245	1.1618	1.9645	1.1741	1.9084	24
37	1.1391	2.0879	1.1503	2.0235	1.1620	1.9635	1.1743	1.9075	23
38	1.1393	2.0868	1.1505	2.0224	1.1622	1.9625	1.1745	1.9066	22
39	1.1395	2.0857	1.1507	2.0214	1.1624	1.9616	1.1747	1.9057	21
40	1.1397	2.0846	1.1508	2.0204	1.1626	1.9606	1.1749	1.9048	20
41	1.1399	2.0835	1.1510	2.0194	1.1628	1.9596	1.1751	1.9039	19
42	1.1401	2.0824	1.1512	2.0183	1.1630	1.9587	1.1753	1.9030	18
43	1.1402	2.0812	1.1514	2.0173	1.1632	1.9577	1.1756	1.9021	17
44	1.1404	2.0801	1.1516	2.0163	1.1634	1.9568	1.1758	1.9013	16
45	1.1406	2.0790	1.1518	2.0152	1.1636	1.9558	1.1760	1.9004	15
46	1.1408	2.0779	1.1520	2.0142	1.1638	1.9549	1.1762	1.8995	14
47	1.1410	2.0768	1.1522	2.0132	1.1640	1.9539	1.1764	1.8986	13
48	1.1411	2.0757	1.1524	2.0122	1.1642	1.9530	1.1766	1.8977	12
49	1.1413	2.0746	1.1526	2.0111	1.1644	1.9520	1.1768	1.8968	11
50	1.1415	2.0735	1.1528	2.0101	1.1646	1.9510	1.1770	1.8959	10
51	1.1417	2.0725	1.1530	2.0091	1.1648	1.9501	1.1772	1.8950	9
52	1.1419	2.0714	1.1531	2.0081	1.1650	1.9491	1.1775	1.8941	8
53	1.1421	2.0703	1.1533	2.0071	1.1652	1.9482	1.1777	1.8932	7
54	1.1422	2.0692	1.1535	2.0061	1.1654	1.9473	1.1779	1.8924	6
55	1.1424	2.0681	1.1537	2.0050	1.1656	1.9463	1.1781	1.8915	5
56	1.1426	2.0670	1.1539	2.0040	1.1658	1.9454	1.1783	1.8906	4
57	1.1428	2.0659	1.1541	2.0030	1.1660	1.9444	1.1785	1.8897	3
58	1.1430	2.0648	1.1543	2.0020	1.1662	1.9435	1.1787	1.8888	2
59	1.1432	2.0637	1.1545	2.0010	1.1664	1.9425	1.1790	1.8879	1
60	1.1433	2.0627	1.1547	2.0000	1.1666	1.9416	1.1792	1.8871	0
′	cosec	sec	cosec	sec	cosec	sec	cosec	sec	′
	61°		60°		59°		58°		

′	32°		33°		34°		35°		′
	sec	cosec	sec	cosec	sec	cosec	sec	cosec	
0	1.1792	1.8871	1.1924	1.8361	1.2062	1.7883	1.2208	1.7434	60
1	1.1794	1.8862	1.1926	1.8352	1.2064	1.7875	1.2210	1.7427	59
2	1.1796	1.8853	1.1928	1.8344	1.2067	1.7867	1.2213	1.7420	58
3	1.1798	1.8844	1.1930	1.8336	1.2069	1.7860	1.2215	1.7413	57
4	1.1800	1.8836	1.1933	1.8328	1.2072	1.7852	1.2218	1.7405	56
5	1.1802	1.8827	1.1935	1.8320	1.2074	1.7844	1.2220	1.7398	55
6	1.1805	1.8818	1.1937	1.8311	1.2076	1.7837	1.2223	1.7391	54
7	1.1807	1.8809	1.1939	1.8303	1.2079	1.7829	1.2225	1.7384	53
8	1.1809	1.8801	1.1942	1.8295	1.2081	1.7821	1.2228	1.7377	52
9	1.1811	1.8792	1.1944	1.8287	1.2083	1.7814	1.2230	1.7369	51
10	1.1813	1.8783	1.1946	1.8279	1.2086	1.7806	1.2233	1.7362	50
11	1.1815	1.8775	1.1948	1.8271	1.2088	1.7798	1.2235	1.7355	49
12	1.1818	1.8766	1.1951	1.8263	1.2091	1.7791	1.2238	1.7348	48
13	1.1820	1.8757	1.1953	1.8255	1.2093	1.7783	1.2240	1.7341	47
14	1.1822	1.8749	1.1955	1.8246	1.2095	1.7776	1.2243	1.7334	46
15	1.1824	1.8740	1.1958	1.8238	1.2098	1.7768	1.2245	1.7327	45
16	1.1826	1.8731	1.1960	1.8230	1.2100	1.7760	1.2248	1.7319	44
17	1.1828	1.8723	1.1962	1.8222	1.2103	1.7753	1.2250	1.7312	43
18	1.1831	1.8714	1.1964	1.8214	1.2105	1.7745	1.2253	1.7305	42
19	1.1833	1.8706	1.1967	1.8206	1.2107	1.7738	1.2255	1.7298	41
20	1.1835	1.8697	1.1969	1.8198	1.2110	1.7730	1.2258	1.7291	40
21	1.1837	1.8688	1.1971	1.8190	1.2112	1.7723	1.2260	1.7284	39
22	1.1839	1.8680	1.1974	1.8182	1.2115	1.7715	1.2263	1.7277	38
23	1.1841	1.8671	1.1976	1.8174	1.2117	1.7708	1.2265	1.7270	37
24	1.1844	1.8663	1.1978	1.8166	1.2119	1.7700	1.2268	1.7263	36
25	1.1846	1.8654	1.1980	1.8158	1.2122	1.7693	1.2270	1.7256	35
26	1.1848	1.8646	1.1983	1.8150	1.2124	1.7685	1.2273	1.7249	34
27	1.1850	1.8637	1.1985	1.8142	1.2127	1.7678	1.2276	1.7242	33
28	1.1852	1.8629	1.1987	1.8134	1.2129	1.7670	1.2278	1.7234	32
29	1.1855	1.8620	1.1990	1.8126	1.2132	1.7663	1.2281	1.7227	31
30	1.1857	1.8611	1.1992	1.8118	1.2134	1.7655	1.2283	1.7220	30
31	1.1859	1.8603	1.1994	1.8110	1.2136	1.7648	1.2286	1.7213	29
32	1.1861	1.8595	1.1997	1.8102	1.2139	1.7640	1.2288	1.7206	28
33	1.1863	1.8586	1.1999	1.8094	1.2141	1.7633	1.2291	1.7199	27
34	1.1866	1.8578	1.2001	1.8086	1.2144	1.7625	1.2293	1.7192	26
35	1.1868	1.8569	1.2004	1.8078	1.2146	1.7618	1.2296	1.7185	25
36	1.1870	1.8561	1.2006	1.8070	1.2149	1.7610	1.2298	1.7178	24
37	1.1872	1.8552	1.2008	1.8062	1.2151	1.7603	1.2301	1.7171	23
38	1.1874	1.8544	1.2010	1.8054	1.2153	1.7596	1.2304	1.7164	22
39	1.1877	1.8535	1.2013	1.8047	1.2156	1.7588	1.2306	1.7157	21
40	1.1879	1.8527	1.2015	1.8039	1.2158	1.7581	1.2309	1.7151	20
41	1.1881	1.8519	1.2017	1.8031	1.2161	1.7573	1.2311	1.7144	19
42	1.1883	1.8510	1.2020	1.8023	1.2163	1.7566	1.2314	1.7137	18
43	1.1886	1.8502	1.2022	1.8015	1.2166	1.7559	1.2316	1.7130	17
44	1.1888	1.8493	1.2024	1.8007	1.2168	1.7551	1.2319	1.7123	16
45	1.1890	1.8485	1.2027	1.7999	1.2171	1.7544	1.2322	1.7116	15
46	1.1892	1.8477	1.2029	1.7992	1.2173	1.7537	1.2324	1.7109	14
47	1.1894	1.8468	1.2031	1.7984	1.2175	1.7529	1.2327	1.7102	13
48	1.1897	1.8460	1.2034	1.7976	1.2178	1.7522	1.2329	1.7095	12
49	1.1899	1.8452	1.2036	1.7968	1.2180	1.7514	1.2332	1.7088	11
50	1.1901	1.8443	1.2039	1.7960	1.2183	1.7507	1.2335	1.7081	10
51	1.1903	1.8435	1.2041	1.7953	1.2185	1.7500	1.2337	1.7075	9
52	1.1906	1.8427	1.2043	1.7945	1.2188	1.7493	1.2340	1.7068	8
53	1.1908	1.8418	1.2046	1.7937	1.2190	1.7485	1.2342	1.7061	7
54	1.1910	1.8410	1.2048	1.7929	1.2193	1.7478	1.2345	1.7054	6
55	1.1912	1.8402	1.2050	1.7921	1.2195	1.7471	1.2348	1.7047	5
56	1.1915	1.8394	1.2053	1.7914	1.2198	1.7463	1.2350	1.7040	4
57	1.1917	1.8385	1.2055	1.7906	1.2200	1.7456	1.2353	1.7033	3
58	1.1919	1.8377	1.2057	1.7898	1.2203	1.7449	1.2355	1.7027	2
59	1.1921	1.8369	1.2060	1.7891	1.2205	1.7442	1.2358	1.7020	1
60	1.1924	1.8361	1.2062	1.7883	1.2208	1.7434	1.2361	1.7013	0
′	cosec	sec	cosec	sec	cosec	sec	cosec	sec	′
	57°		56°		55°		54°		

′	36°		37°		38°		39°		′
	sec	cosec	sec	cosec	sec	cosec	sec	cosec	
0	1.2361	1.7013	1.2521	1.6616	1.2690	1.6243	1.2867	1.5890	60
1	1.2363	1.7006	1.2524	1.6610	1.2693	1.6237	1.2871	1.5884	59
2	1.2366	1.6999	1.2527	1.6603	1.2696	1.6231	1.2874	1.5879	58
3	1.2368	1.6993	1.2530	1.6597	1.2699	1.6224	1.2877	1.5873	57
4	1.2371	1.6986	1.2532	1.6591	1.2702	1.6218	1.2880	1.5867	56
5	1.2374	1.6979	1.2535	1.6584	1.2705	1.6212	1.2883	1.5862	55
6	1.2376	1.6972	1.2538	1.6578	1.2707	1.6206	1.2886	1.5856	54
7	1.2379	1.6965	1.2541	1.6572	1.2710	1.6200	1.2889	1.5850	53
8	1.2382	1.6959	1.2543	1.6565	1.2713	1.6194	1.2892	1.5845	52
9	1.2384	1.6952	1.2546	1.6559	1.2716	1.6188	1.2895	1.5839	51
10	1.2387	1.6945	1.2549	1.6552	1.2719	1.6182	1.2898	1.5833	50
11	1.2389	1.6938	1.2552	1.6546	1.2722	1.6176	1.2901	1.5828	49
12	1.2392	1.6932	1.2554	1.6540	1.2725	1.6170	1.2904	1.5822	48
13	1.2395	1.6925	1.2557	1.6533	1.2728	1.6164	1.2907	1.5816	47
14	1.2397	1.6918	1.2560	1.6527	1.2731	1.6159	1.2910	1.5811	46
15	1.2400	1.6912	1.2563	1.6521	1.2734	1.6153	1.2913	1.5805	45
16	1.2403	1.6905	1.2565	1.6514	1.2737	1.6147	1.2916	1.5799	44
17	1.2405	1.6898	1.2568	1.6508	1.2739	1.6141	1.2919	1.5794	43
18	1.2408	1.6891	1.2571	1.6502	1.2742	1.6135	1.2922	1.5788	42
19	1.2411	1.6885	1.2574	1.6496	1.2745	1.6129	1.2926	1.5783	41
20	1.2413	1.6878	1.2577	1.6489	1.2748	1.6123	1.2929	1.5777	40
21	1.2416	1.6871	1.2579	1.6483	1.2751	1.6117	1.2932	1.5771	39
22	1.2419	1.6865	1.2582	1.6477	1.2754	1.6111	1.2935	1.5766	38
23	1.2421	1.6858	1.2585	1.6470	1.2757	1.6105	1.2938	1.5760	37
24	1.2424	1.6851	1.2588	1.6464	1.2760	1.6099	1.2941	1.5755	36
25	1.2427	1.6845	1.2591	1.6458	1.2763	1.6093	1.2944	1.5749	35
26	1.2429	1.6838	1.2593	1.6452	1.2766	1.6087	1.2947	1.5743	34
27	1.2432	1.6831	1.2596	1.6445	1.2769	1.6081	1.2950	1.5738	33
28	1.2435	1.6825	1.2599	1.6439	1.2772	1.6077	1.2953	1.5732	32
29	1.2437	1.6818	1.2602	1.6433	1.2775	1.6070	1.2956	1.5727	31
30	1.2440	1.6812	1.2605	1.6427	1.2778	1.6064	1.2960	1.5721	30
31	1.2443	1.6805	1.2607	1.6420	1.2781	1.6058	1.2963	1.5716	29
32	1.2445	1.6798	1.2610	1.6414	1.2784	1.6052	1.2966	1.5710	28
33	1.2448	1.6792	1.2613	1.6408	1.2787	1.6046	1.2969	1.5705	27
34	1.2451	1.6785	1.2616	1.6402	1.2790	1.6040	1.2972	1.5699	26
35	1.2453	1.6779	1.2619	1.6396	1.2793	1.6034	1.2975	1.5694	25
36	1.2456	1.6772	1.2622	1.6389	1.2795	1.6029	1.2978	1.5688	24
37	1.2459	1.6766	1.2624	1.6383	1.2798	1.6023	1.2981	1.5683	23
38	1.2461	1.6759	1.2627	1.6377	1.2801	1.6017	1.2985	1.5677	22
39	1.2464	1.6752	1.2630	1.6371	1.2804	1.6011	1.2988	1.5672	21
40	1.2467	1.6746	1.2633	1.6365	1.2807	1.6005	1.2991	1.5666	20
41	1.2470	1.6739	1.2636	1.6359	1.2810	1.6000	1.2994	1.5661	19
42	1.2472	1.6733	1.2639	1.6352	1.2813	1.5994	1.2997	1.5655	18
43	1.2475	1.6726	1.2641	1.6346	1.2816	1.5988	1.3000	1.5650	17
44	1.2478	1.6720	1.2644	1.6340	1.2819	1.5982	1.3003	1.5644	16
45	1.2480	1.6713	1.2647	1.6334	1.2822	1.5976	1.3006	1.5639	15
46	1.2483	1.6707	1.2650	1.6328	1.2825	1.5971	1.3010	1.5633	14
47	1.2486	1.6700	1.2653	1.6322	1.2828	1.5965	1.3013	1.5628	13
48	1.2488	1.6694	1.2656	1.6316	1.2831	1.5959	1.3016	1.5622	12
49	1.2490	1.6687	1.2659	1.6309	1.2834	1.5953	1.3019	1.5617	11
50	1.2494	1.6681	1.2661	1.6303	1.2837	1.5947	1.3022	1.5611	10
51	1.2497	1.6674	1.2664	1.6297	1.2840	1.5942	1.3025	1.5606	9
52	1.2499	1.6668	1.2667	1.6291	1.2843	1.5936	1.3029	1.5600	8
53	1.2502	1.6661	1.2670	1.6285	1.2846	1.5930	1.3032	1.5595	7
54	1.2505	1.6655	1.2673	1.6279	1.2849	1.5924	1.3035	1.5590	6
55	1.2508	1.6648	1.2676	1.6273	1.2852	1.5919	1.3038	1.5584	5
56	1.2510	1.6642	1.2679	1.6267	1.2855	1.5913	1.3041	1.5579	4
57	1.2513	1.6636	1.2681	1.6261	1.2858	1.5907	1.3044	1.5573	3
58	1.2516	1.6629	1.2684	1.6255	1.2861	1.5901	1.3048	1.5568	2
59	1.2519	1.6623	1.2687	1.6249	1.2864	1.5896	1.3051	1.5563	1
60	1.2521	1.6616	1.2690	1.6243	1.2867	1.5890	1.3054	1.5557	0
′	cosec	sec	cosec	sec	cosec	sec	cosec	sec	′
	53°		52°		51°		50°		

′	40° sec	40° cosec	41° sec	41° cosec	42° sec	42° cosec	43° sec	43° cosec	′
0	1.3054	1.5557	1.3250	1.5242	1.3456	1.4945	1.3673	1.4663	60
1	1.3057	1.5552	1.3253	1.5237	1.3460	1.4940	1.3677	1.4658	59
2	1.3060	1.5546	1.3257	1.5232	1.3463	1.4935	1.3681	1.4654	58
3	1.3064	1.5541	1.3260	1.5227	1.3467	1.4930	1.3684	1.4649	57
4	1.3067	1.5536	1.3263	1.5222	1.3470	1.4925	1.3688	1.4644	56
5	1.3070	1.5530	1.3267	1.5217	1.3474	1.4921	1.3692	1.4640	55
6	1.3073	1.5525	1.3270	1.5212	1.3477	1.4916	1.3695	1.4635	54
7	1.3076	1.5520	1.3274	1.5207	1.3481	1.4911	1.3699	1.4631	53
8	1.3080	1.5514	1.3277	1.5202	1.3485	1.4906	1.3703	1.4626	52
9	1.3083	1.5509	1.3280	1.5197	1.3488	1.4901	1.3707	1.4622	51
10	1.3086	1.5503	1.3284	1.5192	1.3492	1.4897	1.3710	1.4617	50
11	1.3089	1.5498	1.3287	1.5187	1.3495	1.4892	1.3714	1.4613	49
12	1.3092	1.5493	1.3290	1.5182	1.3499	1.4887	1.3718	1.4608	48
13	1.3096	1.5487	1.3294	1.5177	1.3502	1.4882	1.3722	1.4604	47
14	1.3099	1.5482	1.3297	1.5171	1.3506	1.4877	1.3725	1.4599	46
15	1.3102	1.5477	1.3301	1.5166	1.3509	1.4873	1.3729	1.4595	45
16	1.3105	1.5471	1.3304	1.5161	1.3513	1.4868	1.3733	1.4590	44
17	1.3109	1.5466	1.3307	1.5156	1.3517	1.4863	1.3737	1.4586	43
18	1.3112	1.5461	1.3311	1.5151	1.3520	1.4858	1.3740	1.4581	42
19	1.3115	1.5456	1.3314	1.5146	1.3524	1.4854	1.3744	1.4577	41
20	1.3118	1.5450	1.3318	1.5141	1.3527	1.4849	1.3748	1.4572	40
21	1.3121	1.5445	1.3321	1.5136	1.3531	1.4844	1.3752	1.4568	39
22	1.3125	1.5440	1.3324	1.5131	1.3534	1.4839	1.3756	1.4563	38
23	1.3128	1.5434	1.3328	1.5126	1.3538	1.4835	1.3759	1.4559	37
24	1.3131	1.5429	1.3331	1.5121	1.3542	1.4830	1.3763	1.4554	36
25	1.3134	1.5424	1.3335	1.5116	1.3545	1.4825	1.3767	1.4550	35
26	1.3138	1.5419	1.3338	1.5111	1.3549	1.4821	1.3771	1.4545	34
27	1.3141	1.5413	1.3342	1.5106	1.3552	1.4816	1.3774	1.4541	33
28	1.3144	1.5408	1.3345	1.5101	1.3556	1.4811	1.3778	1.4536	32
29	1.3148	1.5403	1.3348	1.5096	1.3560	1.4806	1.3782	1.4532	31
30	1.3151	1.5398	1.3352	1.5092	1.3563	1.4802	1.3786	1.4527	30
31	1.3154	1.5392	1.3355	1.5087	1.3567	1.4797	1.3790	1.4523	29
32	1.3157	1.5387	1.3359	1.5082	1.3571	1.4792	1.3794	1.4518	28
33	1.3161	1.5382	1.3362	1.5077	1.3574	1.4788	1.3797	1.4514	27
34	1.3164	1.5377	1.3366	1.5072	1.3578	1.4783	1.3801	1.4510	26
35	1.3167	1.5371	1.3369	1.5067	1.3581	1.4778	1.3805	1.4505	25
36	1.3170	1.5366	1.3372	1.5062	1.3585	1.4774	1.3809	1.4501	24
37	1.3174	1.5361	1.3376	1.5057	1.3589	1.4769	1.3813	1.4496	23
38	1.3177	1.5356	1.3379	1.5052	1.3592	1.4764	1.3816	1.4492	22
39	1.3180	1.5351	1.3383	1.5047	1.3596	1.4760	1.3820	1.4487	21
40	1.3184	1.5345	1.3386	1.5042	1.3600	1.4755	1.3824	1.4483	20
41	1.3187	1.5340	1.3390	1.5037	1.3603	1.4750	1.3828	1.4479	19
42	1.3190	1.5335	1.3393	1.5032	1.3607	1.4746	1.3832	1.4474	18
43	1.3193	1.5330	1.3397	1.5027	1.3611	1.4741	1.3836	1.4470	17
44	1.3197	1.5325	1.3400	1.5022	1.3614	1.4736	1.3839	1.4465	16
45	1.3200	1.5319	1.3404	1.5018	1.3618	1.4732	1.3843	1.4461	15
46	1.3203	1.5314	1.3407	1.5013	1.3622	1.4727	1.3847	1.4457	14
47	1.3207	1.5309	1.3411	1.5008	1.3625	1.4723	1.3851	1.4452	13
48	1.3210	1.5304	1.3414	1.5003	1.3629	1.4718	1.3855	1.4448	12
49	1.3213	1.5299	1.3418	1.4998	1.3633	1.4713	1.3859	1.4443	11
50	1.3217	1.5294	1.3421	1.4993	1.3636	1.4709	1.3863	1.4439	10
51	1.3220	1.5289	1.3425	1.4988	1.3640	1.4704	1.3867	1.4435	9
52	1.3223	1.5283	1.3428	1.4983	1.3644	1.4699	1.3870	1.4430	8
53	1.3227	1.5278	1.3432	1.4979	1.3647	1.4695	1.3874	1.4426	7
54	1.3230	1.5273	1.3435	1.4974	1.3651	1.4690	1.3878	1.4422	6
55	1.3233	1.5268	1.3439	1.4969	1.3655	1.4686	1.3882	1.4417	5
56	1.3237	1.5263	1.3442	1.4964	1.3658	1.4681	1.3886	1.4413	4
57	1.3240	1.5258	1.3446	1.4959	1.3662	1.4676	1.3890	1.4408	3
58	1.3243	1.5253	1.3449	1.4954	1.3666	1.4672	1.3894	1.4404	2
59	1.3247	1.5248	1.3453	1.4949	1.3669	1.4667	1.3898	1.4400	1
60	1.3250	1.5242	1.3456	1.4945	1.3673	1.4663	1.3902	1.4395	0
′	cosec 49°	sec 49°	cosec 48°	sec 48°	cosec 47°	sec 47°	cosec 46°	sec 46°	′

′	44° sec	44° cosec	′	′	44° sec	44° cosec	′	′	44° sec	44° cosec	′
0	1.3902	1.4395	60	21	1.3984	1.4305	39	41	1.4065	1.4221	19
1	1.3905	1.4391	59	22	1.3988	1.4301	38	42	1.4069	1.4217	18
2	1.3909	1.4387	58	23	1.3992	1.4297	37	43	1.4073	1.4212	17
3	1.3913	1.4382	57	24	1.3996	1.4292	36	44	1.4077	1.4208	16
4	1.3917	1.4378	56	25	1.4000	1.4288	35	45	1.4081	1.4204	15
5	1.3921	1.4374	55	26	1.4004	1.4284	34	46	1.4085	1.4200	14
6	1.3925	1.4370	54	27	1.4008	1.4280	33	47	1.4089	1.4196	13
7	1.3929	1.4365	53	28	1.4012	1.4276	32	48	1.4093	1.4192	12
8	1.3933	1.4361	52	29	1.4016	1.4271	31	49	1.4097	1.4188	11
9	1.3937	1.4357	51	30	1.4020	1.4267	30	50	1.4101	1.4183	10
10	1.3941	1.4352	50	31	1.4024	1.4263	29	51	1.4105	1.4179	9
11	1.3945	1.4348	49	32	1.4028	1.4259	28	52	1.4109	1.4175	8
12	1.3949	1.4344	48	33	1.4032	1.4254	27	53	1.4113	1.4171	7
13	1.3953	1.4339	47	34	1.4036	1.4250	26	54	1.4117	1.4167	6
14	1.3957	1.4335	46	35	1.4040	1.4246	25	55	1.4122	1.4163	5
15	1.3960	1.4331	45	36	1.4044	1.4242	24	56	1.4126	1.4159	4
16	1.3964	1.4327	44	37	1.4048	1.4238	23	57	1.4130	1.4154	3
17	1.3968	1.4322	43	38	1.4052	1.4233	22	58	1.4134	1.4150	2
18	1.3972	1.4318	42	39	1.4056	1.4229	21	59	1.4138	1.4146	1
19	1.3976	1.4314	41	40	1.4060	1.4225	20	60	1.4142	1.4142	0
20	1.3980	1.4310	40								
′	cosec 45°	sec 45°	′	′	cosec 45°	sec 45°	′	′	cosec 45°	sec 45°	′

GREEK ALPHABET

Alpha (ăl′fȧ)	Α α	Nu (nū)	Ν ν
Beta (bā′tȧ)	Β β	Xi (ksē)	Ξ ξ
Gamma (găm′ȧ)	Γ γ	Omicron (ŏm′ĭkrŏn)	Ο ο
Delta (dĕl′tȧ)	Δ δ or ∂	Pi (pī)	Π π
Epsilon (ĕp′sĭlŏn)	Ε ϵ	Rho (rō)	Ρ ρ
Zeta (zā′tȧ)	Ζ ζ	Sigma (sĭg′mȧ)	Σ σ or ς
Eta (ā′tȧ)	Η η	Tau (tô)	Τ τ
Theta (thā′tȧ)	Θ θ	Upsilon (ūp′sĭlŏn)	Υ υ
Iota (īō′tȧ)	Ι ι	Phi (fē)	Φ φ or ϕ
Kappa (kăp′ȧ)	Κ κ	Chi (kē)	Χ χ
Lambda (lăm′dȧ)	Λ λ	Psi (psē)	Ψ ψ
Mu (mū)	Μ μ	Omega (ō′mĕgȧ)	Ω ω

USEFUL FORMULAS FOR READY REFERENCES

A = area $\qquad$ V = volume

CIRCLE

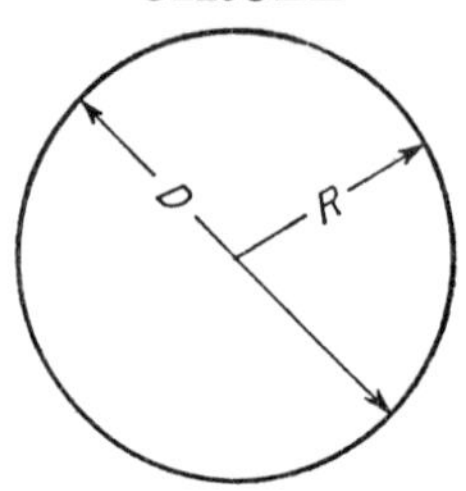

$A = 3.1416R^2$

$A = .7854D^2$

SPHERE

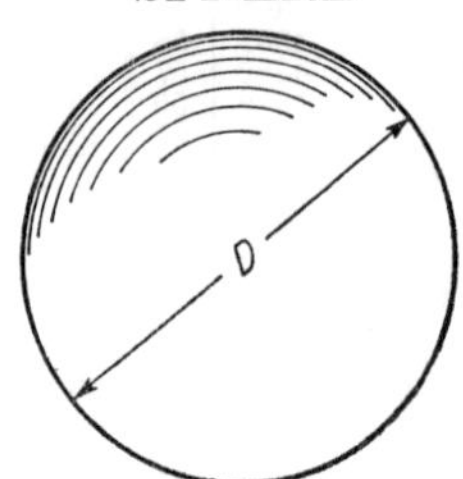

$A = 12566R^2 = 3.1416D^2$

$V = 4.1888R^3 = .5236D^3$

TRAPEZOID

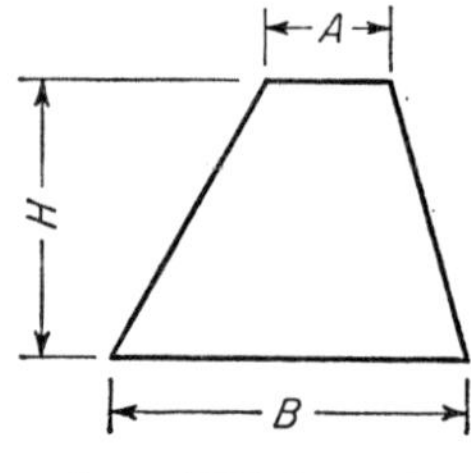

$A = \frac{1}{2}H(A + B)$

CONE

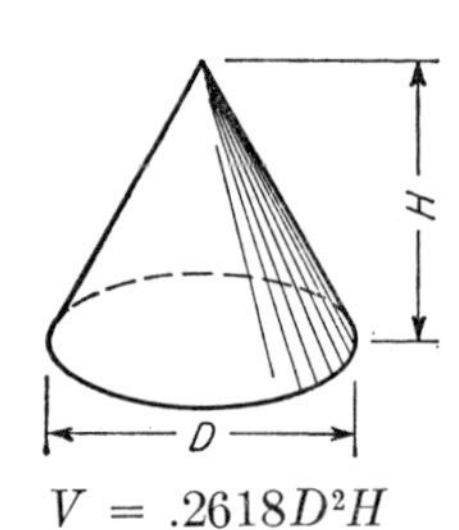

$V = .2618D^2H$

RIGHT CIRCULAR CYLINDER

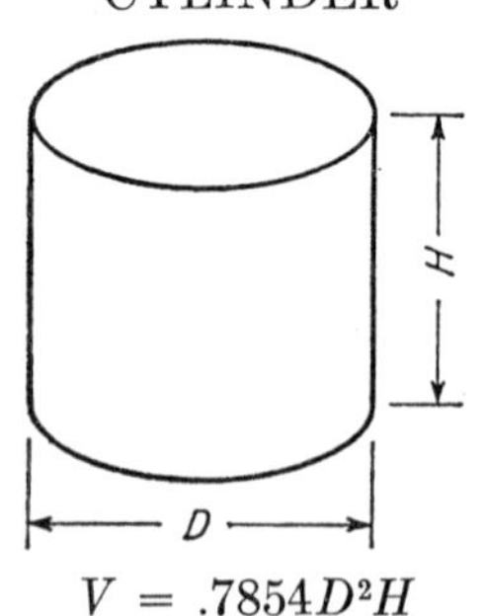

$V = .7854D^2H$

Gallons = $.0034D^2H$ if D and H are in inches.

PARALLELEPIPED

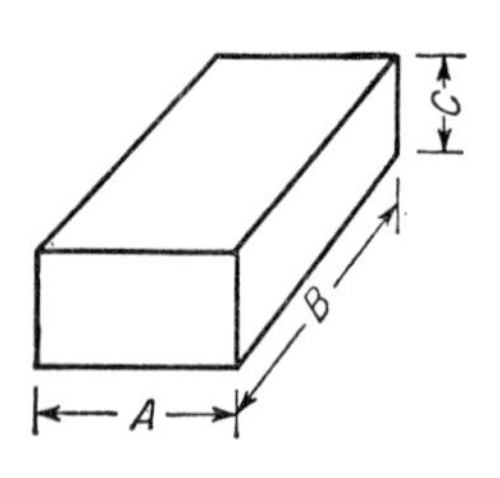

$V = ABC$

Gallons = $.004329ABC$ if A, B, and C are in inches.

HEXAGON

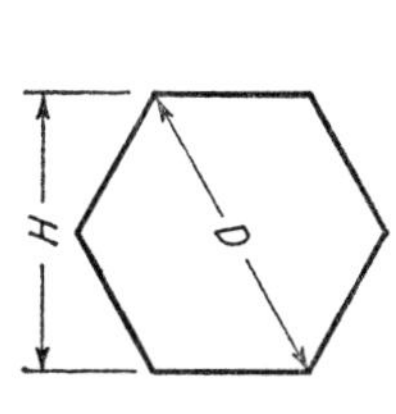

$D = 1.1547H$

KEYWAY

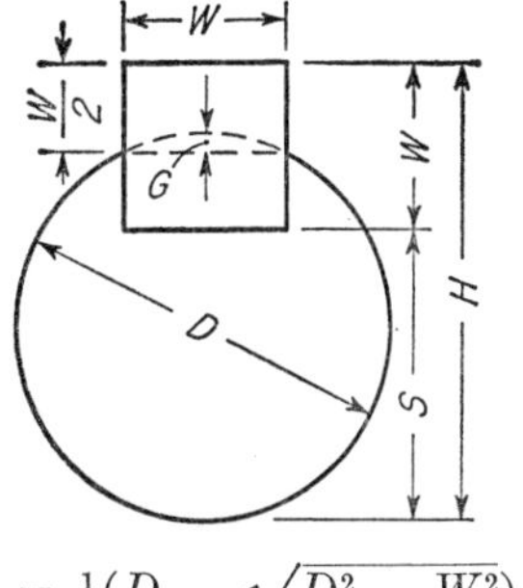

$G = \frac{1}{2}(D - \sqrt{D^2 - W^2})$

$S = D - \frac{1}{2}W - G$

SQUARE

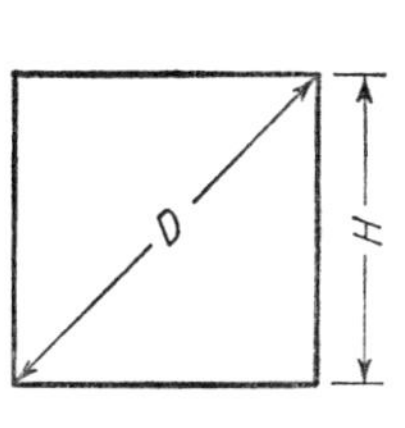

$D = 1.4142H$

CHORD

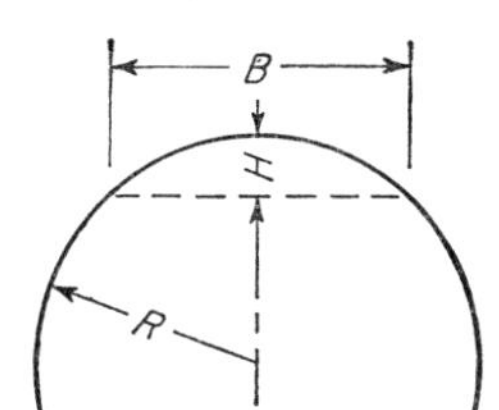

$$R = \frac{B^2 + 4H^2}{8H}$$

STANDARD HEXAGON NUT

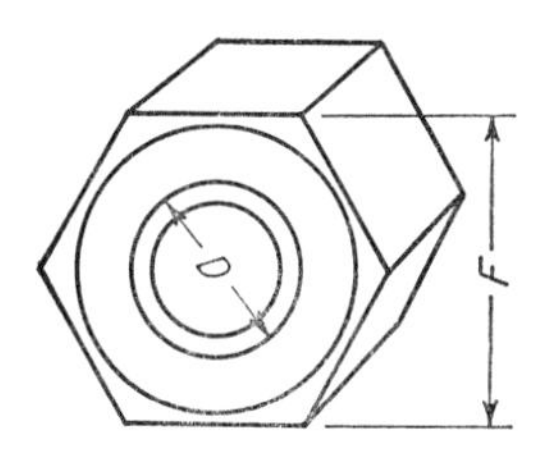

$F = 1.5D + .125$

INSCRIBED CIRCLE

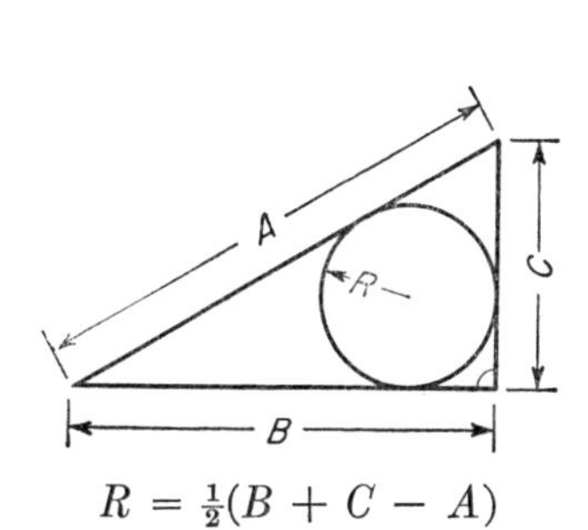

$R = \frac{1}{2}(B + C - A)$

INSCRIBED CIRCLE

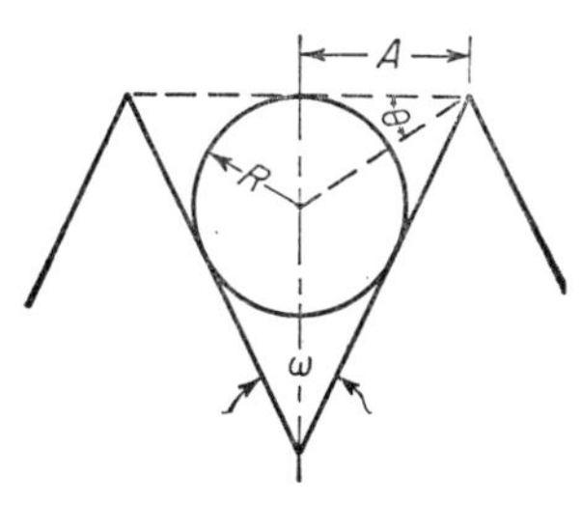

$\theta = \frac{1}{4}(180° - \omega)$

$R = A \tan \theta$

HARDENED AND GROUND STEEL BUSHINGS

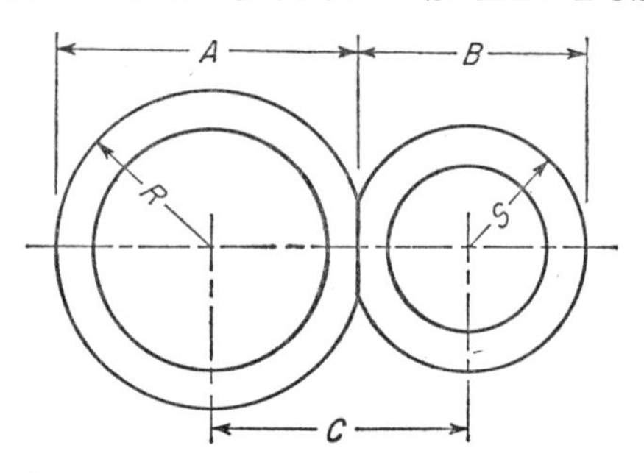

$$A = R + \frac{R^2 + C^2 - S^2}{2C} \qquad B = C - A + R + S$$

DOVETAIL

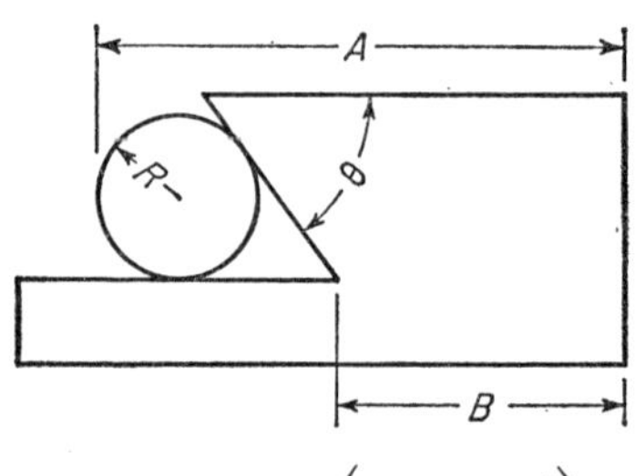

$$A = B + R\left(1 + \cot\frac{\theta}{2}\right)$$

OBLIQUE TRIANGLES—PROJECTION FORMULAS

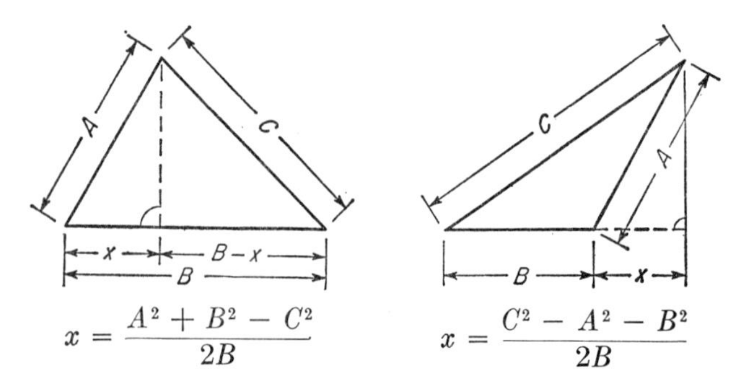

$$x = \frac{A^2 + B^2 - C^2}{2B} \qquad x = \frac{C^2 - A^2 - B^2}{2B}$$

SINE LAW

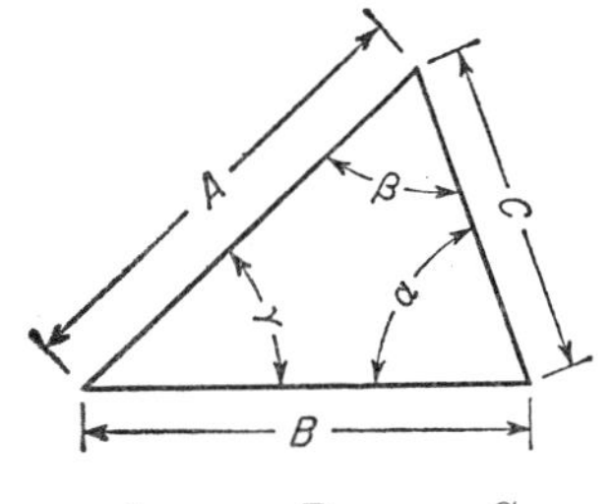

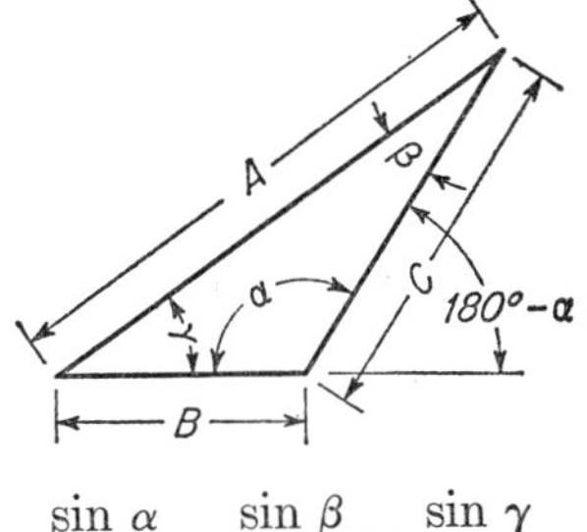

$$\frac{A}{\sin\alpha} = \frac{B}{\sin\beta} = \frac{C}{\sin\gamma} \qquad \frac{\sin\alpha}{A} = \frac{\sin\beta}{B} = \frac{\sin\gamma}{C}$$

$$A = B \sin\alpha \csc\beta = C \sin\alpha \csc\gamma$$
$$B = A \sin\beta \csc\alpha = C \sin\beta \csc\gamma$$
$$C = A \sin\gamma \csc\alpha = B \sin\gamma \csc\beta$$

$$\sin\alpha = \frac{A \sin\beta}{B} = \frac{A \sin\gamma}{C}$$
$$\sin\beta = \frac{B \sin\alpha}{A} = \frac{B \sin\gamma}{C}$$
$$\sin\gamma = \frac{C \sin\alpha}{A} = \frac{C \sin\beta}{B}$$

COSINE LAW

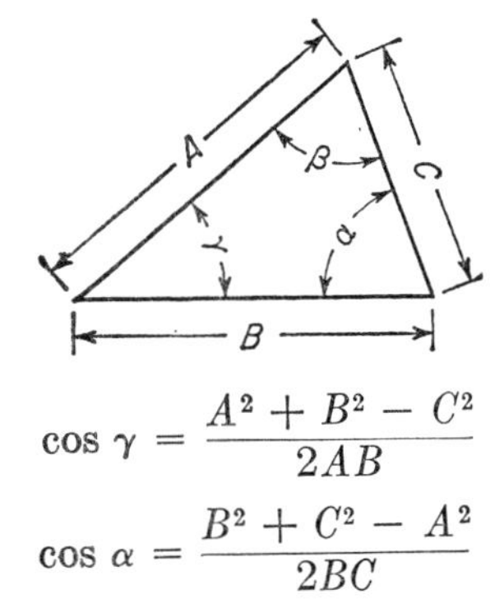

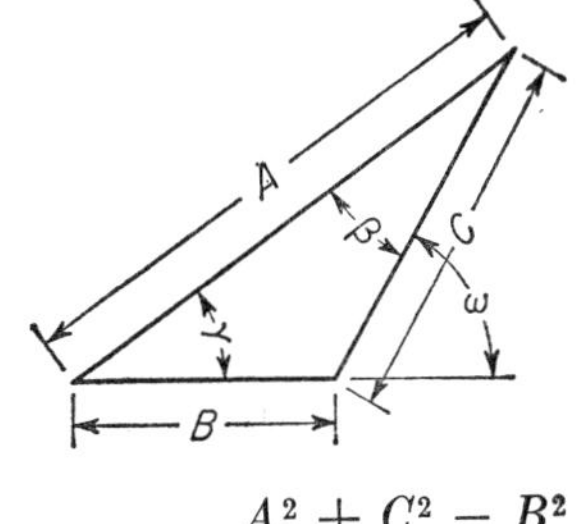

$$\cos\gamma = \frac{A^2 + B^2 - C^2}{2AB} \qquad \cos\beta = \frac{A^2 + C^2 - B^2}{2AC}$$

$$\cos\alpha = \frac{B^2 + C^2 - A^2}{2BC} \qquad \cos\omega = \frac{A^2 - B^2 - C^2}{2BC}$$

ANSWERS

Page	Problem	Symbol	Variable	Answer
17	1	D	748653	6
	2	E	439267	4
	3	F	254273	5
	4	G	532581	6
	5	H	896247	0
	6	J	573862	0
	7	K	7823	4
	8	L	43875	8
	9	M	8236	4
19	1	A	.0982	$\frac{491}{5000}$
	2	B	9.542	$9\frac{271}{500}$
	3	C	.0053	$\frac{53}{10000}$
	4	D	$\frac{932}{1000}$	.932
	5	E	$\frac{31}{10000}$	.0031
	6	F	$\frac{95}{100}$	.95
	1	A	3.4	1.67
20	2	A	7.5323	5.8531
	3	A	11.746	.708
	4	A	.4631	5.9811
	5	A	4.6273	2.4077
21	6	A	2.4285	6.3368
	7	A	2.4285	3.2908
22	1	A	4.3927	38.015
	2	B	8.3576	90.730
	3	C	6.2594	154.46
	4	D	.73826	6.2727
	5	E	.87543	4.2875
	6	F	.46937	2.7999
	7	G	3.4278	23.075
	8	H	7.3492	59.231
	9	J	.93748	10.112
	10	K	1.9	58.872
24	1	B	6.3268	8.2591
	2	C	.85924	7.9892
	3	D	.09387	84.962
	4	E	8.1245	2.3875
	5	F	.83945	1.8458
	6	G	4.3768	1.5994

Page	Problem	Symbol	Variable	Answer
24	7	H	.89537	6.6872
	8	J	9.2843	121.29
	9	K	.07659	.87978
25	10	L	9.3854	.03958
	11	M	4.5876	1.6654
	12	N	.74382	.94319
	13	P	8.2953	.59016
	14	R	.09437	.09287
	15	S	2.4895	2.3557
	16	T	.06382	.43442
	17	U	19	.89473
	18	V	2186	11.566
	19	W	691	.37916
	20	A	.9738	2.3864
	21	A	12.438	1.8545
26	22	A	1.4	1.535
	23	A	.59286	1.3224
	24	A	.4183	1.2517
27	25	A	3.2	.5241
	26	A	2.1	.33
	27	A	2.2589	1.4759
	28	A	6.843	6.652
28	29	A	9.498	1.609
	30	A	1.73	3.78
	31	B	3.705	1.0775
	32	B	.085	11
32	1	N	16	.241
	2	N	11	1.261
	3	N	12	2.487
	4	N	15	3.665
33	5	N	10	.170
	6	N	15	.495
	7	N	12	.632
	8	N	19	.539
	9	N	19	.759
	10	N	17	.837
36	1	L	8	6° 40′
	2	L	4	23° 20′
	3	L	11	37° 55′
	4	L	6	44° 30′
	5	L	10	53° 50′
	6	L	7	63° 35′
	7	L	11	76° 55′
	8	L	7	87° 35′

Page	Problem	Symbol	Variable	Answer
36	9	θ	26°	4° 20′
	10	θ	22°	2° 12′
	11	θ	25°	2° 22′
	12	θ	28°	1° 30′
37	13	U	12	2.137
	14	L	7	57° 35′
40	2	A	52	32
	3	B	28	88°
	4	C	22	5
41	7	D	10	4
	8	E	17	41
	9	F	92	−46
	10	G	19	14
	11	H	26	16
	12	J	24	42
	13	K	14	34
	14	L	45	−15
	15	M	17	−12
	16	N	30	17
	17	P	53	−30
	18	Q	11	−12
	19	R	16	−6
	20	S	17	22
	21	T	8	−9
	22	U	62	−11
	23	V	37	−108
	24	W	42	−21
	25	A	23	−16
43	1	M	78	80
	2	N	12.8	104
	3	L	98	93
	4	K	88	85.5
	5	P	25	121
	6	Q	104	113
	7	R	18	17
	8	A	8	−165
	9	B	2	−56
	10	C	15	388
44	11	D	8	−471
	12	E	20	−5870
	13	F	27	21
	14	G	11	71
	15	H	19	−27
	16	J	20	61

Page	Problem	Symbol	Variable	Answer	
44	17	*S*	17	−390	
	18	*T*	6	85	
54	1	*N*	32	1.3125	
	2	*M*	20	12	
	3	*R*	65	31.787	
	4	*T*	16.75	$10.659	
	5	*S*	11.325	.661	
55	6	*F*	20.2	38.037	
	7	*G*	2.458	.4115	
	8	*K*	67	$\frac{35}{67}$	
	9	*L*	38	21.966	
	10	*H*	38	$\frac{3}{266}$	
	11	*D*	$\frac{8}{11}$	$\frac{10}{11}$	
	12	*C*	53	1.1886	
	13	*P*	57	20.357	
	14	*P*	98	333.2	
57	1	*B*	1280	2596.1	
	2	*C*	63	854.49	
	3	*D*	5.875	1.7279	
	4	*E*	14	24.444	
	5	*G*	105.5	71.990	
	6	*R*	375.5	44.08	
	7	*S*	27.5	2799.2	
	8	*Q*	28600	2.2309	
58	9	*T*	350.5	1602.2	
	10	*H*	.645	.129	
59	1	*N*	21	28.767%	
	2	*L*	225	9.3333%	
	3	*G*	$\frac{2}{9}$	77.777%	
	4	*M*	10	130	
	5	*H*	29	966.66	
	6	No Variable		1.042%	
	7	*F*	13.1	93.893%	
	8	*C*	8.25	$6.64	
	9	*S*	59.75	$36.12	
	10	*T*	.168	.01176	
	11	*N*	48.25	$53.459	
60	12	*D*	.965	.96443	
				Tin	8.19
	13	*A*	9.75	Copper	.4875
				Antimony	.975
				Lead	.0975
	14	*B*	519	Bismuth	259.5
				Lead	129.75

Page	Problem	Symbol	Variable	Answer	
60	14	*B*	519	Tin	64.875
			-	Cadmium	64.875
61	15	*E*	1995	798	
	16	*J*	255	Yellow	63.75
				Green	19.125
				Red	0
				Black	12.75
				Blue	31.875
63	1	*A*	6.7	3.4388	
	2	*B*	3.12	.76923	
	3	*C*	12.1	1.3884	
	4	*D*	6.7	17.42	
64	5	*E*	16.3	.9196	
	6	*F*	11.2	.84625	
	7	*G*	.468	1.8756	
	8	*H*	.406	.50246	
65	9	*J*	1.75	.00102	
	1	*F*	.615	.61536	
	2	*G*	4.625	2.9143	
	3	*H*	31	34.065 %	
	4	*J*	77.2	989.74	
	5	*K*	579.89	26.674	
	6	*L*	165.9	11.115	
	7	*S*	33.4	.57485	
	8	*T*	6900	23.474	
	9	*M*	.545	.71322	
66	10	*N*	3.500	.53485	
	11	*F*	.328	3.5367	
	12	*G*	.663	1.4956	
67	13	*H*	4.25	18.288	
	14	*J*	34	18.307	
	15	*K*	11.25	46.875 %	
	16	*L*	20	$168.00	
	17	*M*	97	3055.5	
	18	*N*	16	$49.12	
	19	*P*	74.75	$51.839	
	20	*Q*	909	3370.0	
71	1	*A*	38296	195.69	
	2	*B*	642934	801.83	
	3	*C*	29 ÷ 43	.82122	
	4	*D*	62895	250.78	
	5	*E*	46.658	6.8306	
	6	*F*	.00547	.07395	
	7	*G*	9.5386	3.0884	

Page	Problem	Symbol	Variable	Answer
71	8	*H*	537.69	23.188
	9	*J*	.00367	.06058
	10	*K*	.36528	.60438
	11	*L*	.05986	.24466
	12	*M*	19.473	4.4128
	13	*N*	.000084	.00916
	14	*P*	.85423	.92424
	15	*P*	91.876	9.5851
	16	*S*	6329.2	79.556
	17	*N*	59.875	7.7378
	18	*P*	6.7982	2.6073
	19	*R*	.26574	.51549
	20	*S*	$\frac{573}{726}$	.88839
	21	*A*	302.68	17.397
	22	*B*	8.762	2.9500
	23	*C*	12.381	3.5186
	24	*D*	21.296	4.6149
72	25	*T*	5.8767	1.9793
	26	*U*	8.9326	1.9924
73	1	*c*	17.8	161.2
	2	*r*	9.8	19.045
74	3	*s*	9	14.345
	4	*B*	8.9	13.6
	5	*n*	21	5.1754
	6	*P*	14	.04764
	7	*N*	62	2.5625
	8	*R*	10	13
	9	*P*	9	.29508
	10	*T*	35	13.221
	11	*B*	9.7	33.184
	12	*B*	8.9	313.22
	13	*N*	27	3.7556
	14	*L*	9.3	1.6468
	15	*D*	21.7	20.434
	16	*N*	43	7.5079
75	17	*H*	8.5	39.1
	18	*D*	5.3	30.151
	19	*H*	8.5	6.0762
	20	*B*	19.9	6.2956
	21	*D*	1.988	1.706
	22	*D*	1.988	.032
	23	*F*	2.487	2.8717
	24	*F*	3.7	4.2723

Page	Problem	Symbol	Variable	Answer
75	25	E	2.225	3.1465
	26	E	2.458	3.4761
76	27	H	1.27	1.8407
	28	H	.687	2.5723
	29	d	.987	1.6055
	30	d	.758	1.262
	31	W	.571	.31525
	32	W	.783	.63325
77	33	D	3.125	3.3976
	34	D	2.375	2.6476
	35	T	9	1.5336
	36	L	11.7	3.3832
92	1	G	14.926	880.63
	2	H	9.25	8.6046
	3	J	20.5	42.983
93	4	T	5.75	178.25
	5	S	4.125	28.875
	6	r	.75	.17617
94	7	U	9.86	14.312
	8	L	20.3	.54133
98	1	G	1.253	7.5857
	2	J	9.5	3.3529
	3	J	9.5	2
99	4	H	3.75	7.0311
	5	T	11.5	1837.1
	6	R	23.75	.14866
100	7	S	77	55
	8	N	.625	111.01
	9	P	2.12	2.4352
	10	L	8.5	7.3995
103	1	E	$\frac{3}{4}$	30
	2	F	9.5	380
	3	G	$\frac{9}{13}$	27.692
	4	H	$\frac{1}{51}$	.78431
	5	J	17	17
	6	K	28	21 or 49
	7	L	38	19
	8	M	48	15
	9	N	92	10
	10	P	65	24

The following four (4) problems have many answers; all of which may be correct. One answer is given.

Page	Problem	Symbol	Variable	Answer
106	1	R	141	$\frac{A}{B} = \frac{6}{21}$ $\frac{D}{E} = \frac{20}{70}$
	2	S	367	$\frac{A}{B} = \frac{3}{21}$ $\frac{D}{E} = \frac{21}{27}$
	3	T	519	$\frac{A}{B} = \frac{2}{27}$ $\frac{D}{E} = \frac{42}{27}$
	4	U	1997	$\frac{A}{B} = \frac{1}{49}$ $\frac{D}{E} = \frac{37}{49}$
120	1	A	6.82	56.7
	2	B	146	47.5
	3	C	7.42	7.99
	4	D	14.44	.493
	5	E	34.7	93.7
	6	F	.317	383
	7	G	4.21	17.7
	8	H	42.1	1770
	9	I	66.4	8.15
	10	J	664	25.8
	11	K	12.6	146
	12	L	28.9	5.00
	13	M	49.4	.267
	14	N	7.42	1.75
	15	P	8.47	17.2
	16	Q	2.18	86.1
	17	R	.132	18.0
	18	S	6.41	999
	19	T	9.68	2.30
	20	U	34.7	.319
	21	V	6.31	2.06
	22	W	5.72	4.12
	23	A	28.6	.0537
	24	B	31.7	2.40
	25	C	19.4	14700
	26	D	94.6	.286
	27	E	462	5.27
	28	F	28.7	36.1
	29	G	30.8	1620
	30	H	43.2	1.03
	31	J	3° 12′	.0558
	32	K	37° 20′	.606

Page	Problem	Symbol	Variable	Answer
120	33	*L*	7° 22′	.129
	34	*M*	30° 15′	.583
	35	*N*	.0831	4° 46′
	36	*P*	.718	45° 54′
	37	*Q*	.215	12° 8′
	38	*R*	.928	42° 52′
	39	*S*	50° 10′	18.9
	40	*T*	39.4	10.6
	41	*U*	43° 12′	603
	42	*V*	4.16	.502
	43	*W*	2° 13′	1770
	44	*A*	14.9	27.9
	45	*B*	41° 18′	107
	46	*C*	27.4	264
	47	*D*	2.14	9° 56′
	48	*E*	87.3	44° 30′
	49	*F*	8.14	6° 6′
	50	*G*	4.61	31° 38′
	51	*H*	82.6	28° 30′
	52	*J*	50.4	37° 30′
	53	*K*	318	5° 49′
	54	*L*	64.2	39° 15′
160	1	*A*	9.125	7.0772
	2	*B*	9.95	6.5039
	3	*C*	10.52	16.684
	4	*D*	11.64	16.911
	5	*E*	7.882	9.8312
	6	*F*	14.75	9.0785
	7	*G*	5.395	10.450
200	13	*A*	1.2	.48109
213	1	*G*	11.062	5.1583
	2	*H*	12.875	12.312
	3	*J*	2.954	11.413
	4	*F*	9.684	4.1105
	5	*E*	15.92	7.2275
	6	*S*	17.23	18.222
214	7	*R*	15.52	6.5877
	8	*T*	3.497	13.050

Since the student is not familiar with interpolation at this time, the next 16 problems have been worked only to degrees and the next smaller number of minutes.

Page	Problem	Symbol	Variable	Answer
218	1	*H*	9.93	55° 26′
	2	*H*	9.93	12.055

Page	Problem	Symbol	Variable	Answer
218	3	G	12.042	50° 45′
	4	G	12.042	7.6204
	5	J	9.67	40° 34′
	6	J	9.67	8.2783
	7	L	5.06	33° 53′
	8	L	5.06	9.0761
	9	U	2.14	60° 33′
	10	U	2.14	4.3525
	11	T	12.4	62° 6′
	12	T	12.4	10.954
219	13	S	3.81	54° 43′
	14	S	3.81	6.5958
	15	R	5.92	56° 40′
	16	R	5.92	3.8936
221	1	G	9.07	39° 24′ 30″
	2	H	4.126	25° 17′ 10″
222	3	T	5.724	48° 59′ 27″
	4	S	9.937	38° 18′ 28″
	5	U	4.623	56° 0′ 6″
	6	R	7.106	41° 47′ 9″
	7	D	10.594	54° 11′ 30″
	8	F	8.521	46° 31′ 17″
224	1	α	38° 56′ 52″	5.9355
	2	α	38° 56′ 52″	4.7975
	3	γ	52° 37′ 16″	19.794
	4	γ	52° 37′ 16″	12.017
	5	δ	58° 19′ 37″	2.1772
	6	δ	58° 19′ 37″	1.1432
	7	ϵ	26° 52′ 29″	4.3273
	8	ϵ	26° 52′ 29″	8.5391
225	9	θ	53° 58′ 45″	13.353
	10	θ	53° 58′ 45″	10.800
	11	ϕ	47° 15′ 8″	7.1466
	12	ϕ	47° 15′ 8″	6.6057
227	1	E	8.463	24° 55′ 41″
	2	F	5.352	6.6396
	3	G	14.87	68° 48′ 58″
	4	D	10.06	11.604
	5	H	6.215	5.7402
	6	J	3.892	29° 32′ 52″
	7	K	7.048	58° 13′ 11″
228	8	L	17.49	36.622
	9	M	5.163	6.3917

Page	Problem	Symbol	Variable	Answer
250	1	J	11.7	35° 19′ 40″
	2	J	11.7	6.6737
	3	U	6.68	7.4529
	4	U	6.68	7.9652
	5	T	9.53	11.682
	6	P	9.97	12.115
253	7	S	22.168	15.8
	8	F	5.5374	4.23
	9	A	3.00	3.1234
	10	B	4.80	98° 33′ 24″
258	1	G	10.9	99° 28′ 18″
	2	G	10.9	31° 3′ 7″
	3	H	18.9	57° 55′ 1″
	4	H	18.9	73° 30′ 58″

INDEX

F

G

H

I

L

M

N

O

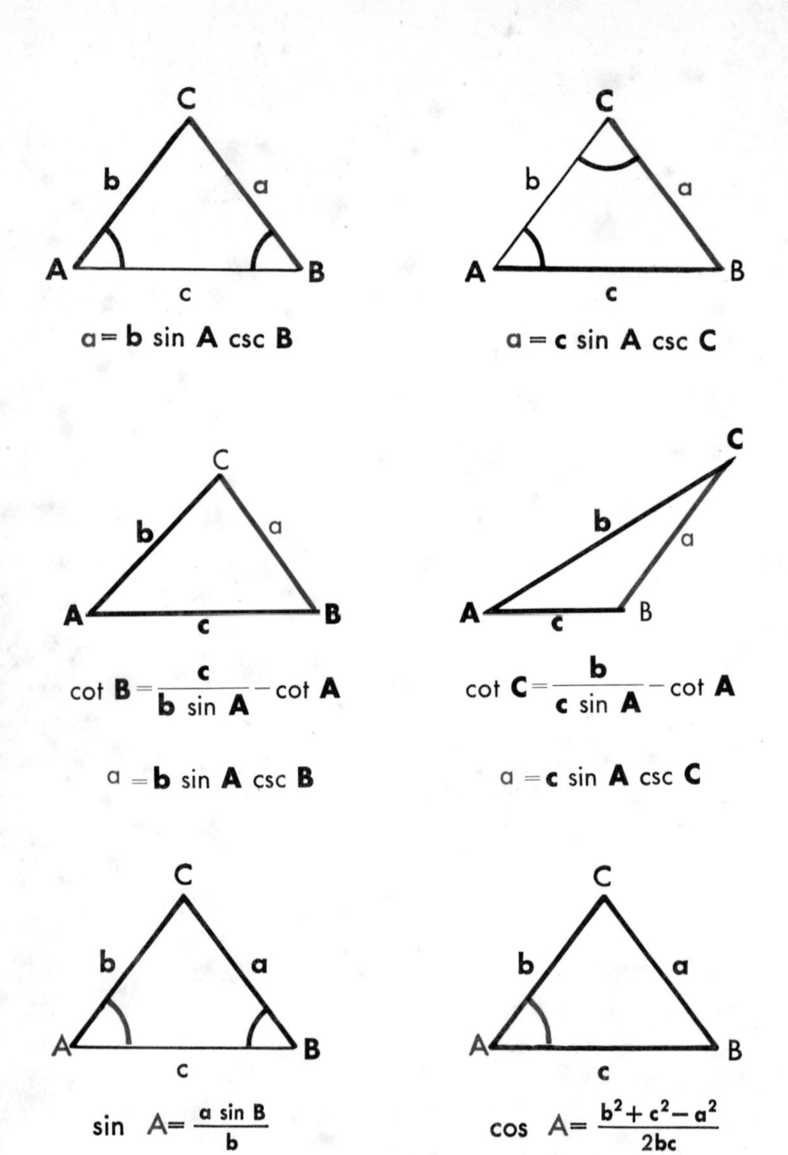

In the above diagrams the unknowns are shown in red. The sides and angles used to determine the unknown are in heavy black lines. The unused sides and angles are in lighter black lines.